Spatial Diversity and Dynamics in Resources and Urban Development

Ashok K. Dutt • Allen G. Noble
Frank J. Costa • Sudhir K. Thakur
Rajiv R. Thakur • Hari S. Sharma
Editors

Spatial Diversity and Dynamics in Resources and Urban Development

Volume 1: Regional Resources

Springer

Editors
Ashok K. Dutt
Department of Geography and Planning
University of Akron
Akron, OH, USA

Allen G. Noble
Department of Geography and Planning
University of Akron
Akron, OH, USA

Frank J. Costa
Department of Geography and Planning
University of Akron
Akron, OH, USA

Sudhir K. Thakur
College of Business Administration
California State University
Sacramento, CA, USA

Rajiv R. Thakur
Department of Geosciences
Missouri State University
West Plains, MO, USA

Hari S. Sharma
Department of Geography
University of Rajasthan
Jaipur, Rajasthan, India

ISBN 978-94-017-7980-7 ISBN 978-94-017-9771-9 (eBook)
DOI 10.1007/978-94-017-9771-9

Springer Dordrecht Heidelberg New York London

Softcover re-print of the Hardcover 1st edition 2015

Printed on acid-free paper

Springer Science+Business Media B.V. Dordrecht is part of Springer Science+Business Media (www.springer.com)

Preface

Natural resources is back on the stage, largely because the human–environment condition has emerged as one of the central issues of the new millennium, particularly as it has become apparent that human activity is transforming nature across several scales in both systemic and cumulative ways. Spatial Diversity and Dynamics in Resources is centrally concerned with the contentious and problematic issues of resources in urban, rural, and peripheral regions of the world. This book is a result of a project initiated by Ashok K. Dutt with his colleagues Allen G. Noble and Frank Costa (Department of Geography and Planning, The University of Akron, Akron, Ohio) to provide a festschrift in honor of Professor Baleshwar Thakur, former Vice-Chancellor of L. N. Mithila University, Darbhanga, Bihar, and Professor of Geography at the Delhi School of Economics, University of Delhi. Baleshwar Thakur has been a collaborator and contributor on several projects initiated at Akron. During the past more than four decades, Thakur has established his national and international credentials as one of the leading exponents of resource management and urban development. He has been recognized for his research contributions by the Association of American Geographer's Regional Development and Planning Specialty Group's Distinguished Scholar Award, besides other recognitions such as the Commonwealth and Fulbright Scholarships, and a Shastri Indo-Canadian Fellowship, among many others. We take this opportunity to thank all the contributors to the volume and the family members of editors in bearing the burden of being away from family responsibilities while working on this project.

Akron, OH, USA	Ashok K. Dutt
Akron, OH, USA	Allen G. Noble
Akron, OH, USA	Frank J. Costa
Sacramento, CA, USA	Sudhir K. Thakur
West Plains, MO, USA	Rajiv R. Thakur
Jaipur, Rajasthan, India	Hari S. Sharma

Acknowledgement

For invaluable editorial and other assistance in the preparation of this volume, we thank the publisher, Dr. Robert K. Doe. Thanks also go to associate editor, Dr. Elodie J. Tronche, and production assistant Naomi Portnoy, for editorial advice and support. The editors thank the contributors for producing the draft and revised chapters on time and without complaint.

Ashok K. Dutt
Allen G. Noble
Frank J. Costa
Sudhir K. Thakur
Rajiv R. Thakur
Hari S. Sharma

Contents

Contributors

Regional Resources

Rajesh Kumar Abhay Department of Geography, Delhi School of Economics, University of Delhi, New Delhi, Delhi, India

Trevor Birkenholtz Department of Geography and Geographic Information Science, University of Illinois at Urbana-Champaign, Champaign, IL, USA

Sanjoy Chakravorty Department of Geography and Urban Studies, Temple University, Philadelphia, PA, USA

Srikumar Chattopadhyay Geographer and Environmental Resource Analyst, Center for Earth Science Studies, Thiruvananthapuram, Kerala, India

Anil K. Chhangani Department of Environmental Science, Maharaja Ganga Singh University, Bikaner, Rajasthan, India

Frank J. Costa Department of Geography and Planning, University of Akron, Akron, OH, USA

Chandrama Dey Sarker School of Engineering and Information Technology, University College, The University of New South Wales at the Australian Defence Force Academy, Campbell, Canberra, ACT, Australia

Alan P. Diduck Department of Environmental Studies and Sciences, University of Winnipeg, Winnipeg, MB, Canada

Ashok K. Dutt Department of Geography and Planning, University of Akron, Akron, OH, USA

Supriya Francis The Energy and Resources Institute (TERI), New Delhi, India

World Sustainable Development Forum, New Delhi, India

D. Fraser School of Engineering and Information Technology, University College, The University of New South Wales at the Australian Defence Force Academy, Campbell, Canberra, ACT, Australia

Archana Gupta Department of Geography, University of Lucknow, Lucknow, Uttar Pradesh, India

Ronald C. Hess Department of Geography-Earth Sciences, Shippensburg University of Pennsylvania, Shippensburg, PA, USA

Ishrat Islam Department of Urban and Regional Planning, Bangladesh University of Engineering and Technology, Dhaka, Bangladesh

Inder Jeet Department of Geography, Maharshi Dayanand University, Rohtak, Haryana, India

X. Jia School of Engineering and Information Technology, University College, The University of New South Wales at the Australian Defence Force Academy, Campbell, Canberra, ACT, Australia

P.P. Karan Department of Geography, University of Kentucky, Lexington, KY, USA

Adil Mohammed Khan Department of Urban and Regional Planning, Jahangirnagar University, Dhaka, Bangladesh

Darin Khongsatjaviwat Department of Political Science and Public Administration, Faculty of Social Sciences, Naresuan University, Phitsanulok, Thailand

L. Lymburner Australian Centre for Remote Sensing (ACRES), Geoscience Australia, Canberra, ACT, Australia

Matthew McCandless International Institute for Sustainable Development, Winnipeg, MB, Canada

Bruce Mitchell Department of Geography and Environmental Management, University of Waterloo, Waterloo, ON, Canada

Allen G. Noble Department of Geography and Planning, University of Akron, Akron, OH, USA

G.K. Panda Department of Geography, Utkal University, Vani Vihar, Bhubaneswar, Odisha, India

Risa Patarasuk School of Life Sciences, Arizona State University, Tempe, AZ, USA

Punyatoya Patra Department of Geography, Aditi Mahavidyalaya, University of Delhi, Delhi, India

George M. Pomeroy Department of Geography-Earth Sciences, Shippensburg University of Pennsylvania, Shippensburg, PA, USA

Ramashray Prasad Bhim Rao Ambedkar College, University of Delhi, Yamuna Vihar, Main Wazirabad Road, Delhi, India

Paul Robbins Nelson Institute for Environmental Studies, University of Wisconsin, Madison, WI, USA

Jayant K. Routray Asian Institute of Technology, Pathumthani, Thailand

Preeti Sachar Department of Geography, Swami Shradhanand College, University of Delhi, Alipur, New Delhi, Delhi, India

Jitender Saroha B. R. Ambedkar College, University of Delhi, Delhi, India

Hari S. Sharma Department of Geography, University of Rajasthan, Jaipur, Rajasthan, India

A. John Sinclair Natural Resources Institute, University of Manitoba, Winnipeg, MB, Canada

Mahendra Bahadur Singh Department of Geography, Banaras Hindu University, Uttar Pradesh, Varanasi, India

Nina Singh Department of Geography, Maharshi Dayanand University, Rohtak, Haryana, India

Ravi S. Singh Department of Geography, Faculty of Science, Banaras Hindu University, Varanasi, Uttar Pradesh, India

Vimalesh Kumar Singh Department of Geography, Banaras Hindu University, Varanasi, Uttar Pradesh, India

Rajiv R. Thakur Department of Geosciences, Missouri State University, West Plains, MO, USA

Rameshwar Thakur Department of Geography, B.R. Ambedkar College, University of Delhi, New Delhi, Delhi, India

Sudhir K. Thakur College of Business Administration, California State University, Sacramento, CA, USA

Swati Thakur Department of Geography, Dyal Singh College, University of Delhi, New Delhi, Delhi, India

Vandana Wadhwa Department of Geography and Environment, Boston University, Boston, MA, USA

Liguo Wang College of Information and Communications of Engineering, Harbin Engineering University, Harbin, China

Urban Development

Swapna Banerjee-Guha Center for Development Studies, School of Social Sciences, Tata Institute of Social Sciences, Mumbai, Maharashtra, India

Ranu Basu Department of Geography, York University, Toronto, ON, Canada

Kathryn J. Bills Department of Geography, Keene State College, Keene, NH, USA

Andrew Boulton Kentucky Community Technical College System, Versailles, KY, USA

Stanley Brunn Department of Geography, University of Kentucky, Lexington, KY, USA

Kevin A. Butler Spatial Analyst Team, ESRI, Redlands, CA, USA

Meera Chatterjee Department of Geography and Planning, University of Akron, Akron, OH, USA

Monalisa Chatterjee Carnegie Institution for Science, Stanford University, Stanford, CA, USA

Frank J. Costa Department of Geography and Planning, University of Akron, Akron, OH, USA

Chris Cusack Department of Geography, Keene State College, Keene, NH, USA

Ben Derudder Department of Geography, Ghent University, Ghent, Belgium

Lomme Devriendt Department of Geography, Ghent University, Ghent, Belgium

José R. Díaz-Garayúa Department of Geography and Geosciences, University of Louisville, Louisville, KY, USA

Ashok K. Dutt Department of Geography and Planning, University of Akron, Akron, OH, USA

John W. Frazier Department of Geography, State University of New York, Binghamton, NY, USA

Jay D. Gatrell Geography and Environmental Studies and Office of Academic Affairs, Bellarmine University, Louisville, KY, USA

Milton E. Harvey Department of Geography, Kent State University, Kent, OH, USA

Norah F. Henry Department of Geography, State University of New York, Binghamton, NY, USA

Satya Prakash Kaushik Department of Geography, Kurukshetra University, Kurukshetra, Haryana, India

Pankaj Lal Department of Earth and Environmental Studies, Montclair State University, Montclair, NJ, USA

Ann Marie F. Murnaghan Department of Environment and Geography, University of Manitoba, Winnipeg, MB, Canada

Allen G. Noble Department of Geography and Planning, University of Akron, Akron, OH, USA

Narendra Kumar Rana Department of Geography, D.D.U. Gorakhpur University, Gorakhpur, Uttar Pradesh, India

Aditi Ranjan Department of Geography, Magadh University, Bodh Gaya, Bihar, India

Neil Reid Department of Geography and Planning, The University of Toledo, Toledo, OH, USA

Gopa Samanta Department of Geography, University of Burdwan, Burdwan, West Bengal, India

Jitender Saroha B. R. Ambedkar College, University of Delhi, Delhi, India

P.R. Sharma Department of Geography, Banaras Hindu University, Varanasi, Uttar Pradesh, India

Annapurna Shaw Public Policy and Management Group, Indian Institute of Management, Calcutta, West Bengal, India

G.N. Singh Department of Geography, Banaras Hindu University, Varanasi, Uttar Pradesh, India

Rana P.B. Singh Department of Geography, Banaras Hindu University, Varanasi, Uttar Pradesh, India

B.R.K. Sinha Department of Geography, Banaras Hindu University, Varanasi, Uttar Pradesh, India

Andres Susaeta School of Forest Resources and Conservation, University of Florida, Gainesville, FL, USA

Christian Tettey Department of Geography, Trent University, Trent, ON, Canada

Rajiv R. Thakur Department of Geosciences, Missouri State University, West Plains, MO, USA

Sudhir K. Thakur College of Business Administration, California State University, Sacramento, CA, USA

A.N. Tiwary Department of Geography, Banaras Hindu University, Varanasi, Uttar Pradesh, India

Frank Witlox Department of Geography, Ghent University, Ghent, Belgium

Part I
Introduction

Chapter 1
Introduction: Regional Resources

Sudhir K. Thakur, Ashok K. Dutt, Allen G. Noble, Frank J. Costa, Rajiv R. Thakur, and Hari S. Sharma

Abstract Natural resources are attributes found in nature such as coal, wilderness, water, soil, and air that can be used as factors of production. A resource is a means to an end. Although resources are plentiful, their distribution is spatially uneven in developing and developed countries. Concomitantly, there is a contrast in the population and resource relationship in relation to regional levels of development. The explosion of population size and its resultant pressure on consumption and depletion of resources is a big question for development analysts and decision makers. Adverse relationships have led to loss of cultivated lands, deforestation, soil erosion, water shortage, groundwater depletion, ecological imbalance, pollution hazards, deterioration of water quality, and environmental degradation. This disparity impacts the gross national product as many governments do not take into account the loss of resources in the measurement of development, thereby inflating the sum of goods and services produced in the economy. Further, resource analysts have advocated an increase in wealth per capita over time to maintain intergenerational social well-being.

Keywords Natural resources • Uneven development • Population and resource relationships • Sustainable development

---

S.K. Thakur (✉)
College of Business Administration, California State University, Sacramento, CA 95819-6088, USA
e-mail: thakurs@saclink.csus.edu

A.K. Dutt • A.G. Noble • F.J. Costa
Department of Geography and Planning, University of Akron, Akron, OH 44325, USA
e-mail: dutt@uakron.edu; agnagn29@gmail.com; usercs7977@earthlink.net

R.R. Thakur
Department of Geosciences, Missouri State University, 128 Garfield Avenue, West Plains, MO, USA
e-mail: rajivthakur@missouristate.edu

H.S. Sharma
Department of Geography, University of Rajasthan, Jaipur, Rajasthan, India
e-mail: pachhora1@yahoo.com

A.K. Dutt et al. (eds.), *Spatial Diversity and Dynamics in Resources and Urban Development*, DOI 10.1007/978-94-017-9771-9_1

Natural resource studies have become a subject of exceedingly high importance in the enrichment of civilized human existence and, ultimately, our biological survival. Mother Earth is rich in diverse natural resources, although their distribution is spatially uneven. There are contrasts in the geographic pattern of distribution of natural resources in both the developed and developing world. The subsistence-level economy of developing countries diverges sharply with highly developed countries in natural resource consumption, resulting in unequal relationship among natural resource and development relationships. The population pressure on natural resources in the two worlds is also significantly different. Natural resource developers and managers are intrinsically concerned with the quantitative explosion in the population size and the concomitant pressure on natural resources. In many parts of the world, this pressure has exceeded the carrying capacity of the resources. Pressure on resources caused by increasing populations and their quest for a high standard of living is growing at a rate that is difficult for the world to sustain. Such trends along with many other factors have resulted in loss of cultivated lands, deforestation, soil erosion, water shortage, groundwater depletion, ecological imbalance, pollution hazards, deterioration of water quality, and environmental degradation.

Many countries are blessed with relatively diverse natural resources, including fertile soil in Argentina, China, India, Nile Valley, Great Plains of the USA; abundant forests in the Amazon basin, Northern Canada, Eurasia; abundant water in North America and humid tropics; rich mineralized rock beds in the USA and former USSR, China, and Australia; energy minerals in the USA, former USSR, Middle East, Venezuela, France, China, and India; and scenic resources in the USA, Canada, Switzerland, and tropical islands. These resources are unevenly distributed, and there are variations in the development of natural resources spatially. The land-to-man ratio is favorable in Australia, Canada, Argentina, USA, Chile, Denmark, and Mexico. Conversely, the land–man ratio is unfavorable in India, Japan, The Netherlands, Egypt, UK, Israel, and China. The developing countries, especially, the humid tropics, are the most densely populated regions of the earth, where by the year 2030 about one third of the world population is expected to reside. The poor productivity in food resources results form the unusual ecological diversity of large parts of the tropics; it is also caused by geologic and climate factors that make tropical soils (Oxisols and Ultisols) deficient in minerals nutrients as compared to middle latitudes.

Soil erosion is occurring in developed and developing countries alike. In the United States and South Africa, soil erosion is a national disaster of the first magnitude. In much of Asia, Latin America, and Africa, change from forest to crop fields is accompanied by increase in runoff. The continued deforestation in the Himalayas and the Aravallis has created serious soil erosion problems in northern India. Soil erosion is making future production more difficult in China. The growth in global population and the expansion of agriculture and industry have put pressure on the world's water resources. Oceania, South America, and North America are relatively water abundant because their per capita water consumption is less than per capita water availability, whereas Asia has the least amount of water available per capita. Forests are important natural resources as they moderate local climates, reduce soil erosion, regulate stream flow, support industries, and provide opportunities for recreation. Latin American and Asian countries have shown an increase in arable areas and decrease in forest coverage. Most European countries are characterized by an increase in the extent of forests and a reduction in arable area.

Mineral and energy resources are central to all economic activities and are perceived as one of the key variables that determine the sustainability of the development process. A region's economic development relies on a variety of energy resources including fossil fuels, hydropower, nuclear energy, biomass fuels, and exotic power resources. Minerals are means for producing energy (atomic minerals, tar shale, oil shale), controlling energy (ferrous and nonferrous minerals), and conserving energy (nonmetals).

In general, natural resources have had a positive effect on the development process, but in some economies it has had negative or less favorable effects on development, especially in Sub-Saharan Africa (Carmignani and Choudhary 2012). Many of the underdeveloped regions in Latin America, Sub-Saharan Africa, Asia, and South Asia are abundant in natural and mineral resources but have not experienced growth compared to resource-poor countries. It is a dilemma that the eastern states of India, rich in coal, iron, aluminum, and groundwater resources, have not shown prosperity relative to the developed western states. What explains this paradox? Perhaps elucidation lies in an alternative explanation called the *resource curse thesis* (Auty 2001, 2007). This concept integrates neoliberal, political, institutional, and environmental approaches that provide an explanation for the fate of mineral-dependent economies. A threat to excessive dependence upon mineral products can result from depletion of such resources, mineral sector marginalization with synthetic substitutes, low responsiveness of primary sector products, and the volatile nature of revenue from this sector.

Further, economic growth and social progress of national economies have been measured by an economic indicator called the Gross National Product (GNP). This is the total sum of marketed goods and services in an economy over a year. This measure of economic progress is misleading because in the process of development natural resources are utilized and consumed as well. Natural resources such as agricultural land, forests, watersheds, groundwater, atmosphere, and ecosystems are utilized; and although some resources can regenerate, their overconsumption leads to depletion and depreciation. Thus, the depreciated asset should be deducted from the GNP to accurately reflect the level of development. Incorrect estimates of GNP can give erroneous signals to policy makers for investment planning and allocation of resources (Miller 1990; Das Gupta 2010). This debate led to the coining of a new measure called the Index of Sustainable Economic Welfare (ISEW). The ISEW is a measure of economic progress that utilizes GNP and accounts for both current environmental issues and long-term sustainable use of natural ecosystem and resources. The much-cited Brundtland Commission on Environment and Development (1987) proposed the idea of sustainable development. This concept implied a process of development in which the need of the current generation does not compromise the ability of production and consumption for future generations. Government planners and decision makers are always vexed with the trade-off between current consumption and intergenerational equity.

Also, the notion of sustainable development has been elaborated as an increase in intergenerational social well-being only if a comprehensive measure of wealth per capita increases over time. National wealth comprises not only manufactured capital, but also knowledge, human capital, and natural capital (Das Gupta 2010). It is necessary for societies to consume natural resources in a manner that allows people and government to invest in natural capital for its regeneration and sustainable use. For example, decision makers have advocated an integrated water resource management (IWRM) approach to manage the complex problem of interstate water conflicts and water management. The IWRM is a process that entails multiple actors to integrate diverse rules and resources in a strategic context for effective water management (Saravanan et al. 2009).

Given this overview, this book is a felicitation honoring Professor Baleshwar Thakur as an academician *par excellence*. The book is a collation of papers on issues resonating with his teaching and research interests over the past 45 years. The papers have been contributed by his friends, colleagues, esteemed teachers, and former students, many of whom have known and worked with him for several decades. The book contains 28 chapters broadly classified around the theme of spatial diversity and dynamics in regional resources, divided into eight parts: Introduction, Methodology, Global Perspectives, Economic Perspectives, Ecological Perspectives, Water Management, Energy and Forest Resources, and Land Cover and Rural Planning.

The introductory part consists of two chapters. The first chapter is an overview to the theme of the book and the second chapter is a review of the academic background and contributions of Professor Baleshwar Thakur to geographic research in general and in particular urban and resource development in developing countries with particular focus on India. The second part on methodology contains two chapters. The first chapter, by **Chandrama Dey Sarker, X. Jia, Liguo Wang, D. Fraser, and L. Lymburner**, discusses flood inundation mapping; in particular, exploring the issue of separating inundated areas from wet areas where trees and houses are partly submerged under water. Experiments were conducted using three different flood events in Australia, leading to satisfactory results. The next chapter, by **Srikumar Chattopadhyay**, posits the importance of a microlevel approach in natural resource management planning with examples at the local level in Kerala.

The third part consists of two chapters concerning global perspectives on natural resources. The first chapter in this part is coauthored by **Supriya Francis and Vandana Wadhwa**. Their analysis concludes that policy makers have a lackluster approach to the effect of climate change on the agriculture sector upon which millions of Indian farmers depend for their livelihoods. The second chapter, by **Bruce Mitchell**, discusses innovative concepts related to environmental, resource, and water management.

The fourth part, covering the economic perspectives of land and natural resources, includes four chapters. The first chapter, by **Sanjoy Chakravorty**, encompasses the evolution of land markets in India before independence. The chapter is written from an economic history perspective to manifest an improved understanding of the role of the state since independence and its ramifications in the current period.

The second chapter, coauthored by **R.C. Hess and G. Pomeroy**, concerns the geographic distribution of land trusts in the United States. Multiple variables such as spatial distribution of land trusts, the number of acres owned, acreage under conservation easement, and the total number of acres protected by all means by the various land trusts are considered for mapping and location quotient analysis. The third chapter in this part, co-authored by **Adil Mohammad Khan and Ishrat Islam**, addresses poverty reduction and social development in Bangladesh. The factors that explain the incidence of regional poverty are lack of electrification, urbanization, number of cooperatives, and industrialization. They suggest a stronger institutional mechanism to address the issue of spatial disparity and policy making. The last chapter in this part, by **Rameshwar Thakur and Swati Thakur**, addresses the topic of mineral resource potential and prospect in Chotanagpur region in eastern India. His analysis suggests that the region has suffered from resource depletion and environmental pollution although the region is endowed with abundant mineral deposits that can contribute to regional development.

The fifth part comprises three chapters under the rubric ecological perspectives of regional resources. The first chapter in this part, by **P.P. Karan**, is an in-depth study of the land, life, and environmental changes in the Himalayas. The second chapter in this part, coauthored by **Paul Robbins and A.K. Chhagani**, contributes to the debate on adaptive management of India's wildlife sanctuaries and suggests the opportunity for the practice of science, democracy, and conservation. The third chapter, jointly authored by **M.B. Singh and V.K. Singh**, describes the food consumption patterns among various ethnic and caste groups in Uttar Pradesh, India.

The sixth part contains nine chapters on water management encompassing issues such as water scarcity, groundwater, floodplain management, water development, and interstate water disputes in India. The first chapter, by **Rajesh Abhay**, utilizes the methods of Thornthwaite and Sullivan to measure water scarcity, alluding to their methodological merits and demerits. **Trevor Birkenholtz** investigates the relationships between groundwater decline and land use changes in Rajasthan. In particular, he highlights the role of markets and social institutions in agriculture decision making. The subsequent chapter by **Inder Jeet** utilizes a political ecological approach to investigate the factors that explain groundwater exploitation in Northwestern India. **G.K. Panda** addresses the impact of floods upon human vulnerability using a quantitative approach. He measures average risk of death and maps vulnerable districts in Orissa, India. The chapter by **Preeti Sachar** assesses water demand for growing various crops in different seasons in the Hindon basin located in Western Uttar Pradesh, India. The next chapter, by **John Sinclair, Alan Diduck, and Matthew McCandless**, alludes to the potentials and pitfalls of small hydropower development in the local areas of the states of Himachal Pradesh and Uttarakhand. The following chapter by **Narendra Rana** utilizes a conceptual approach to address the issue of floodplain management in the Rapti River basin. In a similar spirit, **Archana Gupta** in the subsequent chapter addresses the quality and quantity issue of groundwater, reduction in water table, and water harvesting techniques as a potential solution to recharge aquifers. The last chapter in this part, by

Nina Singh, examines interstate water conflicts in India and offers suggestions for a National Water Commission for mitigating conflict resolution among the States.

The seventh part comprises three chapters around the broad theme of energy and forest resources. The first chapter, by **Jitender Saroha**, examines the nonconventional energy resources in India and further evaluates the achievements, constraints, and measures to reduce gaps between potential and achievements. The next chapter, by **Punyatoya Patra**, examines the concept, characteristics, and perception of joint forest management in India. The last chapter in this part is written by **Ramashray Prasad** and covers community forest and its management in Bhutan.

The last part of the book contains three chapters on land use, land cover, and rural planning. The first chapter, authored by **Risa Patarasuk**, examines the land cover pattern and its distribution along road types in Thailand using the Geographic Information System and remote sensing technology. The change in land cover and loss of forests is a major contributor to local and global environmental change. The second chapter, by **Darin Khongsatjaviwat and Jayant Routray**, examines rural development planning issues in Thailand. The final chapter by **Ravi Singh** concerns with the study of patterns of land use and agricultural change at the district level. There is a need for state intervention as the rural areas are not fully monetized and integrated into the market system, and several constraints affect agricultural development.

References

Auty RM (ed) (2001) Resource abundance and economic development, UNU/WIDER Studies in Development Economics. Oxford University Press, Oxford

Auty RM (2007) Natural resources, capital accumulation, and the resource curse. Ecol Econ 61:627–634

Carmignani F, Choudhary A (2012) The geographical dimensions of the development effects of natural resources. Environ Resour Econ 52:479–498

Das Gupta P (2010) Nature's role in sustaining economic development. Philos Trans R Soc B 365:5–11

Miller J. (1990) A green GNP: taking the environment into account. Dollars & Sense, pp 6–22

Saravanan VS, McDonald GT, Mollinga PP (2009) Critical review of integrated water resource management: moving beyond polarized discourse. Nat Resour Forum 33:76–86

World Commission on Environment and Development (1987) Our common future. Oxford University Press, Geneva

Chapter 2
Baleshwar Thakur: Professional Career and Contributions

Sudhir K. Thakur, Ashok K. Dutt, Allen G. Noble, Frank J. Costa, Rajiv R. Thakur, and Hari S. Sharma

Professor Baleshwar Thakur, 2013

S.K. Thakur (✉)
College of Business Administration, California State University, Sacramento 95819-6088, CA, USA
e-mail: thakurs@saclink.csus.edu

A.K. Dutt • A.G. Noble • F.J. Costa
Department of Geography and Planning, University of Akron, Akron, OH 44325, USA
e-mail: dutt@uakron.edu; agnagn29@gmail.com; usercs7977@earthlink.net

R.R. Thakur
Department of Geosciences, Missouri State University, 128 Garfield Avenue, West Plains, MO 65775, USA
e-mail: rajivthakur@missouristate.edu

H.S. Sharma
Department of Geography, University of Rajasthan, Jaipur, Rajasthan, India
e-mail: prsharma1950@gmail.com

A.K. Dutt et al. (eds.), *Spatial Diversity and Dynamics in Resources and Urban Development*, DOI 10.1007/978-94-017-9771-9_2

Abstract This chapter is a discussion of the growth, evolution, and contribution of Baleshwar Thakur (B.T.) to Indian geographic research. He is a doyen among Indian geographers and has been an acclaimed teacher, scholar, and administrator throughout his illustrious career of four and a half decades. He made immense contributions to the discipline and strived for its popularization both within and outside India. His research career can be divided into two phases: 1964–1980 and the post-1980s. During the first phase he was interested in understanding the long-term processes of urbanization in its regional context in developing countries, focusing on India, using quantitative methods. After accepting the position of Reader in Geography at the University of Delhi, his interests shifted to the study of natural resource management and the history of Indian geography, although he pursued his interest in urban geography as well. His contributions during the past five decades can be subdivided into four broad areas: urban geography, quantitative geography, natural resource management, and history of Indian geography. He continues to be actively engaged in research and publications, and his contributions serve as a model for young Indian geographers aspiring to become scholars.

2.1 Background

Baleshwar Thakur (B.T.), nicknamed '*Bale*' by his parents, is a doyen among Indian geographers. He has worn several hats with distinction and has been an acclaimed teacher, scholar, and administrator throughout his illustrious career of four and half decades. He is among the selected few Indian geographers who transcend the parochial boundaries of academic pursuit. He can be characterized as an erudite knowledge seeker and as a disseminator of knowledge to those who are passionate about learning and understanding the world through the lens of a perceptive geographer. He transmitted immense affection to his students and served on the graduate committees and doctoral dissertations of several dozens of students at Patna and Delhi University. He has an insatiable passion to push the frontiers of academic knowledge beyond its boundaries. Although he retired from the University of Delhi in 2008, yet he spends hours reading journals, engaging in research, visiting the Ratan Tata library at the Delhi School of Economics (DSE), University of Delhi, and meeting with students to advise them on their research projects. His commitment and dedication to research shows that one can retire from a university position but not give up academic life. Throughout his long career, he has been very much a field geographer and has traveled in every nook and corner of the Indian subcontinent.

B.T. was born on July 16, 1943, at Sarabey Village of Madhubai District in North Bihar, in a working-class family. He is the second among three siblings. His father and uncle were influential in his early education that led him to obtain a first place in his matriculation (tenth grade) examination. He received his early school education at Khajauli and his intermediate education at R.K. College, Madhubani, affiliated with Bihar University. In 1960 he was placed in the top ten merit list for his excellent performance in undergraduate studies. During his undergraduate studies, the late Professor Arun K. Dutt (Principal of R.K. College, Madhuhbani, and uncle of the senior editor of this volume, Professor Ashok Dutt) recognized his academic performance and provided a scholarship for his higher education.

2.2 Education and Academic Career

Dr. Thakur obtained first class in both his B.A. Honors (1962) and M.A. (1964) in geography from Patna University, where he was awarded the Merit Scholarship and National Scholarship in Humanities. Soon after completing his master's degree he obtained his first employment as a Lecturer at Ranchi University, in 1964–1965. Simultaneously, he started preparing for the coveted Indian Administrative Service (IAS) examination. One fine morning a chance event changed his trajectory forever. A young and astute professor of psychology (Dr. Jaiballav P. Sinha), who had returned from The Ohio State University in 1965 after obtaining a doctoral degree, convinced the young B.T. of the excellent and world-class academic programs in geography in American universities, and the resulting contribution one could make to knowledge after receiving training from such an acclaimed program. These thoughts mesmerized the young and ambitious B.T., and the very next day he disposed of all the books related to the IAS examination and started applying for a graduate fellowship and admission to academic programs in North America and Great Britain. Professor Ashok Dutt was instrumental in advising and writing recommendation letters for admission to graduate programs in North America. Finally, he was awarded a Commonwealth scholarship to attend a Canadian university and studied for a second master's in geography at the University of Waterloo, Canada, during 1970–1972. He took courses with distinguished urban geographers and resource management specialists. It was there that he became acquainted with Professors Richard E. Preston, Lorne H. Russwurm, and Bruce Mitchell. He worked with the late Professor Lorne H. Russwurn, who became his graduate advisor. At Waterloo University he completed path-breaking work on the topic of hierarchically structured urban places over space and time in southwestern Ontario's complex hierarchical regional systems. This research on urban system dynamics was published in the *Canadian Geographer* journal in 1981 with his advisor as coauthor. Later, he obtained his Ph.D. degree from Patna University, in 1978, on "Entropy Analysis of Changing Urban Patterns in Eastern India" under Professor P. Dayal, former Vice-Chancellor of Magadh University, Bihar. During the 1970s and 1980s his research focused on the long-term processes of urbanization in its regional contexts in developing economies, especially in India.

Subsequently, he was a Visiting Fulbright scholar at the University of Akron, United States (USA), in the summer of 1992 and a Visiting Professor in 2000, where he had the privilege of working under the tutelage of Professor Ashok K. Dutt on "Urban Structure and Processes in India." Professor Ashok Dutt would often introduce B.T. to his colleagues, students, and friends as his 'first-generation student,' as he was indeed his teacher at Patna University during 1960–1962 at the undergraduate level. Further, he has guest lectured at the Geography Department, University of North Dakota, in Grand Forks, USA, during spring of 2005. His epistemological foundations have been influenced by illustrious luminaries in geography, both in India and in North America, such as L.N. Ram, P. Dayal, A.K. Dutt, A.G. Noble, F. Costa, R. Ramachandran, R.P. Misra, Savitri G. Burman (late), A.B. Mukherji (late), V.K. Verma, Gerard Rushton, and Bruce Mitchell. In the past two decades he has cherished his association with the "Akron school of geography and planning" (Thakur 2012b) through which he worked on several book projects and scholarly works with his colleagues at the University of Akron, Ohio (USA).

B.T. was hired as a lecturer in geography at the University of Ranchi (1964–1965) and Patna University (1965–1980), respectively. He relocated and accepted the coveted position of a Reader in Geography at the prestigious and very illustrious Delhi School of Economics (DSE), University of Delhi (1980–1990); then was Professor of Geography, University of Delhi (1990–2005); and was reemployed as Professor of Geography, University of Delhi (2005–2008). He was a Commonwealth Research scholar at the University of Waterloo (1970–1972) and Fulbright Visiting Scholar at the Department of Geography and Planning, University of Akron (1992). He was invited as Visiting Professor at the University of Akron (2000) and Guest Speaker at the Department of Geography, University of North Dakota (spring 2005). He was the Shastri Indo-Canadian Fellow, Department of Geography, York University, Toronto, Canada, during 2004. He was the recipient of several awards and honors, such as University Grants Commission Awards (1969, 1976, 1986), Distinguished Scholarship Award, Regional Development Planning Specialty Group of the Association of American Geographers (AAG) at Washington, D.C. (2005), and Bhoovigyan Samman by Bhoovigyan Foundation, Delhi (2005).

He was ambitious, determined, and truly committed toward achieving the highest accolades in higher education, and worked hard to reach the highest echelon in higher education. He earned a professorship in one of the most reputed institutions of higher education in Asia: the Delhi School of Economics. Although he spent a major part of his academic career living and working in urban areas, his heart was rooted in understanding the rural sector of India. He never missed an opportunity for visiting the rural areas of his home state and thinking about issues pertaining to resource development and the emancipation of the marginal populations of Indian society.

B.T. has been a devout 'Guru' in the proverbial sense of the term and has been a mentor to several generations of students at Ranchi, Patna, and Delhi University. As a mentor, he was conscientious and thoughtful and put much effort into preparing his lectures. His courses had very current readings on the topics of discussion and included readings from cognate fields as well. He had a remarkable ability to appreciate and expose students to alternative debates on the topic, particularly in his conceptual developments in geography courses. He has been an excellent teacher who meticulously prepared his lectures before entering the classroom. He motivated students to do research and spent countless days, months, and years to train them in their research endeavors. His energy, enthusiasm, generosity, and strong commitment to graduate students have produced many budding professionals during the past four decades.

His published works are rich in both ideas and empirical data. Many of his publications were coauthored with younger colleagues and graduate students whom he advised over the years. His students and younger colleagues have been strongly influenced by his inspiration and guidance, and they received an exemplary influence in the pursuit of knowledge. Several of them followed the academic path of going abroad to North American and European doctoral programs to continue seeking higher education and then held academic positions in the USA, Canada, Europe, and Australia. Several others accepted teaching positions at home in regional and national universities. Many of the students competed for coveted positions with the government and in private sectors.

B.T. has accomplished 44 years of teaching and research experience at Ranchi, Patna, Waterloo, Delhi, Akron, and North Dakota. He successfully supervised 21 Doctoral dissertations, 35 Master of Philosophy dissertations, and 62 Master's theses. He was associated with teaching and research at the Department of Geography, Delhi School of Economics, for the longest duration of his career, approximately 28 years (1980–2008), and was the Chairman of the Department during 1996–1999. Later, he also served as the Vice-Chancellor of L.N. Mithila University in Bihar during 2001–2003.

He has been passionate about acquiring knowledge, and his home in Delhi is a testament with the most current research books in geography and social sciences. He spent most of his career engaging in intriguing discussions with his students regarding their research and with colleagues on the frontiers of geographic research. He has a strong interest and an understanding of the philosophy and methodology of geographic debate, enabled by the cerebral power that allowed him to remember details of what he observed, read, and synthesize. He has a phenomenal ability to think systematically, juggling through multiple complex variables and analyzing the cause-and-effect relationship of a spatial phenomenon. During conversations he would often cite references with intricate details. His success can be attributed to the sacrifice, dedication, and congenial atmosphere provided and nurtured by his wife Chanda Thakur. Most of the visitors at his home are students of geography and colleagues who are engaged in the business of promoting the knowledge sector across various disciplines. During visits by students and colleagues, his wife Chanda would provide excellent cuisine, snacks and delicacies in the *Maithili* tradition, which made the academic conversations irresistible, tempting, and oftentimes jovial.

Those who believe in genetics might attribute his influence on his children in pursuing an academic career in geography and related disciplines. His elder son Sudhir earned a doctorate in geography from The Ohio State University and is currently an Associate Professor in the College of Business Administration, California State University, Sacramento; his younger son Rajiv earned a doctorate in geography from Indiana State University and is an Assistant Professor at the University of Missouri State University in Geosciences. Similarly, his elder daughter-in-law, Rajrani Kalra (also his former student at the University of Delhi), earned a doctorate in geography from Kent State University (Ohio). She is an Associate Professor in geography and environmental studies at California State University at San Bernardino. B.T. has nurtured not only a family of geographers, but generations of geographers, as his younger brother is an Associate Professor in Geography at the B.R. Ambedkar College, University of Delhi, and his niece is an Assistant Professor in Geography at Dayal Singh College, University of Delhi. This roster truly is an example of the direct and induced effect of his influence in his immediate and extended family members.

2.3 Professional Contributions

B.T. has been actively associated with and devoted to the cause of Earth Systems Science of the Indian Science Congress Association (ISCA) for more than three decades and has served the Sectional Committee as its Recorder (1990–1992) and as Local Secretary (1997) in the ISCA held in Delhi. His name is familiar among geographers and geologists because of his decade-long association with National Association of Geographers, India (NAGI), as Joint Secretary (1990–1995) and as Secretary General (1997–2002). In addition to holding these illustrious positions, he has been associated with the Association of Geographical Studies, University of Delhi as its Treasurer (1984–1988), Editor (1988–1994), and President (1994–1998). He was also the Vice-President of the Indian Council of Geographers (1990–1994), Institute of Indian Geographers (1995–1997), Indian Regional Science Association (1997–1998), and Institute of Geomorphologists (1998–2000), and Chairman of the Commission of Natural Hazard and Resource Management, NAGI (1993–1998).

His significant contribution in the field of urban geography earned him the position of a member of the International Board of Directors of Asian Urban Research Association (AURA) based at the University of Akron, Ohio, USA. He is also on the editorial board of *Indian Geographical Journal*, *Journal of Region, Health and Care*, *The East-West Geographer*, *The Social Profile*, *Aryabhatta Research Journal of Physical Sciences*, and *Regional Symbiosis*. He has been a member of the Governing Board of A.N. Sinha Institute of Social Studies (1994–1996), Dr. Bhim Rao Ambedkar College (1992–1995), Shivaji College (1997–1999), and Rajdhani College (1999) of Delhi University, as also of the Board of Research Studies of Kurukshetra University, Banaras Hindu University, Chaudhari Charan Singh University, Himachal Pradesh University, and Sri Sahuji Maharaj Kanpur University. He was also a member of the Academic Council, University of Delhi, for a 3-year term (1996–1999).

He has worked extensively for the popularization of geography. He was the first Indian geographer to be interviewed for a video interview on Indian geography ("Geographers on Film") as part of a venture of the 27th International Geographical Union (IGU) meeting in Washington, D.C., during August 1992. He has presented papers at numerous workshops, refresher courses, seminars, and conferences. He has organized four national and international conferences in Delhi: (1) 10th Indian Geographer's Meet, Annual Conference of Institute of Indian Geographers, February 17–19, 1989; (2) 11th Indian Geographical Congress, Annual Conference of National Association of Indian Geographers, India (NAGI), October 3–5, 1989; (3) 29th Annual International Regional Science Conference, January 31–February 2, 1997; and (4) 9th Annual Conference of Indian Institute of Geomorphologists, January 30–February 1, 1998. He continues to attend international conferences in the USA. He has presented papers in organized sessions and panel discussions at the Association of American Geographers (AAG) meetings in Denver (2005), San Francisco (2007), and Las Vegas (2009).

He attended several quantitative method workshops, which expanded and sustained his interests in quantitative geography. Notable among advanced school attendances are the Summer Institute in Quantitative Geography at the University of Gauhati (1973), Summer Institute in Statistics, Department of Statistics, Patna University (May–June 1977, 1978), and the Workshop on Location Analysis for Social Services at the Institute of Development Studies, University of Mysore (August 1978). Later, he was invited as the Resource Person in Quantitative Geography, Department of Geography, Patna University (1986) to deliver the keynote address. He also delivered a number of prestigious lectures, including those at University of Rajasthan, Jaipur (2003), and the Professor R.N. Dubey Memorial Lecture, Allahabad (2008), where he insisted upon interdisciplinary thinking in geography. He has held the view that geographic knowledge is an outcome of its intersection with other social and natural sciences horizontally and not vertically such as physics, psychology, economics, or engineering. His presentations draw ideas from cognate spatial disciplines concerned with space–time processes such as history, geology, regional science, behavioral psychology, and urban sociology.

He has published 35 peer-reviewed journal articles in national and international journals, 70 book chapters, and 21 written and edited books. He has several different projects on which he is currently working. His publications have appeared in the journals of: *Transactions of the Indian Council of Geographers*, *Journal of Bihar Research Society*, *Geographical Outlook*, *Indian Geographical Studies*, *National Geographic Journal of India*, *Geographical Review of India*, *The Deccan Geographer*, *The Canadian Geographer*, *Transactions of the Institute of Indian Geographers*, *Geo Journal*, *Annals of the National Association of Geographers India*, *Regional Symbiosis*, *and Indian Geographical Journal.*

2.4 Major Contributions

His four and one-half decades of research contribution can be classified in four subareas within geography: 'urban geography,' 'quantitative geography,' 'natural resource management,' and 'history of Indian geography.' His publications can be divided into four time periods: 1968–1978, 1979–1988, 1989–1998, and

1999–2011. The remainder of this chapter reviews his research contributions to these broad subfields of geography. His research career can be divided into two phases: 1964–1980 and post-1980s. During the first phase he was interested in understanding the long-term processes of urbanization in its regional context in developing countries, focusing on India and quantitative geography applications. After relocating to the University of Delhi, his interests shifted to the study of natural resource management and the history of Indian geography, although he pursued his interest in urban geography as well.

2.4.1 *Urban Geography*

B.T. received his higher education in geography at the University of Patna and University of Waterloo. At both institutions his focus of study was urban geography. In retrospect, his research follows the evolution of urban geography as a systematic field during the past five decades. He has mainly contributed to four broad areas within the subfield of urban geography: *land use and morphology dynamics*, *internal structure of cities*, *urban structural change*, *and urban system dynamics*. An examination of Fig. 2.1 shows that in the area of urban geography he has published slightly more than half his total publications, 53 % (1968–1978), which increased to 71.4 % (1979–1988) and then declined to 21.4 and 20.3 % during the 1989–1998 and 1999–2011 periods, respectively.

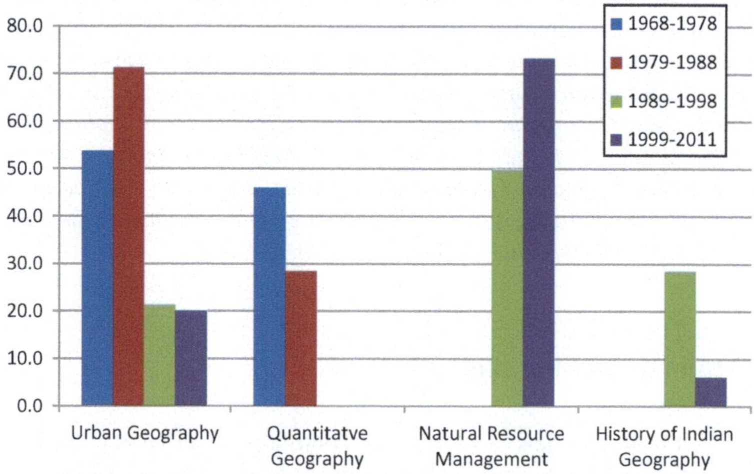

Fig. 2.1 Publication patterns (in %) of Professor Baleshwar Thakur: 1968–2011 (Source: Authors)

2.4.1.1 Land Use and Morphology Dynamics

In the early period the role of environmental determinism was paramount, and this led to a focus on studying the impact of physical features such as site and situation in determining the urban growth. First, he investigated the evolution and ecological structure of three cities in Bihar: Darbhanga, Gaya, and Patna (Thakur 1968a, 1972; Ram and Thakur 1972, 1998), and the spatial structure of Delhi metropolitan region (Rana and Thakur 2000; Thakur and Sagu 2007; Dahiya and Thakur 2007).

Darbhanga City occupies an important place among the towns of the North Bihar Plains. Thakur (1968b) traced the role of successive dynasties and rulers on the evolving urban pattern of the city. In the post-independence period, the city has been characterized by new residential development, ribbon development of government hospitals, and establishment of industrial estates.

The evolution of the city of Gaya cannot be explained by any classical model of land use but by a combination of all three—concentric, sector, and multiple nuclei models. The actual location of land use has been strongly determined by site and physical characteristics rather than cultural factors. Singh et al. (2005) studied the problems of planning in Darbhanga City and suggested the need for social facilities such as posts and telegraphs, police station, cinema house, dispensary, colleges, and children's parks. The folk culture of Mithila region in its geographic, historical, and sociological realms has been addressed by Thakur (2007a). This research focuses upon the unique paintings and folk music in the region. The glory of the paintings is based on both the human and natural resources of the region. Land, soil, water, and rich biotic resources are the bases of its growth. Also, the mental perceptions of the original geometrical design of floor, wall, and paper paintings are the strengths of Mithila women. However, the future and existence of Mithila paintings is threatened by the growing popularity of urban and Western culture.

Further, the sprawl and functional segregation has been largely influenced by religious, transport, and administrative factors. Patna is the capital of the state of Bihar and has been characterized as a great city in peril (Ram and Thakur 1998). The city has a poor location that has inhibited its proper development (Thakur 1972). The implication of a mixed land use pattern has led to a chaotic pattern of unorganized development. The city has an East–West contrast in its morphological and social structure with a directional bias in the growth of population density toward the West. The city needs to provide low-cost housing as well as high-rise apartments in congested areas to provide sustainable metropolitan growth. The condition of infrastructure supply and its maintenance is pathetic, perhaps because of underassessment and the poor collection of taxes, and thus maintenance and repair are in dilapidated condition. Further, land management in the rural urban fringe needs priority as there is frequent demand for zoning changes.

Large metropolitan cities in India are a conundrum of land uses (residential, commercial, industrial, office, agricultural, and open space). In particular the various consumers of usable land compete for land and bid to pay the highest price to put the land to the highest and best possible use. Thakur and Sagu (2007) examined the sociospatial pattern of elite residential colonies in the metropolitan Delhi region during the early 1980s. Their analysis identified 76 elite residential colonies. The colonies were divided into four groups: 'super-elite colonies,' 'elite colonies,' 'marginal elite colonies,' and 'peripheral elite colonies.' The samples of elites were categorized into nine occupational groups, and their absolute and relative distributions across elite residential colonies were mapped. Analysis revealed that location of elite colonies and distances from the core are positively related, as well as that elite colonies are located close to major transport infrastructure lines. The spatial distribution of super-elite colony locations depicts a sector pattern as the major concentration of elite colonies are located in two sectors, one dominated by diplomats and the other dominated by industrial or business households. Most of the elite colonies depict a dominance of elite business or government officials. The marginal elite colonies do not show similar characteristics. The peripheral elite samples do not allow such generalizations. Further, Thakur and Chauhan (2005) made a comparative analysis of Delhi and Mumbai's location of elite residential colonies. Their analysis suggested that in both cities an overall pattern of decentralization and suburbanization of elite colonies and occupational groups has shown a phenomenal growth.

Rana and Thakur (2000) have evaluated land use planning in Delhi, emphasizing commercial areas, the nature of its composition, and its spatial growth. There has been a preponderance of commercial sector growth during 1961–1981, especially in the areas of retail business, wholesale trade, and warehousing and storage facilities. The wholesale and warehousing businesses have a tendency to cluster around retail centers in the central business districts of Delhi. Thus, the objective of decentralization of business activities was defeated. Concomitant to this trend has been the unauthorized conversion of residential buildings into shops or godowns. This trend has led to mixed land uses becoming an inseparable part of the urban landscape in Delhi. The Delhi Master Plans have been criticized because they adopted the old British model of urban planning based on strict zoning regulations and compartmentalized land uses. That model has failed to recognize the more indigenous elements of the urban landscape, that is, informal trade and mixed land uses. However, the Delhi Master Plans did promote some important concepts of commercial land use planning such as land use planning in a regional perspective, introduction and planning of multipurpose building complexes, community centers, and the hierarchical planning of commercial areas on the basis of residential neighborhoods.

The rural–urban fringe is an area of mixed rural and urban land uses and populations between the continuously built-up urban and suburban areas of the central city and rural hinterland. The fringe begins at the point where agriculture land use starts near the city and extends to the point where village land use can be seen. The fringe displays a changing mix of land use and social and demographic characteristics. Dahiya and Thakur (2007) in a study of selected villages around Delhi observed that as the city approaches the urban fringe the behavior and norms of the native people change radically in terms of: household characteristics, occupation structure, economic and social linkages, and values. With urban contact, the Jajmani salaried system is substituted with modern salaried employment, the joint family system is substituted by the nuclear family, the average size and age of the household decreases, and peasants working as domestic or manual workers increase as the city approaches the rural hinterland.

The fringe region is marred with such issues as the location of large-scale urban amenities, the problem of fringe agriculture, the acquisition of land banks for future development, and the social integration of commuters in the urban realm. Thakur (2010a) has intensively studied the structure and dynamics of the urban fringe of Delhi, addressing the issues of spatial spread of the urban fringe in Delhi, impact of urbanization on the natural resources in the fringe of Delhi, and evaluated the programs and policies of Delhi's fringe development. He concludes that the expansion of the city has not been regular and uniform in all directions as a consequence of physical and administrative factors. The directional bias in the spatial spread of Delhi is constrained by its relief features. Khadar is a low-lying area with recent river deposits, and Bangar is composed of older deposits, which cover the northwestern portion of the state. The fringe development is agriculture based in the north, northwest, west, and east, and scenic and luxury areas are based in the south and southwest. The expansion of the city has impacted the natural resources by leading to a sharp fall in the level of groundwater and water quality. Also, fringe development is a dynamic and expansive process, thereby leading to boundary changes.

2.4.1.2 Internal Structure of Cities

The term internal structure is defined as the location, arrangement, and the interrelationship between social and physical elements in the city. The city performs an important role in the location of economic activities by way of site selection, the expression of the layout and transport network, and its central location. An important theme of city structure analysis has been the study of spatial and functional structure of urban business patterns. Commercial ribbons refer to the analysis of city shopping centers on the basis of their location and functional characteristics. Thakur and Dayal (1976) analyzed the commercial ribbons in Patna and classified the business establishments on the basis of their morphological characteristics. The commercial ribbons in Patna were influenced by riverborne trade along the Ganga River and later by railroad development. The physical distance between the old and new ribbons has increased with development. The commercial ribbons may be classified as old, central, new commercial and isolated ribbons on the basis of morphological characteristics. The ribbons have developed as specialized centers of commercial activities: old ribbons as wholesale marts, central ribbons as centers of higher-order shopping goods, new commercial ribbons providing financial and auto-related services, and isolated ribbons as convenience goods center. This specialization supports the hypothesis that as commercial ribbons develop in a certain direction, differences emanate among the old and new ribbons, as the old ribbon specializes in wholesale goods and shopping goods dominate in the newer ribbons.

2.4.1.3 Urban Structural Change

As regional economies develops, the nature of the interaction between economic sectors changes. The economy is primary sector oriented in the beginning, then it becomes manufacturing sector oriented, and finally tertiary sector oriented, leading to quaternary and quinary sector-dominated growth patterns. Thakur (1974a) in this context has explored the relationship between urbanization and economic development. The structural aspect of urbanization concerns the spatial organization of economic activities. Two models predominate this theme: (1) the center–periphery model and (2) the stage model of growth. The first model recognizes that the space economy must be viewed as an interacting system of relationships among developing and depressed regions. This model is more powerful as it articulates the national economic space as an interdependent system of cities. The second model describes the different stages of urban growth. Thakur (1979a) investigated the spatial changes in the concentration of industries in Kitchener-Waterloo during the period 1951–1970. He opined that growth industries do not disperse at a faster rate and that this is devoid of any association with shift from the center of gravity. Also, the female participation rate does not explain the shift from the center of gravity in a statistical sense.

Thakur (1974b) evaluated models of intraurban consumer travel behavior with respect to its ability to reveal the explanatory and predictive power of the models. He utilized four criteria to evaluate the models: 'logical consistency,' 'measurable behavior in real world,' 'dynamic and feedback effects,' and 'easily tested empirically.' He posited that the retail gravity model is a highly generalized model of aggregate human behavior and does not capture the behavior of individual consumers. The central place theory places emphasis upon the nature of physical retail location and becomes important with respect to the analysis of spatial patterns of individual behavior and shopping patterns. He also suggests that micro models of consumer travel behavior hold more promise in explaining intraurban consumer travel behavior.

2.4.1.4 Regional Urban Systems

The term urban system refers to the interdependence and interconnection between urban places. This concept was introduced by Brian Berry as part of his study and application of general systems theory to urban and regional systems analysis. Nations and large regions are organized into a set of urban places and their hinterlands, which exhaust the land area such that goods, services, people, capital, information, and ideas travel up and down the hierarchy through networks connecting a city and its hinterland. A national territory consists of regions that have their own urban systems at the subregional level and are called intraregional systems. Similarly, regions interact with other regions and form interregional systems. As development takes place, the relative role of cities vacillates. Thakur (1978) examined the relationship between urban dynamics and general systems theory and examined the basic concepts related to general systems theory such as 'closed and open systems,' 'entropy and negentropy,' 'stability and instability,' 'equilibrium,' and 'feedback mechanisms.' He pointed to the inappropriate delimitation and definition of boundaries in the empirical study of urban systems. He also suggested the necessity of ensuring the functional wholeness of the urban system in an analysis of spatial interdependence.

Russwurm and Thakur (1981), in their pioneering work, analyzed the functional change and stability patterns in the hierarchic order of central places in Ontario's regional settlement system during 1871–1971. The study area extended from the Western extent of Toronto to beyond the city of London. In their analysis they utilized the Davies index. A spatial analysis suggested that 'surging areas' were characterized by faster growth than their long-term means and were concentrated near large cities. This pattern was similar to 1871 except that the concentration was near Toronto in the latter period. The middle years during 1901–1941 were the most stable. Further, 'slowing places' occurred in the Northern periphery in 1871, from Toronto to Hamilton during 1901–1941 and in the rural area West Kitchener by 1971. The study further stated that the ideas of centrality and nodality in understanding stability characteristics are important. Centrality refers to the surroundings of a place in a region and nodality to the competitive position of a place within the urban system. In aggregate, since 1901 cities have gained and towns have lost with respect to nodality in the Ontario urban system.

Thakur (1968b) identified various types of rural settlements in North Bihar with a view to understanding the relationship between human groups and their natural environments. He identified four types of settlements: 'compact,' 'semi-compact,' 'hamletted,' and 'dispersed, sprinkled, or disseminated settlements.' Further, his interests shifted to the analysis of dynamic aspects of urban settlements. The growth trends and characteristics of urban settlement systems of Bihar for a hundred-year period (1872–1971) were analyzed. The region was delimited into two subregions, the Bihar plains and the Chhota Nagpur plateau. Rank–size relationships depicted a trend toward regularity and orderliness and were dominated by Patna in the Bihar Plains and Jamshedpur-Ranchi in the Chhota Nagpur plateau. The rank–size relationship has not been stable in either region. The position of primacy has not changed in the intervening period in the Bihar plains whereas the position of the most dominant city in the Chhota Nagpur plateau has shifted from Ranchi to Jamshedpur. Application of the entropy method detected changes in urban systems of different physical and economic characteristics (Dayal and Thakur 1982). Thakur (1979b, 1981) further extended his geographic coverage and explored the urban settlement patterns in Eastern India during the hundred-year period (1872–1971) and in North Eastern India during 1901–1971 using entropy analysis. He identified a center–periphery dichotomy in the evolution of the urban pattern in Eastern India. Also, the hypothesis of an increase in uniformity of urban places in North Eastern India been observed to be true.

Thakur et al. (2000) edited the felicitation volume for Professor A.K. Dutt titled *Geographic and Planning Research Themes for the Millennium*, which contains papers on urban and regional planning, economic geography, and social and cultural geography. He also edited a three-volume anthology on City, Society and Planning

(2007) commemorating the life and professional achievements of Professor A.K. Dutt. The first volume focuses upon urban structure and problems ranging from national to international scales. The second volume addresses issues related to social and cultural aspects of human life from different geographic areas around the globe. The third volume draws papers on various aspects of regional planning. Further, he edited another opus with his colleagues at the University of Akron, a volume of scholarly papers honoring the life and contributions of Professor A.G. Noble. The book, titled *Facets of Social Geography: International and National Perspectives* (Dutt et al. 2011), is a collection of scholarly papers on the nature, scope, evolution, and problems in social geography.

2.4.2 Quantitative Geography

In the area of quantitative geography he published a little less than half of his publications, that is, 46 % during the decade 1968–1978, which declined to 28.5 % during the period 1979–1988 (Fig. 2.1). In the early 1970s he was influenced by the quantitative revolution and its impact on geography. His graduate training at the University of Waterloo had a strong impact on his epistemological foundations in the methodological approaches in geography. After he became a faculty member at the University of Delhi in 1980, his interests shifted to conceptual developments in geography and natural resource management issues, and so his contributions in the area of quantitative geography took back stage. His publications in the field of quantitative geography can be classified into three areas: ‘exploratory application in geography,’ ‘sampling and geography,’ and ‘nearest neighbor and entropy analysis.’

His earlier publications concerned exploratory applications of statistical methods in geography in cartography and urban applications. Thakur and Thakur (1982) traced the development of modern techniques in statistical mapping and posited that by combining statistical and cartographic methods complex geographic data can be presented with ease and greater interpretative value. Thakur (1976) reviewed the theory and applications of the Spearman rank correlation coefficient and its limitations with reference to urbanization and central place studies.

Sampling is an important tool in geographic studies often utilized to make inferences about the population when the population is too large for enumeration. Thakur (1974) discussed the types of sampling techniques and the limitations of the application of sampling in geographic research. Random sampling is not suitable when linear trends are present in the data, and the assumption of independence of observations is violated when we analyze the behavior of consumers in a complex cultural region. He evaluated the efficacy of sampling methods for surveying the manufacturing firms in the twin cities of Kitchener-Waterloo, in Ontario (Thakur 1975). Further, he also alludes to the confusion regarding levels of measurement and the use of appropriate statistical techniques. He draws attention to the notion of power efficiency and robustness in inferential statistics (Thakur 1977).

Nearest neighbor analysis is a tool in geographic analysis utilized to determine the spatial arrangement of a pattern of points in a region. The distance of each point to its neighbor and the average nearest neighbor distance of all points are measured. The spacing within a point pattern is analyzed by comparing it with a threshold distance. Thakur (1973) analyzed the spatial pattern of urban places in southwestern Ontario during 1971. He tested the hypothesis that different levels of urban places in the region were uniformly spaced. He also pointed to two inadequacies in the use of nearest neighbor analysis. The first is the interpretation of the *R* statistic, which measures whether the distribution tends toward clustering, randomness, or uniformity. The value of the *R* statistic ranges from 0 to 2.1459 for an ideal distribution, but the values that constitute the measurement of clustering, uniformity, and randomness are not clearly defined. The second problem is the issue of drawing boundaries around the study area.

Entropy analysis is a measure of the amount of uncertainty in a probability distribution of a system subject to constraints. Two concepts central to the entropy method are the notions of macro and micro states. A macro state can be defined as an aggregate frequency distribution of urban places in a region, and the micro state can be defined as the various ways in which these urban places can be distributed in a region corresponding to the same macro state. Thus, entropy approach measures the relationship between a given macro state and the possible micro state that corresponds to it. Thakur (1979b) is the first Indian geographer to apply this approach to explore the urban settlement systems in Eastern India in the five regions: Bihar plains, lower Ganga plains, Chota Nagpur plateau, Utkal coastal plain, and Orissa highland region. He tested the hypothesis if uniformity increased over time in the urban settlements across the region. The analysis indicated an increase in entropy over time across all the regions during the period 1872–1971. In the analysis he also ascertained center–periphery dichotomy in the evolution of urban patterns in Eastern India. The lower Ganga plain was considered as the central region and the others as peripheral regions. In the central region the urban places have shown a trend toward randomness, whereas the peripheral regions have shown a uniform pattern of growth over time. Further, Thakur (1981) extended the analysis to North Eastern India for the period 1901–1971 with the same hypothesis. His analysis confirmed a trend toward greater conformity consistent with the assumptions of the Christallerian and Loschian urban landscape. Thus, the two studies lend support to the thesis of a center–periphery dichotomy in the evolution of the Indian regional urban system. He has suggested the exploration of other regional settlement systems using similar methodologies to understand the dynamics and evolution of regional urban systems in India. He also collated a bibliography on entropy studies in geography that is a wealth of resources for researchers in geography (Thakur 1979c).

2.4.3 Natural Resource Management

B.T. developed an interest in natural resource management (NRM) in the post-1980s. He was particularly influenced by the '*BKW* (Ian Burton, Robert Kates, and Gilbert White) *school of natural hazards*' and the '*MOSS* (B. Mitchell, T. O' Riordan, T.F. Saarinen, and W.R.D. Sewell) *school of PAVE* (*perceptions*, *attitudes*, *values and emotions*)' in his research and teaching endeavors. The BKW school was rooted in the behavioral, perceptual, and institutional responses to flood hazards. The MOSS school was an outgrowth of the BKW school extending the behavioral geographic approach to natural resources management. He was in particular motivated by the coursework he had completed with Professor Bruce Mitchell at the University of Waterloo. During the decades 1989–1999 and 1999–2011 he has mostly published in the area of NRM. Figure 2.1 depicts an increase in publication from 50 % during 1989–1999 to 73.4 % during 1999–2011 in the area of NRM. His publications can be broadly classified in the following three areas: *conceptual developments in NRM, land and scenic resources, and water management.*

2.4.3.1 Conceptual Developments in NRM

Zimmerman (1964) stated "Resources are not; they become." As development takes place, the needs of human being change and with that the means of satisfying the needs change as well. The rapid development of the global economy has put tremendous pressure on the natural resources of the Earth. Thakur (2008) examined the various theories providing an understanding of the interrelationships between population trends, resource use, and its concomitant development impacts. NRM is broadly defined as the as the management of renewable and nonrenewable resources by private, public, and community-based organizations. Thakur (2003) did a substantial review of the theoretical literature examining the decision-making approaches in NRM. In particular he examined the role of perceptions, attitudes, values, and emotions in resource management decision making. He is critical of geographic contributions to decision making in NRM. Several locational, behavioral, economic, and social variables affect the decision-making process in NRM, and the possibility in variations among these relationships is unbounded. Thakur (2012a) delivered the Presidential Address on "Research Perspectives on Resource Management in India" at the Annual Conference of the Institute of Indian Geographers held at Agartala Central University, Tripura. His address analyzed the status of research done in the field of resource management in India in the past 50 years. He has also seen the growth of resource geography in the context of paradigm development in India. This is a very exhaustive review of resource management studies in India and has also presented future directions of research for young geographers.

Further, Thakur (2007b) examined the trends in resource development in India and its implications for sustainable management. He posits that although India has progressed in attaining higher agricultural productivity, this development is restricted to selective crops and geographic areas because of such problems as land resources and their conservation. Much of India's agricultural land is subject to erosion, salinity, leaching, waterlogging, urban encroachment, and unregulated mining of land. India has made tremendous progress in harnessing rivers to distribute water to deficit regions and developing groundwater resources. Although India has a national water policy, problems of availability and adequate distribution exist in urban areas of the country. Deforestation has been the most serious and widespread problem in India. The nation lags in the enforcement of forest regulations, although many efforts have been made toward enhancing forest cover to sustain environmental conservation. Prospects of energy resources will not be bleak if there is continued emphasis on nonconventional energy resources.

2.4.3.2 Land Resources

Land is a basic natural resource upon which most of our productive sustenance is dependent. It is an important factor of production in agriculture, forestry, grazing, fishing, mining, and real estate development. It is an important input to the economic production process. An important theme that has been addressed is the ecological degradation of land. The overarching theme that has been addressed is land encroachment, reclamation of wasteland, degradation of land, land–human relationship, and caste and land ownership patterns.

Thakur (2010b), in his notable R.N. Dubey Memorial Lecture, addressed the theme of the causes and consequences of land degradation in India. In particular he focused upon the spatial and temporal variations of agricultural land and land–man ratio, spatial pattern of land degradation, efforts of reclamation and reforms of land development and management, challenges facing land development, and how effective has been public intervention in mitigating land degradation. He posits that the ratio of land to humans (land–man ratio) varies in accordance with agrarian and social environments. A high land–man ratio is found in the Thar Desert, Chambal badlands, the Malwa Plateau, the Deccan lava plateau, and Karnataka plateau. A low land–man ratio is found in productive areas such as the Indo-Gangetic plains and eastern fringe of Deccan peninsula. This pattern is an outcome of the direct positive relationship between rural population density and intensity of land use. Almost two thirds of the total land area suffers from land degradation and is characterized by soil erosion, desertification, mining, deforestation, salinization, alkalinity, waterlogging, landslides, overgrazing, and physical, biological, and chemical degradation of land. Further, the Government of India has set up the National Wasteland Development Board to enunciate reclamation efforts through a massive program of afforestation. Despite public intervention, the depletion of land resources via such trends as formation of deep gullies in north central India, coastal sandy land on the eastern and western coasts, spreading sand dunes, land affected by shifting cultivation, and derelict land in mining areas adds to the adverse trends. In conclusion his analysis suggests that land degradation trends can be reversed by enhancing access to land, improving tenancy, putting ceilings on land holdings, and redistributing land to the less-endowed classes.

Kapoor and Thakur (2008) analyzed the problem of land acquisition in Delhi. Their analysis suggested land acquisition is an outcome of the processes of growth and urban development in large cities such as Delhi. The expansion of urban land use will curtail agricultural land, and hence the quality of land foregone for urban expansion has to be sacrificed cautiously. Thakur and Bajpai (2008) examined the problem of classification and reclamation of wasteland in Uttar Pradesh. No uniform intervention can be implemented as different types of wastelands exist and thus different reclamations methods are required. They make recommendations for reclamation of saline soil, ravine land, waterlogged land, degraded forest land, and reclamation of uplands.

The land–man relationship and its spatial distribution pattern for East Sikkim revealed interesting trends during 1981 (Thakur and Bhushan 1992). First, agriculture land use is the most predominant in the region. Second, the density of population is the highest around urban centers, along the Tista River and valley bottoms. Third, the spatial variations in the quality of per capita land are largely determined by topographic and climatic conditions as well as ethnic characteristics. Fourth, and last with respect to surplus and deficit land, the Western, Southern, and revenue blocks situated along higher slopes and in the river valleys are surplus in character. Thus, valley bottoms, flat land, and proximity to urban centers have had a significant role in differentiating between surplus and deficit land. The relationship between caste structure and land ownership has been explored (Thakur and Vir 2003). The study revealed the role of the control of rural land resource and the spatial aspects of resource alienation. They further opined caste and spatial expression of rural landscape in terms of distance and direction, land quality, and land security in understanding the reasons for low agricultural productivity in the Central Bihar plain.

2.4.3.3 Water Management

Water management is an interdisciplinary field of study involved with planning, distributing, and the optimal use of water resources. Every competing user of water needs to be satisfied in an ideal world, but the availability of consumable water varies across space and time. Some regions are endowed with excess supply and others are faced with drought conditions. Thakur and Inderjeet (2000) studied the determinants of groundwater depletion in Eastern Haryana. A multivariate analysis revealed that groundwater quality and net annual draft were significant factors leading to the depletion of groundwater in the region. A rapid increase in the cultivation of water-intensive crops, well construction, and pumping machinery has exacerbated the depletion of water levels in the region. This trend advocates public intervention for reversing the loss of groundwater. Thakur and Preeti (2003, 2004) assessed the spatial and temporal availability of water resources in the Hindon basin of Upper Ganga-Yamuna Doab during 1980–1992. They utilized the water balance technique developed by Thornthwaite to calculate the spatial and temporal availability of the water resource. They posited that the availability of water is satisfactory but there are marked spatial and temporal variations in water utilization. Also, there is a need for optimal and efficient utilization of water resources. The efficient utilization of water by small farmers is affected by such factors as water rates, distance of farms from water sources, and the socioeconomic conditions of the farmers. The impact of limestone quarrying on water resources in the Dehradun district was investigated by

Thakur and Kurl (1996). Their objective was to understand the spatial distribution and utilization of water resources for domestic irrigation and industrial use. The analysis suggested wide fluctuations in water discharge, heightened impact of quarrying on water depletion, and ecological degradation.

B.T. has edited a 13-volume opus on *Resource Management in Developing Countries*, published by Concept Publishing Company. A brief summary of the volumes is provided in the next several paragraphs.

Vol. I: Resource Management: Theory and Techniques

This volume gives the practitioner access to the most practical and recent thinking on the subject of natural resources management including theories, models, and paradigms. This immensely stimulating volume is essential reading for those concerned with the resource management process as it operates in both developed and developing countries. It elaborates the concepts of renewable and nonrenewable resources, ecology, and sustainable development and emphasizes approaches to resources management. It also demonstrates the use of Geographic Information Systems and remote sensing techniques in inventory and appraisal of natural resources to improve the quality of decision making as well as in monitoring and evaluation of public policy.

Vol. II: Population, Resources, and Development

Pressures on natural resources caused by increasing populations in developing countries and their quest for a high standard of living are growing at a rate that is increasingly difficult to manage. This volume, therefore, addresses the complex interrelationships between population, resources, and development, including massive quantitative expansion in population size and the concomitant pressure on natural resources, the penultimate problems of human survival, and development opportunities and constraints. It also brings to light new information on the depletion of common property resources and probes into the questions of resource adequacy.

Vol. III: Ecological Degradation of Land

The volume provides a series of essays on various facets of land degradation, focusing on wasteland identifications, distribution, mapping, and remedial measures. It gives special attention to environmental conditions and trends in desertification and takes account of critical issues and the problems emerging thereof. It then examines processes and policies of the loss of agricultural land on the urban fringe and the depletion of common land in tribal areas.

Vol. IV: Land Appraisal and Development

This volume is a collection of empirical studies focusing on methodological approaches to land appraisal of mountainous, desert, and deltaic areas. It provides a comprehensive and in-depth analysis of the dimensions of agricultural development and production potential for the alleviation of poverty. In addition, the volume provides an understanding of relationships between land values, land use planning, and development policies reflecting the changing needs of the urban community.

Vol. V: Soil Deterioration and Conservation

The volume fills a critical void in the soil resource literature by providing the necessary scientific background for a comprehensive understanding of soil deterioration caused by increased human activity, particularly from the users' point of view. It discusses primarily contemporary patterns of soil utilization, estimation of the magnitude and extent of soil deterioration, erosion control, and conservation policy. It also presents a few in-depth case studies on shifting cultivation as a soil management system for subsistence farming in low-population tropical areas.

Vol. VI: Water Supply and Quality

The volume has special significance in the context of current urban water supply and quality problems because of increasing demand for water in the face of rapid industrialization, growing population, and urban migration. The chapters are centered around water supply and demand, the impact of agricultural and industrial activities and waste disposal on water quality, the spatial pattern of drinking water quality, eutrophication, and water balance. The volume also addresses strategies for controlling water pollution in rivers and lakes.

Vol. VII: Water Utilization and Management

In this volume, the researchers present empirical evidence on the environmental, bureaucratic, and socioeconomic constraints in water resources utilization and management. It offers specific strategies for water management in health resort towns, metropolitan cities, canal irrigation, rice cultivation, the dairy industry, refineries, and ponds and tanks. It concludes that both public officials and the general public should have sufficient awareness of both water problems and management options.

Vol. VIII: Groundwater Assessment and Development

Succinct, yet comprehensive in its coverage, this volume focuses on groundwater exploration and development in hard rock and alluvium areas, including the determination of durable and efficient setting of wells, assessing the groundwater potential of the aquifers for management, delineation of the freshwater–saltwater boundary, and rates and magnitude of groundwater depletion and local subsidence. In addition, the volume concentrates on the analysis of spatial and temporal variations in groundwater quality and the economics of irrigation with groundwater.

Vol. IX: Bioresource Depletion and Conservation

This volume contains a wealth of information on the patterns and processes of bioresource depletion. Through empirical evidence, it provides unique insights into the underlying economic, social, and political forces that encourage conversion of forests to other less valuable uses. The book offers an extensive discussion of the strategies to conserve biodiversity, educate the public, and contribute to sustainable development. The book also explores the extraordinarily rich potential for livestock and fisheries development to resolve economic problems.

Vol. X: Environmental Hazards

Based on a wealth of empirical detail, this volume provides a balanced perspective on the occurrence, causes, and impact of environmental hazards on society and natural resources. Included are resource utilization and managerial problems arising from flood, drought, landslip, landslide, riverbank erosion, tropical cyclone, volcano, and earthquakes.

Vol. XI: Scenic Resources and Tourism

This comprehensive volume focuses on scenic resources and tourism development studies conducted by the individual researcher at the microlevel and by tourism development agencies at the macrolevel. It examines the theories in full diversity and manifestation including scenic sites, pilgrimage-based tourism, ecotourism, and agro-tourism, the structure of tourism development and impacts, and some of the major components of the tourism planning process.

Vol. XII: Energy Resources and Environment

This interesting and thought-provoking volume examines the distribution, production, and utilization of both conventional and nonconventional energy resources. It provides new insights into energy demand, problems, and prospects and focuses on energy policies and programs for sustainable development. It explores techniques of mineral resources appraisal, showing their use in mineral conservation and management. In addition, this volume sheds new light on the environmental degradation and planning posed by our resource use.

Vol. XIII: Integrated Watershed Development

Drawing upon empirical studies, this final volume in the series presents an inventory of watershed resources, and their utilization, development, and management, with particular emphasis on identifying problems in humid tropical uplands and river basins. It establishes a mechanism for community participation in decision making in formulating strategies for integrated development of resources to bring enhanced resource productivity on a sustainable basis.

2.4.4 History of Indian Geography

B.T. has published nine papers on the history and development of Indian geography. Approximately 28 % of his publications in this area were published during the 1990s and roughly 6 % during the 2000s. For almost 30 years he taught a course on the 'History of Geographic Thought' and 'Conceptual Developments in Geography' at Patna University and Delhi University, respectively. Geography in the Western world evolved as a field focusing upon the study of geographical phenomena at various scales utilizing man–land relationships, area study, and spatial organization approaches (Taaffe 1974). Currently, the field of geography is undergoing turmoil with a tussle between dominance and survival among multiple approaches such as quantitative versus qualitative, quantitative versus post-positivist, and theoretical versus casual observation.

Thakur (2011) classifies Indian methodological developments in geography into three phases: pre-quantitative, quantitative, and post-quantitative. The first phase existed during the 1930s–1960s, a period when the method of geographic inquiry was purely descriptive of observations such as relief of land use, urban forms, and architecture of settlements. The second phase emphasized fieldwork, with intense focus on data collection and processing of data; this was an expansionary, more diversified, analytical, and scientific phase in the development and application of methodologies in Indian geography. Several alternative statistical methodologies were utilized, such as descriptive statistics, nonparametric statistics, spatial statistics, and multivariate statistics, and an effort toward model building. During the third phase of evolution, that is, the post-quantitative phase, Indian geographers responded to various criticisms made by the different schools within geography such as Marxism, humanism, critical social theory, post-modernism, post-structuralism, and neo-quantification. This phase also comprised the development of spatial statistics and geographical information systems (GIS), which is often called modern spatial analysis, and also the development of qualitative methods and mixed methods approaches.

Thus, Indian geography is characterized by 'abuse of statistical methods,' 'declining field work,' and 'model construction.' The methodological sphere of Indian geography is further characterized by a lack of reliable and robust hypotheses, creative research questions, and appropriate conceptual framework. The adoption of spatial quantitative methods to test spatial hypotheses is on the decline, although Indian geographers have made rapid application of thematic, computer, statistical, and GIS mapping, which has contributed to success in understanding the regional dimensions of Indian geography (Thakur 2011).

In his two-part opus on Indian geography, Thakur (1994a, b) reviewed and critically analyzed the development of Indian geography during the post-independence period. He further provided several directions that have relevance to contemporary Indian society, planning, and policy making. He divides the development of Indian geography into three periods: pre-1950, 1951–1970, and post-1970. During the first phase the primary emphasis was on teaching geography in schools, rapid expansion of the teaching of geography at the undergraduate level in colleges, the formation of geographic societies, and publication of geographic journals to promote research and teaching. During the second period there was greater emphasis upon analytical focus. Further, during the third phase there was expansion in teaching and research as grants from the government increased.

A critical analysis of the development of Indian geography suggests (1) geography is a distinct field of study focusing upon national and regional planning and utilizing maps as a tool of regional analysis, (2) research work on physical geography problems is weak and theoretical contributions are limited, (3) use is increasing in quantitative spatial analysis and synthesis, (4) geographic inquiry suggests a lack of ability to move from micro- to macrolevel generalizations, (5) lack of indigenous model building based on local assumptions and reality, (6) prevalence of an interest in topical division, (7) knowledge regarding spatial organization is given higher attention at the cost of spatial approach, (8) the temporal dimension has been ignored and a static approach favored, (9) excess reliance is placed on secondary as opposed to primary data, (10) limited focus upon systems or a behavioral approach, and (11) to a large extent, research is conducted in isolation in a descriptive style with limited focus upon a theoretical and conceptual framework (Thakur 1994b).

Thakur (2005) traced the development of geography at the University of Delhi at the postgraduate level. He meticulously documented the contributions of the 15 chairpersons and the faculty in fostering and scaling the department to national and international fame in the teaching and advancement of geographic research. In particular, the department has offered specializations in the three areas of (1) urban and regional planning, (2) resources and environment, and (3) social, medical, and political geography. He provides the major milestones from the 'formative stage,' to 'teething years,' 'take-off phase,' via the stage of 'consolidation,' and 'tasks ahead.'

Urban geography is one of the most dynamic and active fields within Indian geography. Thakur and Parai (1993) reviewed the contributions to urban geographic research in India during the 1980s–1990s. Urban geographers have explored in great detail macro-regional urbanization through space–time analysis and empirical studies. Indian geographers have explored challenging topics in intraurban and sociocultural patterns and problems. He opined that only a few geographers have made an effort toward indigenous model building. His evaluation of urban geographic research in India suggests the following areas for future research: (1) analysis of structure, function, and behavior of regional systems, (2) spatiotemporal analysis of periodic markets and rural development, (3) rural service development planning, (4) reformulation of classical land use models to suit Indian conditions, (5) spatial and temporal variations in the urban informal market, (6) emphasis on sociocultural aspects such as language, crime, education, health, and pilgrimage, (7) conservation and management of urban water supply, (8) land management in the urban fringe, and (9) macrolevel national analysis (Thakur and Parai 1993).

Professor Thakur is currently working on various research projects including Urban Structure and Processes in India (with Professor Ashok Dutt) and India: Land, People and Economy (with Professor R.P. Misra). The remainder of this book consists of chapters contributed by Professor Thakur's colleagues, students, and esteemed friends. The book is divided into eight parts comprising 28 chapters covering a wide gamut of topics and themes. As editors we wish to thank the contributors for writing scholarly chapters for this volume.

References

Dahiya BS, Thakur B (2007) Intra-village and inter-village response to urbanization in the rural-urban fringe of Delhi. In: Thakur B, Pomeroy G, Christopher C, Thakur SK (eds) City, society and planning, vol 1, City. Concept, New Delhi, pp 476–93

Dayal P, Thakur B (1982) Settlement systems in Bihar: dynamics of urban settlement systems. In: Alam SM, Gopi KN (eds) Settlement systems in India. Oxford and IBH Publishing, New Delhi, pp 144–167

Dutt AK, Wadhwa V, Thakur B, Costa FJ (2011) Facets of social geography: international and Indian perspectives. Cambridge University Press India, New Delhi

Kapoor A, Thakur B (2008) A geographical analysis of land acquisition for urban development: a study of Delhi. In: Thakur B (ed) Perspectives in resource management in developing countries, vol 3, Ecological degradation of land. Concept, New Delhi, pp 277–295

Ram LN, Thakur B (1972) The ecological structure of Gaya. Geogr Outlook 9:88–99

Ram LN, Thakur B (1998) Patna: a great city in peril. In: Misra RP, Kamlesh M (eds) Million cities of India: growth dynamics, internal structure, quality of life and planning perspectives. Sustainable Development Foundation, New Delhi, pp 635–657

Rana L, Thakur B (2000) Commercial land use planning and development in Delhi. In: Noble AG, Thakur B, Mukerji AB, Costa FJ (eds) Geographic and planning research themes for the new millennium. Vikas Publishing, New Delhi, pp 35–54

Russwurm LH, Thakur B (1981) Hierarchical and functional stability and change in a strongly urbanizing area of Southwestern Ontario: 1871–1971. Can Geogr 25:149–166

Singh VP, Singh DP, Thakur B (2005) Darbhanga City: problems and planning. In: Thakur B, Sinha VNP, Prasad M, Sharma N, Rana P, Mandal RB, Singh RBP (eds) Urban and regional development in India, vol II. Concept Publishing, New Delhi, pp 267–311

Taaffe EJ (1974) The spatial view in context. Ann Assoc Am Geogr 64(1):1–16

Thakur B (1968a) Types of rural settlements in north Bihar. Trans Indian Counc Geogr 5:157–163

Thakur B (1968b) Evolution of the townscape of Darbhanga. J Bihar Res Soc 54:354–368

Thakur B (1972) The central business district of Patna. In: Alam S (ed) Proceedings of symposium on urban land use. National Committee of Geography, Kolkata, pp 77–84

Thakur B (1973) Nearest neighbor analysis as a measure of urban place patterns. Indian Geogr Stud Res Bull 1:59–71

Thakur B (1974a) Urbanization and economic development: the structural contribution. Indian Geogr 14:55–59

Thakur B (1974b) Models of intra-urban consumer travel behavior. Indian Geogr Stud Res Bull 2:62–72

Thakur B (1974c) Sampling methods and the geographer. Indian Geogr Stud Res Bull 4:55–59

Thakur B (1975) Application of sampling methods in an industrial area. Indian Geogr Stud Res Bull 5:75–82

Thakur B (1976) Spearman rank correlation coefficient: a review of technique and applications in geographical research. Geogr Rev India 38:18–28

Thakur B (1977) Problems of measurement, power efficiency and robustness in inferential statistics. Indian Geogr Stud Res Bull 8:51–57

Thakur B (1978) Urban dynamics and general system theory. Indian Geogr Stud Res Bull 11:15–26

Thakur B (1979a) Intra-urban location of manufacturing: a test of the hypothesis of spatial changes. Deccan Geogr 17:499–515

Thakur B (1979b) Urban settlements in eastern India: entropy changes and pattern analysis. Concept Publishing, New Delhi, p 222

Thakur B. (1979c). Bibliography on entropy studies in geography (Vance M, ed) Vance Bibliographies, Monticello, pp 285–320

Thakur B (1981) Entropy analysis of changing urban pattern in northeastern India. Trans Inst Indian Geogr 3:175–181 (reprinted in Mahadev PD (ed) (1986) Urban geography. Heritage Publishers, New Delhi, pp 161–169)

Thakur B (1994a) Indian geography: development trends and prospects: I. Trans Inst Indian Geogr 16:67–85

Thakur B (1994b) Indian geography: development trends and prospects: II. Trans Inst Indian Geogr 16:177–195

Thakur B (2003) Coping with natural hazards in India: a geographer's agenda. Ann Natl Assoc Geogr India 23:1–20

Thakur B (2005) The evolution of geography in the University of Delhi. In: Thakur B (ed) Geography in Delhi, University of Delhi. Manisha Publications, New Delhi, p 1

Thakur B (2007a) The folk culture of Mithila. In: Thakur B, Pomeroy G, Christopher C, Thakur SK (eds) City, society and planning, vol 2, Society. Concept, New Delhi, pp 422–446
Thakur B (2007b) Emerging trends in India's resource development: implications for sustainable development. In: Thakur B (ed) Perspectives in resource management in developing countries, vol 2, Population, resources and development. Concept, New Delhi, pp 1–87
Thakur B (2008) Linking population, resources and development: a review of conceptual approaches. Reg Symbiosis 16:11–40
Thakur B (2010a) The structure and dynamics of the urban fringe of Delhi. In: Dikshit JK (ed) The urban fringe of Indian cities: Professor Jaymala Diddee Felicitation volume. Rawat Publications, New Delhi/Jaipur, pp 172–190
Thakur B (2010b) Depletion and degradation of land resources in India (series editor, L.R. Singh). R.N. Dubey Foundation, Allahabad
Thakur B (2011) Methodological developments and empirical practice in Indian geography. In: Sharma PR, Sharma VN (eds) Research in Indian geography: issues and solutions. R.K. Books, Delhi, pp 215–252
Thakur B (2012a) Research perspectives on resource management in India. Trans Inst Indian Geogr 34(2):1–29
Thakur SK (2012b) Methodological development in social geography. In: Dutt AK, Costa F, Wadhwa V, Thakur B (eds) Facets of social geography: international and Indian perspectives. Cambridge University Press, Delhi
Thakur B, Bajpai M (2008) Capability classification and reclamation of wasteland in Etawah District. In: Thakur B (ed) Perspectives in resource management in developing countries, vol 3, Ecological degradation of land. Concept, New Delhi, pp 323–361
Thakur B, Bhushan PS (1992) Land–man relationship and distribution of agricultural land in East Sikkim. Trans Inst Indian Geogr 14:13–31
Thakur B, Dayal P (1976) Commercial ribbons in Patna: spatial growth and functional structure. Natl Geogr J India 22:164–177
Thakur B, Jeet I (2000) Determinants of groundwater depletion in Eastern Haryana. Indian Geogr J 43:141–145
Thakur B, Kurl A (1996) Impact of limestone quarrying on water resources in the Tons Basin, Dehradun District. Reg Symbiosis 4:91–106
Thakur B, Parai A (1993) A review of recent urban geographic studies in India. GeoJournal 29(2):187–196
Thakur B, Sagu AK (2007) Locational analysis of elite residential colonies in Delhi. In: Thakur B, Pomeroy G, Christopher C, Thakur SK (eds) City, society and planning, vol 1, City. Concept, New Delhi, pp 494–522
Thakur B, Singh Chauhan SB (2005) Growth and characteristics of elite residential colonies in Mumbai and Delhi. In: Thakur B, Sinha VNP, Prasad M, Sharma N, Rana P, Mandal RB, Singh RBP (eds) Urban and regional development in India, vol II. Concept, New Delhi, pp 99–118
Thakur B, Preeti S (2003) Utilization efficiency of irrigation water in the Hindon basin. Indian Geogr J 78(2):77–90
Thakur B, Preeti S (2004) Spatio-temporal study of water balance in the Hindon basin: 1980–92. Reg Symb 12:49–69
Thakur B, Thakur R (1982) Statistical techniques in population and settlement mapping. Geogr Rev India 44:62–77
Thakur B, Vir P (2003) Caste structure, patterns of land ownership and crop security in Ramgarh Block, Bhabua District. In: Sureshwar J, Madhusudan T, Thakur B (eds) Dimensions of society and culture of Bihar: felicitation in honor of Dr. Madneshwar Misra. Academy Press, Allahabad, pp 288–303
Thakur B, Mukerji AB, Noble AG, Costa FJ (eds) (2000) Geographic and planning research themes for the new millennium. Vikas Publishing, New Delhi
Thakur B, Pomeroy G, Cusack C, Thakur SK (2007) City, society and planning. Concept, New Delhi (3 vols)
Zimmermann EW (1964) An introduction to world resources. Harper & Row, New York

Part II
Methodology

Chapter 3
Spectral Unmixing with Estimated Adaptive Endmember Index Using Extended Support Vector Machine

Chandrama Dey Sarker, X. Jia, Liguo Wang, D. Fraser, and L. Lymburner

Abstract The most difficult operation in flood inundation mapping using optical flood images is to separate fully inundated areas from the "wet" areas where trees and houses are partly covered by water: this can be referred as a typical problem, the presence of mixed pixels in the images. A number of automatic information extraction image classification algorithms have been developed over the years for flood mapping using optical remote sensing images. Most classification algorithms generally help in selecting a pixel in a particular class label with the greatest likelihood. However, these hard classification methods often fail to generate a reliable flood inundation mapping because of the presence of mixed pixels in the images. To solve the mixed pixel problem, advanced image processing techniques are adopted, and the linear spectral unmixing method is one of the most popular soft classification techniques used for mixed pixel analysis. The good performance of linear spectral unmixing depends on two important issues: the method of selecting endmembers and the method to model the endmembers for unmixing. This chapter presents an improvement in the adaptive selection of the endmember subset for each pixel in the spectral unmixing method for reliable flood mapping. Using a fixed set of endmembers for spectral unmixing all pixels in an entire image might cause overestimation of the endmember spectra residing in a mixed pixel and hence reduce the performance level of spectral unmixing. Compared to this, application of an estimated adaptive

C. Dey Sarker (✉) • X. Jia • D. Fraser
School of Engineering and Information Technology, University College,
The University of New South Wales at the Australian Defence Force Academy,
Campbell, Canberra, ACT 2600, Australia
e-mail: C.Dey@student.adfa.edu.au; X.Jia@adfa.edu.au; D.Fraser@adfa.edu.au

L. Wang
College of Information and Communications of Engineering,
Harbin Engineering University, Harbin, China
e-mail: wangliguo@hrbeu.edu.cn

L. Lymburner
Australian Centre for Remote Sensing (ACRES), Geoscience Australia,
Symonston, Canberra, ACT 2609, Australia
e-mail: Leo.Lymburner@ga.gov.au

A.K. Dutt et al. (eds.), *Spatial Diversity and Dynamics in Resources and Urban Development*, DOI 10.1007/978-94-017-9771-9_3

subset of endmembers for each pixel can decrease the residual error in unmixing results and provide a reliable output. In this chapter, it has also been proved that this proposed method can improve the accuracy of conventional linear unmixing methods and is also easy to apply. Three different linear spectral unmixing methods were applied to test the improvement in unmixing results. Experiments were conducted in three different sets of Landsat-5 TM images of three different flood events in Australia to examine the method on different flooding conditions and achieved satisfactory outcomes in flood mapping.

Keywords Linear spectral unmixing • Remote sensing • Flood mapping • Endmember selection • Extended Support Vector Machine

3.1 Introduction

In recent years, remote sensing technology has made substantial contributions in every aspect of flood disaster management such as preparedness, prevention, and relief. Remote sensing provides satellite images as the most powerful tools in mapping of the spatial distribution of disaster-related data within a short period of time. Active research has been conducted in recent years to map the extent of a flood by using optical data for reasons of their effectiveness and availability, as well as low cost, and it has been generally observed that in spite of facing difficulties in loss of information from cloud cover or vegetation cover, using optical images is more convenient than microwave images because of the complexities involved in interacting with ground objects and the high costs of the microwave images (Wang 2002). Satellites such as IRS-1C/1D, with the three sensors PAN, LISS-3, and WiFS, NOAA, Landsat 5 Thematic Mapper, and Landsat 7 Enhanced Thematic Mapper provide data for various flood characteristics. However, the great challenges involved in data interpretation for flood mapping are to separate the permanent water bodies from areas under full flood or partly flooded, a typical cause of the existence of mixed pixels in flood images.

Generally, it is assumed that every spot on the ground can be labeled to one and only one category. Such discrete categorization is convenient and challenging by its simplicity, but in reality it is not the accurate portrayal of a real landscape. Because pixels represent a spatial average of all the objects covered, it is inevitable that multiple spectral categories will be included in most of them (Schowengerdt 1997). Therefore, mixed pixels represent a mixture of cover types or information. All natural and man-made surfaces are nonuniform at some level of spatial resolution, and therefore an increase in spatial resolution may reduce the percentage of mixed pixels, but mixed pixels will still occur at the boundary between objects regardless of their size and the sensor resolution (Schowengerdt 1997). Particularly, in relationship to flood scenes, the prevalence of mixed pixels is quite reasonable, given the spread of flood water over the other ground cover types.

To solve the mixed pixel problem (as mentioned in the previous section) in image analysis, advanced image processing techniques has been developed. Among these, linear mixture models are widely used. In this model, each endmember's spectral characteristics are represented by a single spectrum, which may be derived from averaging a set of representative samples. A mixed pixel's reflectance is assumed as a weighted sum of all the endmembers (Keshava and Mustard 2002). More precisely, the basic hypothesis behind the method of linear spectral unmixing is that the electromagnetic radiation that is recorded as the spectral reflectance of a pixel by the sensor is the linear sum of the spectral signatures of pure ground components or endmembers multiplied by their respective ground proportions within that particular pixel. The linear unmixing method only allows estimating the surface fractions of a number of pure spectral components, when a linear mixing occurs within a pixel. Linear mixing results when electromagnetic solar radiation undergoes single scattering to reach the satellite sensor from the ground surface. It also assumes that the incident solar radiation only interacts with a single ground component and does not undergo multiple scattering within the particles of that ground component (Keshava and Mustard 2002). The mathematical explanation of this method is presented in Sect. 3.2.

The spectral mixture analysis method consists of three important areas to be addressed: first, how to model the mixed spectrum; second, how to model each endmember; and third, how to address the method of selection of the endmember.

3.1.1 *Modeling the Mixed Spectrum*

The two most widely used methods to model the mixed spectrum are linear and nonlinear. However, the linear model is more popular in the spectral unmixing method. This model assumes that the surface within a pixel consists of distinct materials with constant spectral properties. More precisely, the incident solar radiation is reflected from the Earth's surface through a single bounce and is not characterized by multiple scattering within a pixel (Mcfeeters 1996). In the linear model, the spectrum of mixed spectra is modeled as a linear combination of the pure spectra of the materials of interest present in a particular pixel weighted by their fractional abundances (Klatt 1994; Mcfeeters 1996). But, on the other hand, in many cases a pixel having a single reflectance between objects in it is merely an assumption of the reality. Generally, there are multiple reflections within a material or between objects within a GIFOV or a pixel, and a nonlinear approach is needed to model the mixing method. In the current experiment, however, the linear model has been applied.

3.1.2 *Modeling Each Endmember*

The second issue concerns how to model each endmember. In the literature of mixture modeling, the dominant distinct ground objects with relatively constant spectral properties are termed the endmembers (Smith 1985). This issue is about the techniques to find the most spectrally pure pixels that typically correspond to endmember spectra representing a primary class or the basic training component, which helps to generate an endmember class. A number of algorithms have been proposed to find the appropriate endmember for spectral unmixing: pixel purity index (PPI), N-FINDER, automated morphological endmember extraction (AMEE) techniques [10], convex cone analysis (CCA), iterative error analysis (IEA), vector component analysis (VCA), and iterated constrained endmembers (ICE) (Schowengerdt 1997). However, the in-class spectral variation is often large, and fractions obtained from the linear mixture algorithm can be negative or not sum to unity because it is inappropriate to expect that the pixels belonging to the same class have the same spectrum. Recently, the potential of the support vector machine (SVM) in providing new understanding of mixed spectral data was investigated. In the current study, an extended version of SVM (u_ESVM) has been developed for spectral unmixing. In this method, the complete set of training samples is used to model the pure pixels of the defined class, and u_ESVM has been applied for both the pure and mixed pixel interpretation.

3.1.3 Selection of Endmembers

Among the three main issues of linear spectral unmixing, selection of endmember class is one of the most important factors. The endmembers generally are collected from image data, and in this way image endmembers can be directly linked with the surface components detectable in the image scene. The number of endmember classes varies based on the spectral difference, number of bands, and spatial resolution. An appropriate number of endmember sets needs to be selected for an accurate unmixing. In addition to the selection of endmembers that are good representatives of ground materials, a wise selection of endmembers for each individual pixel is also necessary for a precise unmixing (Atkinson 2005). However, it is not sensible to expect that all the pixels in the test image contain all the endmembers. Consequently, if a particular endmember class is absent in a pixel, the estimated fractional abundance is not often zero, and large fractional errors are generated because of an excess of endmembers used for that particular pixel. This error can be reduced if an adaptive subset of endmember selection can be made on a per pixel basis where this optimized endmembers set is a more accurate representation for the given pixel (Atkinson 2005). In our previous experiment, we have already demonstrated the proposed method of application of the adaptive subset of endmember index for unmixing improves the unmixing accuracy (see Atkinson 2005 for a detailed illustration). In that experiment, we made the adaptive endmember index including only the two most probable (the two classes with highest and second highest probability for each mixed pixel) classes for mixed pixels to adapt for simplicity in the experiment. In the current paper, we have extended that work by generating the estimated adaptive endmember index file for individual pixels by incorporating all the endmember class (or classes) that reside(s) in that pixel. The methodology is illustrated in the following sections.

3.2 Methodology

3.2.1 Adaptive Estimated Endmember Selection

In the current experiment, an attempt has been made to generate a potential endmember index file for each pixel of the test image. This method only selects those endmember classes that are present in a pixel, and this has made unmixing performance better in the context of accuracy level. Assuming there are P primary classes in an image scene, for a particular pixel, the key task of this proposed method is to select automatically a subset of M number of classes that are relevant for each current pixel. Three factors can be analyzed to eliminate the nonrelevant endmembers to obtain the relevant subset of classes, as mentioned by Jia et al. (2010). First, we can expect that the mixed pixels have low probability of belonging to any defined classes. After sorting the probability of each class for a pixel, if a good gap is observed between the highest and second highest value, the pixel is determined to be a pure pixel. A small gap will indicate a mixed case. Second, for a highly spatially correlated scene, the mixed pixels are found in the boundaries between two neighboring ground cover types. In such cases, initially a thematic map is generated by applying an initial classification based on the multispectral measurement. After that, a window is defined to run on each pixel, and if its neighbors within the window have the same labels, that pixel is pure and there is no need to unmix that. Otherwise, the dominant classes within that window are selected as the resident classes of that pixel. Third, expert knowledge or local knowledge can be applied to eliminate the nonrelevant endmembers for a given case.

These three factors can be combined or applied alone for endmember selection depending on the individual scenes. Here, we have applied a modified version of the first factor to generate an unmixed index matrix to control the subset of classes to apply in the unmixing process. To find the most appropriate classes (or class) for a pixel, we propose to examine the probability of the pixel belonging to each primary class. To generate the probability results, $P(m), m=1,2\ldots P$ (P is the total number of primary classes) for each pixel, the maximum likelihood classifier can be used; this is the most common supervised classification method used in remote sensing image analysis. Each pixel was assigned the number of P(m) based on the total number of primary classes. The $P(m)$ values are normalized and sorted in a descending order. Threshold values are then defined and applied to determine which classes (or class) are contained in that particular pixel. Basically, if $P(m)$ is very small and below the threshold, we believe class m is not present, and if the $P(m)$ value with highest probability within a pixel is greater than a predefined threshold, the pixel is considered as the pure pixel. The determination of threshold value is determined by analyzing the range of the normalized probability values of classes.

Applying these rules, an unmixing index matrix can be formed to indicate which class(es) each pixel contains. The adaptive subset of endmember classes was applied in the linear spectral unmixing method using the extended support vector machine on three different case studies. Results were compared with the results obtained by using a fixed set of endmember classes in unmixing. Moreover, similar sets of fixed and adaptive subsets of endmember classes were applied on the two other most widely used linear spectral unmixing methods, that is, the partial constrained linear spectral (PCLS) unmixing method and the multiple endmember spectral mixture analysis (MESMA). The following section explains the concept of linear spectral unmixing using the extended support vector machine (u_ESVM).

Linear spectral unmixing occurs when the incident solar radiation only interacts once with each ground component, and does not undergo multiple scattering within the particles of that ground component, and the reflectance undergoes single scattering when reaching the satellite sensor from the ground surface. The mathematical representation of the spectral reflectance of a mixed pixel is stated in the following equation (Adams et al. 1986) to facilitate further discussion:

$$r_n = \sum_{m=1}^{M} f_m a_{m,n} + e_n, \quad n = 1, 2, \ldots, N \tag{3.1}$$

where rn, $n = 1, \ldots, N$, is the observed spectrum of any pixel in the nth spectral band of the satellite sensor; fm, $m = 1, \ldots, M$, is the fraction of endmember m in that pixel, and $\alpha m{,}n$ is the spectral reflectance of the mth endmember in the nth spectral band; en refers to the error term in band n, which is the unmodeled portion of the observed reflectance along with any instrumental noise. The linear unmixing problem as observed in Eq. 3.1 represents a system of linear equations that can be expressed in matrix form as

$$\boldsymbol{R}_{ij} = \mathbf{A}\hat{\boldsymbol{f}}_{ij} + e_{ij}, \tag{3.2}$$

where $\widehat{f_{ij}}$ is the estimated $M * 1$ vector of M endmember fractions for the pixel at location (i,j), and $\boldsymbol{A}$ is the $N * M$ endmember signature matrix. The added term eij represents the residual error in the fitting of a given pixel's spectral vector by the weighted sum of M endmember spectra and unknown noise.

One of the most straightforward approaches to solving the linear problems using Eq. 3.2 is by using least squares minimization. Because the number of spectral bands (N) is greater than the number of endmember classes (M), as generally observed in multispectral or hyperspectral images, therefore the linear mixing model becomes an overdetermined set of N equations with M variables. As a result, the matrix $\boldsymbol{A}$ of size ($M \times N$) cannot be inverted. The value of the fractional abundance vector $\widehat{f_{ij}}$ can then be obtained via the pseudo-inverse of matrix $\boldsymbol{A}$ as follows (Gorss and Schott 1998; Khanna et al. 2006; Schowengerdt 1997):

$$\hat{\mathrm{f}}_{ij} = \left(\mathrm{A}^{\mathrm{T}}\mathrm{A}\right)^{-1}\mathrm{A}^{\mathrm{T}} \times \mathrm{R}_{ij} \tag{3.3}$$

The foregoing Eq. 3.3 may, however, produce negative abundance values and their sum may not be in unity and therefore not physically realistic. These issues are a serious limitation of the linear unmixing model. It is inappropriate to expect that pixels belonging to the same endmember class have the same spectrum. This model is inefficient in incorporating in-class spectral ambiguity into the unmixing. Furthermore, poor selection of endmembers, such as an improper number of endmembers used in unmixing, may also cause the generation of unrealistic abundances in this unmixing process. Therefore, with inappropriate handling of the endmember spectral variability by the model and an inappropriate number of endmember selection, the conventional linear spectral unmixing may result in the aforementioned problems (Gorss and Schott 1998). A simple solution is to impose the following constraints to obtain the constrained least square model (Gorss and Schott 1998; Schowengerdt 1997):

$$\sum_{m=1}^{M} f_m = 1 \text{and } 0 \le f_m \le 1 \tag{3.4}$$

For this reason, the conventional linear spectral unmixing method is also termed the fully constrained linear spectral unmixing method.

The concept of ESVM (an extended version of SVM) has been applied for spectral unmixing. In this method, the complete set of training samples is used to model the pure pixels of the defined class. Figure 3.1 illustrates this method with a simple case of 100 % separability between classes of A and B in a two-dimensional case. The support vectors now become the pixel vectors that are on the boundary between pure and mixed regions and are then named as 'just pure' pixels. The two regions above or below the lines formed by the 'just pure' pixels are pure pixels. The region

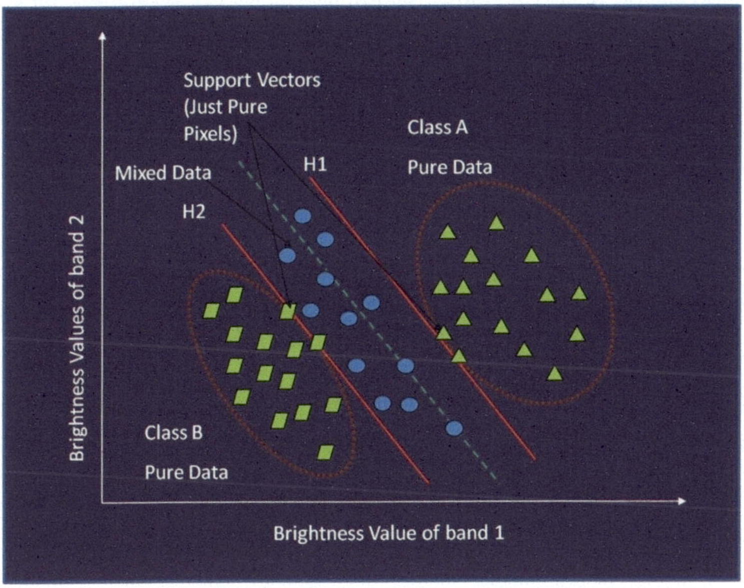

Fig. 3.1 Spectral unmixing with extended support vector machine (ESVM)

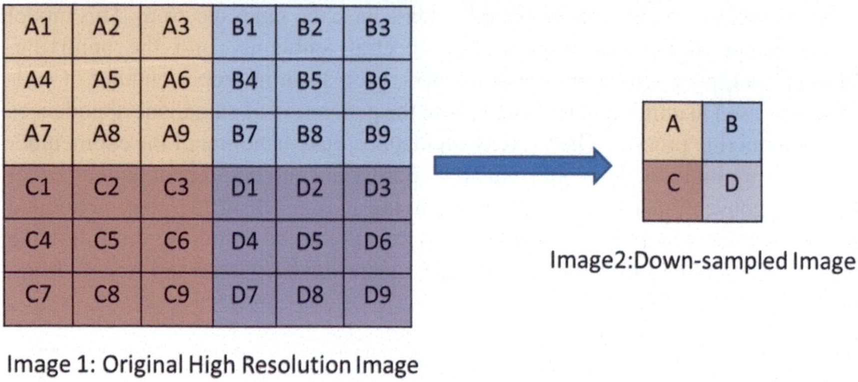

Fig. 3.2 Down-sampling method

between the lines is associated with the mixed pixels. If the spectral mixture is assumed linear, the SVM decision boundary (in the middle of this region) becomes a 50 % mixing line. To perform spectral unmixing, the lines H1 and H2 (as shown in Fig. 3.1) are determined first, and then the distances to the two lines are evaluated. The results represent the fractions of the two classes contained.

For the extended version of the SVM, the margins between the hyperplanes formed by the pixels on the support vectors (see Fig. 3.1) are recognized as the mixed region and the space beyond that is the region consisting of pure pixels only (Jia et al. 2010). Once the discriminant function *f*(**x**) is determined, pure pixels are labeled ω*i*, and the relative mixture proportion of mixed pixels (*βrx*(*ωi*)) for class A and B can then be obtained by the following equation (Jia et al. 2010):

$$\begin{aligned}&\text{if } f(x) \geq 1\\&x \in \omega_A, \text{i.e } \beta_x^r(\omega_A) = 1, \beta_x^r(\omega_B) = 0;\\&\text{if } f(x) \leq -1\\&x \in \omega_B, \text{i.e } \beta_x^r(\omega_A) = 0, \beta_x^r(\omega_B) = 1;\\&\text{if } -1 < f(x) < 1\\&x \in \{\omega_A, \omega_B\},\\&\beta_x^r(\omega_A) = \frac{1}{2}(1 + f(x)), \beta_x^r(\omega_B) = \frac{1}{2}(1 - f(x))\end{aligned} \tag{3.5}$$

where *βxr*(*ωA*) and *βxr*(*ωA*) are the relative mixture proportions of class ω*A* and ω*B* for a particular pixel *x*, respectively. The method is denoted as u_ESVM. For more than two class cases, it is first conducted for each class against the rest of the classes separately to find a pixel's relative mixture proportion, *βxyωi*. An extra step for this case is to normalize the mixture fraction by

$$\beta_x(\omega_i) = \beta_x^{\gamma}(\omega_i) / \sum_{i=1}^{I} \beta_x^{\gamma}(\omega_i) \tag{3.6}$$

It is also important to mention that, for unmixing function, two inputs were required. The first one is the training samples to train the endmember classes. The second one is the constant parameter w to determine the distance between the two hyperplanes and consecutively the mixed pixel region. The changes in the value of this constant also change the accuracy level. Cross-validation is often used to determine the optimal value.

Beside mapping both relatively pure pixels (as binary results and the mixed pixels as fraction results), u_eSVM has the advantage of using a set of training samples to model each primary class instead of a single representative spectrum only. It therefore can accommodate the spectral variations within the primary class. Furthermore, there is no need to impose any artificial constraint to make the results positive, because the unmixing results using u_ESVM are always positive in nature (Jia et al. 2010).

As mentioned earlier, application of the PCLS method was conducted with the help of the ENVI 4.8 software package. This software only applies the sum-to-one constraint in the unmixing process. Mathematically, the implementation of the sum-to-one constraint in Eq. 3.3 can be easily performed by simply by adding another simultaneous equation [the $(N+1)$th] in the unmixing inversion process (Chang 2003; Klatt 1994). This PCLS method was adopted for the comparative study because this method satisfies the requirement of target detection of estimated abundance fractions of the desired target pixel vectors from their surrounding pixel vectors; the unmixing method does not confine the fractional estimates to the range of 0 and 1 as observed in the case of fully constrained linear spectral unmixing, and therefore the PCLS method does not limit the target detection capability of the fractional abundances (Chang et al. 2002). Hence, the evidence of endmembers that are present in a mixed pixel as analyzed by the PCLS method provides a picture more closely resembling the actual ground conditions. In this respect, the PCLS method is acceptable to map the evidence of flood water and non-water endmembers in mixed pixels.

The MESMA method, on the other hand, is based on the concept of a linear spectral mixture approach with the inclusion of more than one linear spectral unmixing model that is initially computed for each pixel in an image. The concept of MESMA was primarily developed by Roberts et al. (1998). Here, in this thesis, MESMA was applied for the subpixel analysis of flood mapping. VIPER tools (an open source software discussed in more detail by Halligan 2008) are used to run this method.

For assessing the accuracy of the fractional abundances derived by running the unmixing operation, two different methods are considered. In the conventional root mean square error assessment, the residual error in the unmixing method is approximated by differences between the estimated and modeled abundances of an endmember class for the entire image. The modeled fractions for the endmembers are generated from the ground truth image. The modeled fractions are considered as the true fractions of the endmember ground components present in each pixel. Therefore, for each class, the average unmixing accuracy is obtained as

$$\mathrm{RMSE}_p = \sqrt{\frac{\sum_{i=1}^{N}\left(f_p(i) - F_p(i)\right)^2}{N}} \tag{3.7}$$

where *fp*(i) and *Fp*(i) are the measured and modeled fraction values of the endmember class p obtained from the unmixing output and the ground truth image, respectively, and N is the total number of pixels in the image. To generate the true endmember fractions, initially the ground truth image is classified with the application of the maximum likelihood classifier, and this classification method assumes that the statistics for each class in each band are normally distributed and calculates the probability that a given pixel belongs to a specific class. To obtain the true fractions of endmember classes, the classified image was spatially reduced (three times for the current experiment) to match the size of the test image. Hence, a pixel value in the reduced classified image actually represents the proportions of the endmember classes present in a corresponding three-by-three matrix of the ground truth classification image. The fraction of each pixel in the reduced classified image therefore is considered as the true fractions of the endmember class and used for accuracy assessment.

3.3 Experiments and Analysis

For the experiment, two different sets of Landsat-5 TM images were used. The dates of satellite images relate to the major flood events in the study areas of the Northern Territory (2008 flood event in the Daly River catchments) and South Australia (the recent 2011 flood event in Innamincka region). For the Innamincka Region, another set of very high resolution World-View2 images (VHR_WV2_IR) was obtained for the experiment. However, this very high resolution image is only used for the analysis and selection of primary classes, not for testing, because, to maintain the parity in the experiments, only the Landsat image sets were applied for all the three cases. Table 3.1 gives the list of data that have been used in this experiment.

One of the main research objectives in this study is to assess the accuracy of mixed pixel analysis. Therefore, ground truth, the true mixture percentages for each pixel, is required for validation of the unmixed results. The true mixture percentages of the pixels were obtained based on the hard classification results on the Landsat-5 TM images. For testing, the down-sampling method was applied to the high-resolution image to generate an artificial low-resolution image in each test experiment. Every pixel in the original (high-resolution) data was assumed to be pure, and the hard classification results were used as ground truth to validate the unmixing

Table 3.1 List of remote sensing images used in the study and their short names

	Landsat-5 TM (30 m)	Landsat-5 TM (30 m)	WorldView-2 (2 m)
Location	Pre-flood imagery	During flood imagery	
Innamincka region	–	25.01.2011	
		TM_IR	
Daly River Basin	10.05.2006	03.03.2008	26.01.2011
	TM_DR	TM_DR	WV2_IR

results obtained from the down-sampled image. By this process, it is possible to avoid the application of different sets of low-resolution satellite images as test images and therefore eliminate error occurring during image registration. Other geometric and radiometric differences between the low- and high-resolution images resulted in large proportions of error being present in estimating the unmix results.

Down-sampling is the most commonly used image operation for image compression, display, and transformation. The method refers to the reduction of the spatial resolution of the image while maintaining the same two-dimensional representation. Down-sampling generally helps in the reduction of the storage requirements of images (Youssef 1999). In Figs. 3.3, 3.4, 3.5, 3.6, 3.7, 3.8, 3.9, and 3.10, the process of the down-sampling method is explained using a simple example. Image 1 in the figure is the original high-resolution image made up of 6 by 6 pixels, which is then

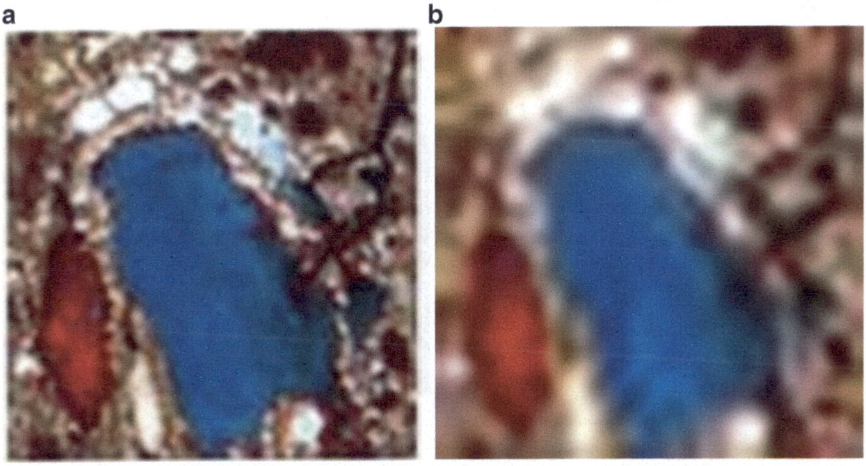

Fig. 3.3 (**a**) Landsat-5 TM image of the Innamincka region (HR_TM_IR). (**b**) Down-sampled Landsat-5 TM image of the Innamincka region (LR_TM_IR)

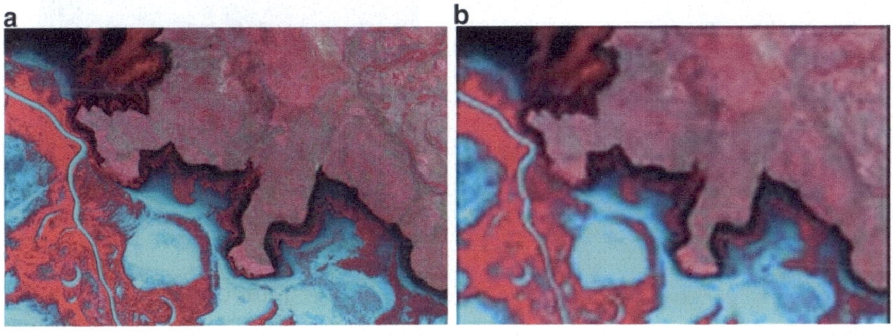

Fig. 3.4 (**a**) Landsat-5 TM image (2008) of the Daly River Catchments area (HR_TM_DR). (**b**) Down-sampled Landsat-5 TM (2008) image of the Daly River Catchments area (LR_TM_DR)

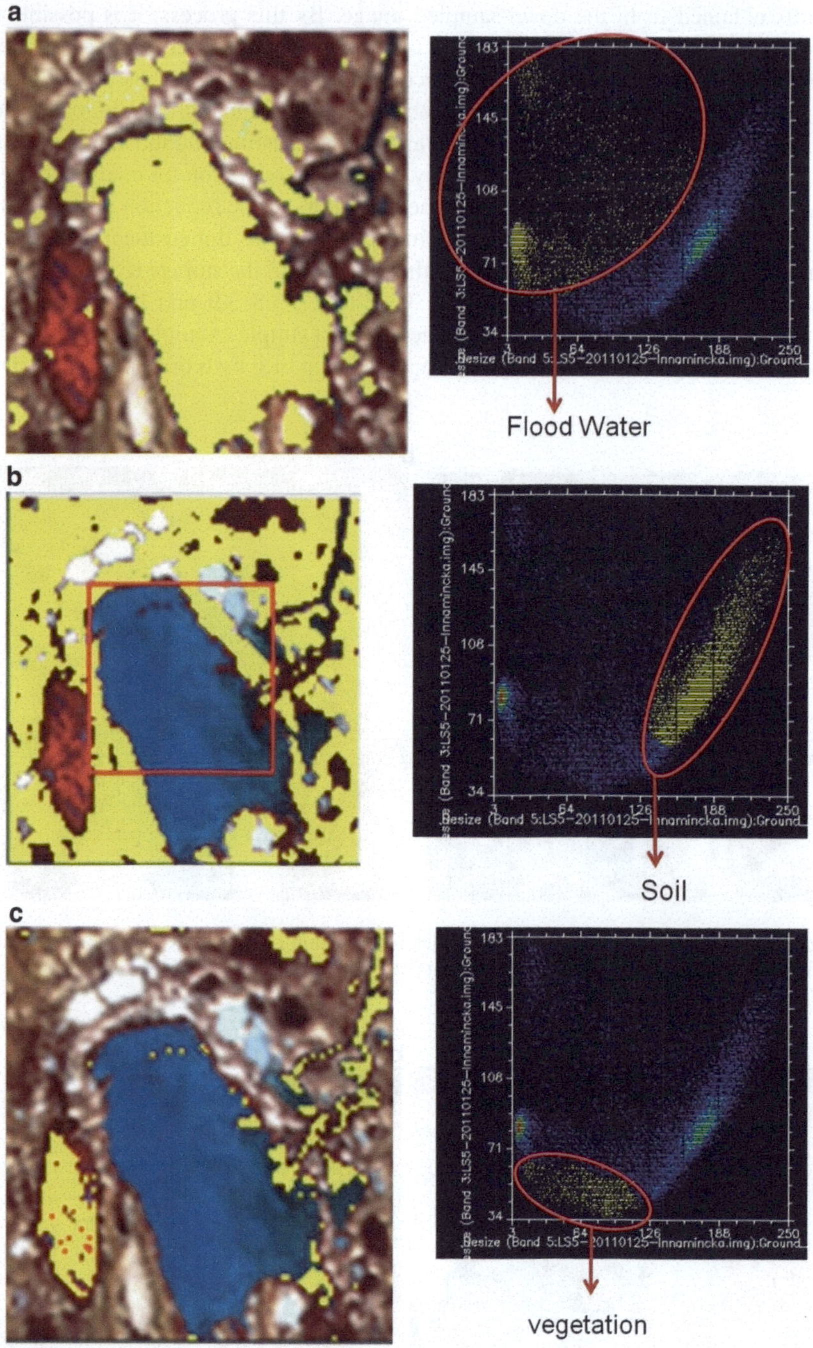

Fig. 3.5 Detection of primary classes based on spectral plot analysis

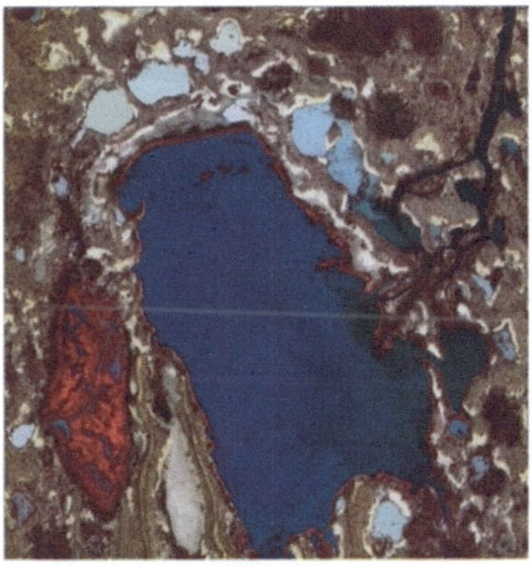

Fig. 3.6 Very high resolution WV2_IR image

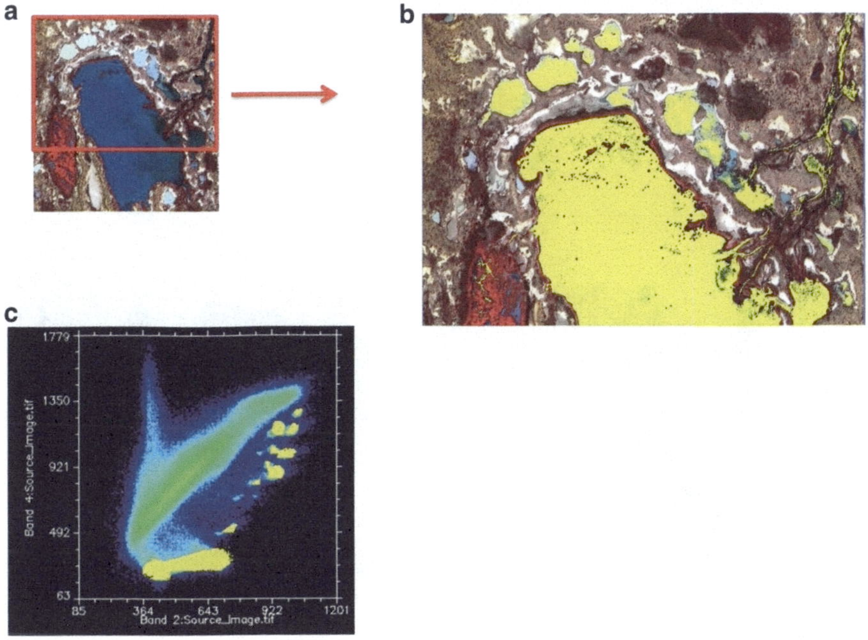

Fig. 3.7 Spectral ambiguity of water class. (**a**) CIR version of VHR_WV2_INA image (**b**) The area within the *red* box in (**a**) (**c**) Spectral plot

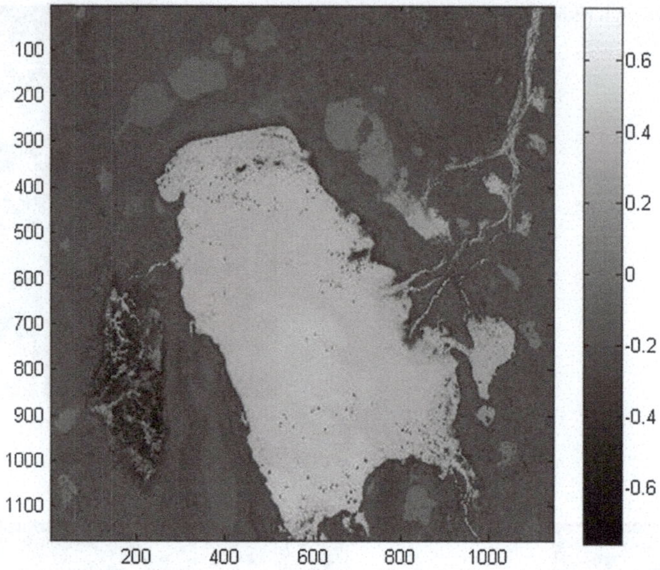

Fig. 3.8 Normalized difference water index (NDWI) image

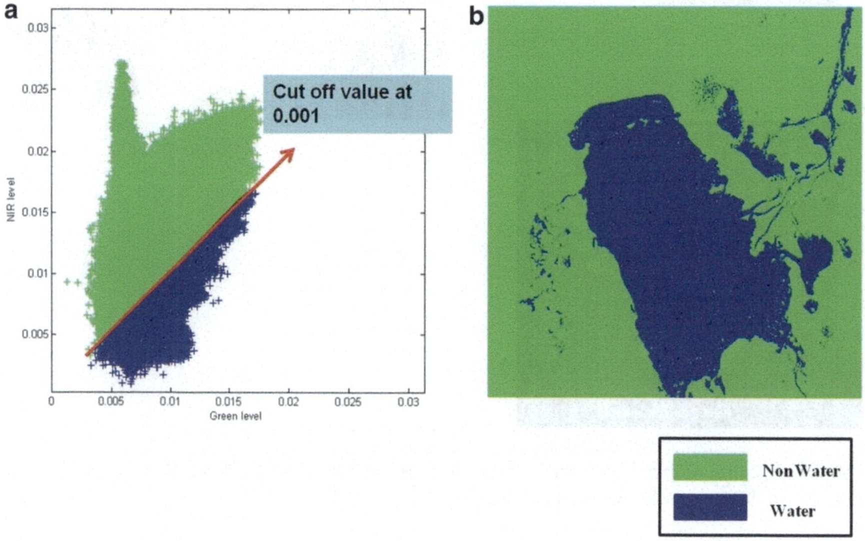

Fig. 3.9 Extraction of water features. (**a**) NIR vs. Green scatter plot (**b**) NDWI with threshold applied

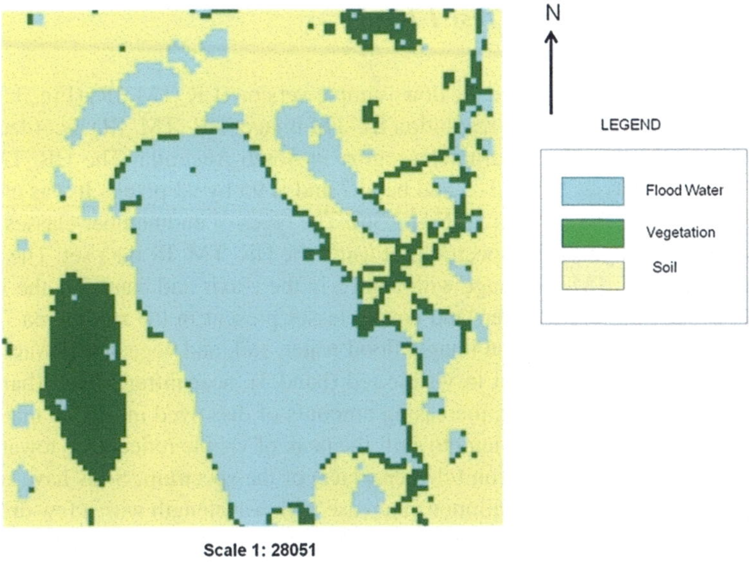

Fig. 3.10 Classification result of HR_TM_IR image

down-sampled nine times. The result is given in Image 2, where A, B, C, and D are the average values of the corresponding nine values in Image 1.

In this study, the down-sampling process was applied to the Landsat 5 TM high-resolution image of the three study areas. For each study area, a subset was selected for experiment. The image was down-sampled by reducing the existing spatial resolution nine times. For the first Landsat image of the Innamincka area, the ground truth image comprises a subset area of 95 by 93 pixels of high-resolution Landsat image. Each pixel of the downsampled test image consists of the mean of every 3 by 3 pixels from the original image. The size of the test image is 31 by 31 pixels (Fig. 3.3a, b). In this respect, it is important to mention that the subset area from the image of Innamincka region constitutes a very small portion as compared to the entire extension of the Landsat image and therefore the subset region may look more pixilated than the subset areas of study 2 and study 3. However, the spatial resolutions of the three images are the same.

From the foregoing two images, it is quite evident that the down-sampled image is more pixilated as compared to the original high-resolution image because of the reduction in spatial resolution. Similarly, in test 2, the subset of high-resolution Landsat-5 TM image of the Daly River basin area comprises 327 by 489 pixels of the original scene and the size of the test image is 109 by 163 pixels (Fig. 3.4a, b).

The entire experiment is illustrated in the following three sections consisting of the three different case studies of floods in Australia.

3.3.1 Experiments for Test 1 Image

The third test was conducted on the downsample version (LR_TM_IR) (Fig. 3.3b) of a subset portion from the high-resolution IR_TM image (HR_TM_IR) recorded during the 2011 flood in the Innamincka region in South Australia. The HR_TM_IR image scene comprises bands 1–5 and band 7 and is 95 by 93 pixels. It was used as the ground truth of the study area (Fig. 3.3a). The types of endmember classes were determined by analyzing the spectral plot using the HR_TM_IR data set. The spectral plot of the HR_TM_IR image with band 5 in the *y*-axis and band 3 in the *x*-axis helped in determining the other land cover classes present in the study area. Three broad classes were found in the image: flood water, soil, and vegetation. Water particles increasingly absorb light in visible red (band 3), near infrared (IR) (band 4), and mid IR (band5). However, increasing amounts of dissolved inorganic materials in water bodies (turbid water) tend to shift the peak of visible reflectance toward the red region from the green region (clearer water) of the spectrum. Soils have reflectance characteristics with a continuous increase with wavelength with a few drops in the water absorption bands at 1.4, 1.9, and 2.7 μm from the presence of moisture. All these drops are almost absent in dry soils and sands. The spectral reflectance curve of vegetation, on the other hand, drops by the water absorption bands at 1.4, 1.9, and 2.7 in the mid-IR range and forms a plateau between about 0.7 and 1.3 μm based on plant cell structure. In the visible range, however, plant pigmentation determines the reflectance structure. Healthy green vegetation therefore absorbs light in the blue and red regions and leaves only green reflection (Richards and Jia 2006). In the spectral plot, the cluster at the center of the plot in Fig. 3.5b shows that the soil and water areas are clearly indicated (see the clusters in the extreme lower left section and at the upper part of the scatter diagram) in Fig. 3.5a, whereas the vegetation forms a cluster just below the water clusters in Fig. 3.5c. However, in this respect it is important to note that the cross plots may not have the capability to show the pure water pixels because of the high spectral variability of water in this image. Therefore, in Fig. 3.5a although all the highlighted areas in yellow were deemed to represent water areas, some of them were not pure water bodies. To select the pure water samples, the normalized differential water index (NDWI) was run on VHR_WV2_IR image (Fig. 3.6).

The area is characterized by desert sands with some patches of water and vegetation. Spectral ambiguity in water-filled pixels is mainly observed in the cross plot of the VHR_WV2_IR image spectra with spectral values of the near-IR (NIR) band (band 4) in the *y*-axis and values of the red band (band 3) in the *x*-axis (Fig. 3.7).

Analyzing the nature of the scatter diagram, it is was clear that the spectral value for pixels showing the presence of flood water contained a broad range of reflectance values in both the spectral regions from low to very high, as observed in the yellow clusters in the upper right to lower left in Fig. 3.7c. Therefore, to separate the non-water from water features, the normalized differential water index image was generated from the VHR_WV2_IR image.

Because of their similar reflectance, some areas with high moisture content could be mistaken for flood water, possibly resulting in overestimation of water surface area (McFeeters 1996). To solve the problem, the normalized differential water index (NDWI) was generated:

$$\mathrm{NDWI} = \frac{\mathrm{Green} - \mathrm{NIR}}{\mathrm{Green} + \mathrm{NIR}},$$

where NDWI refers to the water index image, NIR refers to spectral values of pixels at the near-infrared spectral channel (band 4), and green refers to the spectral values at the visible green spectral channel of pixels from the original image (band 3).

The reflectance in green and NIR wavelengths was used to (1) maximize the typical reflectance of water features by using green light wavelengths; (2) minimize the low reflectance of the NIR by water features; and (3) take advantage of the high reflectance of NIR by terrestrial vegetation and soil features (McFeeters 1996). In the NDWI image (Fig. 3.8), water-filled pixels have positive values and pixels with non-water features have negative values. The NDWI value also ranges between −1 and 1. After analyzing the difference in the NDWI values for water and non-water pixels, different values were attempted to put as a threshold in the NDWI index image to extract the water-filled pixels. Finally, a distinct break has been found with pixel values greater than 0.001 for water and non-water pixels. Pixel values equal to or more than 0.001 proved to contain water and non-water pixels showed values less than 0.001. Therefore, a threshold of 0.001 was applied to extract the water features from the NDWI image (Fig. 3.9). After analyzing the NDWI index image, it was easier to select pixels with pure water reflectance spectra for unmixing.

After a better understanding of the areal coverage of pure water features in the image was gained, the types of endmember classes were determined by analyzing the spectral plot using the HR_TM_IR data set. Therefore, the selection of pure training data from the image (HR_TM_IR) depended on the knowledge gained from both the index image and from the cross-plot analysis. After determining the types of pure ground cover classes, the next most important task was the generation of true fractions from the HR_TM_IR image dataset. The supervised maximum likelihood classifier method (as applied in the previous two experiments) was used to generate the true ground cover fractions. For classification, pure training and testing data were selected from the HR_TM_IR image. Tables 3.2 and 3.3 show the training and test data, respectively. A decision rule was applied to identify the endmember class to which a pixel belonged. The following Tables 3.4 and 3.5 list the mean vectors and covariance matrices, respectively. These class mean vectors and covariance values were applied in the classification function as class signatures. Discriminant functions were then applied as the decision rule to determine the exact class to which a pixel belongs (Richards and Jia 2006).

The classified image (Fig. 3.10) helped to achieve a clearer idea of the distribution of the pure ground components (endmember classes) in the image.

The classified HR_TM_IR image (Fig. 3.10) was down sampled nine times in spatial scale for the generation of ground truth fractions of endmember classes. The fraction images as showed in Fig. 3.11 were obtained from the down-sampling method.

Table 3.2 Classification accuracies of training samples on HR_TM_IR image

Name of endmember class	No. of training data	Producer's accuracy (%)	User's accuracy (%)	Overall accuracy (%)	Kappa statistic
Flood water	487	99.8	100.0	99.8	99.6
Soil	150	100.0	99.4		
Vegetation	178	99.3	99.3		

Table 3.3 Classification accuracies of test samples on HR_TM_IR image

Name of endmember class	No. of test data	Producer's accuracy (%)	User's accuracy (%)	Overall accuracy (%)	Kappa statistic
Flood water	139	100.0	100.0	100.0	100.0
Soil	33	100.0	100.0		
Vegetation	56	100.0	100.0		

These true fraction values were used to assess the accuracy level of the estimated fractional abundances obtained by applying the fixed set of the endmember index file and the adaptive subset of the endmember index file into the unmixing processes of PCLS, MESMA, and u_ESVM. Finally, to initiate the unmixing process, the selected training samples that represented the pure reflectance spectra for each of the endmember classes were collected based on the analysis of the cross plots and the land use classification results of the ground truth image. However, for the requirements of each unmixing method the training samples were used to serve different purposes: for the PCLS method, the mean spectra of each endmember class were applied, whereas for MESMA a few endmember spectra were used. However, for the u_ESVM method the entire training data set was used as an input.

Initially, the estimated fractional values were generated by applying the fixed set of the endmember index file into the test LR_TM_IR image using these aforementioned three methods of unmixing methods, respectively (Fig. 3.12a–c).

In comparison of the amount of errors from the three methods, it has been observed (see Fig. 3.13) that the u_ESVM method renders the least amount of error compared to the other two types of unmixing methods. This advantage occurs because the u_ESVM method has the capability to use the entire endmembers library for primary classes to unmix the pixels that lie within the mixed region as well as to recognize the (relative) pure pixels. In PLCS only, the mean spectrum was used and, in MESMA only, a few endmember spectra were used for mixture analysis.

Now, the new proposed pixel-dependent estimated subset of the endmember index file was generated to apply to those unmixing methods. The mechanism behind the generation of this endmember selection is explained in the Methodology section. The experiment was run on the LR_TM_IR image. Considering each data set, the experiment was run twice based on two different sets of adaptive class index files. The first index file (a_TST) was generated from the ground truth fraction image. The method of generating the true fraction image was previously explained. The results of the estimated fractional abundance obtained from the unmixing

Table 3.4 Covariance matrices of the endmember classes

	Flood water					
Spectral bands	**1**	**2**	**3**	**4**	**5**	**6**
1	311.13298	213.68676	368.2964	367.71144	77.642753	43.026871
2	213.68676	149.63711	251.35989	249.56444	51.615607	28.234919
3	368.2964	251.35989	464.14819	467.84795	106.72089	59.574535
4	367.71144	249.56444	467.84795	481.20213	116.3899	65.404201
5	77.642753	51.615607	106.72089	116.3899	69.608445	36.271867
6	43.026871	28.234919	59.574535	65.404201	36.271867	21.469423
	Soil					
Spectral bands	**1**	**2**	**3**	**4**	**5**	**6**
1	35.42706	23.14638	31.63217	28.32626	24.73281	20.57313
2	23.14638	17.89881	25.70399	22.87996	19.59487	16.55399
3	31.63217	25.70399	47.86495	41.4482	41.69193	33.47531
4	28.32626	22.87996	41.4482	45.09989	38.02088	29.25478
5	24.73281	19.59487	41.69193	38.02088	85.82175	57.72278
6	20.57313	16.55399	33.47531	29.25478	57.72278	45.96842
	Vegetation					
	1	**2**	**3**	**4**	**5**	**6**
1	11.12304	4.550336	11.21477	–28.1387	5.621924	9.879195
2	4.550336	6.871409	2.794094	28.63409	–12.949	–10.575
3	11.21477	2.794094	20.36819	–78.4393	39.80067	36.17356
4	–28.1387	28.63409	–78.4393	663.6201	–173.843	–210.735
5	5.621924	–12.949	39.80067	–173.843	512.3546	284.6664
6	9.879195	–10.575	36.17356	–210.735	284.6664	186.1668

Table 3.5 Mean vectors of endmember classes

	Endmember classes		
Spectral bands	Flood water	Soil	Vegetation
1	131.301848	101.26404	81.33333333
2	82.48459959	58.235955	41.8800
3	88.72073922	84.342697	42.7400
4	44.37577002	93.601124	90.19333333
5	11.02464066	177.51685	97.36666667
6	6.156057495	97.522472	40.7400

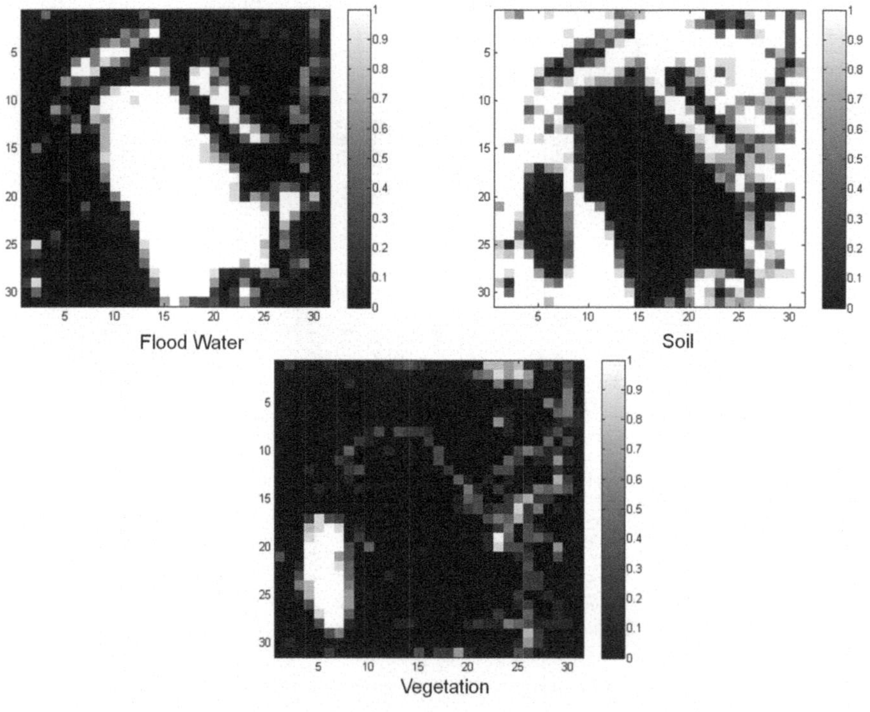

Fig. 3.11 True fractional images obtained from HR_TM_IR image

method with application of a_TST can demonstrate the effectiveness of using the right subset of primary classes to unmix.

The second index file was the pixel-dependent endmember index file, termed the adaptive or estimated set (a_EST) of the endmembers index file. For further convenience, a case of 20 pixels from the probability matrix of the test 1 experiment was selected to show the method of generating the pixel-dependent class index matrix. The probability values for the classes are negative in nature. For facilitating the calculation, the values were normalized (as shown in Table 3.7) within a range of 0 to −1. A value of −0.01 was defined as the maximum threshold l and a value of −0.50 was defined as the minimum threshold value. Table 3.6 represents the probability values of the three primary classes for 20 pixels.

A pixel is considered to be a pure pixel and contains only one class if a class probability value within that pixel is greater than the threshold value of −0.1 and the pixel has then no need to be unmixed. In a pixel a class with probability value less than −0.70 is considered to be the nonrelevant class for that pixel and eliminated. Hence, with the application of thresholds on probability values, an index matrix had been generated that represented the per-pixel subset of classes (Table 3.8).

A similar method was applied to extract the pixel-dependent endmember (primary) class index for the entire image. The resultant fractional abundance images derived by applying the pixel-dependent endmember index file in different unmixing methods are illustrated in Fig. 3.14.

After obtaining the estimated fractional abundances using a_EST, the adaptive true endmember sets (a_TST) were generated from the ground truth image and used

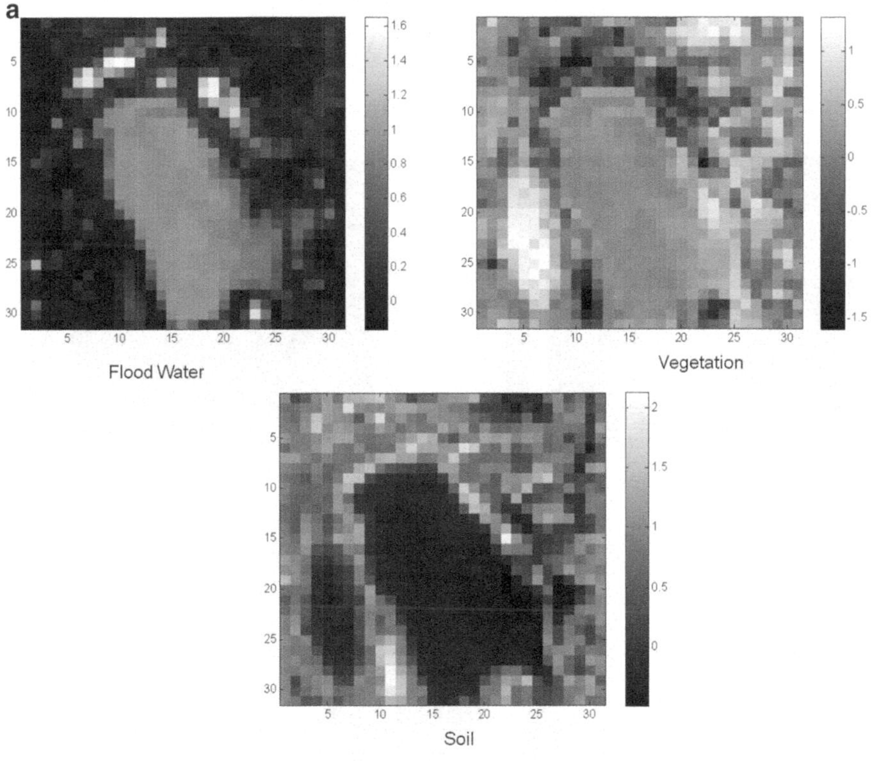

Fig. 3.12 (**a**) Estimated fractional images obtained from the partial constrained linear spectral (PCLS) method. (**b**) Estimated fractional images obtained from the multiple endmember spectral mixture analysis (MESMA) method. (**c**) Estimated fractional images obtained from u_ESVM method

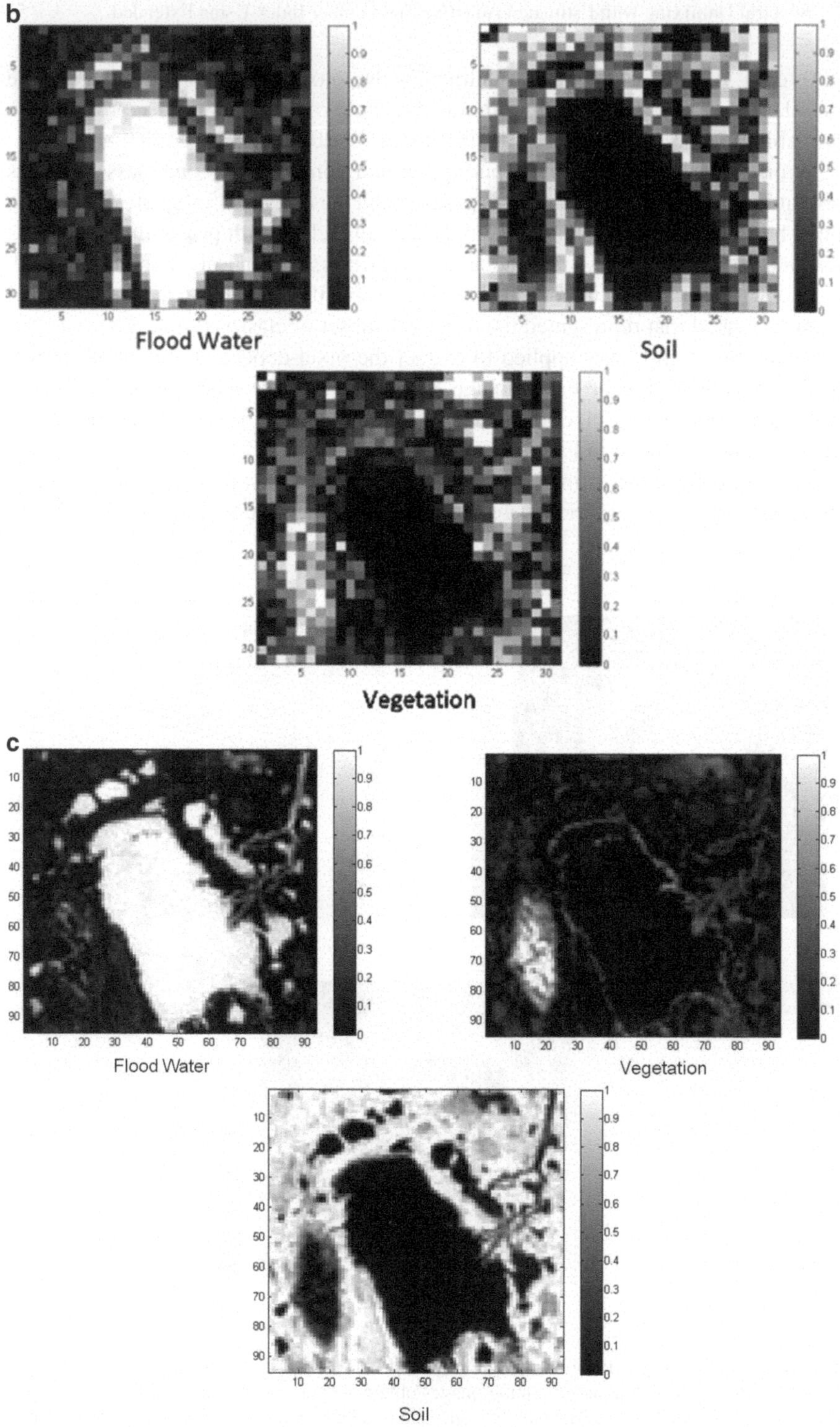

Fig. 3.12 (continued)

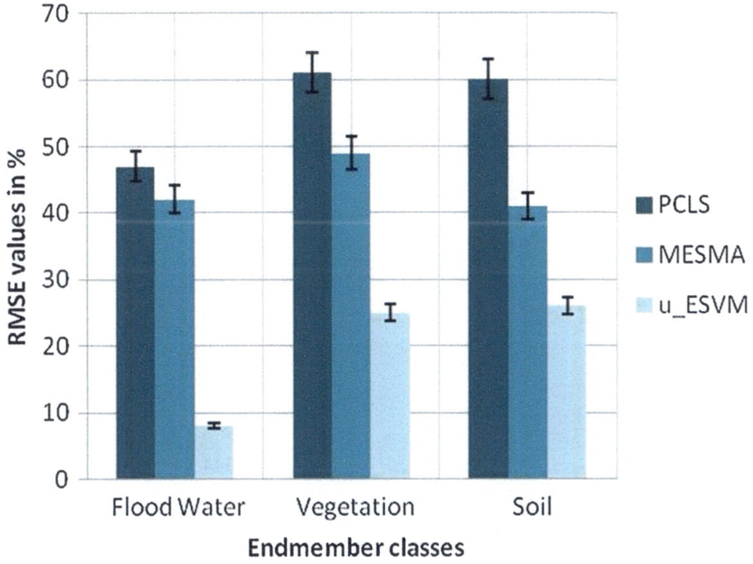

Fig. 3.13 Error assessment of the unmixing methods using the fixed set of endmember index

Table 3.6 Probability values of primary classes in each pixel

No. of pixels	Flood water	Vegetation	Soil
1	−50.13244598	−240.7980304	−495.8884922
2	−19.87511206	−674.3973389	−1238.479589
3	−21.31601147	−585.4990497	−1063.615156
4	−23.41883927	−558.2268685	−1024.550682
5	−18.40685511	−543.3207059	−1095.275169
6	−17.49080924	−662.3377023	−1318.188117
7	−16.81332269	−599.9542628	−1273.585993
8	−20.21841531	−669.2414695	−1417.457131
9	−20.30682625	−668.64095	−1184.084084
10	−210.8525297	−149.1003343	−119.1975891
11	−862.4215656	−685.5199695	−48.47553242
12	−850.6597146	−492.9347917	−29.26265881
13	−201.2553021	−218.368684	−181.2066791
14	−24.89400036	−186.8583966	−618.4377088
15	−119.6067486	−34.21013613	−221.7155883
16	−331.1654651	−149.2524547	−36.14823386
17	−653.2135977	−257.8014693	−17.88901218
18	−682.7598602	−276.8383562	−17.67047553
19	−750.0655929	−289.360029	−26.39714967
20	−283.0345194	−69.90753937	−226.0500062

Table 3.7 Probability values in descending order for each pixel

No. of pixels	Flood water	Vegetation	Soil
1	–0.06372	–0.30604	–0.63024
2	–0.01028	–0.34893	–0.64079
3	–0.01276	–0.35051	–0.63673
4	–0.01458	–0.34755	–0.63787
5	–0.01111	–0.32789	–0.661
6	–0.00875	–0.3315	–0.65975
7	–0.00889	–0.31738	–0.67373
8	–0.0096	–0.31764	–0.67276
9	–0.01084	–0.35698	–0.63218
10	–0.44005	–0.31118	–0.24877
11	–0.54022	–0.42941	–0.03037
12	–0.61963	–0.35906	–0.02132
13	–0.33496	–0.36344	–0.30159
14	–0.02999	–0.22508	–0.74494
15	–0.3185	–0.0911	–0.5904
16	–0.64109	–0.28893	–0.06998
17	–0.70321	–0.27753	–0.01926
18	–0.69864	–0.28328	–0.01808
19	–0.70374	–0.27149	–0.02477
20	–0.48884	–0.12074	–0.39042

Table 3.8 Index matrix of pixel-dependent class subset

No. of pixels	Flood water	Vegetation	Soil
1	1	0	0
2	1	0	0
3	1	0	0
4	1	0	0
5	1	0	0
6	1	0	0
7	1	0	0
8	1	0	0
9	1	0	0
10	1	1	1
11	0	0	1
12	0	0	1
13	1	1	1
14	1	0	0
15	0	1	0
16	0	0	1
17	0	0	1
18	0	0	1
19	0	0	1
20	0	1	1

to unmix the test image using the three unmixing methods. The performance was compared with the case when a fixed set of classes (u_FST) was used for all the pixels. It has been observed that the error is substantially less using the adaptive subset of classes in unmixing (Fig. 3.15). It is also evident from Fig. 3.15 that the estimated index file is not much deviated from the true index file, as the difference in error is not large in most cases. Another most interesting fact is also observed that the error level obtained from PCLS using a_EST file is closer to the minimum possible error of unmixing using a_TST; this occurs because the partially constrained unmixing method (PCLS) is more convenient in terms of applying only unit sum constraints on estimated fractional abundance and therefore the fractional estimations are less customized as compare to the estimates obtained from MESMA and u_ESVM. Hence, in the PCLS unmixing process the generated fractional estimates are able to represent the situation closer to the real conditions that was represented by the true fractional estimates. However, the u_ESVM method with estimated a_ EST file provides more accurate results as compared to MESMA and PCLS.

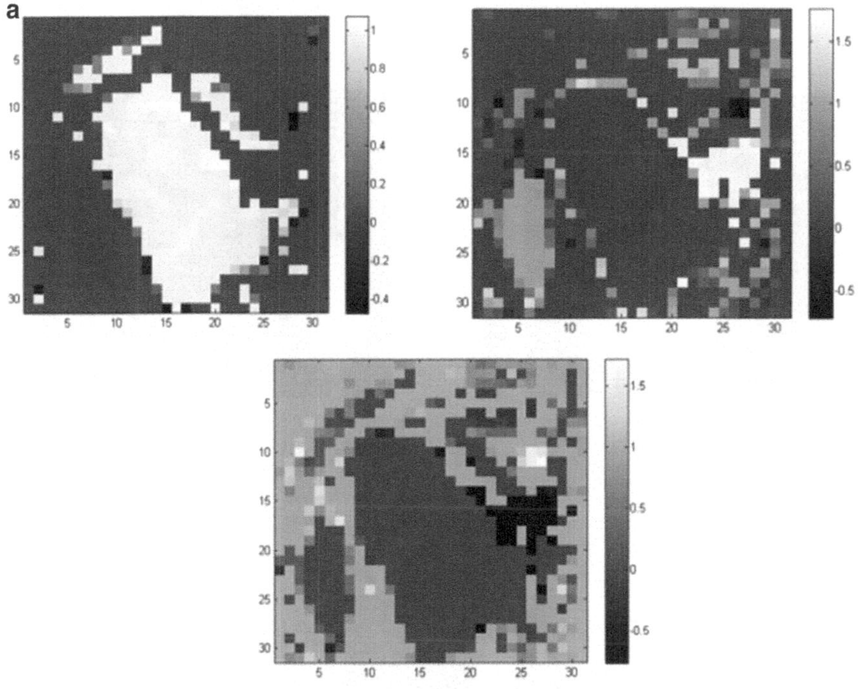

Fig. 3.14 (**a**) Estimated fractional images obtained from PCLS method. (**b**) Estimated fractional images obtained from MESMA method. (**c**) Estimated fractional images obtained from u_ESVM method

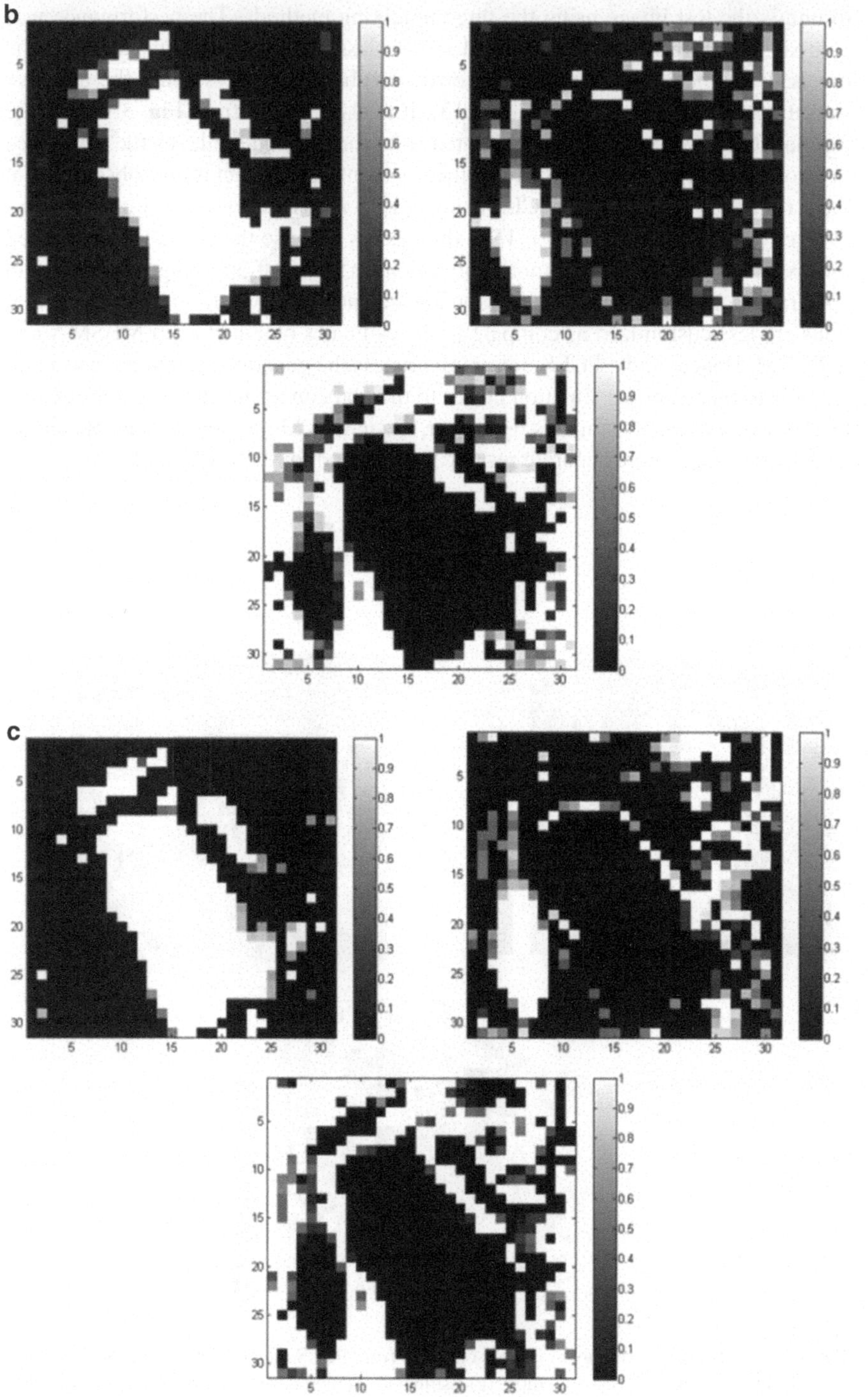

Fig. 3.14 (continued)

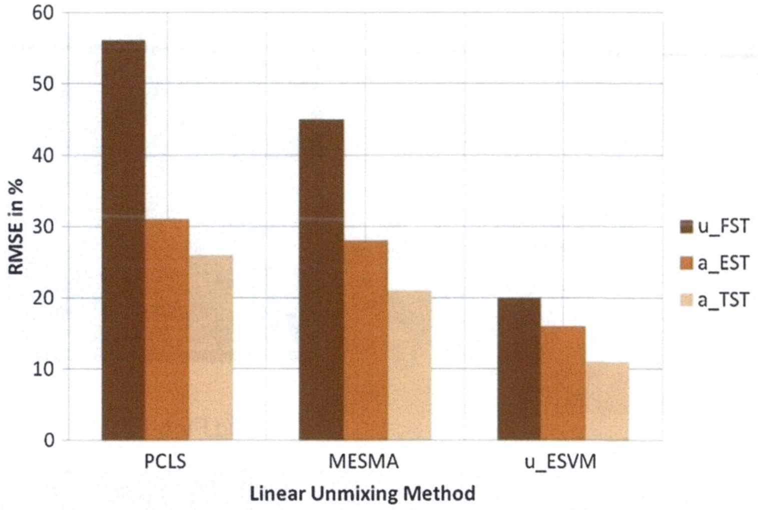

Fig. 3.15 Error comparisons between the outputs obtained from unmixing method using adaptive subset class index files and a fixed set of class for test 1

3.3.2 Experiments for Test 2 Image

For the second data set, the three primary or endmember classes are turbid flood water, pure flood water, and vegetation. To determine the primary classes, spectral plot analysis was performed. The analysis of the cross plot portrays the spectral differences present in the study area. The HR_TM_DR image was used in the cross-plot analysis. The cross plot with reflectance values at band 5 (near IR) in the x-axis and reflectance values at band 3 (visible red) in the y-axis was developed for differentiating the clusters of the pure pixels for the endmember classes by considering the spectral characteristics of water, vegetation, and soil. For example, as in Figs. 3.16a, b, turbid flood water and pure flood water and their corresponding areas in the test image are identified.

From the foregoing figures it is clear that the cluster in the center of the plot represents the region of vegetation (Fig. 3.16c). This cluster represents a vast area in the test image with similar spectral characteristics. The small cluster in the lower left of the plot represents the pure water pixels (Fig. 3.16b), and the cluster in the upper left corner of the plot depicts the region of pure turbid water pixels (Fig. 3.16a). The spectral plot analysis therefore reveals that in the CIR (color infrared) image (Fig. 3.4a) format of the HR_TM_DR and its corresponding data set of the LR_TM_DR image, the area covered with turbid water appears cyan, vegetation and woodlands appear in different shades of red, and the areas covered with pure flood water appear black.

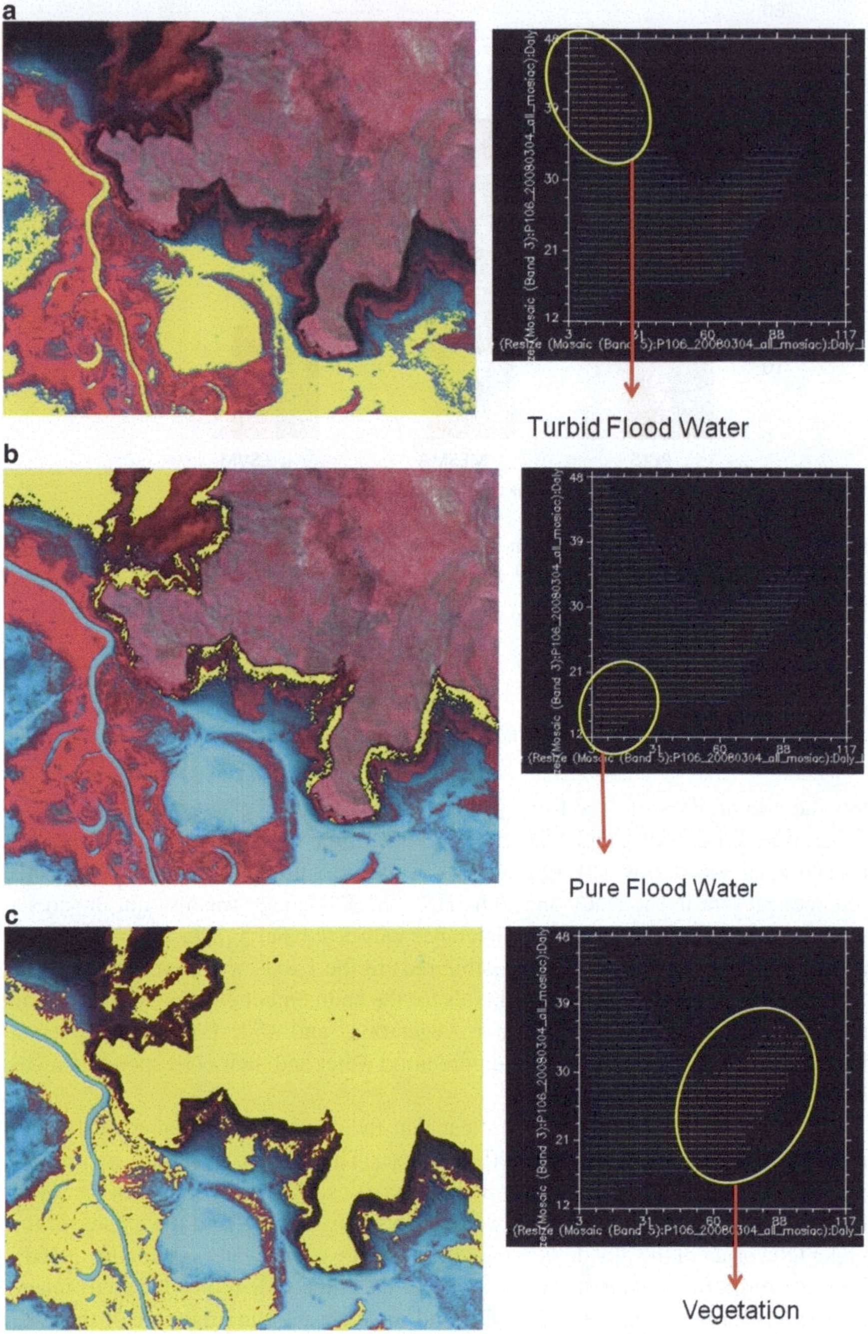

Fig. 3.16 Detection of endmember classes based on spectral plot analysis

The second most important step in this experiment was the ground truth generation of fractional abundances present in each pixel. Hard classification results on the HR_TM_DR image were used to produce the ground truths. The identification of the pure ground cover classes as already discussed helped in selecting the appropriate training and test data to be used in classifying the HR_TM_DR image. The numbers of training and testing samples are given in Tables 3.9 and 3.10, respectively. The supervised maximum likelihood (PML) classification method was used to classify the HR_TM_DR image. Tables 3.11 and 3.12 show the class means vectors and covariance matrices for the endmember classes.

The classification results (Fig. 3.17) for the HR_TM_DR image give a clear picture of the distribution of the pure ground components (endmember classes) in the image. Subsequently, the classified HR_TM_DR image was downsampled nine times lower in spatial resolution to generate the true (ground truth) fractions or endmember classes present in each pixel. The fraction images in Fig. 3.18 show the ground truth fractions for the three endmember classes.

After application of the fixed set of endmember classes (u_FST), the adaptive pixel-dependent endmember sets were developed from the true fraction images (a_TST) and applied on the test image to unmix it into three fractional abundance images based on the three primary classes. After that, the adaptive estimated endmember index file (a_EST) was generated based on the proposed method (as explained in the Methodology section). The generated fractional abundance images using a_EST in PCLS, MESMA, and u_ESVM unmixing methods are listed in Fig. 3.19.

Similar to test 1, a comparative study had been conducted on the test 2 image to measure the error levels of the output fraction images using the u_FST, a_EST, and a_TST index file with the help of the three unmixing methods (Fig. 3.20). An identical condition was observed in the comparative study of the error estimations in

Table 3.9 Classification accuracies for training samples of HR_TM_DR image

Name of endmember class	No. of training data	Producer's accuracy (%)	User's accuracy (%)	Overall accuracy (%)	Kappa statistic
Turbid flood water	1,386	100.0	100.0	100.0	100.0
Pure flood water	358	100.0	100.0		
Vegetation	916	100.0	100.0		

Table 3.10 Classification accuracies for test samples of HR_TM_DR image

Name of endmember class	No. of training data	Producer's accuracy (%)	User's accuracy (%)	Overall accuracy (%)	Kappa statistic
Turbid flood water	612	100.0	100.0	100.0	100.0
Pure flood water	111	100.0	100.0		
Vegetation	312	100.0	100.0		

Table 3.11 Covariance matrices of the endmember classes

Turbid flood water						
Spectral bands	**1**	**2**	**3**	**4**	**5**	**6**
1	1.8147	0.34053	0.74102	0.26484	−0.21849	−0.01108
2	0.34053	0.7841	0.89664	0.40768	−0.16049	−0.02834
3	0.74102	0.89664	2.5296	0.89339	−0.36584	−0.09315
4	0.26484	0.40768	0.89339	2.1218	0.73883	0.17451
5	−0.21849	−0.16049	−0.36584	0.73883	2.3553	0.34581
6	−0.01108	−0.02834	−0.09315	0.17451	0.34581	0.995
Pure flood water						
Spectral bands	**1**	**2**	**3**	**4**	**5**	**6**
1	1.656	0.20145	0.33295	0.11629	0.30196	0.053378
2	0.20145	0.46709	0.24851	0.28434	0.32291	0.061719
3	0.33295	0.24851	0.85571	0.25576	0.52262	0.1491
4	0.11629	0.28434	0.25576	1.5155	1.2316	0.45154
5	0.30196	0.32291	0.52262	1.2316	2.9016	0.70225
6	0.053378	0.061719	0.1491	0.45154	0.70225	0.98265
Vegetation						
Spectral bands	**1**	**2**	**3**	**4**	**5**	**6**
1	3.8792	2.2507	3.882	−7.6628	6.9245	3.4331
2	2.2507	3.0202	3.7553	−2.5866	5.8744	2.6926
3	3.882	3.7553	7.7183	−20.013	12.548	6.5772
4	−7.6628	−2.5866	−20.013	174.3	−21.216	−20.5
5	6.9245	5.8744	12.548	−21.216	48.681	19.64
6	3.4331	2.6926	6.5772	−20.5	19.64	10.137

Table 3.12 Mean vectors of endmember classes

	Endmember classes		
Spectral bands	Turbid flood water	Pure flood water	Vegetation
1	72.89393939	54.81843575	60.68558952
2	38.14718615	18.89106145	27.45960699
3	45.38311688	14.50558659	21.38209607
4	16.53607504	10.34636872	92.46724891
5	7.594516595	8.212290503	62.50327511
6	4.256132756	4.38547486	17.65393013

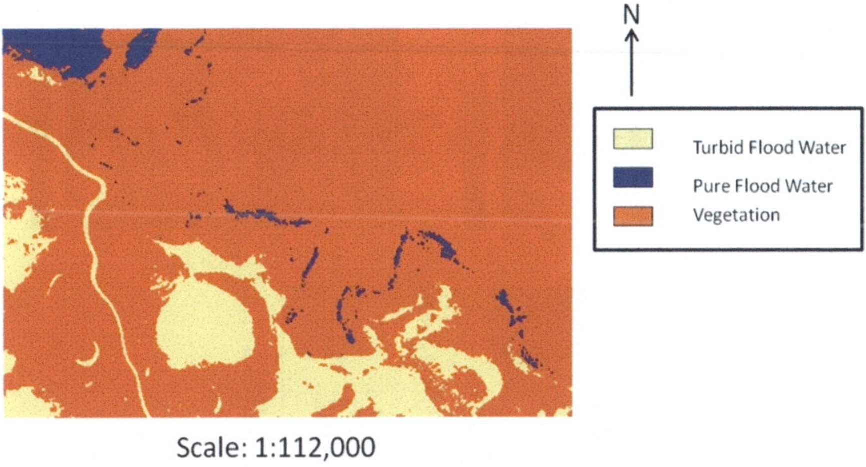

Fig. 3.17 Classified HR_TM_DR image

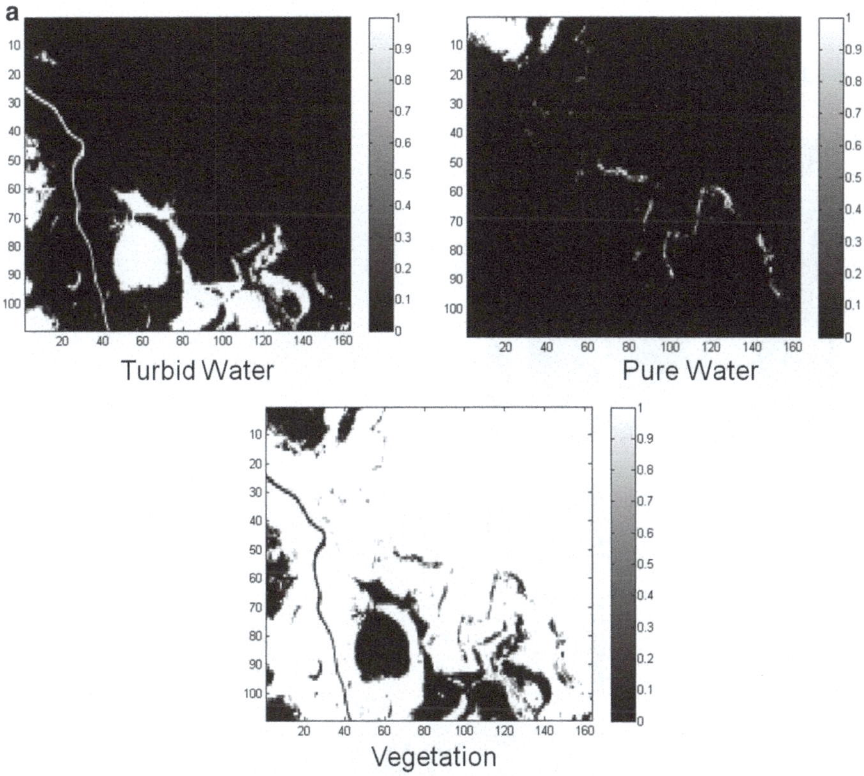

Fig. 3.18 (**a**) True fractional images obtained from HR_TM_DR image. (**b**) Estimated fractional images generated from LR_TM_DR test image

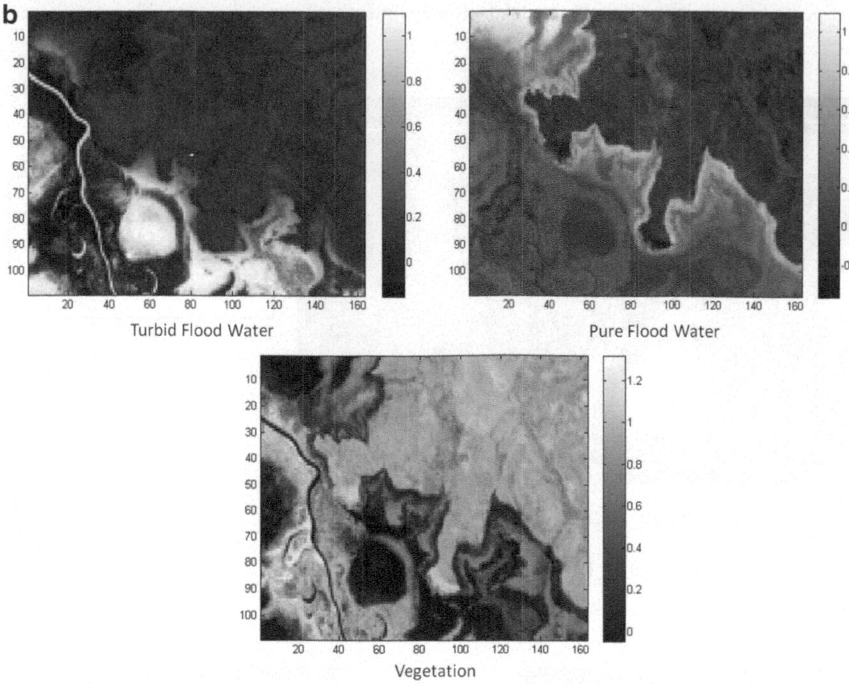

Fig. 3.18 (continued)

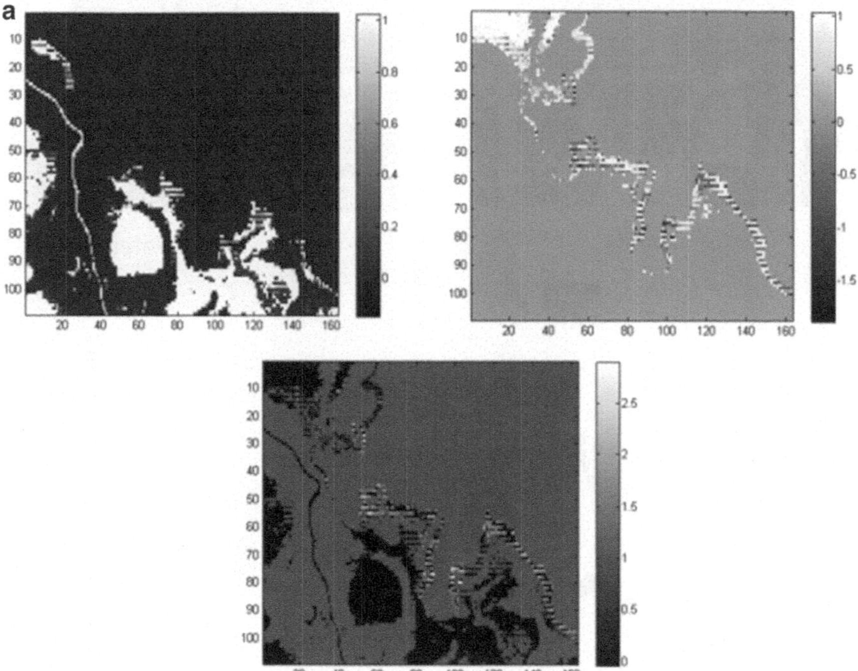

Fig. 3.19 (**a**) Estimated fractional images obtained from PCLS method. (**b**) Estimated fractional images obtained from MESMA method. (**c**) Estimated fractional images obtained from u_ESVM method

Fig. 3.19 (continued)

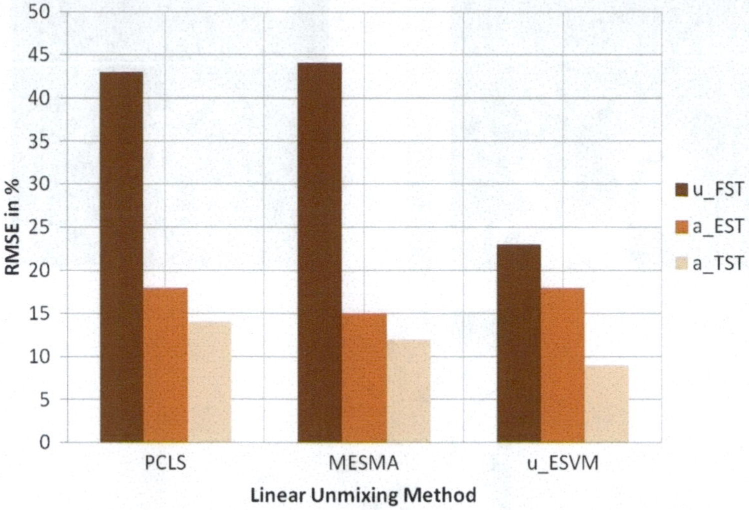

Fig. 3.20 Error comparisons between the outputs obtained from unmixing method using adaptive subset class index files and a fixed set of class for test 1

fractional outcomes in the case of the test 2 image. In this case, also the u_ESVM method responded more effectively in the adaptive pixel-dependent endmember index file.

3.4 Conclusion and Directions

The main concern in this investigation is how to overcome the problem of using a fixed set of primary classes for unmixing to obtain a reliable and efficient flood map. Pixel-dependent subsets of primary classes are introduced in this experiment. It has been proved by evaluating the RMSE of experimental results that the application of the proposed adaptive subset of classes has been able to produce more robust and less erroneous unmixing results over the conventional mixture analysis models. Using the adaptive subset of classes with the u_eSVM method produced more accurate results as compared to the other methods mentioned in this chapter, because primarily only the u_eSVM method is able to include the entire endmember set to consider the in-class spectral variation and also able to distinguish between the mixed and relatively pure pixels during the unmixing process. With the application of an adaptive subset of classes instead of using a fixed set of primary classes for unmixing every pixel in an image, the performance of the u_eSVM unmixing method is enhanced further.

Further investigation can be carried out to merge all the individual fraction images to generate a single thematic image. Several attempts were already recorded in the literature for subpixel mapping based on the assumption of high spatial correlation (Atkinson 2005; Wang et al. 2011). The subpixel mapping results represent the distribution of resident class fractions in each pixel. However, those methods are not easy to incorporate in the case of flood images because of the abrupt changes of land cover types observed in flooded areas resulting from inundation, and also because the land–water boundary is not very distinct in flood images. The development of an appropriate method of subpixel mapping for flood images will be valuable in future research to lead to a precise representation of flood water distribution over an area.

References

Adams JB, Smith MO, Johnson PE (1986) Spectral mixture modeling: a new analysis of rock and soil types at the Viking Lander 1 site. J Geophys Res 91:8098–8112

Atkinson PM (2005) Sub-pixel target mapping from soft-classified, remotely sensed imagery. Photogramm Eng Remote Sens 71(7):839–846

Chang CI (2003) Hyperspectral imaging: techniques for spectral detection and classification. Springer, New York

Chang CI, Chiang SS, Smith JA, Ginsberg IW (2002) Linear spectral random mixture analysis for hyperspectral imagery. IEEE Trans Geosci Remote Sens 40(2):375–392

Gorss H, Schott J (1998) Application of spectral mixture analysis and image fusion techniques for image sharpening. Remote Sens Environ 63:85–94

Halligan K (2008) Viper tools: MESMA. http://www.vipertools.org/?q=content/mesma

Jia X, Dey C, et al (2010) Controlled spectral unmixing using extended support vector machines. In: Workshop on hyperspectral image and signal processing: evolution in remote sensing, Reykjavìk, Iceland

Keshava N, Mustard JF (2002) Spectral unmixing, IEEE Signal Process Mag 19(1):44–57

Khanna RK, Agarwal CK, and Pravin K (2006) Remote sensing and GIS applications in flood management, Indian disaster management congress, National Institute of Disaster Management, New Delhi, India

Klatt JW (1994) Error characterization of spectral products using a factorial designed experiment. Chester F. Carlson Centre for Imaging Science. Royal Military College of Canada, Kingston

Mcfeeters SK (1996) The use of the normalized difference water index (NDWI) in the delineation of open water features. Int J Remote Sens 17(7):1425–1432

Richards JA, Jia X (2006) Remote sensing digital image analysis: an introduction. Springer, Berlin

Roberts DA, Gardner M, Church R et al (1998) Mapping chaparral in the Santa Monica Mountains using multiple endmember spectral mixture models. Remote Sens Environ 65(3):267–279

Schowengerdt RA (1997) Remote sensing: models and methods for image processing. Academic Press, St. Louis

Smith MO, Johnson PE, Adams JB (1985) Quantitative determination of mineral types and abundance from reflectance spectral using principal component analysis. J Geophys Res 90:C797–C804

Wang L, et al (2011) Sub-pixel mapping based on sub-pixel to sub-pixel spatial attraction model. Institute of Electrical and Electronics Engineers (IEEE)—International Geoscience and Remote Sensing Symposium (IGARSS), 593–596, Vancouver, Canada

Wang Y, Colby JD, Mulcahy KA (2002) An efficient method of mapping flood extent in a coastal flood plain using Landsat TM and DEM. Int J Remote Sens 23(18):3681–3696

Youssef A (1999) Image downsampling and upsampling methods. In: Conference on imaging, science, systems, and technology (CISST), Las Vegas, June

Chapter 4
Natural Resource Management: Why Does Micro Matter? Experience from Kerala

Srikumar Chattopadhyay

Abstract Natural resources management has witnessed perceptible changes in recent years, as there are symptoms of degradation causing irreversible damage to the natural resource base and loss of production potential even as society is striving for sustainable development. This chapter argues that it is necessary to introduce a microlevel approach for achieving prudent management practices in the case of natural resources, citing specific examples from an experiment initiated in Kerala.

Keywords Natural resource management • Panchayat • Kerala

4.1 Introduction

Prudent management of natural resources, particularly land, water, and biomass, forms a key element in planning for sustainable development. Overexploitation, misuse, and inappropriate use of resources and indiscriminate human intervention in the biophysical framework has pushed the Earth's systems outside a stable environmental state (Rockstrom et al. 2009). On average, human populations are withdrawing resources at a rate 20 % faster than the Earth can renew and consequently are depleting ecological assets (Wackernagel et al. 2002; Jorgenson 2003). The existing natural resource management system is being questioned as there are signs that this high degree of exploitation is resulting in irreversible environmental damage. The Millennium Ecosystem Assessment report (2005, p. 16) observed that 'changes in ecosystem have contributed to substantial net gains in human well being and economic development, but these gains have been achieved at growing costs in the form of degradation of many ecosystem services, increased risks of nonlinear changes and the exacerbation of poverty for some groups of people.' The challenge is to reorient policies, institutions, and practices in our natural resource use and reduce negative trade-offs. Sustainable resource management must be an integrated and

S. Chattopadhyay (✉)
Geographer and Environmental Resource Analyst, Center for Earth Science Studies, Thiruvananthapuram, Kerala, India
e-mail: srikumarc53@gmail.com

A.K. Dutt et al. (eds.), *Spatial Diversity and Dynamics in Resources and Urban Development*, DOI 10.1007/978-94-017-9771-9_4

interdisciplinary process aiming at interdependencies between institutions, environmental dynamics, economic processes, applied technology, and dominant cultures in managing and administrating natural resources (Rammel et al. 2007). But how do we develop such management practices? Our attempt in this chapter is to argue that this is largely achievable within the context of broad-based sustainability following a microlevel approach. In broad-based sustainability the emphasis is not on any fixed definition or ideal, but rather on 'evolutionary processes of improving the management of systems, through improved understanding and knowledge' (Cary 1998, p. 12, c.r.; Rammel et al. 2007). This concept is significant as the potential of evolutionary thinking in resource management is being stressed in recent years because natural resource management cuts across disciplinary boundaries and warrants a socioecological system analytical approach. There is a growing realization that natural systems and socioeconomic systems are governed by self-organization and co-evolutionary dynamics and therefore are mutually dependent, sharing characteristics of complex adaptive systems that express large macroscopic patterns emerging from local small-scale interactions (Rammel et al. 2007). The microlevel approach facilitates understanding of local-level interactions.

This chapter is organized in eight sections. Section 4.2 discusses the changing nature of natural resource management systems. It also discusses the need for such changes in the context of a paradigm shift in development processes and society's quest for sustainable development. Section 4.3 presents the macro–micro debate. Section 4.4 spells out Kerala's natural resource base. Section 4.5 describes microlevel resource zones in Kerala, top sequences at various altitudinal zones, and micro zones and production potentials. Section 4.6 presents Kerala's initiatives in microlevel resource management. It elaborates on execution of the program, mapping module, field examples, and the extension of Panchayat resource mapping (PRM). Implication of this initiative is discussed in Section 4.7. The eighth section concludes the chapter.

4.2 Changing Nature of the Natural Resource Management System

The science of natural resource management was shaped under the premise of utilitarian values that view nature as merely a storehouse of raw materials. Resources were thought to be valuable if they could be used to create wealth (Holling et al. 1998). Overemphasis on economy and technology has led to the exploitation of natural resources as commodities and development of management systems accordingly. Commodification is not limited to just natural resources but was even extended to people's own labor (Polanyi 1980). This conventional, disciplinary, and reductionist view is found to be inadequate as manifested through resource depletion, irreversible change in many cases, and cascading environmental impacts of indiscriminate resource use. The ecosystem dimension of resources is frequently overlooked, which perhaps has created the serious environmental problems that are threatening the economic base of society.

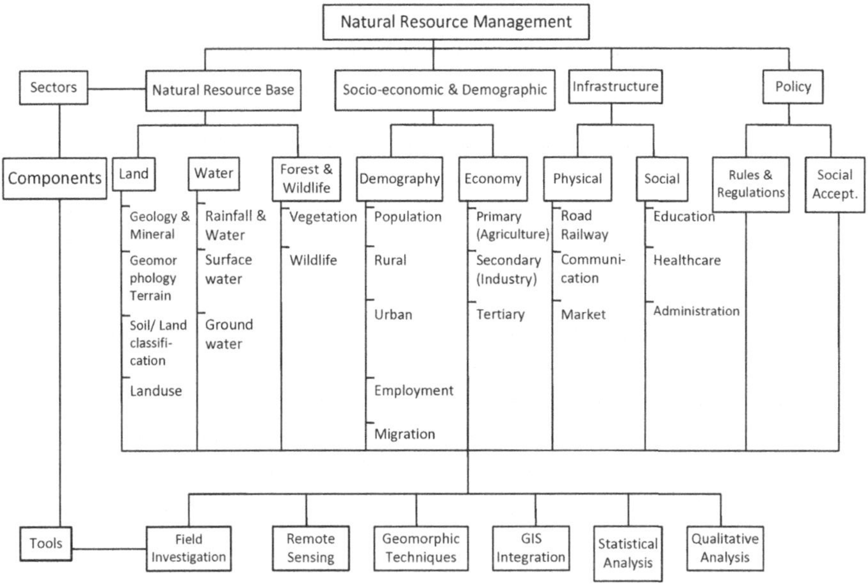

Fig. 4.1 Integrated framework of natural resource management studies, including tools

The alternate form of resource management now being advocated draws from the systems approach and evolutionary theory. It is fundamentally interdisciplinary and combines historical, comparative, and experimental approaches at scales appropriate to the issues (Holling et al. 1998). There are four basic domains of natural resource management: (1) the natural resource base: the land, water, air, forests, and wildlife, or the ecological domain; (2) socioeconomic and demographic characteristics; (3) infrastructural and technological issues; and (4) institutional and policy issues (Fig. 4.1). Each of these subsystems is governed by its own set of rules and internal dynamics. Mutual interactions among these subsystems are scale dependent, multidimensional, and co-evolutionary. The reductionist approach hitherto followed for natural resources management might have to be replaced by a co-evolutionary perspective emphasizing a "natural resource management system as hierarchically arranged mosaics of co-evolving social, technological and environmental processes or elements" (Rammel et al. 2007, 12). A socioecological approach is suggested "to relate management practices based on ecological understanding to the social mechanisms behind these practices, in a variety of geographical settings, cultures and ecosystems" (Berkes et al. 2003, p. 4).

This paradigm shift in the concept of natural resource management, in view of degradation, overuse, and growing stress on the natural resource base, should strive to pair utilization with conservation. Further, it should internalize the conservational aspects within the gamut of the development process, thereby contributing to sustainability. In other words, the focus of natural resource management is on sustainability encompassing ecological, economic, and social concerns. Avoiding irreversibility is an important step in the present approach of natural resource management and can serve as a useful link between ecological science and public policy decisions on resource use (Anonymous 1989). The complex ecological interdependence of various components of natural resources dictates a threshold or resilience limit, crossing of which will change the system irreversibly. This limit needs to be recognized while using resources, and policy may be laid down appropriately. The natural resource management program must rely on adequate information on ecosystems, proper understanding of the ecosystem processes, and the status of resources use, resource demand, existing load, environmental health, and impacts. Participation of the stakeholders and maintaining diversity are part of this exercise in changing perspectives. The co-evolutionary approach strives for an adaptive system in which the ecological system and the social system interact and feed each other. This approach is manifested at the resource user level where individuals and the community directly interact with natural resources within the boundary of an ecosystem. This concept can be captured with full ramifications within an eco-zone or micro-watershed, a functional ecosystem entity. The issue of governance is important in this context of social ecological functions because societal actions impact natural resource systems, thereby changing the natural process and also being affected by the changes. A microlevel natural resource analytical frame is given in Fig. 4.2.

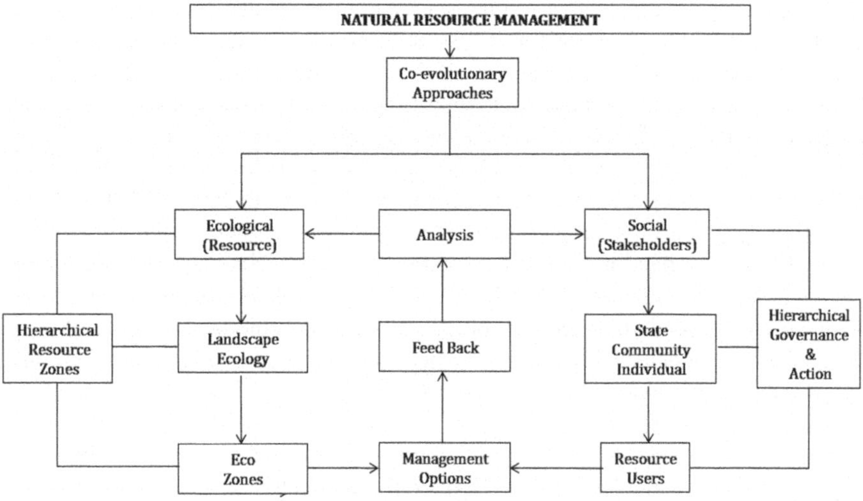

Fig. 4.2 Natural resource management: framework for microlevel analysis

4.3 Macro–Micro Debate

The macro–micro debate in resource management and the planning process is age old. There are schools proposing alternate views and different positions regarding the relationship between resource management and planning. Centralized planning and sectorial approach are often contested because they have not yielded desired results. Local-level planning and a development framework have emerged as an important approach for regional planning. In 1992, the 73 and 74 constitutional amendments under the articles of 243 G and 243 W, respectively, have specifically empowered local self-governments to take up the responsibilities of preparing and implementing locally developed plans for economic development and social justice in their respective jurisdictions (Chattopadhyay et al. 2004).

This renewed emphasis on local-level initiatives opened a significant opportunity for the people, professionals, planners, and administrators to engage and cooperate at the village level. To be positioned as an effective system for developing local-level economy, the decentralized process must address the issues of local resources, resource use, microlevel environment, stakeholders, scope of institutionalization, and people's participation (Chattopadhyay 2003). It is also well realized that macrolevel and aggregate data are not adequate for the purpose of drawing plans relevant at the local level. The emergent paradigm for balanced living with the environment brings together decentralization, reliance on the local resource base, democracy, and diversity: this will give way to an inclusive holism, open system thinking, and diverse options instead of reductionism, linear thinking, and standard solutions (Chambers 1994). It is now well accepted that macro studies and analyses are required to develop the overarching frame or broad contour of management, whereas microlevel analysis is important to understand the processes. Nonavailability of data at the ground level is a major problem faced by the concerned people. Mukherjee and Bandyopadhyay (1993) discussed this point elaborately in a review of panchayat activities in West Bengal and pointed out the data gap at the microlevel for drawing up meaningful panchayat plans under decentralized development, for which West Bengal is well known.

4.4 Natural Resource Base of Kerala

Kerala's performance in the sphere of social and human development has been remarkable during the past couple of decades. Natural resource endowment has been considered as an important contributor underlying the Kerala model of development or Kerala's development process (Chattopadhyay and Franke 2006; Chattopadhyay 2005). The natural resources, particularly land, water, and biomass, provide the foundation for a productive sector, which is the base for any economic development and growth.

The geologic setup, high physiographic variations within a short distance, low latitudinal position and limited longitudinal extension, topographic diversity even at the microlevel, a long coastline (560 km), and location with respect to sea and mountain have endowed the state with a rich and diversified natural resource base. About 58 % of the total area of Kerala lies below an elevation of 100 m elevation. In all, 22 geomorphic units have been identified, and there are four districts, each having as many as 14 units (Chattopadhyay and Chattopadhyay 1995). Indian Council for Agricultural Research (ICAR) has classified the soils of Kerala into 35 categories (Kerala State Land Use Board 1995). There are 13 agro-climatic zones with varying production potential (Government of Kerala 1974). On average, Kerala receives more than 3,000 mm rainfall annually. Precipitation increases from the coast to the foothills, recording 4,510 mm in Nariyamangalam in the western foothills of the Western Ghats and also from south to north. There are places all along the foothills that receive rain every afternoon as in tropical rainforest conditions. The eastern part of the Western Ghats bordering Tamil Nadu is a rain shadow area where precipitation is reduced considerably, the least being 470 mm in Chinnar (leeward side of the Western Ghats). There are 9 stations that each receive an annual average rainfall of more than 3,500 mm and 11 stations that record less than 2,000 mm in a year (Simon and Mohankumar 2004). Rain occurs in every month, and the bulk of the rainfall occurs during southwest and northeast monsoons. Diurnal and seasonal variations of temperature are low (5–7 °C) in major parts of Kerala. The orographic effect on climate is well pronounced. As average duration of sunshine is more than 10 h in a day, production seasons prevail throughout the year.

The state is well drained by 44 rivers. Many of these rivers have been dammed in the upstream section for irrigation and hydroelectricity projects, impounding 30 reservoirs whose combined area is 296 km^2 (square kilometers). A 1-km length of a main river channel is sustained by 12 km^2 of area compared to an all-India average of more than 1,100 km^2. Total annual water yield in Kerala is 70,165 million m^3 (cubic meters) (Government of Kerala 1974). On average, a 1-km^2 area yields 1.81 million m^3 of water, of which 60 % is usable (Nair and Chattopadhyay 2005). The replenishable groundwater resource is 6,840 million m^3, and the overall stage of development is 43 %, as stated by the Central Ground Water Board, Government of India. According to Kudumbashree Mission, there are 46,337 ponds covering 275 km^2 in area Government of Kerala (2005).

Natural vegetation covers the rugged slopes, hills, and plateaus in the eastern part of Kerala. The main vegetation types are evergreen, semi-evergreen, moist deciduous, dry deciduous, subtropical montane, and temperate shola. Government statistics indicate that the total forest area in 1991–1992 was 11,223 km^2, of which 7,860 km^2 was dense forest, 1,538 km^2 plantations, and the remaining 1,825 km^2 was given to various government and semi-government agencies (Government of Kerala 1994).

A transect from the coast to the Western Ghats crest shows land use change along with altitude from predominantly coconut up to 100 m, rubber from 40 to 500 m, tea and coffee up to 1,000 m, and cardamom above 1,000 m of altitude. The microlevel sequence is rice/seasonal crops in the valley, mainly coconut in the lower side slope, mixed tree crops with sparse settlements in the upper side slope, and predominantly settlement with mixed tree crops along the ridges. Upper slopes and ridges in the higher altitudes are covered by agricultural plantations, forest plantations, and natural vegetation. The state enjoys a diversified cropping pattern, evident from the fact that none of the districts accounts for more than 20 % for any major crop except for Palakkad, which accounts for 28.57 % of the total area under food crops, and Kottayam for rubber (20.19 %). About 18 % of the state's forest area was under plantations as of 1992, as compared to 12.6 % in 1973. Teak is the main plantation crop, covering slightly less than 50 % of the total plantation area. Forest products include logs, poles, charcoal, firewood, sandalwood, honey, reeds, and bamboo. Minor forest products and medicinal plants are regularly collected from the forests by the tribal people.

The foregoing discussion highlights that Kerala has three distinct natural resource-based productive sectors, namely, fisheries (marine and back water), agriculture, and forestry, distributed throughout the state, and providing a diversified resource base at the microlevel. This ecological framework, coupled with institutional interventions such as land reforms, has provided the people in the state better access to resources in terms of both location and possession. Adequate water availability both in space and in time facilitates intensive land use and also settlement dispersal. The general productivity level of the state is high. Long growing seasons add to its agricultural productivity. High biomass production, the high assimilating capacity of the biosphere, and very high species diversity combine to increase the production base of the state, giving Kerala a higher carrying capacity than other Indian states. Kerala State does not suffer from large-scale calamities and thus there are minimum negative externality effects. All these natural advantages, which we consider as part of the ecological security, have placed the state up front to utilize opportunities at the individual and community level. These advantages have to be protected and enhanced by efficient use with limited injury to the resource base and environment to maintain the level of Kerala's development and to sustain it at the higher level.

4.5 Microlevel Resource Zones in Kerala

The state's microlevel variation emerges from landscape diversity, mainly from the great local variations in geomorphology and soil types. Kerala's landscapes follow geographic hierarchy theory as established by various empirical studies in other parts of the world (O'Neil et al. 1989). This hierarchy can be established by carving the landscape according to the drainage basins that are governed by hydrological processes. A hierarchical order of the drainage basins could easily be identified. A main drainage basin can be divided into sub-watersheds, mille watersheds, and micro watersheds. Kerala has 44 major river basins, 151 sub-watersheds (100–500 km^2), 960 mille watersheds (10–100 km^2), and more than 2,000 micro watersheds (<10 km^2) (Chattopadhyay 2002).

Taking note of the landform conditions and geomorphic processes, such as fluvial or coastal conditions operating in a given area and the specific actions that are involved therein, it is possible to identify the microlevel zones and their specific characteristics. It is these micro-zone characteristics that influence the production base.

The topography of Kerala is mostly undulating, that is, each small area can be characterized by valley, slope, and ridge formations. This arrangement is known as a 'topo sequence,' the smallest unit of a landscape. The valleys can be broad, narrow, flat bottomed, or sloping. Similarly, the slope joining the valley and ridge can have a steep or gentle gradient. The ridge can be elongated, narrow, or wide. Whatever the case, the basic element of any landscape unit is this topo sequence. Similar microlevel units can be identified in the coastal plain and other low-lying areas. Ridges (sandy ridges) and runnels (low-lying areas between two sandy ridges) are the normal topo sequences in the coastal plain.

4.5.1 Topo Sequence at Various Altitudes

Macro- and micro zones interact to produce a further set of variations in that the micro zones arrange themselves differently at different altitudes as one goes east from the Kerala coastline (Chattopadhyay and Franke 2006). Typical combinations are ridge and runnel arrangements in the coastal plain; levees and back swamps in the fluvial plains; valleys, mesa side slopes, and mesa surfaces in the lateritic hard crust area; flat-bottomed valleys, side slopes, and ridges/mounds in parts of the midlands; valleys, lower slopes, upper slopes, and ridges in the midlands and foothills; and narrow valleys, steep side slopes, and narrow ridges in the hilly areas. The general topo sequence in most parts of Kerala, as indicated earlier, has either three or four components: valleys or plains, lower slopes, upper slopes, and ridges. Rice and other seasonal crops are cultivated in the valleys. There is also widespread use of the valleys for settlements and tree crops. The lower slopes are planted mostly in tapioca, areca nut, and coconut. They occur as ribbons around the valleys; the upper slopes typically have settlements, coconut, and other fruit trees. Ridges are given over to settlements, roads, and other structures. Coconut trees are also planted on the ridges.

With an increase in elevation, the micro-zone pattern changes. The ridges and upper slopes are covered with plantations of rubber and cashew whereas the lower slopes have settlements and coconut trees. At even higher altitudes, within the forest area, grasslands mainly occupy the ridges. The change of crops from valley to ridge is an indicator of changes in water availability and soil composition within the topo sequence. Spanning the entire state, these microlevel variations in ecological characteristics create enormous opportunities for resource management.

4.5.2 Micro Zones and Production Potential

Each of these microlevel units or micro zones, also known as terrain components, has a different production potential (Chattopadhyay and Chattopadhyay 1995). Nutrient-laden sediment and water from the ridges and slopes move down to the valleys, and thus the valleys are replenished and maintain their fertility. Kerala's high water availability, topography, and soil character allow the valleys to be normally given over to planting seasonal crops such as rice, whereas tree crops, which have deep root penetration, are planted on the slopes and ridges. The Panchayat resource mapping program (PRM) introduced in Kerala in 1990 (Center for Earth Science Studies 1991) attempted to delineate these microlevel variations to enable the panachayats to plan for more appropriate land, water, and biomass management with involvement of the stakeholders. We briefly describe this program here and try to highlight its uniqueness in the matter of natural resource management at the microlevel.

4.6 Kerala's Initiatives in Microlevel Resource Management

In spite of the state's richness, there are symptoms of stress at various levels, particularly in the form of deforestation, soil erosion, loss of soil fertility, overexploitation of riverine resources such as sand and clay, salinization, depletion of the groundwater table, and water pollution. The scenario is further complicated by the high concentration of population in the state and occasional natural disasters offsetting the development process. To arrest further degradation and enhance the resource base, it is important to turn to microlevel initiatives. Introduction of decentralized planning following the three-tiered Panchayat Raj system in Kerala was believed to be needed as macrolevel interventions failed to address microlevel issues. A program known as Panchayat Resource Mapping (PRM) was introduced in Kerala by the early 1990s to help planning from lower levels.

4.6.1 Execution of the Program

The Centre for Earth Science Studies, in collaboration with the Kerala State Land Use Board (KSLUB) and Kerala Sastra Sahitya Parishad (KSSP), initiated the project "Panchayat Level Resource Mapping for Decentralized Planning with People's Participation" in 1991. The Department of Science and Technology (DST) of the Government of India sponsored the pilot project to establish the methodology and create a cadastral-level database for Natural Resource Data Management System program of DST. It is a novel venture to map natural resources with the participation of local volunteers, duly trained and guided by professionals. Resource maps were prepared in 1:12,500 scale and in cadastral scale (1:3,960). Table 4.1 presents the

Table 4.1 Activities for executing panchayat resource mapping (PRM) program

Phase/type of activity	Details
Environment building	Interaction with the panchayat and NGO. Campaigning in the panchayat, preparing ground for initiating the program. Formation of a PRM committee. Conducting PRA as entry program.
Training materials and guidebook	Preparation of materials for training volunteers such as handouts, sample maps, and other required materials. Guidebook for master volunteers and other professionals.
Selection of volunteers	Selection of at least five volunteers from the local people in each ward/village/hamlet/unit area of 1–2 km^2
Training of volunteers	(A) Training of master volunteers, who will lead the team and coordinate the work at the panchayat level, preferably in a R&D institute
	(B) Training of other volunteers in the respective panchayat
Mapping by volunteers	Land use and asset mapping in cadastral scale follows training of other volunteers. Consolidation and replotting of mapped data continue.
Mapping by professionals	Mapping of landform, surface material/soil, water potential, and other relevant details in desired scale (1:12,500 or in cadastral scale for smaller area).
Finalization of maps	Finalization of all maps, preparation of short writeup, and preparation of environmental appraisal map for discussion at the panchayat.
Action planning	Presentation of maps in the panchayat and developing an action plan map through interaction with the panchayat. Appraising PRM committee/panchayat about the resources, potentials, and problems of the panchayat.

Source: Compiled from CESS (1991), Chattopadhyay et al. (1999)

various steps followed to execute this program in Kerala. Although it started as a DST-supported pilot program, in the course of time it emerged as a major program for covering all the panchayats in the state. The Government of Kerala issued required orders to this effect (Center for Earth Science Studies 2000). It is now mandatory for all the panchayats to prepare such resource maps and work out their development plans. Kerala has 999 panchayats and each of these panchayats has prepared a development report containing details of their local resources, although resource maps are yet to be completed for all the panchayats.

4.6.1.1 Mapping Module

This mapping module envisaged as a minimum-needs module consists of seven or eight maps, namely, landform, surface material, depth to bedrock/thickness of weathered mantle, potential areas of water availability, land use, assets, and environmental appraisal for land use planning or an action plan (Chattopadhyay et al. 2004). The activities follow:

1. Land use and asset mapping by the volunteers after due training on the 1:3,960/1:7,920 scale cadastral map and collection of details about all water

sources (well/pond/tube well/bore wells) in a structured format. Volunteers also maintained a field dairy recording the details of the survey plot, including problems narrated by the residents.

2. Preparation of relief, landform, surface material, and water availability maps by the Earth (project) Scientists on a 1:12,500 scale through field survey and collection of data at 250-m intervals (16 samples/km^2).
3. Preparation of environmental appraisal map through collation of both sets of data by the Earth Scientists and working out an action plan map through interaction with the panchayat people.
4. Giving all the maps with a brief write-up to the respective panchayat.

Table 4.2 shows the utility of the basic resource maps generated for a panchayat. Initially, volunteers also collected household-level socioeconomic data; however, this was subsequently discontinued.

4.6.1.2 Field Example

This field example describes resource mapping in a panchayat of North Kerala to provide an idea of the type of maps prepared and the kind of information available therein. Onchiyam panchayat, with an area of 9 km^2, is located in Kozhikode district. It is one of the 25 panchayats considered for PRM under DST sponsorship. Maps prepared are relief, landform, surface material, water availability, land use, assets, and environmental appraisal. One additional map shows micro-watershed boundaries, traced during a 'drainage mapping program' attempted by KSSP volunteers under the guidance of CESS and CWRDM scientists. It was a map needed by the panchayat to develop a watershed management plan under watershed-based development schemes. This map was also necessary to understand the problem of waterlogging and the scope of providing required drainage canals to drain the area.

There are eight types of landform units in this panchayat. Surface materials are duricrust, laterite soil, riverine alluvium, and coastal alluvium in addition to hard laterite crust and occasional rock outcrops. A water survey indicated that there were a total of 84 ponds, of which 8 each had an area greater than 300 m^2. The relatively elevated areas have problems of water availability, although water supply is a problem in the coastal plain as well. Tree crops composed of coconut, cashew, and other fruit trees dominate land use pattern. Only 22 ha of area was under rice cultivation as of 1991 compared to 164 ha in 1975–1976. The field investigation indicated further decline in area under rice cultivation; it was only 3 ha in 1996, and at present all the paddy fields have been diverted to other uses. A comprehensive table indicating existing conditions, problems, and suggestions for intervention in each of the mapping units has been prepared (Table 4.3).

Each of these eco-zones has distinct characteristics. Problems of sea erosion, gully erosion, sheet erosion, waterlogging, salinity, and reclamation are found to be unique to particular zones. Thus, zone-specific intervention could be worked out for better resource management. The land use types had overlapping characteristics, but

Table 4.2 Resource maps and their utility

Thematic maps	Information collected	Utility of the map
Landform	Geomorphic units, slope categories and direction of slope, erosion-prone areas	Land capability assessment, action plan for soil conservation, land use planning, and carrying capacity assessment
Basement configuration	Lithology, depth to bedrock, and geologic structure	Assessment of the thickness of weathered or unconsolidated materials, assessment of water-holding potential, erosion, and landslide management
Water availability	Perennial and seasonal areas, depth to water table, location of springs, approximate water potability	Identification of potential sources areas, development of ponds and springs, guideline for optimum well depth and location, planning for conjunctive use of water resources
Surface material	Rock outcrop, cover material soil type, soil texture, soil thickness/depth, soil drainage	Assessment of productivity, soil stability, geohydrological investigation availability of minor minerals (sand, clay, rock)
Land use	Existing land use, plot-wise details of crops, sites of paddy field reclamation, distribution of wastelands, marshes/ wetlands, water bodies	Assessment of existing land use pattern, computation of area under different land use categories/crops, formulation of sustainable land use pattern, planning for land use consolidation
Assets	Ward boundary and administrative divisions, roads, railways, canals, all educational infrastructures, schools, colleges, hospitals, dispensaries, power lines, transformers, markets, ration shops, recreational facilities, postal and telegraph/telephone facilities, ponds, springs, wells, and other water structures	Assessment of spatial distribution of all infrastructural facility, help preparing estimates for construction and engineering purposes, delineation of gaps and underdeveloped areas
Environmental appraisal	Erosion and landslide proneness, waterlogging, water deficit areas, wastelands, flood-prone areas, extent of salinity infestation, degraded land, reclamation/land conversion, deforested areas., and any other problems	Environmental hazard management, land use planning in conformity to natural setting base to initiate discussion at local level for action planning
Action plan	All problem areas warranting intervention such as ponds to be desilted, drainage requirement, low productivity zone, erosion-affected areas	Preparation of action programs, project preparation, priority assessment, and planning

Source: Chattopadhyay (2003)

Table 4.3 Microlevel eco-zones in Onchiyam panchayat, Kerala

Micro-zone Sl. no.	Unit description	Area (km^2)	Soil	Water availability	Land use availability	Problems
1.	Ridge/hill crest with laterite	1.36	Gravelly loam	Low	Coconut, mixed tree crops, laterite quarry	Shallow soil, water shortage
2.	Steep and moderately steep slope with laterite	0.75	Gravelly clay loam	Moderate	Coconut, settlement, mixed tree crops	Gully erosion, water shortage
3.	Gentle to moderate side slope with laterite	2.38	Loam to gravelly clay loam	Moderate to high	Coconut, mixed tree crops, tapioca, settlement, arecanut, banana, vegetable	Sheet erosion
4.	Valley floor with ravine alluvium	1.27	Silty clay loam and loam	High	Coconut and vegetable settlement	Waterlogging and land reclamation
5.	Coastal plain with coastal alluvium	1.68	Sandy loam	High	Coconut and settlement	Waterlogging and salinity
6.	Sandy ridges and runnels	1.03	Sandy	High	Settlement and coconut in runnel	Excessive drainage, salinity, and waterlogging
7.	Beach	0.20	Sand	High	Open space	Sea erosion
8.	Tidal marsh	0.39	Sandy clay		Coconut	Reclamation

Source: CESS (2000); Panchayat Resource Mapping: Final report, p 25

the range of crop variation within short distances is significant. All the paddy fields in this panchayat have been made on land reclaimed by draining and filling tidal marshes. Reclaiming the land in this way leads to waterlogging in the unit of the valley floor with riverine deposits. This association was seen more clearly when micro-zones four and five were identified. Waterlogging prevents the soil from recuperating, interferes with transportation links, and facilitates mosquito reproduction in some areas, leading to health problems. The problem of water shortage and waterlogging within the same panchayat is a manifestation of the high degree of differentiation of the landscape system. As many as seven micro-watersheds could be identified in this single panchayat. Because the different zones are so close spatially, the interventions required for combating waterlogging in one zone can contribute to increasing the water availability in the water shortage zones by specific watershed management actions. Waterlogging during a monsoon, salinity intrusion, lack of an irrigation system, scarcity of drinking water and sanitation facilities, dereliction/silting of ponds, and landscape disfigured by abandoned laterite quarries, soil erosion, coastal erosion, drying up of wells, and lack of maintenance of roads are some of the problems noted from the survey and further corroborated by local people. Interventions such as excavation of ponds, aquaculture in marshy areas, vegetative bounding/barrier, and water harvesting are some of the suggestions made during discussions with the panchayat members and local people.

4.6.1.3 Extension of PRM

Bharat Gyan Vigyan Samithi (BGVS), a New Delhi-based volunteer organization, adopted RPM methodology to implement drinking water and sanitation programs. It was implemented in selected Community Development Blocks of West Bengal, Bihar, Orissa, and Tamil Nadu under the Rajiv Gandhi National Drinking Water Mission. The program was implemented with certain modifications such as use of only cadastral scale for preparing all thematic maps and focusing the survey on drinking water and sanitation requirements of villages. The States of Madhya Pradesh, Karnataka, Pondicherry, and Tripura are utilizing this technique for various land and water resource management projects. The Department of Rural Areas and Employment of the Government of India recommended PRM as one of the methodologies to conduct resource inventories for watershed management and wasteland development. Mapping proved to be an important tool for mobilizing the people. There were instances in Bihar where interest in literacy among the tribal groups could be generated following resource mapping. Therefore, PRM has potential to pave the way for larger interventions at various levels.

4.7 Implication for Natural Resource Management

The emerging trend in natural resources management is to emphasize microlevel intervention following diversity of the resource base and to involve stakeholders in the management process. The PRM internalizes both these objectives and therefore is emerging as a useful planning tool (Thomas Isaac and Franke 2000; Chattopadhyay 2002). Specific study indicated that the panchayats with resource maps were well placed to work out their plan, especially for land and water management (Chattopadhyay et al. 1999). The program could imbue a new culture and new thought processes in the sphere of natural resource-based planning. The common people became aware of the fact that resource documentation and mapping are vital for local-level planning. In many places the actual users of the land participated in the mapping exercise. They could bring in some of their traditional practices while discussing the action plan, and this gave them a higher level of confidence, which has long-term implications for capacity building and resource management with stewardship. A sense of leadership and working together for the common good came to characterize the volunteers while participating in the program. Perhaps PRM is the beginning of a paradigm shift in natural resource management. In the emerging information age, decentralized management of basic resources, grounded in detailed location-specific information, may turn out to be the most desirable option (Gadgil et al. 1998). This program has also the potential to institutionalize the process of resource management at the microlevel. The microlevel approach will also bring forward the spatial dimension of resource distribution, interlinkages of basic resources, and need for integrated policy in natural resource management.

4.8 Conclusion

Natural resource management is experiencing a paradigm shift. The thrust is on a microlevel and integrated approach. The components of resources are ecologically interdependent and the interrelationship is nonlinear. A straight-jacket commodity approach appears to be counterproductive. Knowledge-based decisions warrant a detailed spatial database in micro scale and a systematic analytical frame. The policy may be oriented accordingly to cater to this emergent need. This chapter has demonstrated, with examples from Kerala, that there are significant microlevel variations in land, water, and land use. These variations need to be captured to optimize production potential and evolving suitable management strategies. The Kerala experiment on microlevel resource management, starting with resource mapping at the panchayat level, is a novel attempt. It has the potential to contribute significantly in evolving a microlevel methodology in natural resource management. The recent emphasis on accounting for diversity and participation is well addressed in this experiment in Kerala.

Acknowledgments The author thanks the Director of CESS for extending facilities and permission to prepare this chapter.

References

Anonymous (1989) Resource management law reform: sustainability, intrinsic values and the needs of future generations. Working Paper No. 24. Ministry of Environment, Wellington

Berkes F, Colding J, Folke C (eds) (2003) Navigating social ecological systems: building resilience for complexity and change. Cambridge University Press, Cambridge

Cary J (1998) Institutional innovation in natural resource management in Australia: the triumph of creativity over adversity. Abstract of the conference "Knowledge generation and transfer: implications for agriculture in the 21st Century". University of California - Berkeley, Berkeley

Centre for Earth Science Studies (1991) Panchayat level resource mapping: an approach paper. CESS, Thiruvananthapuram

Centre for Earth Science Studies (2000) Panchayat level resource mapping: final technical report. CESS, Thiruvananthapuram

Chambers R (1994) Participatory rural appraisal: challenges, potentials and paradigm. World Dev 20(10):1437–1454

Chattopadhyay S. (2002) Panchayat resource mapping: a base for watershed management and development. Experience from Kerala. In: Vijayanand SM (ed) Watershed management practices in India. Western Ghats Cell, Planning and Economic Affairs Department. Government of Kerala, Thiruvananthapuram

Chattopadhyay S (2003) Towards Gram Swaraj: West Bengal and Kerala experiments. Anasakti 1(1):93–115

Chattopadhyay S (2005) Ecological security as a base for social security: context Kerala. Geogr Rev India 67(2):109–125

Chattopadhyay S, Chattopadhyay M (1995) Terrain analysis of Kerala. Technical Report No. 1/95. Science Technology and Environment Department, Government of Kerala, Thiruvananthapuram

Chattopadhyay S, Franke R (2006) Striving for sustainability: environmental stress and democratic initiatives in Kerala. Concept, New Delhi

Chattopadhyay S, Krishnakumar P, Rajalekshmi K (1999) Panchayat resource mapping to panchayat level planning in Kerala: an analytical study. Discussion Paper No. 14, KRPLLD. Centre for Development Studies, Thiruvananthapuram

Chattopadhyay S, Sasidharan CK, Mathai J (2004) Panchayat resource mapping: an experiment from Kerala. In: Ravindra Kumar GR, Subhash N (eds) Silver jubilee compendium: earth system science and natural resources management. CESS, Thiruvananthapuram, pp 401–416

Gadgil M, Hemam NS, Mohan Reddy B (1998) People, refugia and resilience. In: Berkes F, Folke C (eds) Linking social and ecological systems: management practices and social mechanisms for building resilience. Cambridge University Press, Cambridge, pp 30–47

Government of Kerala (1974) Report of the committee on agro-climatic zones and cropping patterns. State Planning Board, Thiruvananthapuram

Government of Kerala (1994) Agricultural statistics of Kerala, 1992–1993. Department of Agriculture, Thiruvananthapuram

Government of Kerala (2005) Economic review 2004. Kerala State Planning Board, Thiruvananthapuram

Holling CS, Berkes F, Folke C (1998) Science, sustainability and resource management. In: Berkes F, Folke C (eds) Linking social and ecological systems: management practices and social mechanisms for building resilience. Cambridge University Press, Cambridge, pp 342–362

Jorgenson AK (2003) Consumption and environmental degradation: a cross-national analysis of the ecological footprint. Social Problems 50(3):374–394

Kerala State Land use Board (1995) Land resources of Kerala State. Government of Kerala, Thiruvananthapuram

Mukherjee N, Bandyopadhyay D (1993) New horizons for West Bengal's panchyats: a report for the government for West Bengal. Government of West Bengal, Kolkata

Nair KN, Chattopadhyay S (2005) Water resources of Kerala: issues and case studies. Centre for Development Studies, Thiruvananthapuram

O'Neill RV, Johnson AR, King AW (1989) A hierarchical framework for the analysis of scale. Landsc Ecol 3(2):193–205
Polanyi K (1980) The great transformation. Beacon Press, Boston
Rammel C, Stagl S, Wilfing H (2007) Managing complex adaptive systems: a co-evolutionary perspective on natural resource management. Ecol Econ 63(1):9–21
Rockström J, Steffen W, Noone K et al (2009) A safe operating space for humanity. Nature (London) 461(7263):472–475
Simon A, Mohankumar K (2004) Spatial variability and rainfall characteristics of Kerala. Proc Indian Acad Sci (Earth Planet Sci) 113(2):211–221
Thomas Isaac TM, Franke RW (2000) Local democracy and development: people's campaign for decentralized planning in Kerala. Leftword, New Delhi
Wackernagel M, Schulz NB, Deumling D, Linares AC, Jenkins M, Kapos V, Monfreda C, Loh J, Myers M, Nogaard R, Randers J (2002) Tracking the ecological overshoot of the human economy. Ecology 99(14):9266–9271. doi:10.1073/pnas.14203 3699

Part III
Global Perspectives

Chapter 5
Natural Resource-Based Livelihoods in the Context of Climate Change: Examining the Stance of Decision Makers in India

Supriya Francis and Vandana Wadhwa

The Earth has enough resources to meet people's needs,
but will never have enough to satisfy people's greed.

(M.K. Gandhi)

Abstract Climate change varies across different regions and leads to changes in many meteorological elements such as rainfall, temperature, sea level, and various extreme events. These changes not only affect natural and human systems independently but also their interfaces, thus changing ecosystems and production, diversity, and the functionalities of livelihoods. People engaging in primary economic activities such as farming, fishing, or forestry comprise a large proportion of the Indian population and are the most vulnerable to changes in weather patterns as they depend directly on these eco-resources for their livelihood. Therefore, it is highly essential to address the issue of climate change to ensure the steadiness of their sources of livelihood. The words of the decision makers thus become very important; any decisions regarding future development should take into account climate change and its impact on livelihoods, and correspondingly, the words of decision makers at the national and local level should reflect cognizance of this relationship. This chapter discusses the stance of the country's decision makers regarding this relationship through the means of text-based pragmatic analysis that examines decision makers' speeches and statements for their level of concern regarding the climate change–livelihood preservation relationship. It highlights the disparity between the people's

S. Francis (✉)
The Energy and Resources Institute (TERI), New Delhi, India

World Sustainable Development Forum, New Delhi, India
e-mail: supriyafrancis@gmail.com

V. Wadhwa
Department of Geography and Environment, Boston University, Boston, MA, USA
e-mail: vandanaw@verizon.net

A.K. Dutt et al. (eds.), *Spatial Diversity and Dynamics in Resources and Urban Development*, DOI 10.1007/978-94-017-9771-9_5

reality and the knowledge base of the experts, which are certainly available to these decision makers, and what really seems to concern the decision makers. The study finds that references to livelihood preservation are at best implicit when present, but are largely conspicuous by their absence in text.

Keywords Climate change • India • Pragmatic analysis

5.1 Introduction

Climate change has become a major buzzword in development discourse because of its implications for almost all aspects of development; changing environmental conditions will impact entire ecosystems, with major implications for health and well-being and on social and economic productivity (Intergovernmental Panel on Climate Change) (IPCC 2007). Climate change especially affects the people and communities whose livelihoods depend largely upon resources acquired through their natural environment, because their activities are highly climate sensitive (e.g., agriculture, fisheries, and forestry, which are dependent upon temperature regimes, land quality, water availability, and precipitation: see Ward and Shivley 2012 for review of relevant literature).

The impact of climate change on natural resource-based livelihoods is especially sharp in the global South, where a large proportion of the population depends on them for their sustenance, and where the political economies have not yet built up the resiliency to withstand significant shifts in resource availability. Africa and Asia are already feeling the shifts and unpredictability that are associated with climate change (Mendelsohn et al. 2006; Schneider et al. 2007; Ward and Shivley 2012). Agrarian livelihoods are among the most affected because of their high dependence on climatic variables; more than half of India's billion-plus population are engaged in agriculture, a significant proportion of which is primarily rain fed (Dar 2011; Food and Agricultural Organization of the United Nations [FAO] 2012), contributing to livelihood insecurity.

The question that arises, when the availability or predictability of such natural resources are threatened, and the human dimension of the impact on livelihoods becomes of great concern, is then how are decision makers addressing the matter of livelihood preservation of the multitude that depends on such resources? In fact, what is their stance toward the relationship between climate change and livelihood security? This study takes the first step toward answering these crucial questions by using pragmatic analysis (PA) on textual evidence from media reports, speeches, and addresses. This chapter highlights the disjoint between the people's experience on the ground and the pronouncements of the decision makers, despite the access of the latter to the knowledge base of experts from relevant disciplines. As the analysis shows, issues of the livelihood security of those dependent on natural resources are at best addressed implicitly but are largely conspicuous by their absence in explicit text. The significance of such an analysis derives from its ability to measure political rhetoric against its relevance to on-ground realities.

We seek to qualify at the outset that the scope of this study does not include debates on climate change (e.g., causes, extent) and the degree of its impact on resource-based livelihoods. We assume that climate change is occurring, and is and will continue to affect natural resource-based livelihoods based on the ample evidence provided in existing literature, studies, and reports emanating from academia, governments, think tanks, and the media, as cited in the study context. The primary purpose of this study is to demonstrate the value of the little-utilized methodology of PA in resource geography and development. We particularly highlight how it can be used to examine development-related discourse within contemporary contexts of (1) the challenge of climate change occurring at the global level, and (2) preserving natural resource-based livelihoods in the global South, using India as a case in point.

Further, for the purposes of this study, the natural resources we refer to are those immediately related to climate and weather conditions (temperature, precipitation, extreme events such as droughts and floods), and the primary livelihood of concern here is agricultural, because it is both highly climate sensitive and accounts for a large proportion of the livelihoods in India. Finally, we would also like to qualify that we do not seek to deem the position taken by the decision makers as wrong or right, but only seek to analyze their explicit and implicit contents. This exercise also does not seek to interpret the decision makers' perspectives, but rather, the readers' and our own (per Duffy 2008).

5.2 Study Context: Livelihoods and Climate Change in India

To provide a comprehensive context for this study, it is important to touch upon (1) the definition of livelihoods and, relatedly, of employment and jobs; (2) climate change and its impact on natural resource-based livelihoods, (3) the situation in India regarding climate change and current/potential effects on natural resource-based livelihoods. This section provides brief notes on these aspects to facilitate the creation of a larger context for the analysis. Finally, a fourth subsection provides the discourse-specific context of development paradigms and national climate change policies adopted by Indian leaders.

5.2.1 Livelihoods

In development literature, the underpinning of any understanding of sustainable livelihoods is the one provided by Chambers and Conway (1992). Based on this seminal work, Carney [1998; four cited in International Institute of Sustainable Development (IISD) 2003] defined livelihoods as follows:

> ... compris[ing] the capabilities, assets (including both material and social resources) and activities required for a means of living. A livelihood is sustainable when it can cope with and recover from stresses and shocks and maintain or enhance its capabilities and assets both now and in the future, while not undermining the natural resource base.

This definition is widely used in sustainable livelihoods literature and work and is adopted by us in the present work. However, for the present study, it is also important to differentiate between the concept of 'livelihoods' and the terms 'employment' and 'jobs.' A background note prepared by the United Nations Development Program (UNDP) as a contribution to the 35th session of the Commission for Social Development (theme: Productive Employment and Sustainable Livelihoods) explicitly demonstrates that 'employment' constitutes a subset of the more holistic concept of 'sustainable livelihood' (see Lawrence and Singh 1997). The note mentions the adoption of this differentiation in various developing countries, including India. The UNDP and Laotian joint Human Development Report (Ministry of Planning and Investment, and UNDP 2009) also provides a useful set of definitions for the two terms as follows:

> Employment: This refers to the engagement of a person in an economically meaningful activity. A person could be self-employed or be employed for a wage/salary. Words like 'job' and 'employment' are interchangeably used.
>
> Livelihoods: It is the whole process of how households earn a living. The process is not restricted to income from one or two main activities, but includes self-provisioning, resources obtained from commons, incomes from migration, repatriation of funds, etc., by members of the household. The word 'livelihoods' is commonly used for agrarian set ups.

These definitions are valuable references against which to evaluate the understanding and point of view of decision makers regarding livelihoods, as is demonstrated in the study analysis.

5.2.2 Climate Change

Climate change has attracted much attention from the international development community, experts, and decision makers alike. In the hundred years spanning 1906–2005, global surface temperatures increased by 0.74 °C ± 0.18 °C. It is estimated that the twenty-first century will see an increase in both the frequency and intensity of high temperature and intense rainfall events, heat waves, and extreme drought (IPCC 2007; World Bank 2008). Such extreme events and unpredictability in climatic conditions are expected to disproportionately hurt the most vulnerable, that is, those who depend directly on the natural environment for their livelihoods. Changes in meteorological elements such as rainfall, temperature, sea level, and the occurrence of extreme events affect natural and human systems independently, and in concert with other major factors to change ecosystems and production. These changes in turn impact diversity and functionalities of livelihoods by disrupting access to 'natural capital' (IISD 2003; Mendelsohn et al. 2006).

Meteorological observations since 1950 suggest that changes in both magnitude and frequency of some extreme weather and climate events in some global regions are already observable (Climate and Development Knowledge Network (CDKN) 2012; Schneider et al. 2007; World Bank 2008). As mentioned, Africa and Asia are not only expected to bear the brunt of climate change consequences but are already beginning to experience climatic shifts, extreme events, and the associated livelihood insecurity that occurs when large proportions of the population are directly reliant on ecosystem resources (IISD 2003; IPCC 2007; Ward and Shivley 2012). Asia is experiencing trends of increased surface temperatures, greater precipitation variability, and greater intensity and frequency of extreme events such as typhoons, heat waves, floods, and mudflows, as well as an unsurprising trend of decreased crop yields in some countries (Cruz et al. 2007). South Asia, specifically, has seen an increase in the number of warm days and nights, and increase in tropical cyclones that affect coastal economies, both urban and rural (CDKN 2012).

5.2.3 Impact of Climate Change on Agriculture in India

Almost 600 million people comprise India's agricultural population, which accounts for 51 % of its labor force. Sixty percent of India's land area is under cultivation, of which 60 % is dependent on rain-fed agricultural practices, as is 60 % of livestock farming (Dar 2011; FAO 2012). Therefore, not only is a large proportion of the Indian population dependent upon agriculture as a source of livelihood, much of this livelihood is based on access to water, suitable land, and amenable climatic conditions, and is thus vulnerable to external shocks and threats to natural capital. In fact, the International Climate Risk Index ranked India the seventh most vulnerable country in the world to extreme climate events (see Kim 2011).

Climate change is one such threat that is proving to be immediate reality rather than an abstract concept in many regions of the country. Parts of Northeast India have experienced decreasing trends in average annual precipitation while the Northwest has seen reverse trends during summer months for the past few decades; annual mean temperatures and number of hot days per year have trended upward, which greatly affect crop yields (Cruz et al. 2007). Droughts and floods have become more frequent, tropical cyclone activity has become more intense, and the Indian monsoon has become much more unpredictable; many of these events have resulted in crop failure, loss of income, and loss of livelihood (CDKN 2012; Cruz et al. 2007; India Meteorological Department 2009).

Chattopadhyay and Hulme (1997) concluded from analyzing meteorological trend data spanning 1940 to 1990 that Central and Southern India showed warming trends over this period, and the country as a whole experienced increasing temperatures during the post-monsoon season. Conversely, both pan-evaporation (an integrated measure of various climatic elements such as temperature, precipitation, humidity, wind, and solar radiation) and potential evapotranspiration have decreased for the country as a whole, a trend likely to continue into the future (Chattopadhyay and Hulme 1997; Dhar and Mazumdar 2009). All these elements significantly impact agriculture, the economy, and millions of livelihoods.

Occurrence of off-seasonal natural disasters, such as drought and flooding, and patterns of change in seasonal precipitation and temperature regimes have been observed at subregional and local scales as well. The Indian monsoon is a climatic element most perceptibly affected by climate change. The southwest monsoon season (June–September) showed significantly decreasing precipitation trends from

1901 to 2003 in parts of Jharkhand, Chhattisgarh, and Kerala States, and Kerala was also found to experience large intraregional differences in rainfall trends during different seasons (Guhathakurta and Rajeevan 2006; Pal and Al-Tabbaa 2009). The National Action Plan on Climate Change (Ministry of Environment and Forests 2008) also provides a comprehensive report on observed and potential changes in surface temperature, rainfall, extreme events, and impact on sea levels and the Himalayan glaciers.

For the past several years, the media have also carried several reports on the human impact of climate change and its effect on rural lives and livelihoods. More than 200,000 farmer suicides have been reported since 1997; causes point to poverty, rise in cost of agrarian inputs, and fall in crop prices, but all this has been exacerbated by crop failures, mainly caused by droughts (Report sought on India farm suicides 2011). This and other repercussions are poignantly reported by Ramesh and Nelson (2009):

> A succession of droughts, compounded by flash floods in recent years, have destroyed crops and ruined the soil, leaving farmers in debt to loan sharks. The growing numbers who have committed suicide to escape the shame has attracted concern. But less attention has been paid to farmers handing wives and daughters to prostitution.
>
> Sangeeta, a farmer's wife from the Bundelkhand area, which straddles the massive states of Uttar Pradesh and Madhya Pradesh [was sold by her husband to a pimp] for a month to raise 2,500 rupees to settle a debt….
>
> The disclosure of a trade in women emerged as an Oxfam report detailed how climate change had affected Indian agriculture.

Floods are also equally destructive, as recounted in a report by Sengupta (2007):

> ….unusually heavy rains submerged [village] Puchaldini's fields, destroying crops, drowning cattle and goats and killing 10 people, part of a death toll of 160 across southern and western India over less than four days….The deluge turned this village into a living example of India's chronic vulnerability to the rains, which come too heavy in some years and not at all in others, destroying lives and livelihoods and sending ripples through the economy. (By Thursday, the monsoon death toll hovered near 500 across India, according to news agencies.)

Those who have lived their entire lives dependent on nature are familiar with its regular course and are now personally encountering the signs of climate change. As reported by Pallavi (2009):

> Dhodabai who lives in the Bhimashankar area of Maharashtra's Pune district is respected for her deep knowledge of weather phenomena and her ability to coherently present her observations. According to Dhondabai Asawle, 80-plus, 'Till about the 1970s, the rainfall pattern in the area could be predicted like clockwork. We planned our agricultural activities around the time-table of *nakshatras* (stars). And till I was in my 40s, I do not remember the timetable ever failing us.'

Coastal areas are already experiencing the consequences of sea-level rise, as Edwards (2005) reports:

> Tulsi Khara can sense something serious is happening. For more than 70 years she has lived on the Ganges delta in India, scraping a living off the land. But now, most of her precious two hectares has disappeared under rising waters.

[Khara]: 'We have lost our livelihood,...Displacement and death are everywhere here. The land is shrinking and salty water gets into our fields. Why is nature turning so violent?'

Given the foregoing study context, it is important to examine the stance of decision makers on an issue that is already altering the lives and livelihoods of so many in the country. Our methodology, analysis, and findings are presented in the following sections.

5.2.4 *Climate Change and Development: The Indian Context*

It would be useful to bear in mind that the decision makers were likely operating within the climate prevalent during the study period of 1992–2009. During this time frame, 'sustainable development,' 'climate change,' and neoliberalism were already well accepted as established paradigms in development processes, as is well encapsulated by Drexhage and Murphy (2010, p. 2):

> ...climate change has become the de facto proxy for implementation of the sustainable development agenda; but the framework of the climate changes negotiations are not always the appropriate forum for broader strategic discussions of sustainable development.
>
> While sustainable development is intended to encompass three pillars, over the past 20 years it has often been compartmentalized as an environmental issue. Added to this, and potentially more limiting for the sustainable development agenda, is the reigning orientation of development as purely economic growth. This has been the framework used by developed countries in attaining their unprecedented levels of wealth, and major and rapidly developing countries are following the same course.

Also important to the context is the position of Indian decision makers in international climate change negotiations and its domestic policies regarding the same. At, and immediately following, the landmark United Nations Conference on Environment and Development (UNCED) in 1992, a global North–South divide was palpable on climate change policies. Northern countries view the global south's growth potential as a likely source of emissions and seek to impose caps, while the southern countries see such impositions as a "neo-colonial attempt to interfere with their development," and "North–South suspicions [regarding an equity divide]...has aroused considerable domestic pressure, particularly in India, not to compromise on basic equity perceptions or 'give in' to any Northern demands." (Paterson and Grubb 1992, pp. 298–299).

India has since stuck to the demand that the 'equity principle' (per capita rather than total emissions as a basis of mitigation) should underlie all international climate change mitigation policies. Additionally, efforts in such a direction must be accompanied by knowledge and technology transfer and aid from industrialized countries (for example, clean development mechanisms), and must not compromise the Indian economic growth trajectory (see Badrinarayana 2011; Harrabin 2007, section 'Crucial time' para. 9–16; Jha 2009; Sathaye et al. 2006). The following discussion sheds light on the Indian position on climate change policy:

> Public policy on climate change officially therefore continues to be guided by the need to eradicate poverty and develop economically. The Government of India maintains that "the most important adaptation measure to climate change is development itself"[1]. This approach can be seen in the National Action Plan on Climate Change (NAPCC) which seeks to promote development objectives that yield 'co-benefits' that address climate change but are not solely aimed at mitigation or reducing emissions. (ICP n.d. b, India and Climate Change: Introduction, para. 3)

In the last few years, India has imposed voluntary caps on emissions (Dessler and Parson 2010; Jha 2009), but these are still in line with the foregoing 'equity' and 'economic growth' priorities. The following methodology allows us to explore the stance of Indian decision makers on livelihoods as related to the foregoing contexts of climate change and related policies.

5.3 Methodology: Pragmatic Analysis

A type of discourse analysis, "Pragmatic Analysis refers to a set of linguistic and logical tools with which analysts develop systematic accounts of discursive political interactions. They endeavor to identify the full range of inferences that a reader or hearer would make when encountering the locutions of an author or speaker, considered in context." (Duffy 2008, 168). In other words, pragmatic analysis (PA) considers not only what is actually said, but also another layer of 'implied' or inferred meaning of the words. PA is particularly suited to examining political rhetoric, because it is often predicated on getting across both 'explicit and implicit,' and 'direct and indirect,' meanings of speech (Wilson 1990).

According to Duffy (2008), PA follows a number of precise 'operational' steps, enabling the systematic analysis of discourse, which he enumerates as follows:

1. Positing a set of background assumptions: this step requires the analyst/s to qualify the "pre-understanding" necessary for the reader to understand the discourse in its context. Duffy (2008) recommends that these specifications of 'background assumptions' be 'noncontroversial' to avoid conflicts and debatable points at the outset of the study.
2. Make initial specification of text's implicit contents: this step comprises the explicit enunciation of the inferences drawn from the text. These inferences should also be obvious enough to any reader of the text containing the discourse to be analyzed.
3. Dialogical argument analysis: involves the use of the foregoing step on each 'move and countermove' in the argument of the discourse. If the discourse does not include multiple parties (as in this study), this step is skipped.

[1]Text in quotation was cited by source and double-checked by authors to emanate from the National Action Plan on Climate Change (Ministry of Environment and Forests, 2008, p. 12).

4. Semantic network representation: incorporates steps 1 and 2 as well as the explicit contents of the text (the actual words), which together form a denotional and connotational matrix that can lend itself to analysis. This matrix is then used to for specifying an "action theorem" that proposes a particular result from the semantic network—a result with which the analyst/s wish the reader to concur.

Proving the action theorem: the theorem must be validated by proving it follows logically from the semantic network. If this cannot be accomplished, the analyst/s regards the theorem as 'underspecified' and seeks to build up materials incorporated in steps 1 and 2.

Sensitivity analysis: propositions made thus far within the semantic network are removed from it one by one. Any such removal that does not affect the validity of the action theorem is considered dispensable. All other propositions are retained.

Syllogism construction: the propositions retained from step 6 are utilized to build a model that frames the analyst/s' perception of what a reader would interpret from the text. At this point, the PA is concluded, but its highly systematized nature allows for rigor through critical examination.

5.3.1 Data and Sources

For the present study, the text used for the analysis was sourced from statements made by political decision makers, either in news media (newspapers, interview transcripts, or press releases) or in addresses made at international and national/regional/local forums. To gain an understanding of Indian decision makers' stances on livelihoods and climate change, we conducted a search in the Lexis Nexus database for newspapers, interview and speech transcripts, and wire service relays from major world and national/regional publications containing the terms "climate change," "livelihoods," "India," and "Minister" (the last as proxy for national and state-level decision makers). This last term encompasses decision makers such as the Prime Minister of India, Ministers of State, Chief Ministers (de facto heads of States), and Cabinet Ministers. The timeframe chosen for the study was based on climate change policy-related milestones: the starting date was January 1, 1992, the year of the United Nations Conference on Environment and Development (UNCED), popularly known as the Rio Summit, where 172 governments participated and which ultimately led to the 1997 Kyoto Protocol. The end date was December 31, 2009, just as the Copenhagen Climate conference concluded, which was also the month preceding the start of this project. The database search yielded a total of 212 items, of which 60 were relevant, unduplicated items that were utilized for the final analysis. Only those items were used that placed climate change and livelihoods in the same discussion and context. However, because such items using both terms in the same context were scarce, references to the natural environment and its elements were also considered in the analysis as implicit references to factors affecting natural resource-based livelihoods.

5.3.2 Hypotheses and Research Questions

The primary research question, as mentioned in the introduction, was 'What is the stance of Indian decision makers toward the climate change and livelihood security relationship?' Apart from this, some other research questions of interest arose during the analysis, which are addressed in the concluding section of this chapter.

As typical of exercises in PA, no a priori assumptions were made or hypotheses posited regarding possible stances or findings. The study was intended simply to be an exploration into the positions of Indian decision makers regarding the important connections between climate change and livelihoods, using PA as a demonstration tool for facilitating such an analysis. 'Climate change' in this exercise includes references to occurrence of extreme events, climate change consequences, mitigation, and adaptation measures. 'Livelihoods' refers to natural resource-based activities that are directly affected by climate (e.g., agriculture, fisheries, forestry), and that fit the definition provided in the study context.

5.4 The Analysis: Reading Between the Lines

The semantic representation that lends itself to analysis must first make clear the background assumptions. This is a very important step in PA because the method is so highly context specific (Wilson 2005). Following Duffy's (2008) recommendations to keep these assumptions simple and straightforward, we specify that the analysis-specific background assumptions are as follows:

1. The Ministers, as elected officials, or in the case of Cabinet Ministers, appointees of elected officials, represent India, and speak and act in the interest of its citizens. The latter part is assumed here to be true in theory, to avoid any controversies—whether or not they actually act in the interest of India and its citizens is a debate outside the purview of this study.
2. By dint of their position, each of the Ministers whose words are presented here hold some degree of decision-making power; Ministers at various levels (state level, national level, and the Prime Minister) are either directly responsible for decision making or are at least actively engaged in decision-making processes at their respective levels of representation.
3. The Ministers have the necessary access to the knowledge base of experts, scientists, and policy makers in various relevant fields such as climate change, agriculture, and forestry. The knowledge base is made available to them through the media and interactions with experts, as well as the various reports produced by government and nongovernment entities—for example, as can be seen in the following media report (Global warming will hit India hard, report warns 2005):

   Global warming will push temperatures in India up by 3°–4°C by the turn of the century, hitting agriculture and infrastructure, a joint India–U.K. study said yesterday.

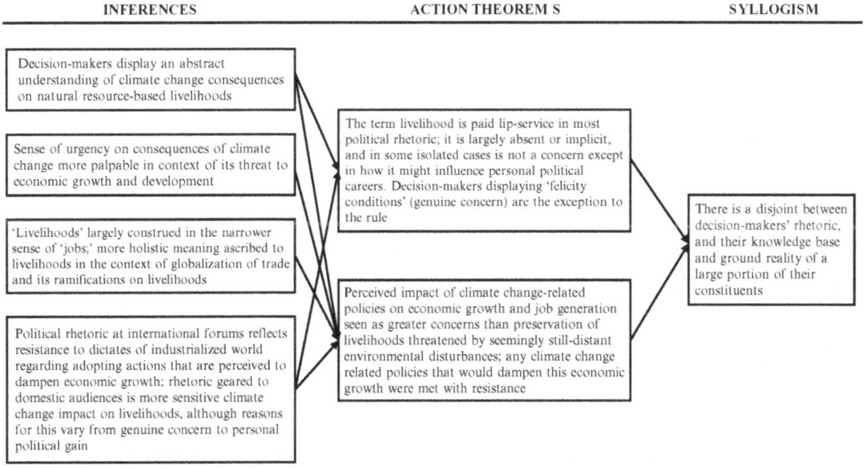

Fig. 5.1 Inferences, action theorems and final model (syllogism) emerging from pragmatic analysis of Indian decision-makers' stance on climate change impact on natural resource-based livelihoods

> Rainfall will increase substantially in many areas, resulting in problems with food supply, and affecting the livelihoods of much of the population as well as spreading diseases such as malaria, said the report released by Indian Environment Minister A. Raja and British Deputy Minister for trade and investment Ian Pearson.

The National Action Plan on Climate Change has been also available as a comprehensive resource on climate change, its consequences, and required actions since its release in June 2008.

Given these background assumptions and the context provided before, the textual data is more amenable to interpretation. In Duffy's (2008) scheme of PA, the implicit meanings inferred by the analysts should be expressly put forward as a second step of the analysis, and qualifies that these are not necessarily beyond the 'mundane.' We present some of the inferences we drew from the explicit words of decision makers on their concerns regarding climate change and livelihoods, providing some examples from the (explicit) text to support the analysis. These inferences as well as the action theorems and final syllogism that follow from the semantic network are also presented in Fig. 5.1.

First, we found that the understanding of the consequences of climate change on livelihoods is certainly present, but in an abstract sense for most decision-makers. Talk about ramifications on livelihoods is largely paid lip-service, with decision makers' rhetoric largely referring to the existence of consequences, but no detailed statements regarding concrete steps for addressing this concern. For example, the report on the meeting of heads of state at the 14th South Asian Association for Regional Cooperation (SAARC) Summit in April 2007 shows that most decision makers are aware of the important link between climate change and livelihoods. Yet, immediate concrete action even at this Summit is limited to commissioning a panel of experts:

> The heads of the States...expressed deep concern over global climate change and the consequent rise in sea level and its impact on the lives and livelihoods in the region. They agreed to commission a team of regional experts to identify collective actions in this regard, the declaration said. (SAARC Pledge to Fight Terror 2007)

More evidence supporting this inference is found in other examples of explicit text provided in the rest of this chapter, but also in observations from other sources such as the India Climate Portal (ICP), a public resource set up to promote Indian leadership on climate change (see ICP n.d. a, Policy Brief: Towards Regional Cooperation, para. 5)

> While climate change has been part of the agenda right from the 5th SAARC Summit in 1990, even a 2007 ministerial meet in Dhaka and the [ensuing] 'SAARC Action Plan on Climate Change' yielded no concrete results. Action has been slow coming and pledges to implement the Dhaka Action Plan (2009–2012) have yet to be initiated.

Second, matters related to climate change are not seen as much as an immediate threat to environment and its resources (thus to natural resource-dependent livelihoods) as they are to economic growth and to development. Preserving and promoting India's rapid economic growth recurred repeatedly in the rhetoric in the context of climate change, in keeping with the Indian position presented in the study context, while livelihoods often took a secondary place, even when explicitly mentioned.

For instance, at the launch of the National Action Plan on Climate Change (NAPCC) on June 30, 2008, Prime Minister Manmohan Singh specifically mentioned the direct connection between climate change and its consequences on livelihoods, but only in one reference, whereas the emphasis on economic growth was more noticeable. He said:

> Without a careful long-term strategy, climate change may undermine our development efforts, with adverse consequences across the board on our people's livelihood, the environment in which they live and work, and their personal health and welfare.... The time has come for ... ecologically sustainable development [but] our people have a right to economic and social development and to discard the ignominy of widespread poverty. For this we need rapid economic growth". (India PM unveils plan to fight climate change 2008)

While the Prime Minister follows up with "But I also believe that ecologically sustainable development need not be in contradiction to achieving our growth objectives," (ibid.) it is clear from constant references by him and other decision makers that growth objectives are a greater priority. In fact, just the year before, the Group of Eight (G8) Summit in Germany, where climate change, sustainability, the environment, and carbon emissions were some of the items on the agenda, India, along with some other nations, refused to support compulsory caps on greenhouse gas (GHG) emissions as a climate change mitigation measure. The emphasis on economic growth as a nonnegotiable factor was clear in the Prime Minister declaration before leaving for this Summit: "It is a fact that more and not less development is the best way for developing countries to address themselves to the issues of preserving the environment." (Wax 2007).

A few months after the launch of the National Action Plan on Climate Change, the Prime Minister addressed the nation on the anniversary of Indian Independence and said:

> Our ancient Himalayas are under environmental threat. If the Himalayan glaciers recede, the flow of water in our sacred rivers will go down. Climate change can disrupt our economy in several ways. Some of our coastal areas could be submerged. Our monsoon pattern may change. We need long-term solutions to such threats. Our Government has come forward with a NAPCC. This plan shows how each of us must adapt our ways of working and living and how we must treat our natural resources, so that our carbon emissions remain within reasonable limits and our environment is protected. (India PM unveils plan to fight climate change 2008)

Again, the first emphasis here is on the impact of climate change on the economy; the reference to livelihoods is indirect and refers to climatic changes that will disrupt them, and to adaptation. Although mitigation through limiting GHG is mentioned, other rhetoric stresses that such "reasonable limits" will simply be anything not exceeding emissions by the developed countries: "I have already declared, as India's Prime Minister, that despite our developmental imperatives our per capita GHG emissions will not exceed the per capita GHG emissions of the developed industrialized countries" (India PM unveils plan to fight climate change 2008).

Perhaps the best example of this emphasis on economic growth is in the words of then Foreign Minister S.M. Krishna in his address to the UN General Assembly in 2009. While his address touched on various issues, on the topic of climate change, he stated: "Poverty alleviation and livelihood security are central imperatives for India. For this accelerated economic growth and energy security are critical drivers" (Krishna says India wants to resolve all issues peacefully with Pak 2009).

A third inference that we drew relates to both the foregoing inferences and concerns the decision makers' understanding of the term 'livelihood.' We found that apart from a few instances as those just given, whenever there was mention of climate change and related issues such as mitigation, the rhetoric not only tilted toward growth as a livelihood generator, but also that it reflected a more limited conceptualization of livelihoods as 'jobs.' For example, in a 2007 interview with Mike Williams of BBC News, Finance Minister P. Chidambaram was asked how India's rapid growth goals and related increases in energy consumption could be justified given consequent impact on environment and climate change issues. His reply:

> Growth is the best antidote to poverty. Growth gives incomes to people who are employed, throws up jobs for those who are not employed. Therefore growth is imperative…. (Chidambaram 2007)

In fact, the more holistic conceptualization that has greater application to resource-based agrarian livelihoods was more likely to appear in rhetoric touching upon economic issues. This is apparent later on in the same interview, where Chidambaram mentions livelihoods, but in the context of trade issues rather than climate change:

> [GDP share of] agriculture will move down maybe 3 or 4 percent [in the next few decades]. But please remember the bulk of the working population is dependent on agriculture which is why it's important to pay attention to agriculture…. Agriculture in India today is a livelihood issue, which is why when we go to the WTO [World Trade Organization] we do not look upon agriculture as simply a trading issue, it is a livelihood issue; we have to protect the livelihood of millions of people.

Interestingly, the same 'talking point' of livelihoods as related to trade issues was echoed by Prime Minister Manmohan Singh at the 12th Business Roundtable Meeting organized by The Economist:

> Industrialized nations should unilaterally bring down their emission levels and not use the issue as a bargaining tool in multi-lateral trade talks,… I hope [climate change] is not introduced as an additional conditionality in trade negotiations as it will only complicate the negotiations…. it must be recognized that for us agriculture is not just a business, but a way of life and a major source of livelihood (Keep climate out of WTO: PM 2007).

Thus, the concern with maintaining economic growth in the face of climate change issues is paramount, but concern for livelihoods does not seem to garner similar attention unless it is in an economic context as opposed to climate change. Very importantly, also evident in much of the rhetoric is strong resistance toward adopting mitigation measures that might threaten the pace of economic growth, and decision makers frequently appear loath to be seen as 'toeing the line' set by the industrialized nations or the traditional 'powers that be' (also see Paterson and Grubb 1992). In fact, a decade ago, MSN Mid-Day (India hits out at developed nations 2002) reported the following remarks from then Prime Minister Atal Bihari Vajpayee at the International Climate talks at New Delhi, India:

> Asking the advanced nations to pump more funds to enhance capacity building in developing countries, Vajpayee said developing countries do not have adequate resources even to meet their basic human needs. "Climate change mitigation will bring additional strain to the already fragile economies of the developing countries [and will affect our efforts to achieve higher GDP growth rates to eradicate poverty speedily][2]", he said….Describing [sustainable] agriculture, food and nutritional wellbeing and weather-related economic loss as among the key areas of adaptation, he said "there is a need for strengthening the capacity of developing countries in coping with extreme weather events, which are increasing in frequency and severity due to climate change."

Another example of this resistance is contained in the BBC Interview with P. Chidambaram already cited (Chidambaram, 2007); after justifying the reason for rapid growth targets for India, the former Finance Minister said:

> When the rest of the world, the developed world, was growing nobody asked them why are you consuming so much energy, nobody asks them today why didn't you slow down and why didn't you consume less energy. The point is we have a right to grow, just as much as the US and Europe had a right to grow in the 19–20th Century.

[2]Text in brackets is provided to create better context, and is taken from original statements by former Indian Prime Minister Atal Bihari Vajpayee at the High Level Segment of the Eighth Session of Conference of the Parties to UN Framework Convention on Climate Change, New Delhi, 30 October, 2002.

Overall, our inferences point to the largely absent or cursory connections between climate change and livelihood concerns in decision makers' rhetoric and the greater significance of economic growth and jobs, along with resistance to dictates from developed nations that were perceived in any way to curtail the Indian economic growth trajectory. However, there were some exceptions to these general trends where the words of decision makers show genuine concern and understanding (or felicity conditions: see Duffy 2008, p. 174) of livelihood issues as related to climate change. For example, Supriya Sule, a Member of Parliament, Lok Sabha wrote in an editorial for *The Hindu* (Sule 2010):

> While we, the common people, might think that climate change is something that only the people living in the coastal regions and politicians have to worry about, the truth is quite the opposite. Everything that we depend on for our day today sustenance is directly affected by how the climate evolves over time. Be it the wheat and rice crops that get ruined due to untimely rains, thereby pushing up the prices, or the healthcare situation on the country that gets worse due to the increasing case of vector-borne diseases....While global temperatures continue to rise, glaciers melt and ultimately people's livelihoods and lives are lost.

Another case in point is seen in the following report (Sengupta 2007):

> Jairam Ramesh, the [then] Indian State Minister for Commerce, offered a sobering assessment. He cited a startling new study by a government economist, Arvind Virmani, who concluded that as much as 45 percent of the variation in India's gross domestic product over the last 50 years can be explained by the fluctuations in rainfall.
>
> To Ramesh, this was a warning call to a nation gripped by its own economic success. 'Many Indians seem to be living in a cocoon of self-delusion,' he said. 'Nine percent growth and world power are all for real. But if the monsoon fails, the Indian economy may show some resilience, but Indians suffer.'

However, sometimes the political motivations of local leaders are all too explicit and it is easy to infer that in some cases, political interest is a greater if not the sole concern than the livelihoods of the agrarian masses. For example, the same report (Sengupta 2007) continues about the impact of the monsoon:

> It is not for nothing that Indian politicians watch the monsoon with a keen eye.
>
> 'I track it every day,' said Manvinder Singh, a member of Parliament with the opposition Bharatiya Janata Party. His desert belt district in the western state of Rajasthan was submerged in three days of fluke rains last July, killing 109 people and leaving thousands marooned.
>
> This year, the forecast was good, and Singh had hope. 'The most important thing about a good monsoon is that the mood is good,' he said. 'And anything which creates a good mood is always good for politics.'

Thus, our analysis brings us to an additional aspect related to the first inference; in some cases, rhetoric not only reflected cursory concern regarding climatic affects on livelihoods but also preserving and promoting of personal political interests.

5.4.1 The Action Theorems and Syllogism

According to Duffy (2008), the semantic network should facilitate the development of 'action theorems' (set of propositions flowing logically from the inference that should appeal to reader/s interpretations). The action theorem should then be crystallized into the final step of constructing a syllogism (analysts' model of the reader's interpretation of text).

From the semantic network representation, we posit two action theorems. The first action theorem is that with few exceptions, the impact of climate change on natural resource based livelihoods is largely paid lip-service, and in some isolated cases is not a concern except in how it might influence personal political careers. The second action theorem is that perceived impact of climate change-related policies on economic growth and job generation were seen as greater concerns than preservation of livelihoods threatened by some seemingly still-distant environmental disturbances; any climate change-related policies that would dampen this economic growth were met with resistance. This action theorem also finds support from outside the semantic network representation, as demonstrated in the fourth subsection of the study context.

The syllogism that follows is that there is a disjoint between decision makers' rhetoric on the one hand, and their knowledge base and ground reality of a large portion of their constituents on the other hand. The rhetoric is largely concerned with impact on economic growth and development, and in line with the national position on climate change, whereas their knowledge base and actual ground reality are largely related to impact on access to natural capital. The stance of decision makers belies their greater concern for preserving economic growth and job generation (with a seeming conflation of livelihoods and jobs), while concern for the security of natural resource-based livelihoods is implicit at best and largely conspicuous by its absence in text.

5.5 Future Research Directions and Concluding Remarks

Although the primary research question of this study has been explored and answered, two secondary questions arose during the analysis, which we attempt to answer here. The first was: How long after the 1992 Rio Summit did climate change and livelihoods become a topic that appeared in decision makers' rhetoric? From our search of relevant data items, it appears that it took a full decade for this issue to become a significant part of Indian decision makers' speeches and addresses at various forums. The second question was: Does the point of view of decision makers change according to the nature of the forum? For example, are positions taken at international forums different from that at local forums? This question arose because we noted that rhetoric prioritizing, economic growth, and resisting policies from the developed world that might potentially curb this growth were largely apparent at international forums, but that the decision makers' stance became more sensitive to climate change–livelihood connections when the forum changed from the international to the local level. In effect, decision makers seemed mindful of their audience and the audience's reception of their words: economic growth concerns were played on at international forums where such rhetoric and resistance to the global north is perceived favorably by the Indian public (see, for example, Paterson and Grubb 1992;

Boykoff 2010; Jha 2009), but are not as apparent in local-level rhetoric as is concern with the environment that makes livelihoods possible.

Future research in this vein would be valuable, including finding and researching larger semantic networks[3] to answer more effectively all the foregoing questions, as well as for additional exercises such as matching rhetoric against on-ground efforts and extending the scope of the study to include an argument analysis of climate change negotiations to examine decision makers' priorities. The latter proposed research direction would be particularly valuable given that many new schemes and programs have been talked about by various decision makers to handle the issue of livelihood security in the face of climate change. For example, "empowering local panchayati raj institutions [five-member councils] and making them aware about the impact of the changes in the livelihood issues" (Sharma 2009) and "training the rural poor to be climate managers and … appointing a climate change coordinator for every village panchayat (council) in the country." (Pope 2008). Because effective implementation of such programs remains undetermined, it is here that PA can be used to gauge rhetoric versus action.

Another useful exercise would be to analyze the words of experts to examine their implicit allegiances and priorities. For example, Noble laureate and Intergovernmental Panel on Climate Change (IPCC) Chairman R.K. Pachauri warned of consequences of ignoring climate change and advocates awareness and appropriate, sustainable agricultural practices (Climate change may reduce India's crop production—Nobel laureate 2007):

> 'Basically, yields of some crops like wheat, rice and pulses will go down. We have got evidence on decline in the productivity of wheat in the country. It is high time farmers should know why their yields are not growing,' he noted. …Pachauri said farmers have to realize that they cannot take natural resources for granted. They should be aware of water scarcity, which is likely to grow in future. …'Also, farmers perhaps need to change their cropping pattern and agricultural practices to adapt to climate change.'

However, Dr Akhilesh Gupta, Advisor Scientist and Coordinator of National Climate Change Program of the Government of India, paints a less dire picture:

> [He] described the hype overmelting of glaciers as pressure tactics being used against India by developed countries to reduce carbon emissions.
>
> 'Many studies have shown that glaciers in Jammu and Kashmir had not receded at all and were static as compared to the previous available data. We cannot ignore the threat, but we should not panic and abandon development either. We still have time to adopt more sustainable models,' he said. (Verma 2009)

The present research is thus significant in that it has also brought up more questions that should be explored, such as those presented here. As for its significance as it currently stands, the syllogism or final model emanating from our analysis reveals the disjoint between the decision makers' emphasis on economic growth and the

[3] It is, however, important to remember that the context-specific nature of Pragmatic Analysis means that it is not amenable to automated analytical procedures; thus, Duffy (2008) cautions against overly large corpora for analysis.

realities of the rural masses—in conclusion, we leave readers with the following example that ably demonstrates this phenomenon:

> Two very different recent scenes from India: At a power breakfast between many of the country's corporate leaders and top economic officials, the former Finance Minister Pranab Mukherjee declared that India had 'weathered the storm' of the global economic crisis and was witnessing 'green shoots' in industry and services that signaled a return to more rapid growth by next year.
>
> Hundreds of kilometers away in this farming village in the southern Indian state of Andhra Pradesh, weeds were the only green shoots sprouting in the black soil that belongs to the widow Chandli Bai. Her field went 12 weeks without rain during India's annual monsoon season before showers finally arrived on August 23, splattering down too late onto the dry dirt. Her summer crop of lentils was stillborn in the ground.
>
> 'We eat once a day,' said Mrs. Bai, 65, explaining how she and her family had survived the lack of rain. (Yardley 2009)

The contrast between the power breakfast and the one-meal-a-day existence is jarring, but real. Pragmatic analysis provides the ability to policy makers and regional planners, among others, to examine whether decision makers' statements reflect the thought processes and actions that are required within a particular context, in this case, ensuring the livelihoods of those dependent on natural capital assets, when these very resources are jeopardized by climate change.

References[4]

Badrinarayana D (2011) India's constitutional challenge: a less visible climate change catastrophe. In: Richardson BJ, Le Bouthillier Y, McLeod-Kilmurray H, Wood S (eds) Climate law and developing countries: legal and policy challenges for the world economy. Edward Elgar, Northampton, pp 63–83

Boykoff M (2010) Indian media representations of climate change in a threatened journalistic ecosystem. Clim Change 99:17–25. doi:10.1007/s10584-010-9807-8

Carney D (1998) Sustainable rural livelihoods: what contribution can we make? Department for International Development, London

Chambers R., Conway G. (1992) Sustainable rural livelihoods: practical concepts for the 21st century. IDS discussion paper 296. IDS, Brighton

Chattopadhyay N, Hulme M (1997) Evaporation and potential evapotranspiration in India under conditions of recent and future climate change. Agric For Meteorol 87:55–73

Chidambaram P (2007) Indian finance minister: full transcript. (M. Williams, Interviewer). BBC News, Feb 5

Climate change may reduce India's crop production: Nobel laureate (2007 Dec 18). BBC Monitoring South Asia – Political

Climate and Development Knowledge Network [CDKN] (2012) Managing climate extremes and disasters in Asia: lessons from the SREX report. CDKN. Accessed 14 Jun 2012 from www.cdkn.org/srex

Cruz RV, Harasawa H, Lal M, Wu S, Anokhin Y, Punsalmaa B, Honda Y, Jafari M, Li C, Huu Ninh N (2007) Asia. In: Climate change 2007: impacts, adaptation and vulnerability. Contribution of Working Group II to the Fourth Assessment Report of the Intergovernmental Panel on Climate

[4]When a website or other source not indicated, all references are data items retrieved from the Lexis Nexus as indicated in the Methodology section.

Change (Parry ML, Canziani OF, Palutikof JP, van der Linden PJ, Hanson CE, eds). Cambridge University Press, Cambridge, pp 469–506

Dar WD (2011) Agriculture yearbook 2011: revitalizing rainfed agriculture for India's second Green Revolution. International Crops Research Institute for the Semi-Arid Tropics (ICRISAT)

Dessler AE, Parson EA (2010) The science and politics of global climate change: a guide to the debate, 2nd edn. Cambridge University Press, Cambridge

Dhar S, Mazumdar A (2009) Impact of climate change under the threat of global warming for an agricultural watershed of the Kangsabati River. Int J Environ Sci Eng 1(3):153–162

Drexhage J., Murphy D. (2010) Sustainable development: from Brundtland to Rio 2012. IISD. http://www.un.org/wcm/webdav/site/climatechange/shared/gsp/docs/GSP1-6_Background%20on%20Sustainable%20Devt.pdf. Accessed 23 June 2012

Duffy G (2008) Pragmatic analysis. In: Klotz A, Prakash D (eds) Qualitative methods in international relations: a pluralist guide. Palgrave Macmillan, Basingstoke

Edwards R. (2005) G8: everyone agrees global warming is wrecking our planet... everyone except those who can stop it: two-page special report on the Summit which could change the planet. The Sunday Herald, Jun 19, p 12

FAO (2012) FAO statistical yearbook 2012: world food and agriculture. United Nations Food and Agriculture organization. http://www.fao.org/docrep/015/i2490e/i2490e00.htm. Accessed 6 July 2012

Global warming will hit India hard, report warns (2005) The Globe and Mail, Sept 9, S2. Canada

Guhathakurta P, Rajeevan M (2006) Trends in the rainfall pattern over India. National Climate Centre, India Meteorological Department, Pune

Harrabin R (2007) How climate change hits India's poor. BBC News–South Asia, Feb 1. http://news.bbc.co.uk/2/hi/south_asia/6319921.stm. Accessed 22 Aug 2012

ICP (n.d. a) Policy brief: towards regional cooperation—SAARC and the Thimpu declaration on climate change 2010. http://www.indiaclimateportal.org/component/option,com_policybrief/view,policybriefdetail/id,4. Accessed 22 Aug 2010

ICP (n.d. b) India and climate change: introduction. http://www.indiaclimateportal.org/India-and-Climate-Change. Accessed 22 Aug 2012

IISD (2003) Livelihoods and climate change combining disaster risk reduction, natural resource management and climate change adaptation in a new approach to the reduction of vulnerability and poverty. International Institute for Sustainable Development (IISD), Winnipeg

India hits out at developed nations (2002) MSN mid-day, Oct 30

India Meteorological Department (2009) 2009: the warmest year since 1901. India meteorological department. http://www.imd.gov.in/doc/warm2009.pdf. Accessed 22 Feb 2012

India unveils first climate strategy (2008) Modern power system, Aug 6, p. 6

IPCC (2007) Climate change 2007: synthesis report. Contribution of Working Groups I, II and III to the Fourth Assessment Report of the Intergovernmental Panel on Climate Change (Core Writing Team, Pachauri RK, Reisinger A, eds). IPCC, Geneva

Jha PS (2009) Indian public perceptions of the international climate change negotiations. In: Michel D, Pandya A (eds) Indian climate policy: choices and challenges. Stimson Regional Voices, Washington, DC, pp 31–36

Keep climate out of WTO: PM (2007) Hindustan Times, March 13

Kim SY (2011) Public perceptions of climate change and support for climate policies in Asia: evidence from recent polls. J Asian Stud 70:319–331. doi:10.1017/S0021911811000064

Krishna says India wants to resolve all issues peacefully with Pak (2009) Hindustan Times, September 27

Lawrence J, Singh N (1997) Productive employment and poverty eradication: how can livelihoods be more sustainable? Background note for UNDP's contribution to the 35th Session of the Commission for Social Development entitled "Productive employment and sustainable livelihoods." UNDP. http://www.undp.org/sl/Documents/documents.htm. Accessed 5 July 2012

Mendelsohn R, Dinar A, Williams L (2006) The distributional impact of climate change on rich and poor countries. Environ Dev Econ 11(02):159–178

Ministry of Environment and Forests (2008) *National Action Plan on Climate Change (NAPCC)*. Government of India. http://moef.nic.in/downloads/home/Pg01-52.pdf. Accessed 20 May 2012
Ministry of Planning and Investment, and UNDP (2009) Employment and livelihoods: the 4th human development report. http://hdr.undp.org/en/reports/national/asiathepacific/lao/name,20698,en.html. Accessed 5 July 2012
Pal I, Al-Tabbaa A (2009) Trends in seasonal precipitation extremes: an indicator of 'climate change' in Kerala, India. J Hydrol 367:62–69
Pallavi A. (2009) 'Stars don't foretell anymore.' Down to Earth, Dec 31. http://www.downtoearth.org.in/node/2574. Accessed 20 Mar 2010
Paterson M, Grubb M (1992) The international politics of climate change. Int Aff 68(2):293–310. Accessed 25 Aug from JSTOR
Pope C. (2008). Cool panchayats. The Huffington Post, March 28
Ramesh R, Nelson D (2009) Gandhi's party scales down as India's poor hit the wall. The Age, September 16, p 9. Melbourne, Australia
Report sought on India farm suicides (2011) BBC News India, Dec 21. http://www.bbc.co.uk/news/world-asia-india-16281063. Accessed 12 July 2012
SAARC Pledge to Fight Terror (2007) Hindustan Times, April 4
Sathaye J, Shukla PR, Ravindranath NH (2006) Climate change, sustainable development and India: global and national concerns. Curr Sci 90(3):314–325
Schneider SH, Semenov S, Patwardhan A, Burton I, Magadza CHD, Oppenheimer M et al (2007) Assessing key vulnerabilities and the risk from climate change. In: Parry ML, Canziani OF, Palutikof JP, van der Linden PJ, Hanson CE (eds) Climate change 2007: impacts, adaptation and vulnerability. Contribution of Working Group II to the Fourth Assessment Report of the Intergovernmental Panel on Climate Change. Cambridge University Press, Cambridge, pp 779–810
Sengupta S. (2007) Indian economy lives and dies by monsoon. The International Herald Tribune, July 6, p. 2
Sharma A. (2009) Draft policy paper on forestry ready, climate change a core issue. Indian Express, November 20
Sule S. (2010) Climate for change (op-ed). The Hindu, February 9, 2010
Verma H. (2009). Melting glaciers: situation not alarming yet, say experts. Indian Express, October 29
Ward GE, Shivley PS (2012) Vulnerability, income growth and climate change. World Dev 40(5):916–927
Wax E. (2007) A sacred river endangered by global warming: glacial source of Ganges is receding. The Washington Post, July 17, p. A14
Wilson J. (1990) Politically speaking: the pragmatic analysis of political language. Blackwell, Oxford
Wilson D (2005) New directions on research on pragmatics and modularity. LINGUA 115(8):1129–1146. doi:10.1016/j.lingua.2004.02.005
World Bank (2008) Development and climate change: a strategic framework for the World Bank Group—Technical report. International Bank for Reconstruction and Development, The World Bank. http://climatechange.worldbank.org/overview/strategic-framework-development-and-climate-change. Accessed 15 June 2012
Yardley J. (2009) As India grows, rural areas left in the dust. The International Herald Tribune, September 5, p 4

Chapter 6
Innovations and Challenges Related to Resource and Environmental Management

Bruce Mitchell

Abstract Innovations are necessary if resource and environmental management are to evolve and improve. In this chapter, a critical review is provided for concepts that have already shifted, or have the potential to enhance, the practice of resource and environmental management. *Resilience* highlights that management approaches should facilitate capacity of systems to tolerate disturbances without flipping into an entirely new regime. As a result, resilience questions some basic beliefs associated with *sustainable development. Adaptive management* emphasizes that we always face uncertainty and thus should expect to encounter surprises. Thus, we need to be able to adapt, ideally through *active adaptive management.* To adapt, we need to create capacity for *social learning*, whether by individuals, groups, or organizations through single-, double-, and triple-loop learning. *Collaborative approaches* serve as a fundamental component of an ecosystem approach and require structures and processes to incorporate insight from stakeholders. *Governance* reminds us that resource and environmental management should not focus on technical issues, but also need to consider how decisions are and should be made, as well as the nature of relationships and the implications of power differentials. All these concepts are pertinent regarding *environmental justice*, which directs us to ensure no group of people experiences disproportionate costs from resource and environmental management decisions.

For water management, *integrated water resource management* (IWRM) has become a means frequently used to address the water and land, surface and groundwater, and upstream and downstream relationships associated with interconnected terrestrial and hydrological systems. Implementation challenges are being experienced with IWRM and require attention. The concept of *virtual water* has been developed to reduce the vulnerability of water-scarce regions to meet food requirements but also raises other challenges. The *soft path approach* has emerged as a complement to traditional supply-and-demand management approaches. It requires us to reconsider not only how but why we use water for some services. And, finally, more attention is needed regarding *water ethics* if a more principled approach to water development and management is to be achieved.

B. Mitchell (✉)
Department of Geography and Environmental Management, University of Waterloo, Waterloo, ON, Canada
e-mail: mitchell@uwaterloo.ca

A.K. Dutt et al. (eds.), *Spatial Diversity and Dynamics in Resources and Urban Development*, DOI 10.1007/978-94-017-9771-9_6

Keywords Resilience • Adaptive management • Social learning • Collaborative management • Environmental justice • Integrated water resource management • Virtual water • Soft path approach • Water ethics

6.1 Introduction

I first met Baleshwar Thakur when he arrived at the University of Waterloo as a Commonwealth Scholar in the mid-1970s to study for, and complete, his Master's degree. Although I was not his advisor, we shared some common interests. We stayed in contact after he became a faculty member in the Department of Geography at the University of New Delhi. I also had the privilege of writing the Foreword for the second volume in the 14-volume work he edited, The *Perspectives in Resource Management in Developing Countries* (Thakur 2007), and one of my articles was included in the first volume (Thakur 2003).

A very special experience was my being hosted by Thakur as a visitor to the University of New Delhi in January 2006, where I lectured to geography postgraduate students, following my attendance at the 93rd Indian Science Congress in Hyderabad. Thakur had arranged for me to be an invited speaker at the 93rd Congress.

To celebrate Thakur's interests in urban geography, and in resource assessment and management, this chapter examines some major innovations, along with challenges, in resource management. The innovations have relevance for both developing and developed countries. I first consider five initiatives applicable to all aspects of resource and environmental management, followed by four initiatives focused upon water, my major research focus. My remarks are based on personal experience, as well as three recent publications (Mitchell 2010, 2011; Dearden and Mitchell 2012).

6.2 Resource and Environmental Management

In this section, attention is directed to the concepts of resilience, adaptive management and social learning, collaborative management, governance, and environmental justice.

6.2.1 Resilience

The Resilience Alliance is a network of researchers seeking to understand change in social and ecological systems. Many view Holling as the pioneer (Holling 1973; Gunderson et al. 1995; Gunderson and Holling 2002), although others also have major roles (Carpenter et al. 2001; Gunderson and Prichard 2002; Berkes et al. 2003; Walker et al. 2004; Ostrom 2005; Westley et al. 2006).

> Ecosystem resilience is the capacity of an ecosystem to tolerate disturbance without collapsing into a qualitatively different state that is controlled by a different set of processes. A resilient system can withstand shocks and rebuild itself when necessary. Resilience in

social systems has the added capacity of humans to anticipate and plan for the future. (Resilience Alliance 2009)

Resilience can be degraded or lost as a result of various factors, including (1) loss of biodiversity, (2) toxic pollution, (3) inflexible and closed institutions, (4) incentives encouraging unsustainable use of resources, and (5) emphasis on production and increased efficiency causing loss of desirable redundancy. In addition to these factors, resilience can be negatively affected by pervasive but slowly changing variables such as climate, land use, nutrient stocks, human values, and policies. In contrast, resilience is enhanced or protected when diversity is nurtured.

Walker and Salt (2006, p. 14) extend the foregoing ideas with three key insights. First, contemporary efforts to achieve 'sustainable natural resource management' too often model or assume average conditions regarding incremental growth and ignore the likelihood of major disturbances or disruptions. Furthermore, these efforts normally strive to optimize specific aspects of a system (e.g., food production, wood production from forests) in isolation from other interconnected components. Second, contemporary practice usually aims to increase efficiency and optimize performance of ecological systems parts that deliver economic benefits but ignores negative secondary effects and feedback leading to changes in the overall, larger system. Third, given these two challenges, they argue that "Resilience thinking is about understanding and engaging with a changing world. By understanding how and why the system as a whole is changing, we are better placed to work with change, as opposed to being a victim of it" (Walter and Salt 2006, p. 14).

The implications of this shift in perspective from sustainable resource management to a resilient approach are profound. "Any proposal for sustainable development that does not explicitly acknowledge a system's resilience is simply not going to keep delivering the goods (or services). The key to sustainability lies in enhancing the resilience of social-ecological systems, not in optimizing isolated components of the system" (Walker and Salt 2006, p. 9).

6.2.2 Adaptive Management and Social Learning

"Expect the unexpected, and learn from it—this is the underlying principle of adaptive management" (Noble 2004, p. 442). Holling (1978) and Walters (1986) conducted some of the earliest work that underlies the current approach to adaptive management. Their early research used computer modeling to simulate different natural system conditions and to identify possible interventions. The intent was to treat management interventions as experiments in which hypotheses about system response to initiatives would be posed, specific initiatives would be introduced, changes in system structure and behavior would be monitored, and adjustments would be made. An overriding premise was that uncertainty and surprise should be expected. Once that is accepted, it follows that the understanding and prescriptions of scientists and managers will often be incomplete or incorrect, requiring willingness to reflect on experience to learn.

Adaptive management has been applied (Lee 1993) and assessed (McLain and Lee 1996; Allan and Curtis 2005; Armitage et al. 2007; Jacobson et al. 2009), and much has been learned. Noble (2004) and Allan and Curtis (2005) provide good overviews of adaptive management, and the following is based on their work. The basis premise for adaptive management "… is that environmental policy and management practices must be designed with uncertainty in mind. This means that resource and environmental managers must attempt to manage uncertainty by designing policies and practices in such a way that they are capable of dealing with unexpected events and actually benefit from them" (Noble 2004, p. 443). Six core ideas underlie adaptive management: (1) it favors action, (2) accepts and benefits from uncertainty, (3) allows discretion, (4) seeks resilience, (5) provides feedback, and (6) facilitates learning and integration.

Three different types of adaptive management exist (Allan and Curtis 2005, p. 415). *Evolutionary* adaptive management involves "undirected learning from random experience, or trial and error learning." In contrast, *passive* adaptive management is more directed, by "focusing on the implementation of an historically informed best practice or policy, followed by a review of that implementation." The third version, *active* adaptive management, is significantly different, with a focus on learning rather than only on implementation. In other words, managers implement policies and strategies to be able to test hypotheses. The active approach "may use a range of treatments–practices designed to achieve strategic goals–to test the hypothesis that "best" practice is just that." The key features of passive and active adaptive management are shown in Table 6.1.

Active adaptive management is the ideal to which resource managers should aspire. However, various obstacles have to be overcome. Perhaps the most challenging is the reluctance of many managers to acknowledge openly when initiatives have not been effective or successful and major modifications are needed. Most reward systems do not positively recognize individuals or organizations who continuously state that what they are doing has not been as effective as they had hoped for, but they are learning, and believe the next time they will do better. A corollary is that relatively few managers are prepared to acknowledge openly that the change and complexity of the systems they are responsible to manage are beyond their

Table 6.1 Characteristics of passive and active adaptive management

Passive adaptive management
1. Active culture of reflection, involving effective evaluation, rewards for reflection, and effective communication to all project participants, and
2. Mechanisms to incorporate learning into planning and management
Active adaptive management
1. The two characteristics of passive adaptive management, as well as:
2. Management activities designed explicitly to test hypotheses through ecosystem-scale experiments,
3. Complexity recognized and embraced,
4. Mechanisms included for multi- or interdisciplinary investigation, as well as multi-stakeholder engagement, and
5. Social learning strongly emphasized

Modified from Allan and Curtis (2005, p. 415)

understanding and competence, and so they must move forward in a trial-and-error mode, incrementally improving their management practices.

Social learning is also emerging as a significant topic for research and practice (Fazey et al. 2007). As noted by Diduck (2004, pp. 504), and then reinforced in Dearden and Mitchell (2012, 167), various theories provide a foundation, and the 'theory of action' is one of the most compelling. From this perspective, learning is a process to identify and correct errors such as (1) matched intentions and outcomes, and (2) mismatched intentions and outcomes.

Three types of learning have been identified. The first, *single-loop learning*, occurs when matches between intentions and outcomes align, or when mismatches can be resolved by a change of behavior. To illustrate, Diduck (2004) uses the analogy of a thermostat in a home. When the thermostat detects the temperature is too hot or too cold, it acts to turn the heat either off or on. Action occurs because the thermostat receives information (room temperature) and facilitates action (heat off or on). When a mismatch is corrected by first questioning the behavior and then making an adjustment, *double-loop learning* occurs. Continuing with the thermostat example, before turning the heat off or on, the thermostat would question the temperature setting, and perhaps change the setting before commanding the heat to be turned off or on. A third type of learning also could occur. That is referred to as *triple loop*, and occurs when different behavior is prescribed. For example, if it is too cool, close all windows or put on a sweater.

Social leaning can occur at the level of individuals, organizations, communities, and nations. Double- and triple-loop learning reflect 'out-of-the-box' thinking necessary for innovation (Dearden and Mitchell 2012, p. 167). Social learning, whatever its form, is essential if capacity and competence are to be created to expedite active adaptive management.

6.2.3 Collaborative Management

Collaboration has been identified by Cortner and Moore (1999) as one of four fundamental principles for ecosystem management. The other three are (1) holistic, integrated science, (2) socially defined goals and objectives, and (3) adaptable institutions. "Collaboration can be defined as an approach to addressing natural resource and public policy problems to which stakeholders build consensus and work jointly on involving complex problems. … Collaboration entails the creation of stakeholder groups to review information, share analysis, identify objectives, develop an agreed strategy, and implement actions" (Margerum 2007, pp. 135–136).

Self-organizing collaborative groups created and managed by stakeholders "may be more capable of implementing an ecosystem-based approach than are agency-sponsored groups" (Bonnell and Koontz 2007, p. 154). Beyond the normal challenges any group faces related to ecosystem-based management, locally initiated collaborative groups must handle other basic issues, including defining a mission and purpose; developing organizational structures and procedures; managing finances and raising funds; recruiting, hiring, and managing staff; and developing leadership (Bonnell and Koontz 2007, p. 154). These are not trivial matters. Depending on the circumstances, at different times citizen-based, agency-based, and mixed partnership models can be appropriate. They indicate that a citizen-based model may not be appropriate when community depth is minimal as related to issues needing attention (Moore and Koontz 2003).

The increasing complexity and uncertainty associated with resource and environmental management have made the "traditional isolated, hierarchical organizations and decision-making processes … increasingly inappropriate and ineffective" (Lurie and Hibbard 2008, p. 431). In contrast, effective approaches require "flexible, adaptive institutions and networks of organizations and interests with fluid boundaries between authoritative decision makers and the communities in which they are imbedded in order to respond to changing knowledge and issues over time. Network characteristics requiring dynamic rather than rigid institutions include self-organization, horizontal structure, and voluntary participation" (Lurie and Hibbard 2008, p. 432). This view is reinforced by Mendis-Millard and Reed (2007, p. 543) who argue that local communities can do much more than facilitate instrumental improvements: they also can provide critiques of community capacity as well understanding about what is practical to achieve.

Various possible and positive outcomes have been identified through a collaborative approach. Conley and Moote (2003) have noted the following: resource management practices implemented, environmental conditions improved, enhanced social capital and legimacy realized, efficiency promoted and resources expanded, and priority resource issues focused upon at appropriate geographical scales (Gensknow 2009, pp. 411–412). Reed (2007, p. 30) identified further benefits noting that collaborative approaches help to create "practices that respect local circumstances, skills and concerns."

At the same time, researchers and practitioners appreciate that locally based collaborative approaches are not a panacea. As Bradshaw (2003, p. 138) observed, the potential innovation from collaboration will not be realized, "if newly empowered communities are not credible in their management of local resources or have sufficient capacity to do so. That is, communities must display a genuine desire to steward local resources in the interests of all stakeholders–including future generations and nonlocals–and have sufficient capacity to manage the resource base in order to achieve adequate and stable returns." Parochial and narrowly focused, short-sighted decisions by locally based groups are a real possibility and therefore a liability. In addition, developing the necessary trust among and between stakeholders can take many years, and, in some cases, may never be established (Lurie and Hibbard 2008, p. 433).

6.2.4 Governance

Governance has become a significant focus of research intended to improve management capacity (Kooiman 2003; Adger and Jordan 2009). As Tropp (2007: 21) has observed, "… governance is about the processes of making choices, decisions and trade-offs. Governance addresses linkages and processes between and within organisations and social groups involved in decision-making, both horizontally across sectors and between urban and rural areas, and vertically from local to international." Tropp, along with others, also highlights that 'governance' is not the same as 'government' as it needs to include civil society and the private sector.

> Fundamentally, the goal of improving governance is to build flexibility and resilience into the governance structure, thus enhancing the ability to adapt and effectively address current and future challenges. This contrasts with traditional centralized approaches to decision making and management; yet is increasingly recognized both globally and nationally as *the* [emphasis in original] priority. (Brandes and Curran 2009, p. 2).

As an example and reflecting the forgoing belief, Robins (2007) analyzed decentralized governance arrangements for integrated watershed management across Canada. In pursuing this type of governance research, she reflects the view of the World Water Assessment Programme (2006, p. 75): "When it works, decentralization has many benefits: it can allow for a democratization of decision-making through improved stakeholder inclusiveness, transparency and accountability. Appropriately implemented, it can empower people, particularly those lacking the social and political clout and financial means to have a voice and take part in decisions that define their livelihood opportunities." Such research draws upon and extends the ideas discussed in the previous subsection.

6.2.5 Environmental Justice

As noted by Mitchell (2011, p. 61) and Dearden and Mitchell (2012, pp. 167–168), the U.S. Environmental Protection Agency (US EPA 1997) defines environmental justice as "… the fair treatment and meaningful involvement of all people regardless of race, color, national origin, or income with respect to the development, implementation, and enforcement of environmental laws, regulations and policies" (www.epa.gov/compliance.basics/ej.html). 'Fair treatment' directs that no group of people should bear a disproportionate share of negative environmental consequences from industrial, commercial, or municipal operations, or from the implementation of federal, state, local, or tribal policies or programs (US EPA 1997).

> "The environmental justice movement emerged in response to environmental and social inequities, threats to public health, unequal protection, differential enforcement, and disparate treatment received by the poor and people of color. It redefined environmental protection as a basic right."

Bullard observed that environmental justice "embraces the principle that all communities are entitled to equal protection and enforcement of environmental, health, employment, housing, transportation, and civil rights laws and regulations that have an impact on the quality of life" (Bullard: www.ourplanet.com;/imaversn/122/bullard.html).

The concept of environmental justice emerged from a protest in 1982 regarding a hazardous waste landfill site in Warren County, North Carolina, USA. The protesters opposed establishment of a landfill site for PCB-contaminated soil to be removed from 14 different places in the state and be taken to the Warren County site. The landfill site was adjacent to a small, low-income community and the residents were predominantly African-American. The protesters argued that the siting decision highlighted such hazardous facilities were too often being located in areas in which the dominant inhabitants were minorities and/or low-income people.

Environmental justice is not confined to 'local issues' (Bryant 1995, 2007b; Pellow and Brulle 2005; Pellow 2007). Because of increased restrictions on disposal of toxic wastes in developed countries, combined with growing opposition to toxic waste sites, governments and private waste management companies have been seeking alternative sites in other countries. In Bryant's (2007a, p. 1) view, the target countries have been "the politically and economically less powerful nations of the world." The attraction for governments of receiving nations is substantial payments to receive toxic wastes and opportunity to create employment opportunities in building and operating the waste sites. Given such examples, and others in Clapp (1994a, b, 2001), advocates of environmental justice have argued that it is not an acceptable solution to resolve national or domestic problems in developed countries by moving the contentious material or facility to a developing country.

Two examples illustrate the diversity of research related to environmental justice. Holifield (2009) examined the conflict associated with a wood treatment plant initially operated by the St. Regis Paper Company within the Leech Lake Indian Reservation in Minnesota (USA). The company and its successors conducted research that showed contamination in groundwater was being successfully remediated. In contrast, local people, supported by external experts, argued that the subsequent company's modeling and monitoring of groundwater dynamics were fundamentally flawed. Holifield concluded that the successful challenge was the result of a combination of local and external people challenging the science underlying the remediation.

A different type of research examined the way in which environmental justice has been introduced into Scotland, beginning in 1999 (Slater and Pedersen 2009). They found that 'social justice' is a particularly distinctive feature. Indeed, in Scotland, environmental justice was defined, by the Scottish Environmental Protection Agency, as 'the aspects of social justice to which environmental protection relates the most' (Slater and Pedersen 2009, p. 802). However, "despite all the rhetoric, the actual translation of policy to legislative change has in fact been rather small. Those that have been implemented have, with a few exceptions, been due to international commitments rather than a grassroots approach to reform by the Scottish Executive" (Slater and Pedersen 2009, p. 809). This conclusion reinforces the role and value of international agreements that encourage governments to have their policies and practices conform with them.

Environmental justice provides considerable scope for geographic research, particularly because of the long-standing interest related to siting decisions regarding LULUs (locally unwanted land uses) and NIMBYs (not in my backyard), which draw upon the traditions of spatial and environmental analysis (Agyeman et al. 2009).

6.3 Water Management

In the following subsections, four concepts are reviewed: integrated water resource management, virtual water, soft path approach, and water ethics.

6.3.1 Integrated Water Resource Management

The concept of integrated water resource management (IWRM) has attracted attention around the world as practitioners strive to use an ecosystem or holistic approach to address issues related to an array of matters: (1) water, land, and other resources are interconnected, and their relationships need to be considered; (2) upstream and downstream relationships need attention; and (3) surface and groundwater systems need to be considered together. Furthermore, often there are challenges from vertical and horizontal fragmentation among and between public agencies at different spatial scales. These conflicts require coordination and collaboration to overcome the reality of 'silos' and 'stovepipes' leading agencies to pursue only their own legal responsibilities and interests. "IWRM has unquestionably become one of the mainstream initiatives discussed by governments. The major challenge remains its effective implementation in the field" (Rahaman and Varis 2005, p. 20).

IWRM is not a brand-new concept, as 'comprehensive river basin management' emerged in the 1960s, and was preceded by initiatives such as the Tennessee Valley Authority during the 1930s that sought to plan and mange for various aspects of land and water systems. Today, the most frequently cited definition of IWRM is that from the Global Water Partnership (2000, p. 22), "… a process which promotes the coordinated development and management of water, land and related resources, in order to maximize the resultant economic and social welfare in an equitable manner without compromising the sustainability of vital ecosystems."

Integrated river basin management "can be understood as the management of all surface and subsurface water resources of the river basin in its entirety with due attention to water quality, water quantity and environmental integrity" (Jaspers 2003, p. 91). He also added "a participatory approach is followed, focusing on the integration of natural limitations with all social, economic and environmental interests," a view endorsed by Sabatier et al. (2005).

Reservations have been expressed about IWRM, especially regarding the capacity of public agencies to implement it effectively, and the rationale for not considering relationships with other resources, such as energy. "In the real world, integrated water resources management, even in a limited sense, becomes difficult to achieve because of extensive turf wars, bureaucratic in-fighting, and legal regimes (like national constitutions) even within the management processes of single resources like water, let alone in any combined institution covering two or more ministries which have been historic rivals." Biswas (2004, p. 254, 2008a, b).

IWRM continues to be applied and assessed. "There is no simple solution for implementing the concept of IWRM; it depends on the particular framework and institutions related to the water resources in a country" (Blanco 2008, p. 91). Grigg (2008, p. 290) also has argued 'institutional barriers' provide a significant challenge, and while "There is no easy way to overcome the institutional barriers, … the concept of "governance" is widely recognized as a key to the solution." Grigg's comment reinforces the importance of 'governance,' discussed in an earlier section of this chapter. Other evaluations are provided by Shrubsole (2004), Garcia (2008), Varis et al. (2008), and Saravanan et al. (2009).

Opportunity exists for research to clarify the definition and interpretation of IWRM, to identify barriers that hinder effective implementation, and to generate constructive suggestions about how to either improve IWRM or offer a constructive alternative.

6.3.2 Virtual Water

The negative consequences of water shortages for food production and other economic activities, and the possibility that such shortages could lead to regional conflict or wars, were the triggers for the concept of 'virtual water.' Virtual water was conceived by Tony Allan (2011) to allow water-scarce countries to reduce problems by importing food requiring significant water inputs.

Warner and Johnston (2007, p. 65) have explained the basic principle associated with virtual water:

> … water-stressed regions of the world can ensure water and food security by actively seeking out a strategy which reduces domestic agricultural production of basic foodstuffs, increases reliance on the global food trade system to secure domestic requirements, and, in turn, increases the availability of water for non-agricultural development needs, reduces environmental stress and avoids senseless fighting over water.

By adopting a virtual water strategy, and importing the foodstuffs that requiring high amounts of water input, governments in water-stressed or water-deprived countries could meet food security needs despite a limited water endowment.

A virtual water strategy raises at least five challenges (Dearden and Mitchell 2012, p. 395). First, if a nation decided to reduce significantly domestic production of high water-demanding crops, and import those crops from other countries, many local farmers could lose their livelihoods, leading to other social problems. Second, a relatively poor nation would be unlikely have the necessary foreign currency to buy food products requiring significant water inputs. Third, a government may not want to depend on other countries for food needs. Referring to China as an example, Liu et al. (2007, p. 86) observed that "For food security, the government pays more attention to food self-sufficiency than to water use efficiency. Food self-sufficiency is overwhelmingly favoured by the Chinese government, which regards reliance on international food markets as a threat to domestic security." Fourth, adoption of virtual water could obscure in-country water shortages and lead to delays of necessary action to address that limitation. And fifth, despite its attractiveness, virtual water should never become the only criterion for determining water policies (Wichelns 2010).

For his work in developing the concept of virtual water, Professor Tony Allan from King's College, London, received the Stockholm Water Prize in August 2008, which includes an award of $150,000 (US). Professor Allan's citation praised his work regarding virtual water, and noted: "The improved understanding of trade and water management issues on local, regional and global scales are of the highest relevance for the successful and sustainable use of water resources."

6.3.3 Soft Path Approach

As highlighted in Dearden and Mitchell (2012, pp. 392–394), water needs can be met through various approaches (Gleick 2003; Brooks 2005; Brandes and Brooks 2006). The best known is *supply management*, a technological strategy to find or create new water supply sources when demand exceeds present supply. *Demand* management, in contrast, aims to alter human behavior to reduce water use. Common demand management methods are full cost and peak pricing of water, regulations to restrict water use at specific times of a year or day, and incentives to encourage conservation behavior. Both supply and demand management can and should be used together.

The soft path challenges basic patterns of consumption (Brooks et al. 2009). Brandes and Brooks (2006, p. 209) explain that although demand management emphasizes the question of 'how' to do the same with less water, the soft path raises the question of 'why' water is even used for a function. Furthermore, Brandes and Brooks (2006, p. 9) raise key questions.

> Why do we use water to carry away our waste? Demand management would urge low-flow toilets, but waterless systems are available, perhaps not for homes (because of the need for regular maintenance), but certainly for larger buildings.
>
> Why do we use half the potable water that is piped to a house in the summer is for watering lawns and gardens – and sidewalks. Demand management would urge more efficient sprinklers with automatic shut-offs, maybe even water restrictions. The soft path goes further: recycling water from bathtubs and washing machines or, better yet, drought-resistant greenery that requires little or no watering once it is established. (Brandes and Brooks 2006, p. 9)

Four basic principles underlie the soft path approach (Brandes and Brooks 2006, pp. 10–13). (1) Water is treated as a service rather than as an end in itself. Thus, the end is not to flush toilets or irrigate crops, but to dispose of wastes or to grow food. (2) Ecological sustainability is basic. Ecosystems are legitimate users of water, and also are one foundation of economies. Thus, the amount of water required to meet environmental needs is subtracted from what is available from any supply source under consideration to satisfy human needs. (3) Quality of delivered water is matched to an end-use requirement. Thus, this view results in wastewater from one use becoming the supply for another use with less stringent quality needs. (4) Determine the desired future condition, and plan back to the present. Attention is directed not to the most probable future, but instead on the most desirable future. An outcome does not assume present uses of water will be the same in the anticipated desirable future.

6.3.4 *Water Ethics*

Some suggest that in the twenty-first century water will be equivalent to what oil was at the end of the twentieth century in terms of being a strategic resource. Some even suggest that regional wars can be expected because of water shortages. For many needs, there is no substitute for water, and many needs and interests are involved as groups seek to obtain a share of water.

Attention to development of 'water ethics' on which to base water management decisions is striking by its scarcity. Some relevant reports have been published by UNESCO, such as Selbourne (2000). However, literature on water ethics is very modest, creating an opportunity for research and reflection.

An ethic usually is a statement of principles or values to guide behavior (Matthews et al. 2007). However, they recognize that a set of ethics cannot resolve all dilemmas. Indeed, different ethics may lead to different outcomes (e.g., be efficient, be equitable). Notwithstanding such difficulties, Matthews et al. (2007, pp. 350–353) offered six 'imperatives' for a new water ethic: (1) meet basic human needs to enhance equity today and for the future; (2) safeguard ecosystems by allocating sufficient water resources; (3) encourage efficiency and conservation of water resources; (4) establish open and participative decision-making processes; (5) respect system complexity and emphasize precaution; and (6) seek multiple sustainability benefits from water-centered initiatives.

There is considerable scope, need, and opportunity for work focused on water ethics, given growing water security issues, the number of nations already categorized as 'water stressed,' and additional numbers of countries expected to become water stressed by 2025. Opportunities exist to partner with agencies, ranging from local to international scales, to test the practicality of such ethical imperatives when making water management decisions.

6.4 Implications

There is a long tradition of innovation related to resource and environmental management by geographers and others. For example, Gilbert F. White's insight that structural and nonstructural adjustments need to be used together challenged conventional wisdom about floodplain management and led to a new approach that has been adapted around the world (White 1974, 2002; Burton et al. 1978; Kates and Burton 1986; Hinshaw 2006). Although White truly was a giant in terms of research and influencing the practice of water management, the innovations identified in this chapter continue the tradition of critically reflecting on resource management practice and identifying new directions. Such innovative and constructive contributions exemplify the approach taken by Baleshwar Thakur throughout his career. We thus celebrate his distinguished career, and encourage those who follow to aspire to the same high standards and constructive contributions.

References

Adger WN, Jordan A (eds) (2009) Governing sustainability. Cambridge University Press, Cambridge, MA
Agyeman J, Cole P, Haluza-Delay R, O'Riley P (eds) (2009) Speaking for ourselves: environmental justice in Canada. UBC Press, Vancouver
Allan C, Curtis C (2005) Nipped in the bud: why regional scale adaptive management is not blooming. Environ Manage 36:414–425
Allan T (2011) Virtual water. I.B. Taurus, London/New York
Armitage D, Berkes F, Doubleday N (eds) (2007) Adaptive co-management: collaboration, learning and multi-level governance. UBC Press, Vancouver
Bakker K (2007) Introduction. In: Bakker K (ed) Eau Canada: the future of Canada's water. University of British Columbia Press, Vancouver, pp 1–20
Berkes F, Colding J, Folke C (eds) (2003) Navigating social-ecological systems: building resilience for complexity and change. Cambridge University Press, Cambridge
Biswas AK (2004) Integrated water resources management: a reassessment. Water Int 29:248–256
Biswas AK (2008a) Current directions: integrated water resources management: a second look. Water Int 33:274–278
Biswas AK (2008b) Integrated water resources management: is it working? Int J Water Resour Dev 24:5–22
Blanco J (2008) Integrated water resources management in Colombia: paralysis by analysis? Int J Water Resour Dev 24:91–101
Bonnell JE, Koontz TM (2007) Stumbling forward: the organizational challenges of building and sustaining collaborative watershed management. Soc Nat Resour 20:153–167
Bradshaw B (2003) Questioning the credibility and capacity of community-based resource management. Can Geogr 47:137–150
Brandes OM, Brooks DB (2006) The soft path for water in a nutshell. Friends of the Earth, Ottawa; University of Victoria, POLIS Project, Victoria
Brandes OM, Curran D (2009) Setting a new course in British Columbia–water governance reform options and opportunities. POLIS Discussion Paper 09-02. University of Victoria, Victoria
Brooks DB (2005) Beyond greater efficiency: the concept of water soft paths. Can Water Resour J 30:83–92
Brooks DB, Brandes O, Gurman S (eds) (2009) Making the most of the water we have: the soft path approach. Earthscan, London
Bryant R (ed) (1995) Environmental justice: issues, policies and solutions. Island Press, Washington, DC
Bryant R (2007a) International chronology of Environmental justice, accessed on-line at http://www-personal.umich.edu/~bbryant/iejtimeline.html. Accessed 20 Sept 2009
Bryant R (ed) (2007b) Environmental advocacy: working for economic and environmental justice. Bunyan Bryant, Ann Arbor
Burton I, Kates RW, White GF (1978) The environment as hazard. Oxford University Press, New York
Carpenter SR, Walker B, Anderies JM, Abel B (2001) From metaphor to measurement: resilience of what to what? Ecosystems 4:765–781
Clapp J (1994a) Africa, NGOs and the international toxic waste trade. J Environ Dev 3(2):17–46
Clapp J (1994b) The toxic waste trade with less-industrialized countries: economic linkages and political alliances. Third World Q 15(3):505–518
Clapp J (2001) Toxic exports: the transfer of hazardous waste from rich to poor countries. Cornell University Press, Ithaca
Conley A, Moote MA (2003) Evaluating collaborative natural resource management. Soc Nat Resour 16:371–386

Cortner HJ, Moore MA (1999) The politics of ecosystem management. Island Press, Washington, DC
Dearden P, Mitchell B (2012) Environmental change and challenge: a Canadian perspective, 4th edn. Oxford University Press, Don Mills
Diduck A (2004) Incorporating participatory approaches and social learning. In: Mitchell B (ed) Resource and environmental management in Canada: addressing conflict and uncertainty. Oxford University Press, Toronto, pp 497–527
Fazey I, Fischer J, Sherren K, Warren J, Reed N, Dovers S (2007) Adaptive capacity and learning to learn as leverage for social-ecological resilience. Front Ecol Environ 5:375–380
Garcia LE (2008) Integrated water resources management: a 'small' stop for conceptualists, a giant step for practitioners. Int J Water Resour Dev 24:23–36
Genskow KD (2009) Catalyzing collaboration: Wisconsin's agency-initiated basin partnerships. Environ Manag 43:411–424
Gleick PH (2003) Global freshwater resources: sort-path solutions for the 21st century. Science 302:528–534
Global Water Partnership, Technical Advisory Committee (2000) Integrated water resources management. TAC Background Paper No. 4. Stockholm: Global Water Partnership Secretariat
Grigg NS (2008) Integrated water resources management: balancing views and improving practice. Water Int 33:279–292
Gunderson LH, Holling CS (eds) (2002) Panarchy: understanding transformations in human and natural systems. Island Press, Washington, DC
Gunderson LH, Prichard L (eds) (2002) Resilience and the behaviour of large-scale systems. Island Press, Washington, DC
Gunderson LH, Holling CS, Light SS (eds) (1995) Barriers and bridges to the renewal of ecosystems and institutions. Columbia University Press, New York
Hinshaw RE (2006) Living with nature's extremes: the life of Gilbert Fowler white. Johnson Press, Boulder
Holifield R (2009) How to speak for aquifers and people at the same time: environmental justice and counter network formation at a hazardous waste site. Geoforum 40:363–372
Holling CS (1973) Resilience and stability of ecological systems. Annu Rev Ecol Syst 4:1–23
Holling CS (1978) Adaptive environmental management and assessment. Wiley, Chichester
Jacobson C, Hughey KFD, Allen WJ, Rixecker S, Carter RW (2009) Toward more reflexive use of adaptive management. Soc Nat Resour 22:484–495
Jaspers FGW (2003) Institutional arrangements for integrated river basin management. Water Policy 5:77–90
Kates RW, Burton I (eds) (1986) Geography, resources, and environment: selected writings of Gilbert F. White, vol 1. University of Chicago Press, Chicago
Kooiman J (2003) Governing as governance. Sage, London
Lee KN (1993) Compass and gyroscope: integrating science and politics for the environment. Island Press, Washington, DC
Liu J, Zehnder AJB, Yang H (2007) Historical trends in China's virtual water trade. Water Int 32:78–90
Lurie S, Hibbard M (2008) Community-based natural resource management: ideals and realities for Oregon Watershed Councils. Soc Nat Resour 21:430–440
Margerum RD (2007) Overcoming locally based collaboration constraints. Soc Nat Resour 20:135–152
Matthews C, Gibson RB, Mitchell B (2007) Rising waves, old charts, nervous passengers: navigating toward a new water ethic. In: Bakker K (ed) Eau Canada. University of British Columbia Press, Vancouver, pp 335–358
McLain RJ, Lee RG (1996) Adaptive management: promises and pitfalls. Environ Manag 20:437–448
Mendis-Millard S, Reed MG (2007) Understanding community capacity using adaptive and reflexive research practices: lessons from two Canadian biosphere reserves. Soc Nat Resour 20:543–559

Mitchell B (ed) (2010) Resource and environmental management in Canada: addressing conflict and uncertainty, 4th edn. Oxford University Press, Don Mills
Mitchell B (2011) LULUs, NIMBYs, and environmental justice. In: Fung T, Lesbivel SH, Lam K-C (eds) Facility siting in the Asia-Pacific: perspectives on knowledge production and application. Chinese University Press, Shatin, pp 57–83
Moore EA, Koontz TM (2003) A typology of collaborative watershed groups: citizen-based, agency-based, and mixed partnerships. Soc Nat Resour 16:451–460
Noble B (2004) Applying adaptive environmental management. In: Mitchell B (ed) Resource and environmental management in Canada: addressing conflict and change. Oxford University Press, Toronto, pp 442–466
Olsson P (2007) The role of vision in framing adaptive co-management processes: lessons from Kistianstads Vattenrike, southern Sweden. In: Armitage D, Berkes F, Doubleday N (eds) Adaptive co-management: collaboration, learning, and multi-level governance. University of British Columbia Press, Vancouver, pp 268–285
Ostrom E (2005) Understanding institutional diversity. Princeton University Press, Princeton
Pellow DN (2007) Resisting global toxics: transnational movements for environmental justice. MIT Press, Cambridge
Pellow DN, Brulle RJ (eds) (2005) Power, justice and the environment: a critical appraisal of the environmental justice movement. MIT Press, Cambridge, MA
Rahaman MM, Varis O (2005) Integrated water resources management: evolution, prospects and future challenges. Sustain Sci Pract Policy 1:20
Reed MG (2007) Uneven environmental management: a Canadian perspective. Environ Manag 39:30–49
Resilience Alliance. (2009). *Resilience*, accessed at http://www.resiliencel.org/576/php. Accessed 20 Sep 2009
Robins L (2007) Nation-wide decentralized governance arrangements and capacities for integrated watershed management: issues and insights from Canada. Environments 35(2):1–47
Sabatier PA, Focht W, Lubell M, Trachtenberg Z, Vedlitz A, Matlock M (eds) (2005) Swimming upstream: collaborative approaches to watershed management. MIT Press, Cambridge, MA
Saravanan VS, McDonald GT, Mollinga PP (2009) Critical review of integrated water resources management: moving beyond polarised discourse. Nat Resour Forum 33:76–86
Selbourne L (2000) The ethics of freshwater use: a survey. UNESCO, Paris
Shrubsole D (ed) (2004) Canadian perspectives on integrated water resources management. Canadian Water Resources Association, Cambridge, ON
Slater A-M, Pedersen OW (2009) Environmental justice: lessons on definition and delivery from Scotland. J Environ Plan Manag 52:797–812
Thakur B (ed) (2003) Perspectives in resource management in developing countries, volume 1, resource management: theory and technique. Concept Publishing, New Delhi
Thakur B (ed) (2007) Perspectives in resource management in developing countries, vol 2. Population, resources and development. Concept Publishing, New Delhi
Tropp H (2007) Water governance: trends and needs for new capacity development. Water Policy 9(2):19–30
US Environmental Protection Agency (1997) About environmental justice. Available at: http://epa.gov/swerops/ej/aboutej/html. Accessed 20 Sept 2009
Varis O, Rahaman MM, Stucki V (2008) The rocky road from integrated plans to implementation: lessons learned from the Mekong and Senegal River basins. Int J Water Resour Dev 24:103–121
Walker B, Salt D (2006) Resilience thinking: sustaining ecosystems and people in a changing world. Island Press, Washington, DC
Walker B, Holling CS, Carpenter S, Kinzing A (2004) Resilience, adaptability and transformability in social-ecological systems. Ecol Soc 9(2): 5. Online at http://www.ecologyandsociety.org/vol9/iss2/art5/ (last accessed 18 March 2015)
Walters CJ (1986) Adaptive management of renewable resources. Macmillan, New York

Warner JF, Johnston CL (2007) 'Virtual water' – real people: useful concept or prescriptive tool? Water Int 32:63–77

Westley F, Zimmerman B, Patton MQ (2006) Getting to maybe: how the world has changed. Random House, Toronto

White GF (ed) (1974) Natural hazards: local, national, global. Oxford University Press, New York

White GF (2002) Autobiographical essay. In: Gould P, Pitts FR (eds) Geographical voices: fourteen autobiographical essays. Syracuse University Press, Syracuse, pp 341–364

Wichelns D (2010) Virtual water: a helpful perspective, but not a sufficient policy criterion. Water Resour Manag 24:2203–2219

World Water Assessment Programme (2006) Water: a shared responsibility. The United Nations World Water Development Report 2. United Nations Educational, Scientific and Cultural Organization, Paris; Berghahn Books, New York

Part IV
Economic Perspectives

Chapter 7
Inherited Land: The Evolution of Land Markets and Rights Before Independence

Sanjoy Chakravorty

Abstract Land is the most important resource in India and, as a result, always in contention. This chapter takes a long-term view of land markets and land rights—as they evolved through pre-colonial and colonial regimes (with an emphasis on the latter)—to contextualize some of the fundamental struggles over land in independent India. The maximization of land revenue was the primary objective of all pre-independence states, from the Mughals and Marathas to the East India Company and British Crown regimes. There were significant regional variations in the operationalization of these policies—from the more sustainable raiyatwari system used in south and west India to the harsher and more extractive zamindari system used in the east and north—variations that influence agriculture, urbanization, and the political economies of these regions until today. Independent India thus inherited a complex and geographically diverse system of land markets, rights, and fragmentations, created through several centuries of peasant domination and misery, and is still engaged in the task of mitigating and coming to terms with that inheritance.

Keywords Land markets • Land rights • Land acquisition • Land revenue • Pre and post-independent India

7.1 Motivation

Conflicts over the acquisition of agricultural land have been much in the news in the past 5 years. The conflicts have become so widespread and the issue so important that a new Land Acquisition, Rehabilitation and Resettlement law was created in 2013 to replace the land acquisition legislation in existence, the colonial period

---

This paper is excerpted and edited from Chaps. 5, 6, and 8 from the author's book *The Price of Land: Acquisition, Conflict, Consequence* (Chakravorty 2013). The book was written with support from Temple University (a sabbatical) and the Center for the Advanced Study of India, University of Pennsylvania.

S. Chakravorty (✉)
Department of Geography and Urban Studies, Temple University,
Philadelphia, PA 19122, USA
e-mail: sanjoy@temple.edu

A.K. Dutt et al. (eds.), *Spatial Diversity and Dynamics in Resources and Urban Development*, DOI 10.1007/978-94-017-9771-9_7

Land Acquisition Act of 1894. That is, a law that had ostensibly served its purpose for more than 115 years—primarily to acquire agricultural land for nonagricultural use—could no longer do the job. If one colonial law was found to be useful by the independent Indian state for more than 60 years, the same cannot be said about other colonial laws and policies on agricultural land. The very first acts of independent India's parliament were focused on reforming and changing those colonial conditions. The First, Fourth, and Seventeenth amendments to the constitution created the legal basis for key land reforms: the abolition of intermediaries and land ceilings. In other words, the land system that independent India inherited from the colonial raj was considered unacceptable in some domains (requiring major new land reform laws) but convenient in at least one major domain (land acquisition).

It is probably not an exaggeration to argue that the most fundamental elements of India's political economy were based on agricultural land in the initial decades after independence, land reforms on the one hand, and on the other hand large-scale acquisitions for massive projects (from Bhakra Nangal, Hirakud, and Damodar Valley to Raurkella, Bhilai, and Durgapur) that together affected the lives of dozens of millions of agriculturalists. These fundamental transformations took place on a system of land markets and rights created by the colonial raj. This chapter is an attempt to understand the evolution and details of the land system inherited by independent India to better contextualize the actions of the state after independence, actions whose consequences continue to reverberate in contemporary India.

7.2 The State and Land

Land markets operate under rules set down by the state. In general, the state has the power to dictate:

- Land use, or the activity that takes place on a given parcel. Broadly, the possible uses include agriculture, industry, commerce, infrastructure, and housing. Each of these broad categories can be subdivided. In agricultural settings, at the extreme, the state has been known to decide what specific crop will be grown on a specific land parcel.
- Division of land output between the producer of the output and the state. This is simply the rate of taxation. In agricultural settings, there is evidence (discussed in this paper) that this rate has been as high as 75 % (or three-fourths of production) for long periods, even as high as 90 % on occasion.
- Land ownership, which includes the meaning of ownership itself (that is, the rights that are associated with ownership, at the individual, family, or community level), who can own what, and who can own how much. Included in this power is the ability to change ownership from one entity to another (individual to individual, individual to state, community to state, etc.).

It is necessary to be careful about the meaning of "ownership" of land. The word has a clear and unambiguous meaning in the contemporary world. To own something suggests the right to do what one pleases with it—use it or waste it, give it away, or sell it. These complete rights of ownership have, however, never been associated with land: this is true even today, even in societies (such as the United States) that are identified with the glorification of private property rights. One of the main reasons for constrained ownership rights is the existence of "externalities" or "spillovers," which are the public effects of private uses of land. The effects of the activities that take place on a land parcel do not necessarily stop at the border of that parcel. Therefore, it is possible to generate negative externalities or negative spillover effects on those who do not own that parcel, hence do not derive benefits from it, but are subject to what is done with it. There can be positive externalities too, which are usually framed in the language of "public purpose"; for example, a piece of land can be used for a highway rather than a paddy field, and thereby generate utility for the public at large rather than a private entity. The contemporary infrastructure of land use law and eminent domain law has been created to handle the positive and negative externalities of land ownership, at least nominally.

Therefore, there are constraints on land ownership in contemporary market societies that otherwise recognize and protect private ownership rights in land. As we move back in time we find weaker and weaker rights associated with ownership until, in India, we reach a point where the concept of private ownership of land did not exist at all, or existed only for a select few, the chosen ones favored by the state. There were some rights for all, to be sure, such as use rights (the right to cultivate and keep a share of the produce, or to graze cattle) and community rights (typically in forests and water bodies), but these were not private rights of ownership in anything close to the contemporary meaning of the word.

What one does not own, one cannot sell. What is not sold cannot also be bought. There is no market for that good. As we step back in time, we find weaker and weaker "land markets" till we reach a point where they did not exist in any meaningful sense. Therefore, as we try to gain an understanding of the evolution of the state–peasant–land relationship during the past 300 plus years, it is useful to think of two overarching themes:

- The gradual creation of land markets and the establishment of private property rights in land, and
- The tug of war between the power of the individual or citizenry and the power of the state to determine the rights attached to land ownership.

Hence, the right to private property in land is the key variable, not only the right itself, legally constituted by the state, but limits on the right, also legally defined by the state. Can an ordinary citizen own land? If yes, what can he do with it? And just as important, what can the state do with it? This study of the past leading up to the present is largely an attempt to understand the historical formation of this basic right, the right to private ownership of land.

A useful way to understand the meaning of an individual right to land is to understand what the state can do—that is, its dictates on land taxes and land uses and land takings. The specific dictates change over time, but in general it is possible to identify long-term regimes, periods that are at least several decades long, during which these dictates operate within narrow margins. The dictates of the state also change over space, or the territorial units over which regimes have authority. In other words, if we want to understand the evolution of the relationship between the Indian peasantry and his land, we need to have knowledge of the different state regimes that have controlled that relationship. We also have to recognize that during any given time period a number of spatial regimes have existed within the boundaries of the modern map with which we are all familiar. This paper is a study of spatial land regimes in pre-independence India. It provides a very brief account of pre-colonial regimes in North and South India, followed by a longer account of the colonial regimes (especially the zamindari and raiyatwari systems).

7.3 Before Colonization

The eighteenth century was the period of transition from pre-colonial to colonial regimes. So, as far as this discussion is concerned, the pre-colonial period refers to the seventeenth and the first half of the eighteenth century; the colonial period refers primarily to the second half of the latter, all the nineteenth century, and into the first half of the twentieth century, a period of approximately 180 years.

Seventeenth-century India is commonly thought of as Mughal India. The size of the Mughal Empire was not fixed throughout the century, but it covered all of North India during the entire period, and increasingly more of the Deccan, reaching an apogee in the northern parts of the modern states of Tamil Nadu and Kerala under Aurangzeb. The other great power during this period were the Marathas, who were a potent force in western, central, and southern India in the period spanning the decline of the Mughals and the ascendance of British colonial power. Therefore, in the pre-colonial period, we need to discuss the state–peasant–land relationship for North and South India separately. They were similar in many respects, but different enough that it may be possible to argue that the long shadow of pre-colonial India continues to influence development patterns in these two regions to date.

7.3.1 The Mughals

There is no question that the power of the Mughal regime was based on revenues from peasant agriculture. This regime was a typical instance of the cyclical argument justifying imperialism: the greater the amount of land revenue, the larger the army that can be supported; the larger the army, the larger the land that can be controlled and the greater the revenue. By definition, successful empires control large land areas. The principal governmental problem is how to most effectively raise the largest land revenue possible from ever larger territories.

The system instituted by the Mughals relied on jagirdars and zamindars (or mansabdars) for revenue collection. The jagirdars were few in number (they were considered the equivalents of princes or nobles), in the mere dozens; the zamindars/

mansabdars were more numerous, numbering several thousand. Irfan Habib (1982, p. 242) writes: "It has been estimated that in 1646, a mere sixty-eight princes and nobles at the top claimed 36.6 per cent of the jama' of the empire; the next 587 officials claimed nearly 25 per cent. On the other hand, the remaining 7,555 mansab-holders claimed between a quarter and a third of the revenues." Jagirdars had very short tenures, averaging between 2 and 4 years at a single location. That is, they had powerful but transferrable positions. Mansabdaris, on the other hand, were inheritable. The jagirdars and mansabdars were responsible for collecting revenue (mostly in cash, some in produce) from the peasantry. The unit of revenue collection was the mansabdar or zamindar rather than the individual cultivator.

Irfan Habib (1982, p. 238) writes: "Amidst the complexity of arrangements for assessment and collection, one major aim of the Mughal administration still stands out in bold relief: the attempt at securing the bulk of the peasant's surplus." How much was that? The rates varied somewhat from administration to administration and on the specific land rights of a specific peasant, but in general, the direct tax was around half the peasants' output. On top of that there were a number of additional levies, imposts, and official's fees, which typically added up to another quarter of the output. The transferable jagirdars almost always looked for opportunities to extract additional revenues for themselves, harshening what has been called a system of reckless exploitation of the peasantry. Aurangzeb imposed an additional head-tax on non-Muslims (jizya), which came to about a month's salary for farm wage laborers. As a result, the Mughal regime extracted no less than three-quarters of a peasant's output, frequently more than that.

Consider the example of village Akahera in Pargana Rinsi (in the jagir of Raja Bishan Singh) described by R.P. Rana (2006, p. 26): "The peasants of this village produced 16,000 mans of grain in the kharif season of 1665. Out of this gross produce, the peasants paid 8,000 mans to the jagirdar as land revenue (mat). From the remaining 8,000 mans, the peasants paid other cesses which totalled 4,500 mans. They were left with 3,500 mans… It may be noted that most of these parganas later became the centre of the Jat revolts. These figures clearly show that the peasants had paid 78 per cent of their total output as revenues to the jagirdar."

What if the peasants refused to pay? Irfan Habib (1982, p. 240) again: "The collection of revenue was enforced by severe methods. Nonpayment of revenue was deemed equivalent to rebellion. While eviction was not unknown as a punishment, the more usual method seems to have been imprisonment and torture of the headmen, followed by the massacre of the adult male population and enslavement of women and children." Despite such severe punitive measures, sporadic insurgencies were common. Even more common was the mass-scale abandonment of farms, sometimes by entire villages.

R.P. Rana identifies several possible causes for the decline of the Mughal Empire. The principal ones include "Hindu reaction" to Aurangzeb's excesses (argued by Jadunath Sarkar 1952); the failure of the jagir system, especially the insecure tenure of jagirdars (Satish Chandra 1982); the "Shivaji factor" and the demoralization of the Mughal army the further it entered the Deccan (Pearson 1976); and cultural failure and technological stagnation (Athar Ali 1975). But ultimately Rana indicts the debilitating revenue extraction system. He writes (2006, p. 24): "As oppression increased, the number of absconding peasants grew, cultivation declined and peasants took to arms giving birth to rural uprisings of varying intensity. Consequently the empire fell prey to the wrath of an impoverished peasantry."

7.3.2 The Marathas

The revenue collection systems and rates in South India were more diverse because, in contrast to Mughal North India, the region was never under the rule of a single empire. By the end of the fifteenth century, after the fall of the Vijaynagar kingdom, there arose the Nayak kingdoms of Ikkeri, Mysore, Senki, Thanjavur, and Madurai, and the Bijapur and Golconda kingdoms further north. The Malabar Coast had several small kingdoms such as Calicut and Cochin. Most of these were squeezed between the expansionary drives of the Mughals from the north and the Marathas from the west.

A good sense of the condition of the peasantry in pre-colonial South India may be derived from the significant scholarship available on the reign of Shivaji, arguably the most significant Maratha ruler. Shivaji's revenue collection apparatus generally took less of the peasant's output than the Mughal system, between one third and two fifth of the produce directly, plus several other cesses and taxes that may have added up to another one fifth of the output, and what may be more important, did so using the raiyatwari system in which the peasant, as opposed to the village or the zamindar, was the unit of collection. The burden of taxes in the territories Shivaji controlled but did not rule directly was considerably higher, and included chauth (one fourth) and sardeshmukhi (one tenth) on the output. As a result, peasants in these subjugated lands paid taxes that were equivalent to Mughal rates. It is possible that these taxes allowed Shivaji to be more generous to the peasants he ruled directly.

The exploitations that were associated with the jagirdari system of the north were less in evidence under the Marathas as they sought to minimize the power of the traditional local revenue collectors, the Deshpandes and Deshmukhs. Hiroshi Fukazawa (1983, pp. 254–256) writes: "…the Zamindari system of the north Indian type…was generally absent in the Deccan…" but goes on to note that "…yet many large inamdars…and the like had to be maintained and created…for such was not only the custom of the time but political and social necessities compelled medieval rulers to admit and rely on the landed interests."

It is possible to summarize the conditions of the pre-colonial period with the following observations:

- The state had little interest in controlling what was grown on the land. Peasants grew what they wanted on "their" land, although the meaning of ownership and property rights were variable and significantly different from what is commonly understood today.
- The state had a large say in land ownership rights. It defined different types of use rights (with associated taxation levels), and directed more expanded rights

and control at the top of the hierarchy. Officers of the state controlled rights and uses at the level of the peasantry, who also had various forms of collective ownerships.

- The primary interest of the state was to extract revenue from the land. The overall rate of revenue extraction was typically around 75 % of the output. Mughal North India had higher rates than South India, a region that never came within a single ruler's ambit.
- Mughal North India used a jagirdari-zamindari system that was harsher for peasants than the raiyatwari system used for much of the time in much of the south. These different systems have had very long term consequences.

7.4 Colonial Regimes

There is a vast quantity of research on the condition of agriculture and the Indian peasantry under British colonization. This brief synopsis cannot do justice to anything close to all the findings, and especially the arguments between historians on the effects of colonial rule: Did colonial rule basically continue the patterns of pre-colonial governance? Was it more injurious to the welfare of its subjects than what went before? Did the institutions of colonial governance create immiserization and pauperization of the Indian peasantry? In B. Chaudhuri's (1983, pp. 86–87) summary of the argument:

> ...the nationalists who assumed an increasing rural impoverishment blamed it mainly on certain aspects of the British land revenue administration, such as the high pitch of land revenue demand, the insistence on its payment in money and its relative inflexibility which, by preventing accumulation of agricultural capital, inevitably impoverished agriculture. Later writers... emphasized other factors, such as the establishment of private property in land, the creation and proliferation of a class of 'parasitic' landlords, the increasing burden of rent and rural indebtedness. All these...caused large-scale alienations of peasant holdings, with the result that peasants ceased to be 'self-possessing, self-working and self-sufficient' producers and increasingly depended for their subsistence on agricultural wage labour and sharecropping...According to the opposing point of view...the old agrarian society was far from egalitarian; the considerable redistribution of landed property rights brought about by the British revenue laws only marginally affected the old system of land control at the village level.

It is, indeed, quite difficult to draw an objective and narrow conclusion about the effects of close to 200 years of rule. The difficulties arise for a number of different reasons: the variation in governance patterns in different regions of India; the variation in the objectives of the colonial state over this very long period; the variation in ideologies and global economic conditions, which mattered more and more over time; and the variation in the responses of the Indian elites and commoners to the different patterns of governance over time.

Also, it is obvious that the colonial state at the end of the eighteenth century (at which point much, but not all, of India was under direct or indirect control of the

East India Company) was far different from the one that left India in 1947. It is useful to divide this period into four segments:

- Early Company Raj (1765–1830), a period that began after the battle of Buxar gave the Bengal Diwani to the Company, after which the mercantile firm based in London gained control of a very large land mass and tried to figure out ways to best meet its objectives.
- Late Company Raj (1831–1857), a period during which the continuous warring for control of the land had ended, and as a result there was a reasonably settled governance system (with significant regional variation, of course) that came to an end with the Sepoy Rebellion (or the First War of Indian Independence).
- Early Crown Raj (1857–1918), covering the first decades of direct rule by the British crown and parliament which ended more or less with the first world war and the first inklings of the end of empire.
- Late Crown Raj (1919–1947), covering the period between the world wars, including the global economic depression (with profound effects in India), and ending with formal independence.

7.5 Company Raj

What were the issues uppermost in the minds of the leadership of the Early Company Raj, the accidental imperialists? In Dharma Kumar's (1983, pp. 214–215) words:

> As the British extended their rule over south India, with its bewildering variety of land systems, they were faced with three sets of problems. Their most urgent need was for land revenue, and here the high rates charged by their immediate predecessors were very helpful. Then there was the question of whom to settle with for the land revenues. Should the land revenue be taken directly from individual cultivators or from cultivators as a group, and if so, represented by whom? Should contracts be made with intermediaries—the 'poligars,' or the officials and others established as zamindars during Muslim rule, or the adventurers, foreign and native, spawned by the turmoil of the eighteenth century? And finally, as the Company turned into a government, it had to consider wider issues of land law and rights. How should it arbitrate between different interests? On the whole, the government prized political stability above abstract notions of justice; it preferred not to disturb the traditional distribution of powers and rights, if only it could discover what that was, unless its own interest suffered.

As David Washbrook writes (1981, p. 650), there is no doubt that "land remained overwhelmingly the single most important source of wealth and the base of production." So, in simple terms, the issue at hand for the Company Raj was how to maximize revenue from land—about whose distribution and management the Company had little knowledge—with the minimum of effort. This question was not being considered in an ideological vacuum. The leaders of the Company Raj were men of commerce. They believed in markets and trade backed by the rule of law (remember, Adam Smith's *Wealth of Nations* was published in 1776, mere years into the East India Company's adventures in the subcontinent). Anand Swamy writes (2011, p. 138): "much of the discussion was conducted in terms that would be familiar to contemporary economists: secure property rights and contract enforcement, and, more generally, good governance, would promote investment, trade, and economic growth. The Company would (it was argued) provide this essential support for economic development far better than the despotic and mutually hostile regimes that had preceded it."

Reality turned out to be murkier than theory, of course, and expedient too. In the event, the Company Raj generally attempted to use the regional land revenue extraction systems that predated their control. To keep this exposition brief and manageable, I focus on two regions—the east and the south—and their respective "ideal type" revenue systems—zamindari and raiyatwari. (A third system of revenue collection, called mahalwari, was also used by the British. It was a mix of the zamindari and raiyatwari systems and used over a limited area, mainly in Punjab. I will not complicate these discussions with more detail on the mahalwari system.)

The Company's "experiments" with governance (with the perspective of time, these initial policies do feel like experiments) began in the east, in the Bengal Diwani, which the Company Raj acquired in 1765. In much of the east (present-day West Bengal and Bangladesh, Bihar, and Odisha), the system predating Company control was the Mughal zamindari system. For the sake of simplicity we can think of this as the "landlord" system of revenue extraction.

The raiyatwari system, as discussed earlier, is sometimes referred to as a "non-landlord" system because the regime attempted to collect revenue directly from the peasant (or raiyat). This term is not strictly accurate because the pre-colonial south had a variety of kingdoms with a variety of revenue collection mechanisms. For instance, in many subregions of present-day Kerala (for example, Travancore and south and central Malabar), there existed a peasant-labor system that resembled slavery more than anything. In addition, there were populations that were never cultivators. Brahmins, who were not allowed to touch the plow, always used tenant farmers. Also, "there was a large group who were born into agricultural servitude and could rarely emerge from it" (Kumar 1983, p. 216).

Nevertheless, the zamindari-raiyatwari systems can be generalized to north/east and south/west India. In the following paragraphs I describe these two systems and their consequences, with most of the attention focused on the initial phase during which the basic outlines of the remaining decades were established.

In the east, at first, the Company Raj, similar to typical "weak regimes," attempted to collect taxes through "revenue farming," using intermediaries and 5-year assessments, etc., but faced nonpayment and corruption (such as tax collectors who absconded with their collections). In 1770 a massive famine killed about one third of the population of Bengal, which raised serious questions about the sustainability of the tax collection system. There were several peasant uprisings, notably in Rangpur (1783) and Birbhum (1788–1790).

Then, in 1793 Lord Cornwallis created the Permanent Settlement in the east, a policy that froze land revenue for the region for perpetuity. The policy document declared that "at the expiration of the term of settlement no alteration will be made in the assessment which they have respectively engaged to pay, but they and their heirs and lawful successors will be allowed to hold their estates at such assessment for ever" (Guha 1963, p. 11). Henceforth, the Company could not increase its revenue demand from the zamindars. It declared that the zamindars were not mere revenue collectors, as in the past, but proprietors of the land. Agricultural land became property in ways that it had never been before, an object that could be bought and sold; this is the critical move that began the formation of a proto land market in India.

The Permanent Settlement policy seems outlandish from a contemporary perspective, and at the time it was justified by arguing that the zamindars would not only pay their taxes more readily, but with a fixed and known revenue burden would become prudent managers of the land and invest in raising productivity. Most historians agree that all the "free market" talk aside, the main reason for the Permanent Settlement policy was its simplicity. The Company was too weak at that time to implement anything too complicated.

The short-term effect of the Permanent Settlement was to unleash chaos in the east. The revenue rate was fixed at 90 % of the rent, which was an extremely heavy burden. The Company administrators generally refused to be flexible, even when there were crop failures. When zamindars could not pay, their estates were sold on auction. There were large-scale defaults (for example, more than half the 3,000 estates in Odisha defaulted and were auctioned within the first 15 years after the settlement), and "new men" emerged to become a new landlord class. Many came from the merchant society of Calcutta, men who had made their money from trade in salt and abkari (alcohol), but mostly they came from the Company administration itself, men who had "networks" and "insider information" that they used to acquire land.

The sale price of an estate at that time barely equaled the annual revenue demand from it. It should be noted, however, that by the early twentieth century, the revenue rate had effectively dropped below 20 % of output and the price of estates was manifold higher, as much as 15 or 20 times the revenue demand. In other words, the Permanent Settlement created a land market where none existed before; it was volatile in the early decades and zamindaris were precarious; but over time, capable zamindars emerged (such as in Burdwan), and agricultural land became a valuable commodity that was exchanged between people with access to capital and information.

In the south and west (in the Madras and Bombay Presidencies), the raiyatwari system was established principally under the leadership of Thomas Munro, a believer in "utilitarianism," who became governor of the Madras Presidency in 1820. Before that, in the latter part of the eighteenth century, the Company had been using various systems of village headmen and middlemen to collect revenue, and between 1802 and 1807 had actually established the Permanent Settlement of zamindar-based revenue collection over large areas under Company control. Although by 1822 the Company had decided that the raiyatwari system would be introduced everywhere, in 1830 about 30 % of the Madras Presidency was under the zamindari system. Even then, all raiyatwari regions were not alike, and the Company used local large landlords (mirasdars) of various kinds for revenue collection.

In some senses the village hierarchies in the south appear to have been flatter, at least in collective decision making at the village level on water distribution and community assets, but in others, such as caste roles, the hierarchies in the south appear to have been steeper. The cultivator still paid heavy taxes. In zamindari lands he retained between one fifth and one third of his produce. In raiyatwari lands there was great variation in revenues extracted in practice, and only some of the variation came from whether a land was "wet" (irrigated) or "dry." In some fortunate regions (such as Coimbatore and Thanjavur) the cultivator could retain half or more of his produce; in others, such as North Arcot, the cultivator might retain as little as a one quarter or one fifth of his crop.

By 1830 the essential features of the state–peasant–land relationship had been established all over the subcontinent. The hapless and tax-burdened peasant was in evidence everywhere—east, south, north, and west—although some pockets of relative peasant prosperity existed in the south and west. In general, the revenue burden on the peasantry was no better or worse than during the Mughal and Maratha periods.

The great churning that had been taking place at the top of the Indian revenue-extracting class—thousands of zamindars ruined, "new men" in their place, selected poligars and mirasdars given power, others not—had largely settled down. Interventions in the social system had begun, which essentially reified the Brahminical view of caste and varna hierarchies, which eventually "made the nineteenth century the Brahmin century in Indian history"; see Washbrook (1981, p. 653), who goes on to suggest that this may explain why the twentieth century was "anti-Brahmin." These social interventions, through the legalization of "Hindu family law" and marriage and inheritance norms, had a great influence on the distribution of land by the time the colonial order was over.

Of primary importance here is the comparison of "Hindu family law" with English customary or common law on land inheritance. In the latter, a system of primogeniture prevailed, in which land was inherited by the oldest male child of a household. In so-called Hindu law, land was equally inherited by all male children. This law was not a major problem when land was plentiful (if not necessarily abundant), at least in large parts of the country, until late in the nineteenth century. But once the population started to expand rapidly, the law or custom of equal inheritance led to significant levels of land fragmentation, a problem that is fundamental to many of the contemporary conflicts.

Under the Late Company Raj (1830–1857), the governing order as described remained fundamentally unchanged. There were changes, of course, but they were incremental rather than revolutionary. One key change was the growth of cash crops. Opium was a reliable commodity with stable prices; it was grown primarily to export to China, and large areas of Bihar and the United Provinces turned to its cultivation. Indigo prices were fickle and farmers were unwilling to cultivate it, so it was grown mainly by European landowners using coercive tactics. This was one instance of the "plantation economy" system, in which swaths of land originally used for growing food or kept fallow were turned over to the production of cash crops (often, as in the case of indigo, over the protests of the cultivators). Other major plantation crops introduced by the Late Company Raj were tea (in Assam in the 1830s) and coffee (in Kerala and Tamil Nadu in the 1840s). Overall, all these cash crops together covered a relatively small proportion of the total land under cultivation.

The late Company Raj period was a very profitable one for the Company despite the fact that much of the period (starting in the late 1820s and going into the early 1850s, almost uninterrupted) was deflationary for agricultural products. There were several reasons for the profitability.

First: Wars for control of the subcontinent had ended. True, peasant uprisings took place frequently, but they were isolated events and were crushed quickly and without hesitation. Kathleen Gough (1974) estimates that there were 77 peasant uprisings in colonial India. Many took place outside the period under review (such as the Tebhaga Andolan in Bengal and the Telengana Uprising in Andhra in the 1940s), but many took place during the rule of the Company Raj (such as the Moplah Rebellion of 1836) or were caused by its actions (such as the Indigo Rebellion in 1859–1860). *Second*: More and more marginal land was brought under cultivation. Chaudhuri notes (1983, p. 136): "While at the time of the Permanent Settlement barely 30 to 35 per cent of the available land was cultivated, the percentage by the end of the nineteenth century was seldom less than 75 to 80, except in some districts such as Nadia, Bankura, Birbhum and Champaran." *Third*: The revenue demand continued to be very high. For instance: in Puri, between 1842 and 1852, a peasant had to pay over 90 per cent of his rice production as rent; in Assam, between 1824–1825 and 1849–1850, the revenue demand increased by over 480 per cent; in the south, as noted earlier, the demands were variable but as high as 80 per cent of an individual peasant's production in some places. There were several other issues of importance—the increasing commercialization of agriculture and fragmentation of land being the most significant. I discuss these issues in the next section.

7.6 Crown Raj

The Sepoy Rebellion in 1857 led to significant transformations in the Raj. Not only was authority officially taken from the Company and assumed by the Crown, but there were major ideological changes that affected governance. It is not possible, in this space, to detail the many significant changes then taking place in British society and thought, which inevitably guided British actions in India, but it is useful to recognize that Britain itself was a society undergoing great and fundamental change.

Britain was in the full swing of industrialization; Manchester and Birmingham were the factories for the world, and the slums and tenements there and in London were new and, for many, shocking phenomena (later these came to be called "shock cities"). *The Communist Manifesto* of Marx and Engels was published in 1848 and Wordsworth's romantic masterpiece *The Prelude* in 1850. Empire and industry were creating great wealth, but it sat cheek by jowl with numbing visible poverty; ideas on free enterprise, individualism, and rationality competed with notions of social justice and collective action. The polarities that are our daily information inputs today originated in those tumultuous and transformative decades in England.

I do not mean to suggest that the Crown Raj was a kinder, gentler system than the Company Raj. Exploitations and atrocities abounded. Among the first acts of the new Raj, after the brutality of the end of the Sepoy Rebellion, was the suppression of the Indigo Rebellion in central Bengal. The Deccan riots of 1875 were put down just as mercilessly. Ideas about British racial, cultural, and moral superiority imbued the everyday actions and long-term thinking of the Raj.

But the Sepoy Rebellion was a jolt that forced the Crown Raj to think afresh about the size of the land revenue burden and the rights of raiyats and tenant farmers and sharecroppers, that is, the vast majority of the Indian population which worked the land but had little or no "ownership" right to it. David Washbrook writes (1981, p. 685): "The consequences which the British Indian civil service, at various times, claimed to fear most from a competitive capitalist conquest of agriculture were a decline in the land revenue, a link up between the wrath of dispossessed peasants and the emergent nationalist movement and a general collapse of political order leading to mass revolt." To simplify, there were two contradictory ideas at the fore: to maximize revenue from land at the same time that conditions were made better for the actual cultivators.

The conditions for the actual cultivators had been getting worse. The commercialization of Indian agriculture had taken a big step with the creation of land markets during the first decades of the Company Raj (as already discussed). The process continued with the gradual conversion of all revenue payments to cash and later with the expansion of credit markets. In Bihar, for instance, the old batai system (a division of the actual crop) was replaced by the danabandi system (a division of the estimated crop) at the same time that zamindars increasingly demanded payment in cash at market rates at their time of choosing. This demand was compounded by the problem of indebtedness to moneylenders.

Chaudhuri (1983, p. 144) quotes one Mr. Metcalfe, the commissioner of Patna, as writing that "the agriculturist regards a village without its moneylender as an abnormal state [of] things." The village moneylender was needed all the time because the tenant farmer, who had no ownership rights to the land and had to pay a significant majority of his output to the zamindar, got little or no help from the zamindar in acquiring the inputs necessary for production (tools, seeds, water). The landless cultivator had to endure legal coercive tactics by the zamindar, including imprisonment, banishment, and physical beatings (much of which would qualify as torture in contemporary ethics). The zamindar in turn had the right to extract revenue but not any responsibility for providing input into production. The moneylender provided credit, especially when times were better (that is, commodity prices were high), and less so when times were worse (especially during the years between the two world wars that saw a global economic depression and falling commodity prices in India).

The debt situation was bleakest where markets were most established—where it was legal to transfer ownership of small land parcels (that is, the cultivator had some proprietary rights and could use land as collateral for credit), and the commodity market was commercialized and profitable rather than subsistence based and marginal (that is, the peasants were more credit worthy). In such situations, "the moneylender's concern was no longer limited to acquiring profits as a financial intermediary, but was directed increasingly to the acquisition of and speculation in land" (Chaudhuri 1983, p. 146). We should note that the proprietary rights that some tenants had acquired had come about as a result of the dilemma faced by the Crown Raj after the Sepoy Rebellion. The tug of war between zamindar rights (which made revenue extraction easier) and tenant rights (which made the populace more manageable) swung this way and that and was expressed in laws and court cases (some legendary) throughout the Crown Raj regime.

We should also note (ironically) that the very expansion of tenant rights led to expanded markets in land. That is, the first land for which a market was created was estate land, for example, large zamindaris such as the Burdwan estate that had more than 3,000 separate parcels. The expansion of tenant rights now began to create a market for small parcels of land. This market was not, however, supported with a formal credit market or a banking system, which led to the growth of the informal credit system of moneylenders and, almost inevitably, significant quantities of distress sales by indebted small cultivators. Finally, it should be noted that during the 1930s even the moneylenders were in trouble. The Kisan movement had generated "class hatred" and several of the vilest moneylenders were murdered. With the general depression in agricultural prices, creditors by the thousands were unable or refused to pay. And a number of legislations to curb the power of moneylenders were enacted, such as The Bengal Moneylenders Act (1934), The Bengal Agricultural Debtors Act (1936), and the Bihar Moneylenders Acts (1938 and 1939).

The growth of tenant rights and credit markets led to increasingly complex hierarchies of tenancy. The basic feudal system had the zamindar on top and the tenant farmer below him. But this basic feudal system was infused with what has been called sub-infeudation, in which the original tenant was frequently an absentee tenant, where the actual cultivation was done by subtenants or sharecroppers or other landless labor. It was not uncommon to have subtenants of subtenants, whereby between the actual cultivator and the zamindar there were several layers of intermediaries. This condition of sub-infeudation and multiple intermediaries between grower and revenue collector would become one of the most serious problems that had to be tackled at independence.

Overall, however, there appears to be little doubt that at least during the early Crown Raj, the small cultivator in the raiyatwari regions of the south and west fared better than in the zamindari regions of the east and north. The third quarter of the nineteenth century is generally seen as a period of growing prosperity in the raiyatwari lands, until the devastating famine of 1876–1878. Tenant farmers kept half the crop in dry lands, and less (between one fifth and one third) in the more productive irrigated areas. The distribution of land was unequal but remained remarkably stable from the 1850s through the 1940s, but this is not to suggest that all was well in the raiyatwari lands. Physical intimidation and violence were routinely used for revenue collection in the mid-1800s. Tenant farmers routinely fell into debt traps, as in the zamindari lands, especially later in the inter-war years of the twentieth century, and clashes with moneylenders were common. The caste system was solidified, and marks of "social superiority" permeated all social and economic interactions.

These ground-level incremental changes took place in the context of large structural shifts in the colonial economy, especially during the late Crown Raj period, which had major implications for the state–peasant–land relationship. First, land revenue was no longer the driver of colonial policy. Second, a significant growth in population created more pressure on the tenant and landless class. Let us consider both phenomena.

Between 1840 and 1880 the share of land revenue in total revenue fell by about one third, from about 60 to about 43 %. Between 1880 and 1920 this share fell by about one half, from 43 to 23 %. Land revenue was simply not as important as it used to be for the Crown Raj. The same became true from the perspective of the revenue payer: in the south, the share of land revenue in total agricultural output fell to less than 10 % by the late 1890s and by the time of the inter-war years it was as low as 4 or 5 %. In the east this share fell below 20 % by the First World War and dipped below 10 % during the 1930s. This change made agriculture more profitable and agricultural land more desirable and valuable.

At the same time, a very significant demographic transition had begun. Between 1800 and 1900, India's population was virtually unchanged. There were small spikes in the 1860s and 1880s, but these were reduced by high mortality rates, usually caused by famine, in the decades immediately following. But from the 1921 census began a long upswing in total population, from 251 to 361 million in the 1951 census, a growth of 110 million people, or more than two fifths of the original population. With little new land available that could be brought under cultivation, with Hindu family law proscribing the division and subdivision of land from generation to generation, this population increase led to increasing land fragmentation and demand for land.

These two factors—the declining significance of the revenue demand from land and the rising significance of demand for land from a growing population—became the driving forces of the evolving market for agricultural land. In combination with the increased protection of tenant rights (which intensified the sub-infeudation process discussed earlier and created more agents or participants in land transactions), these factors drove up the value of land and created land markets of increasing complexity.

7.7 Land Acquisition

It is interesting to note that in everything that has been discussed so far, in all the readings and scholars that have been cited (and those that have not), the 1894 Land Acquisition Act is not mentioned even once. It seems to have been irrelevant, certainly in relationship to the larger structural issues of taxes and rights, but it was also not an act that affected large numbers of people. The colonial state was not a developmental state; it did not undertake many large development projects that required a large amount of land. Hence, the colonial land acquisition act was not used as much by the colonial regime as it came to be in independent India.

The origins of eminent domain law in India can be traced to the colonial state's need to create infrastructure to facilitate the movement of goods and people and enable commerce; these first "public works" were typically canals and roads, later came railways, mines, and irrigation schemes, and even later came factories and other business establishments. One of the main reasons to use eminent domain, then and now, is to get the needed land cost-effectively. Another, possibly even more important reason, is to get it quickly by avoiding protracted negotiations with numerous small landholders (including "holdouts" or owners waiting for better offers) and sidestepping the legal problems of sorting out the considerable ambiguities about who "owns" what. Therefore, eminent domain was and remains essential to collate multiple properties and to own them "free and clear" of legal encumbrances. The justification for the taking has always been "public purpose," an all-encompassing term whose meaning and ambit have been debated in the courts from the very beginning.

The first law relating to land acquisition was Regulation I of the Bengal Act of 1824. It enabled the state to acquire land for public purpose at a "fair and reasonable" price. When the first seven sections of Regulation I were extended to all lands within the town of Calcutta by Act I of 1850, it included a rule that a declaration by the Governor of Bengal that land was needed for a public purpose was sufficient evidence that the purpose was public. Act XX of 1852 and the Amending Act I of 1854 maintained the state's absolute authority to determine public purpose. Act VI of 1857 repealed all existing laws on land acquisition and compensation and laid out a new policy for all of British India. A secretary to a local government could now decide what land was required for a public purpose. The compensation award of arbitrators, for which the law did not provide any guidelines, could not be challenged except on the grounds of corruption or misconduct, nor were the arbitrators required to explain or justify their method of calculation. Act X of 1870 addressed some of these problems by providing detailed rules for the assessment of compensation and creating a system through which the power of the assessors, who were often incompetent and corrupt, was diluted. In between, Act XXII of 1863 made the first provision for the government to acquire land on behalf of private persons or companies (Aggarwala 2008; Fish 2011).

The culmination of all these legislations was the Land Acquisition Act of 1894 (LAA), which provided the definitive framework for determining (a) the conditions under which land may be acquired and (b) a methodology for compensating for the acquisition. It is worth noting that these two remain the core issues on land acquisition even today (Chakravorty 2011).

7.8 The Colonial Land Legacy

How do we evaluate this long and complex period in Indian history? One can take a nationalist perspective and argue that colonialism denuded the country of its wealth and dignity, imposed alien values, and left it poorer than it would have been had it been left to its own devices. No doubt these views are correct from a national perspective. Britain was enriched with the surplus transferred from India. The British Empire outside the subcontinent was conquered and maintained by the British Indian Army, which by the early twentieth century consumed 40 % of the Indian budget. New institutions such as the judiciary brought in new values (such as protection of private property rights) and enshrined into law some "invented traditions" of India (Manu's laws, the "Hindu" family laws). There were interventions in every sphere, always based on self-interest and often based on ignorance.

But if we narrow our scope to the peasantry and land, it is not obvious that the colonial system created outcomes that were inferior to what existed before (and presumably would have continued without the intervention of colonialism). On the question of taxes there is little doubt that the colonial regimes were almost never more burdensome than the pre-colonial regimes. In the initial decades, through much of the Company Raj period, the revenue rates were high, more so in the zamindari areas than in the raiyatwari areas, but during the Crown Raj period, as a proportion of the total income from land, these declined significantly. In the provision of public goods—such as irrigation and transportation—the evidence appears to favor the colonial regimes. In the provision of public safety and the conditions for safe travel and commerce, the colonial regimes were superior, certainly to the tumult of the eighteenth century. Famines were frequent and devastating (several took the lives of millions of people, sometimes as much as a quarter or a third of the affected region), but were they more frequent or more devastating than famines in pre-colonial regimes? Very doubtful.

The fundamental difference between the pre-colonial and colonial regimes was the creation of a proto land market. The process began with the Permanent Settlement which created a market in estates; by the end of the colonial period, with increasing rights granted to tenant farmers (in both zamindari and raiyatwari lands) incipient markets had been created for small land parcels. One can view this development through an ideological lens—as many appear to have done—and conclude that because markets are inherently inferior to "traditional" non-market transaction systems, the colonial interventions led to unquestionably inferior outcomes.

Alternatively one can take the view that "ownership" rights, however weak and tenuous, are superior to "traditional" rights, which, more often than not, are fickle and arbitrary. That is, a peasant is likely to have more power in a market system than in a non-market feudal system. The image of the pre-colonial wholesome, communitarian, organic village society is a romantic myth. The proto market system in land with which the colonial regimes ended had very serious problems (of inequality of land distribution, peasant indebtedness, and largely because of population pressure without commensurate technological change, increasing fragmentation and marginality), but the question we should focus on is this: what rights did the peasant have relative to the established authorities, the zamindars and mirasdars, and above them, the colonial authority?

The impoverished and embattled Indian peasant of the pre-colonial and early colonial periods had little bargaining power and few avenues of resistance. They abandoned farms when it was still possible to do, when the land abundance of the pre-colonial period had not yet been transformed into the land scarcities of the late colonial regime. They rose up in isolated rebellions when conditions became intolerable and they were able to organize under good local leadership. They randomly attacked some zamindars and moneylenders in less organized, more sporadic ways.

But none of these forms of resistance led to any systematic rights until gradual changes were legislated (and supported by the judiciary) in the late Crown Raj. As a result, when the colonial regime ended, the Indian peasant had more agency, and a greater ability to act in his own interest, than ever before. There are many possible explanations for why this happened (I have discussed some of the most plausible ones earlier), but those explanations matter less than the fact that this expansion in peasant and tenant rights did happen.

The importance of individual rights in land has been given a new perspective by scholars who have compared the very long term effects of the zamindari and raiyatwari systems, the former with fewer rights and greater exploitation than the latter. The original work was done by Banerjee and Iyer (2005), who investigated differences in agricultural productivity between zamindari and raiyatwari lands in independent India. Kapur and Kim (2006) extended their analysis to all the twentieth century, including the last decades of the colonial regime. These scholars argue that the different tax and property right regimes in these regions created fundamentally different incentive structures for individual cultivators, which led, in the twentieth century, to significantly better development outcomes and patterns in the raiyatwari regions despite the fact that the zamindari regions had superior agricultural land.

How different were the agricultural outcomes? Although in 1901 the productivity levels in both regions were on par, by 1931 the raiyatwari lands had 22 % higher productivity. This difference narrowed in the two decades after independence, but after the Green Revolution, by 1981, the raiyatwari lands had 26 % higher agricultural productivity (and higher levels of urbanization and industrial production). Scholars see this as an outcome of regional differences in colonial policy. But, given (as shown earlier) that colonial policy was pragmatic and largely a continuation of pre-colonial revenue systems, it is possible that the long shadow of Mughal India can be discerned in the economic life of the nation even today.

7.9 Conclusion

In general, there is little doubt that the political economy of the independent nation was shaped in fundamental ways by colonial policies on land. It is important to understand that the original goal of the colonizers was to extract land revenue. For well more than a century it remained the primary and overwhelming objective and, therefore, the thrust of all colonial policy. Industrial colonialism came later, by which time the land policies were firmly entrenched. This was what independent India inherited. The inequalities and deprivations of the colonial systems, especially zamindari, required large-scale reforms—first in the form of the abolition of intermediaries, followed by land ceiling laws and tenancy reform laws. These were among the first actions taken after independence—specifically through the First,

Fourth, and Seventeenth Amendments to the Constitution. The effects of these various reforms have been debated at length and the conclusions are sobering. The abolition of intermediaries was the only unequivocally successful reform; tenure and ceiling reforms have generally had adverse effects in most states (with the exception of Kerala and West Bengal, the only two states that undertook these reforms seriously) and have increased land inequality and fragmentation and decreased agricultural productivity (Chakravorty 2013; Ghatak and Roy 2007).

What the independent Indian state also inherited from the colonial period, an inheritance that was not reformed because it was convenient for the Indian state, was the law on land acquisition. Because the new nation required very large quantities of land to launch its modernization and industrialization policy—for dams, irrigation, steel plants, power plants, townships, and so on—it retained unchanged the Land Acquisition Act of 1894 (which was amended twice, in 1962 and 1984, both times to enlarge the power of the state). This law had been used relatively sparingly by the colonial state. It was independent India that used this colonial act massively for its own purpose. What the nation is facing now is the fallout of post-colonial India's use of colonial legislation on land acquisition.

Land was viewed through much of Indian history as a source of revenue. These revenues sustained the ruling classes, and the desire to maximize revenue was the source of almost all conflict between regimes. That fundamental fact of India changed in the twentieth century as the significance of land revenue declined in relationship to revenue from industrial production and trade, and it has disappeared entirely in independent India. Land struggles are no longer over revenue, largely because of the extremely low productivity and income from agriculture. Instead they are about land use, as a modernizing and urbanizing nation attempts to take land from an intensively fragmented and marginalized agricultural class. This is the story of land in India today.

References

Aggarwala OP (2008) Commentary on the Land Acquisition Act, 8th edn. Universal Law Publishing, New Delhi

Athar Ali M (1975) The passing of empire: the Mughal case. Mod Asian Stud 9:386–396

Banerjee A, Iyer L (2005) History, institutions and economic performance: the legacy of colonial land tenure systems in India. Am Econ Rev 95:1190–1213

Chakravorty S (2011) A lot of scepticism and some hope. Econ Polit Weekly 41:29–31

Chakravorty S (2013) The price of land: acquisition, conflict, consequence. Oxford University Press, New Delhi

Chandra S (1982) Medieval India: society, jagirdari crisis and the village. Macmillan, Delhi

Chaudhuri B (1983) Agrarian relations: Eastern India. In: Kumar D, Desai M (eds) The Cambridge economic history of India. Volume 2, c. 1757–c. 1970. Cambridge University Press, Cambridge

Fish C (2011) Land acquisition for special economic zones in India. Unpublished Master's thesis. Temple University, Philadelphia

Fukazawa H (1983) Agrarian relations: western India. In: Kumar D, Desai M (eds) The Cambridge economic history of India, vol 2, c.1757–c.1970. Cambridge University Press, Cambridge. doi:10.1017/CHOL9780521228022.005

Ghatak M, Roy S (2007) Land reform and agricultural productivity in India: a review of the evidence. Oxf Rev Econ Policy 23:251–269

Gough K. (1974) Indian peasant uprisings. Econ Polit Weekly (special number, August)

Guha R (1963) A rule for property in Bengal: an essay on the idea of the permanent settlement. Moulton and Co., Paris

Habib I (1982) The systems of agricultural production: Mughal India. In: Raychaudhuri T, Habib I (eds) The Cambridge economic history of India, vol 1. Cambridge University Press, Cambridge

Kapur S., Kim S. (2006) British colonial institutions and economic development in India. Paper W12613. National Bureau of Economic Research

Kumar D (1983) Agrarian relations: South India. In: Kumar D, Desai M (eds) The Cambridge economic history of India. Volume 2, c. 1757–c. 1970. Cambridge University Press, Cambridge, UK

Pearson MN (1976) Shivaji and the decline of the Mughal empire. J Asian Stud 35:221–235

Rana RP (2006) Was there an agrarian crisis in Mughal North India during the late-seventeenth and early-eighteenth centuries? Soc Sci 34:18–32

Sarkar J (1952) History of Aurangzeb, vol III, 3rd edn. Orient Longman, Calcutta

Swamy AV (2011) Land and law in colonial India. In: Ma D, Van Zanden JL (eds) Long-term economic change in Eurasian perspective. Stanford University Press, Palo Alto, pp 138–157

Washbrook DA (1981) Law, state and agrarian society in colonial India. Mod Asian Stud 15:649–721

Chapter 8
The Geographic Distribution of Land Trust Activities in the United States: An Analysis Based on 2005 National Land Trust Census Report Data

Ronald C. Hess and George M. Pomeroy

Abstract This study is concerned with the geographic patterns of land conservation strategies employed by the 1,667 land trusts included in the Land Trust Alliance's 2005 National Land Trust Census. The spatial distribution of land trusts, the number of acres owned, acreage under conservation easement, and the total number of acres protected by all means by the various land trusts are considered, and these are in turn mapped, analyzed, and discussed: this is done first by utilizing sum totals and then again by using location quotients to find patterns of concentration relative to the United States. Possible causal factors for the spatial distribution patterns found such as per capita income and population are also examined. Several three-dimensional visualizations are presented to offer a unique perspective as well as to facilitate an increased understanding of certain patterns of concentration than could be achieved from traditional mapping techniques and location quotient tables alone.

Keywords Land trust • Geographic distribution • Location quotient • Per capita income • Spatial visualization • USA

8.1 Background

Air and water are considered common property resources that cannot be owned because they provide benefits to all. Land is different in that it can be owned, but not so different in that it can provide many public benefits regardless of whom owns it. Obviously, the public derives benefits from certain types of land, such as farmland and ranchland that produce food, forestland that provides wildlife habitat as well as adding to our oxygen

R.C. Hess (✉) • G.M. Pomeroy
Department of Geography-Earth Sciences, Shippensburg University of Pennsylvania, Shippensburg, PA, USA
e-mail: rhess3@gmu.edu; gmpome@ship.edu

A.K. Dutt et al. (eds.), *Spatial Diversity and Dynamics in Resources and Urban Development*, DOI 10.1007/978-94-017-9771-9_8

supply, and wetland that improves water quality. Perhaps less obvious, however, are the benefits gained from simply the sheer beauty of scenic landscapes, historic places, and open space. Negative externalities that go along with development, particularly sprawl development, are the loss of many of these public benefits. Although land use regulation may seem like the obvious choice to protect these public benefits and manage growth, such measures have had only limited success as many areas do not have zoning and thus comprehensive plans go unenforced (Daniels 2004). For reasons such as these, there has been a shift known as 'the land trust' toward another way of protecting these benefits now and for future generations (Daniels 2004).

Land trusts are nonprofit organizations that conserve land in perpetuity for public benefit. That is, they protect land from development by acquiring it through fee simple purchases, as donations, or by acquiring only specific development rights associated with the land through purchase or donation without actually taking ownership of the land itself, also known as a conservation easement (Brewer 2003). Tax breaks may be awarded to property owners who donate a conservation easement to a non-profit (501C3) land trust (Yuan-Farrell et al. 2005). Strangely, however, the holder of a conservation easement does not have the right to develop the land, because conservation easements differ from traditional easements, such as those held by utility companies, which allow the holder a specific use of the property; conservation easements simply deny the owner of the land certain uses (Gustanski and Squires 2000). This provision also creates a burden for land trusts to monitor the land in perpetuity and to ensure that the terms of the conservation easement are kept throughout the generations, because once the specific rights associated with the conservation easement are severed from the bundle of usual property rights, they remain with the conservation easement, and the land trust that holds such, regardless of changing land ownership or time passed (Land Trust Alliance 2009).

Land trusts, which have been around since 1891, exist on several geographic scales, and conserve various land types. They remained relatively few in number before 1950, with only 53 in existence at that time (Trustees History 2009). By 1980, they had increased in number to 431, an eightfold increase (Brewer 2003). Today, there are 1,700 private land trusts with more than two million members that have collectively conserved 37 million acres of land in the United States (Land Trust Alliance 2009): this equates to an area roughly the combined size of Pennsylvania, Maryland, and Delaware (U.S. Census Bureau 2009). Most of these 1,700 private land trusts are local in scope, operating at the level of several counties to small villages. Others, such as the Trustees of Reservations, which also has the distinction of being the first land trust, operates at the Massachusetts State level. Still others, such as The Trust for Public Land (2009), which conserves urban, rural, and historic sites for public enjoyment, and The Conservation Fund (2009), which conserves a variety of land types in all 50 states, are national in scope. International Land Trusts exist as well; these include the Nature Conservancy (2009), which focuses on ecological land conservation and operates in more than 30 countries, and Ducks Unlimited (2009), which conserves wetlands and waterfowl habitats throughout North America (Brewer 2003).

8.1.1 *The Land Trust Alliance*

The Land Trust Alliance (LTA), based in Washington, DC, is an umbrella organization that unites land trusts by providing operational standards intended to ensure that a land trust endures over the long term. The Alliance also advocates favorable tax policies as well as providing publications, training, and support (Land Trust Alliance 2009). Key components in the LTA operational standards, to which most land trusts have agreed, are rules of conduct, increased budget requirements, and stewardship endowments, necessary for long-term compliance monitoring. Although some land conservation organizations consider themselves land trusts, the Land Trust Alliance has a very specific definition that it uses to determine if a land conservation organization is actually a land trust: "A land trust is a nonprofit organization that, as all or part of its mission, actively works to conserve land by undertaking or assisting in land or conservation easement acquisition, or by its stewardship of such land or easements" (Land Trust Alliance 2009).

8.2 Literature Review

Existing studies concerning land trusts and conservation easements tend to be of an informational nature or concerned with specific topics such as public access, goals, legislation, tax benefit, permanence, or another specific purpose (Brewer 2003; Lieberknecht 2009; Kiesecker et al. 2007). Yet, surprisingly, even at the national level no studies were found that examined the geographic patterns of land trust activities such as the spatial distribution of a land trust, acres owned, acres under conservation easement, and total acres protected by land trusts. Studies of such patterns could provide clues or links to causal factors, thus improving our understanding and perhaps lead to additional studies.

Although no specific studies exploring the geographic patterns of these land trust activities were found, two studies had some relevance. A study by Yuan-Farrell et al. (2005) used statistical analysis to search for factors contributing to the spatial distribution of conservation easements in California, and conservation easement preference patterns were mapped. Although only regional in nature, the study does address patterns of conservation easements, albeit based more on contributing factors. Another study by Wikle (1998) identified spatial patterns in concentration of membership among The Nature Conservancy, Natural Resources Defense Council, and the Wildlife Fund throughout the U.S. Using county-level location quotients, rankings, and mapping, membership patterns among these organizations were examined. The study found similar strong membership concentrations in the Northeast, West, and Rocky Mountain states and similar weaker membership concentrations in the Midwest and South. Based on these concentration patterns, conclusions were drawn that included the suggestion that higher levels of income and education correspond to higher membership ratios, and similar patterns may exist for other environmental organizations. Using similar techniques, this research aims to reveal the geographic patterns and concentrations of specific land trust activities, as well as to consider possible causal factors contributing to such patterns.

8.3 Data

The Land Trust Alliance conducts a census every 5 years that tracks trends in private land conservation at the national level. This census records various attributes connected with private land conservation such as amounts and type. Some of the main items surveyed, and the main items of interest in this study, include the number of land trusts and the year they were founded, the number of acres owned by land trusts, the number of conservation easements acres, and the total number of acres protected by all means. Other items, although not an exhaustive list, include the number of board members, operating budget, whether the land trust practices stewardship, whether the land trust seeks to acquire land or conservation easements, and the primary land type sought for conservation.

These data were collected by the Land Trust Alliance during the first 8 months of 2006 by a survey that was conducted online and by mail of 1,840 land conservation organizations. Respondents were instructed to provide information up to the last day of 2005. Based on the Land Trust Alliance's definition of land trusts mentioned previously, 173 of these organizations were determined not to be land trusts, leaving 1,667 land trusts in the final census report. Of these, more than 940 responded. Additional information was supplemented beyond the survey results by other means such as telephone, e-mail, and land trust support centers and the use of previously collected data in the absence of new data.

In addition to the census dataset, other data were gathered to facilitate the study objectives, including additional boundary files: "counties," and "cities and towns," which were obtained from the GIS Data Depot and published by the U.S. Geologic Survey in 1999, and "zipcodes," shapefiles obtained from the U.S. Census Bureau 2007 TIGER/Line® Shapefiles. These data were necessary to attach the survey data to spatial features. Additionally, per capita income by state from 1980 to 2005 was downloaded in table form from the U.S. Department of Commerce, Bureau of Economic Analysis, and recent population statistics that included the year 2005 were downloaded in table form from the U.S. Census Bureau to facilitate some comparative analysis.

8.4 Methods, Results, and Discussion

The 2005 National Land Trust Census Data were provided in Microsoft Access format and were directly imported into an ArcGIS geodatabase. Data were joined to cities and towns, zip codes, counties, metropolitan statistical areas, regions, and states, layers using a series of joins, sum totals, data exports, and other traditional methods. Income and population data were also imported and joined using similar methods.

8.4.1 *Land Trusts: Density*

A dot density map showing land trusts per state in 2005 was generated as a first step to look for general density patterns (Fig. 8.1). This map clearly shows that the highest densities of land trusts are found on the East and West coasts with the Northeast having the highest density overall and California, while not quite as dense, clearly having the highest density among the Pacific States as well as the entire West. What also stands out on this map is the relatively few land trusts present in the northern central U.S. as well as the central portion of the U.S. in general. Also of note is the low density of Alaska, which although not shown to scale covers an area larger than Texas.

8.4.2 *Geographic Patterns*

Next, three choropleth maps were produced to search for factors that might be associated with the spatial pattern of land trusts. These, as most maps produced in this project, use a quantile classification scheme consisting of three classes. This particular classification scheme was chosen because it is easy to understand conceptually, facilitates easy comparison even when different color schemes are used, and as 51

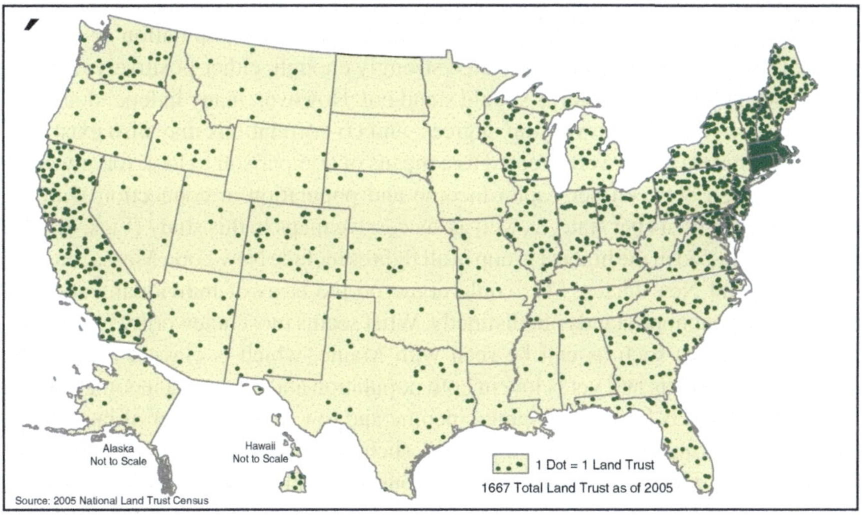

Fig. 8.1 Density of U.S. land trusts in 2005

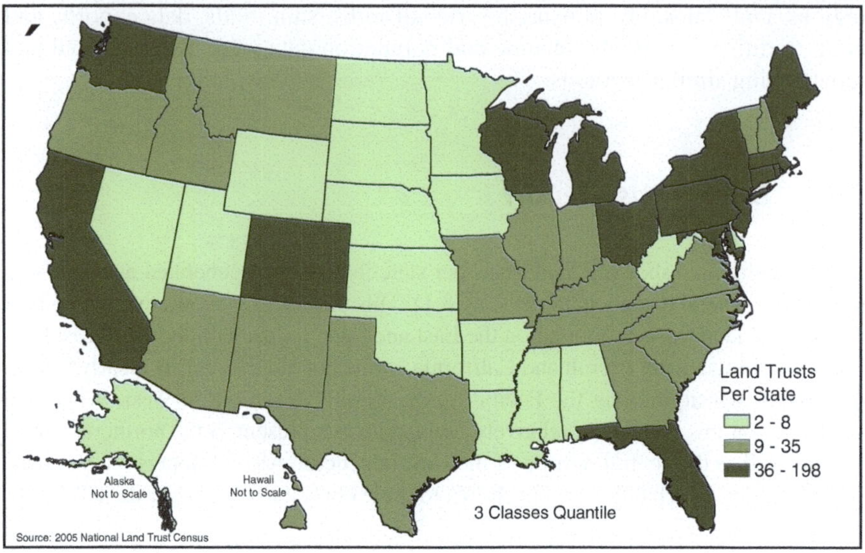

Fig. 8.2 Number of land trusts per state

divides equally by 3, 17 areal units fit perfectly into each tertile. This classification scheme is similar to the one in the study by Yuan-Farrell et al. (2005), which used choropleth maps with three classes to show the spatial pattern of densities for conservation easements in California. The first of these choropleth maps shows the number of land trusts per state (Fig. 8.2).

Obvious patterns exist on this map, which compares well with the dot density map shown previously as they both represent the same data. This pattern suggests a strong association between other factors such as income or population or perhaps others. If only two factors are correlated strongly enough, either positively or negatively, such comparable patterns should stand out. However, many independent variables may be correlated to varying degrees, and co-correlations may also exist.

Two additional maps were created as a means of comparison to look for any obvious association between per capita income and population in connection with the number of land trusts per state, as well as successive maps in this study (Figs. 8.3 and 8.4). States classed in the highest group in all figures include New York, Massachusetts, California, and New Jersey. One could argue on the basis of individual states that patterns exist but none stands out distinctly. What seems most noteworthy is the obvious dissimilarities such as can be seen with Maine, which is classed high in the number of land trusts and yet is low in both population and per capita income. Alaska on the other hand is high in per capita income and low in the number of land trusts. Sometimes striking similarities are present, such as the one mentioned at the beginning of the preceding paragraph, but no obvious associations were found. Yet, factors such as population can be examined in other ways, such as the concentration of land trusts as compared to population, a concept that is addressed later.

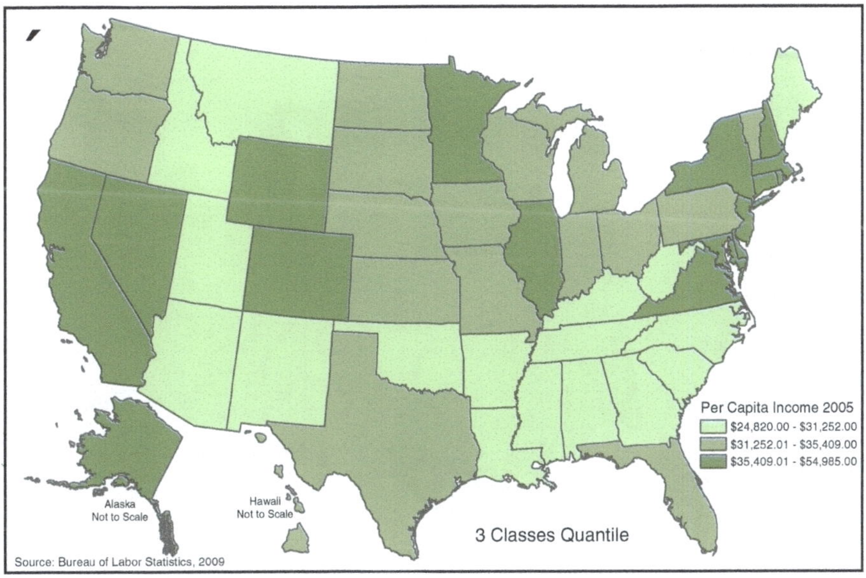

Fig. 8.3 Per capita income per state, 2005

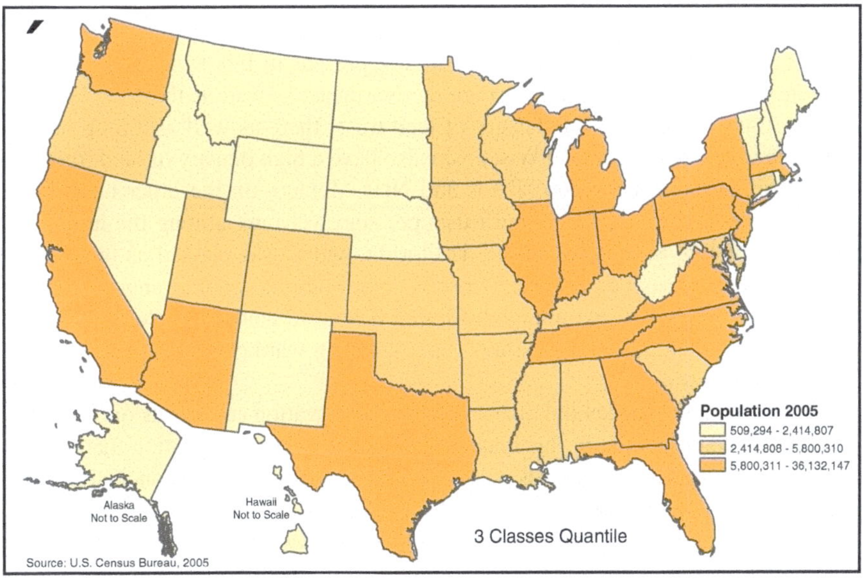

Fig. 8.4 Population per state, 2005

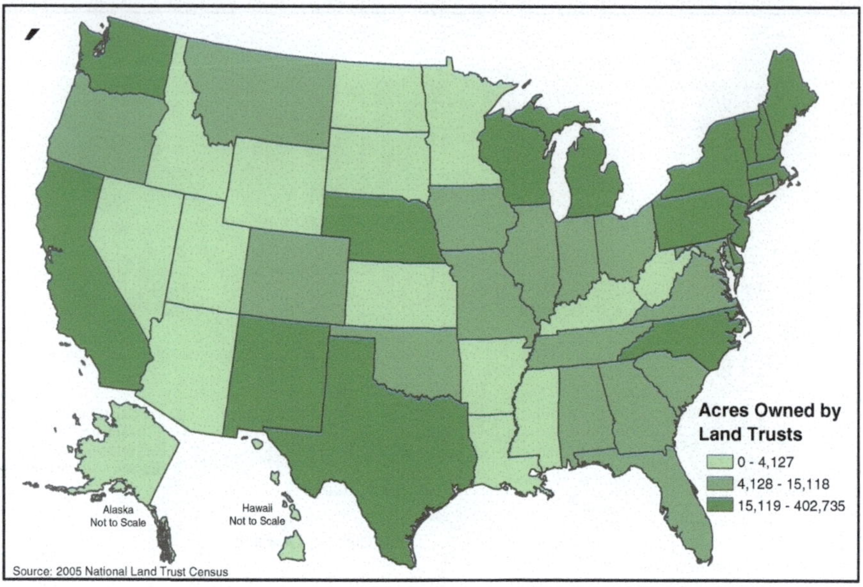

Fig. 8.5 Acres owned by land trusts

To consider acres owned by land trust, another map was created (Fig. 8.5). Not surprisingly, certain recognizable patterns are present in this spatial distribution. Many acres are owned by land trusts in the Northeast as well as the New England states because there is a high density of land trusts there as well as a long history with such trusts. Michigan and Wisconsin also have a high density of land trusts, as do California and Washington. Texas and New Mexico, on the other hand, have somewhere between 9 and 35 land trusts per state yet rank among the highest in acres owned (Fig. 8.2). Also notable is Florida, which was classed as high in the number of land trusts yet in acres owned is somewhere in the midrange classification. One possibility is that although Florida has more land trusts, less acreage is owned, which could also be a result of a preference toward conservation easements over land acquisition.

Figure 8.6 shows the spatial distribution of conservation easements only. Florida is shown in the midrange in this map as well, so the initial suggestion that Florida has more land trust and less acres probably is the better of the two conclusions suggested. The spatial distribution in Fig. 8.6 shows a cluster of states: Iowa, Missouri, Illinois, Indiana, and Oklahoma are classed in the lowest range for conservation easement acres, yet these same states are in the midrange for acres owned. Other states such as Arkansas and Kansas were also part of this cluster of states with the fewest conservation easement acres, but were also in the lowest class with respect to acres owned. Montana, which is shown in the highest class, is also unexpected as this state has only 14 land trust and many more acres under conservation easement than under ownership, suggesting a possible preference for conservation easements

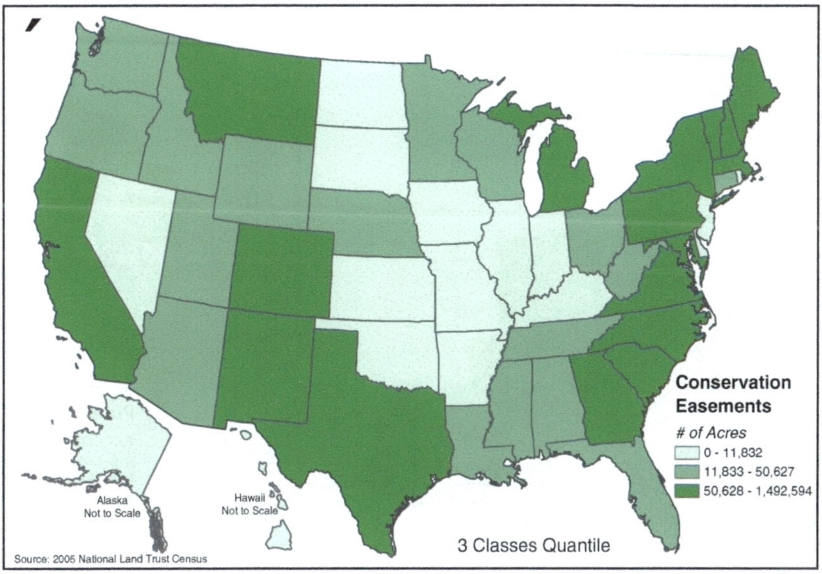

Fig. 8.6 Acres held in conservation easement by land trusts

here as well (Figs. 8.1 and 8.5). This conclusion seems to hold when looking at the total acres protected by land trusts, as Montana is also in the highest class there as well (Fig. 8.7).

When looking at total acres protected, it is hard not to notice the strip-like pattern that appears or seems to divide the country in half. This pattern coincides well with the dot density map shown earlier (Fig. 8.1), suggesting that not only are there few land trusts in these states, few acres are protected as well. Another pattern that has held for each three-classed choropleth map shown thus far is that Massachusetts, New York, and California have been in the highest class on each (Figs. 8.2, 8.3, 8.4, 8.5, 8.6, 8.7), which means that these three states were highest in the number of land trusts, per capita income, population, number of acres owned, number of acres under conservation easements, and total number of acres protected by all means.

Noticing that these three states were highest in income, one could argue that there appears to be a distribution pattern reflecting a spatial bias toward the New England and Mid-Atlantic states and California on many of the maps produced thus far. In light of this, it seemed prudent to create a map that showed the spatial distribution of the ratio of acres owned by land trust to per capita income (Fig. 8.8). A table was also created and ratio values were ranked in three reverse-order quantiles (Table 8.1).

This map is almost identical to the map created for acres owned. In fact, the only two states with a different classification are Idaho and Colorado, whose classifications have been reversed. It was tempting to claim that there must be some association between the variables “acres owned” and “per capita income, but a closer examination

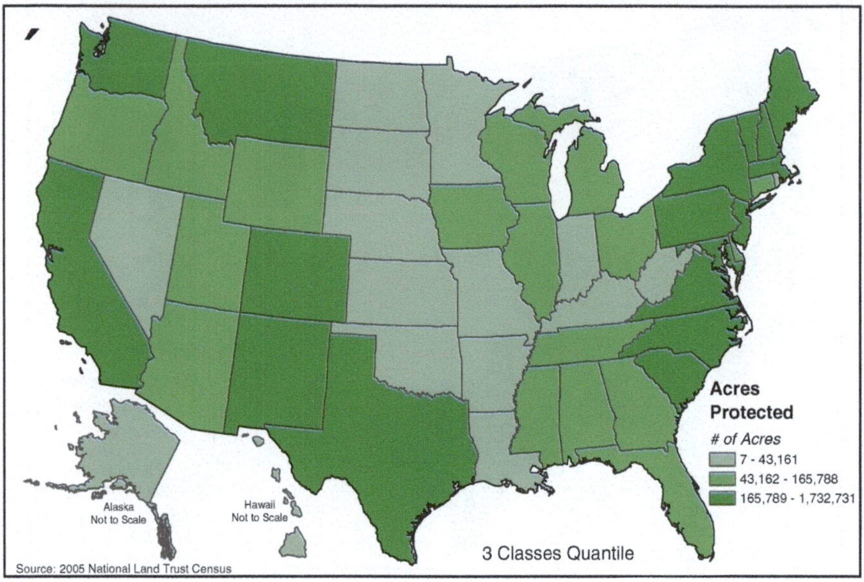

Fig. 8.7 Total acres protected by land trusts through all means

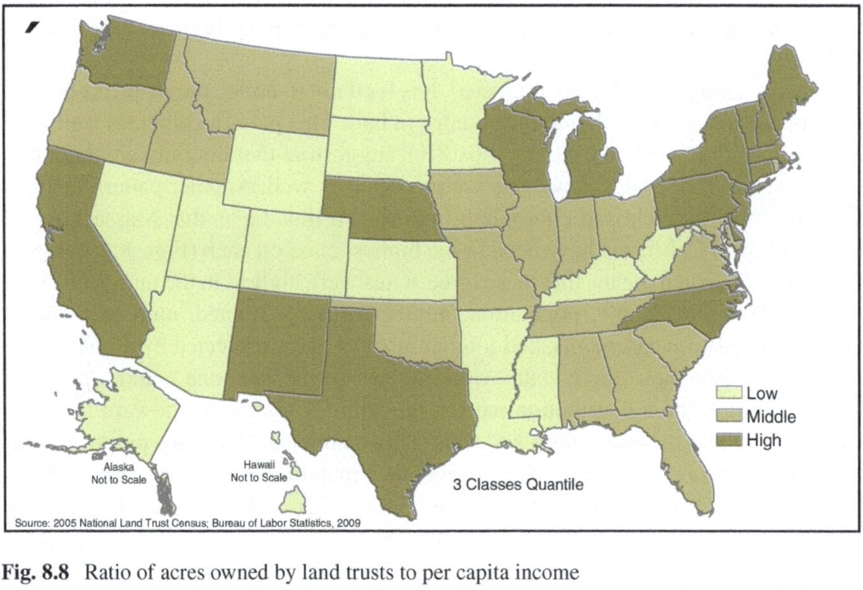

Fig. 8.8 Ratio of acres owned by land trusts to per capita income

Table 8.1 Acres owned by land trusts in state 2005 divided by per capita income, 2005, ranked in three reverse-order quantiles

Highest quantile			Middle quantile			Lowest quantile		
State	Ratio value	Rank	State	Ratio value	Rank	State	Ratio value	Rank
New Mexico	14.5686	51	Rhode Island	0.4182	34	Colorado	0.1271	17
California	8.2188	50	Tennessee	0.4159	33	Utah	0.0799	16
New York	4.1538	49	Iowa	0.3836	32	Kentucky	0.0604	15
Maine	2.6966	48	Indiana	0.3755	31	Minnesota	0.0581	14
Massachusetts	2.6754	47	Ohio	0.3504	30	Arkansas	0.0379	13
Vermont	2.0765	46	Missouri	0.3317	29	Arizona	0.0373	12
New Hampshire	2.0756	45	Virginia	0.3161	28	Mississippi	0.0309	11
New Jersey	1.2649	44	Montana	0.2992	27	Alaska	0.0297	10
Connecticut	1.0529	43	Florida	0.2906	26	West Virginia	0.0267	9
Pennsylvania	1.0153	42	South Carolina	0.2706	25	Wyoming	0.0180	8
Michigan	1.0063	41	Oregon	0.2318	24	Nevada	0.0137	7
Texas	0.8809	40	Illinois	0.2134	23	Louisiana	0.0104	6
North Carolina	0.7593	39	Oklahoma	0.1794	22	South Dakota	0.0089	5
Washington	0.6464	38	Alabama	0.1696	21	Hawaii	0.0084	4
Delaware	0.5861	37	Georgia	0.1664	20	Kansas	0.0056	3
Wisconsin	0.5399	36	Idaho	0.1466	19	District of Columbia	0.0000	2
Nebraska	0.5158	35	Maryland	0.1277	18	North Dakota	0.0000	1

Source: National Land Trust Census (2005), Land Trust Alliance (2009)
Ranked in three reverse order quantiles

of the differences in ratio values between the extremes in the highest class from Table 8.1 prompted concern that, perhaps as a result of per capita income having such a narrow range of values, the classes were not much affected. The only way to settle this was to check the two variables to see if they were correlated. Although statistical analysis was not intended to be part of this study, an exception was made here. A Pearson's correlation performed on the two variables showed the correlation coefficient to be 0.094, although results from both the chi-squared and the Kolmogorov–Smirnov test showed the data were not normally distributed. A Kendall tau rank correlation nonparametric test was then performed, as this test does not require normally distributed data. The result was a correlation coefficient of $0.25p_{<0.01}$, which is very low.

8.4.3 *Location Quotients to Reveal Patterns of Concentration*

Another way to determine spatial distribution patterns is through the use of location quotients. Location quotients are simply the percentage of the activity in the local region divided by the percentage of the activity in the base region, which shows the concentration of that activity with respect to the base region. Location quotients were computed using the following equation:

$$LQ = \frac{A_l / B_l}{A_{us} / B_{us}}$$

where A_l = the total activity in the state (for purposes of this study, these activities included the number of land trusts, acres owned, easement acres, and total acres protected), B_l = the comparative base of the state (population and area are used), A_{us} = the total activity of the entire region (U.S.), and B_{us} = the base of the entire region to which to compare the activity. Location quotient (LQ) values greater than 1 indicate that an activity is more concentrated in the state than in the U.S., values equal to 1 indicate that the activity is equally concentrated in the state as in the U.S., and values less than 1 indicate that the activity is less concentrated in the state than in the U.S. To identify any states that were close to being equal in concentration to the U.S. but may not have ratio values of exactly 1, modified breakpoints were used for mapping purposes. The three classes used were manually defined: in order of lowest to highest these were 0.00–0.80, 0.81–1.20, and 1.21 or greater.

Using location quotient values to compare the concentrations of spatial patterns of specific land trust activities, namely, acres owned, acres with conservation easements, and total acres protected by all means, is similar to the way Wikle (1998) used location quotients in his analysis of spatial concentration of membership patterns for three environmental organizations. In addition, in this study location quotients were also used to show the spatial pattern of the concentration of land trusts compared with population.

Figure 8.9, using location quotients, shows the concentration of the number of land trusts to population. Many of the New England States show as more concentrated than the nation as a whole. With Massachusetts and Connecticut, this is most likely the result of the sheer high number of land trusts. Maine, in fact, has the highest concentration as it has many land trusts and ranks low in population (Fig. 8.4, Table 8.2). Alaska, which has a low number of land trusts, but a high number in comparison to its population, is also more highly concentrated than the nation. What is most interesting, however, is the four Mountain States that are in the highest class, which gives the illusion of a spatial pattern that is more weighted in this region. The New England States all have ratios much higher, however, Montana being the only Mountain State with a ratio greater than 2 (Table 8.2).

Another way to show these geographic patterns of concentration is using ArcScene to add another dimension. Height can be shown proportional to the location quotient values, which allows one to really see the differences in concentration. Figure 8.10 shows the concentration of acres owned by land trusts using area as the base. The pattern displayed for the highest class shows many of the same states that were in the highest class in Fig. 8.5, with the exception of Washington, Texas, and North Carolina.

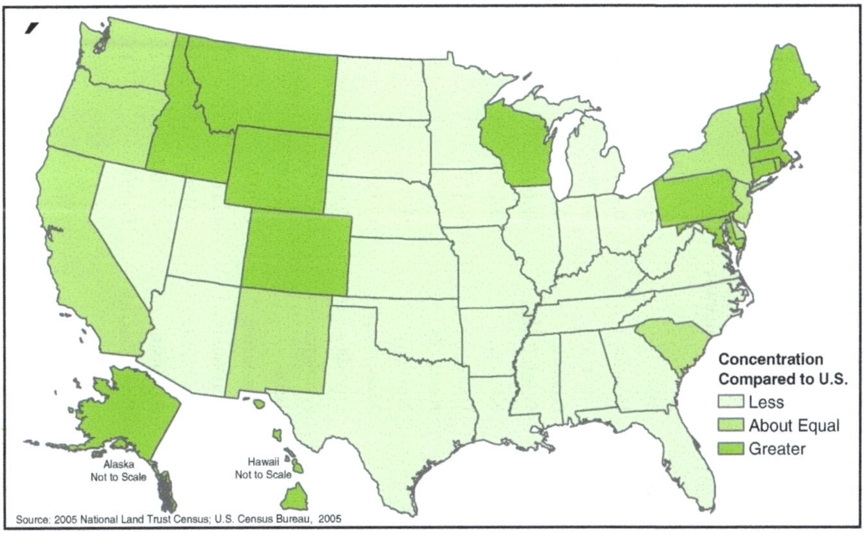

Fig. 8.9 Location quotients for the number of land trust per state using population as the base. Concentration determined using location quotients. Location quotients calculated as ratio of land trusts per state to population of state divided by the ratio of land trusts per U.S. to population U.S. Classes manually defined: LQ<0.81 (concentration <U.S.), LQ 0.81–1.20 (concentration ≈ U.S.), LQ>1.20 (concentration > U.S.) (*Source*: 2005 National Land Trust Census, Land Trust Alliance (2009))

The Northeast states really stand out along with Michigan, California, and New Mexico. This figure, however, shows just how much more concentrated many of these Northeast States are, particularly the New England States, as evidenced by their skyscraper-like appearance. Massachusetts and Rhode Island have ratios greater than 30 (Table 8.3). This technique would also have proved useful for the study by Wikle (1998) for membership concentration among three environmental organizations.

Figures 8.11 and 8.12 were also produced, showing the geographic pattern of concentration for conservation easement acres and total acres protected by all means. The figures each show a region of the East from South Carolina to Maine that is mostly high in concentration. In fact, it becomes clear just how weighted the pattern of geographic concentration really is. From the location quotient values, Maine ranks the highest with respect to conservation easement acres with Vermont coming in second, and Vermont ranks highest in concentration for total acres protected with Maine ranking second (Tables 8.4, 8.5). An uninterrupted region from South Carolina to Maine appears to dominate both maps. This technique is clearly powerful and would have proved useful to Wikle (1998) for membership concentration patterns revealed.

Table 8.2 Location quotients: land trusts in state per state population 2005 divided by land trusts in United States per U.S. population, 2005, ranked in three reverse-order quantiles

Highest quantile			Middle quantile			Lowest quantile		
State	Location quotient	Rank	State	Location quotient	Rank	State	Location quotient	Rank
Maine	11.4369	51	Washington	1.0180	34	Illinois	0.4876	17
Vermont	9.9886	50	South Carolina	1.0029	33	Iowa	0.4795	16
Rhode Island	7.7655	49	Oregon	0.9767	32	Georgia	0.4704	15
Connecticut	6.4837	49	California	0.9744	31	South Dakota	0.4583	14
New Hampshire	4.7509	47	New Jersey	0.8770	30	Kansas	0.4535	13
Massachusetts	4.4739	46	New York	0.8311	29	Utah	0.4320	12
Montana	2.8505	45	New Mexico	0.8299	28	Kentucky	0.4261	11
Hawaii	1.9521	44	West Virginia	0.7829	27	Nevada	0.3682	10
Alaska	1.8755	43	Michigan	0.7730	26	Mississippi	0.3652	9
Maryland	1.8097	42	Indiana	0.7655	25	Florida	0.3598	8
Wyoming	1.7457	41	Virginia	0.7519	24	Alabama	0.2731	7
Wisconsin	1.7022	40	Ohio	0.6825	23	Texas	0.2489	6
District of Columbia	1.6149	39	North Carolina	0.6553	22	Minnesota	0.2079	5
Idaho	1.4931	38	Arizona	0.6287	21	Louisiana	0.1575	4
Colorado	1.4483	37	Tennessee	0.6262	20	Oklahoma	0.1504	3
Pennsylvania	1.3590	36	Missouri	0.5825	19	Arkansas	0.1280	2
Delaware	1.0540	35	Nebraska	0.5055	18	North Dakota	0.0000	1

Source: National Land Trust Census (2005), U.S. Census Bureau (2005), Land Trust Alliance (2009)

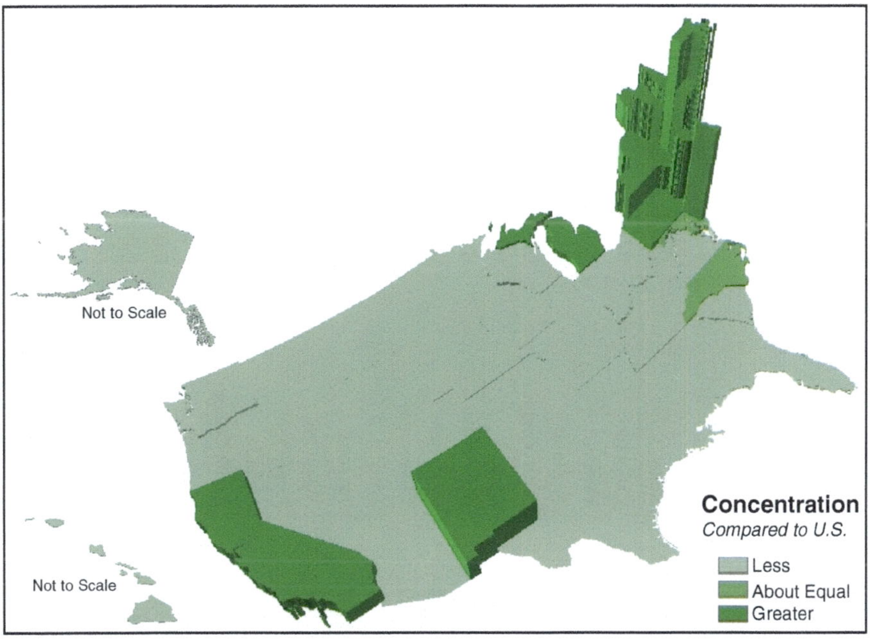

Fig. 8.10 Concentration of acres owned by land trusts. Location quotients for the number of acres owned per state using area as the base (Classes manually defined: LQ < 0.81 (Concentration < U.S.), LQ = 0.81–1.20(Concentration ≈ U.S.), LQ > 1.20, (Concentration > U.S.). State heights extruded proportional to actual LQ value (*Source*: National Land Trust Census (2005), Land Trust Alliance (2009))

8.5 Further Discussion

When considering activities throughout the United States for geographic patterns, a quick generalization can easily be made. Obviously, there is uneven distribution with the East far more heavily weighted than the rest of the country. This view especially holds true when examining the geographic patterns of concentrations among the various land trust activities as mapped and from location quotient values. The central portion of the country shows a lack of land trust activity, which was also confirmed from location quotients that showed their relative concentrations, using population as the base for the number of land trusts, and area as the base for acres owned, conservation easement acres, and total protected acres. In the West, California is in the highest class in 11 of the 12 choropleth maps produced. When viewed in terms of concentration, patterns for acres owned, conservation easement acres, and total acres protected by land trusts, several of the Mountain States such as Montana, Colorado, and New Mexico appear (Figs. 8.10, 8.11, 8.12). Also worthy of note is that Massachusetts was in the highest class in every figure produced. Geographic patterns of sum total data by state such as the number of acres owned, conservation easement acres, and total acres protected show Texas and New Mexico in the highest class there. For

Table 8.3 Location quotients: acres owned by land trusts in state per area of state divided by acres owned by land trusts in the United States per area of U.S., ranked in three reverse-order quantiles

Highest quantile			Middle quantile			Lowest quantile		
State	Location quotient	Rank	State	Location quotient	Rank	State	Location quotient	Rank
Massachusetts	30.5224	51	Indiana	0.6791	34	Colorado	0.0975	17
Rhode Island	30.4592	50	Tennessee	0.6771	33	Hawaii	0.0953	16
Delaware	22.2576	49	Virginia	0.6415	32	Kentucky	0.0899	15
Connecticut	21.2985	48	Ohio	0.5815	31	West Virginia	0.0631	14
New Hampshire	18.1242	47	South Carolina	0.5233	30	Utah	0.0556	13
New Jersey	15.5246	46	Nebraska	0.4720	29	Minnesota	0.0541	12
Vermont	15.1708	45	Iowa	0.4639	28	Arkansas	0.0405	11
New York	7.2941	44	Florida	0.3641	27	Mississippi	0.0346	10
New Mexico	6.9632	43	Missouri	0.3190	26	District of Columbia	0.0319	9
Maine	5.5162	42	Illinois	0.2882	25	Arizona	0.0209	8
California	4.0614	41	Texas	0.2277	24	Wyoming	0.0142	7
Pennsylvania	1.6444	40	Alabama	0.2011	23	Louisiana	0.0118	6
Michigan	1.2116	39	Georgia	0.1859	22	Nevada	0.0093	5
Maryland	1.1529	38	Oregon	0.1613	21	South Dakota	0.0076	4
North Carolina	0.9957	37	Oklahoma	0.1583	20	Kansas	0.0047	3
Washington	0.7161	36	Montana	0.1257	19	Alaska	0.0039	2
Wisconsin	0.6801	35	Idaho	0.1042	18	North Dakota	0.0000	1

Source: National Land Trust Census (2005), Land Trust Alliance (2009)

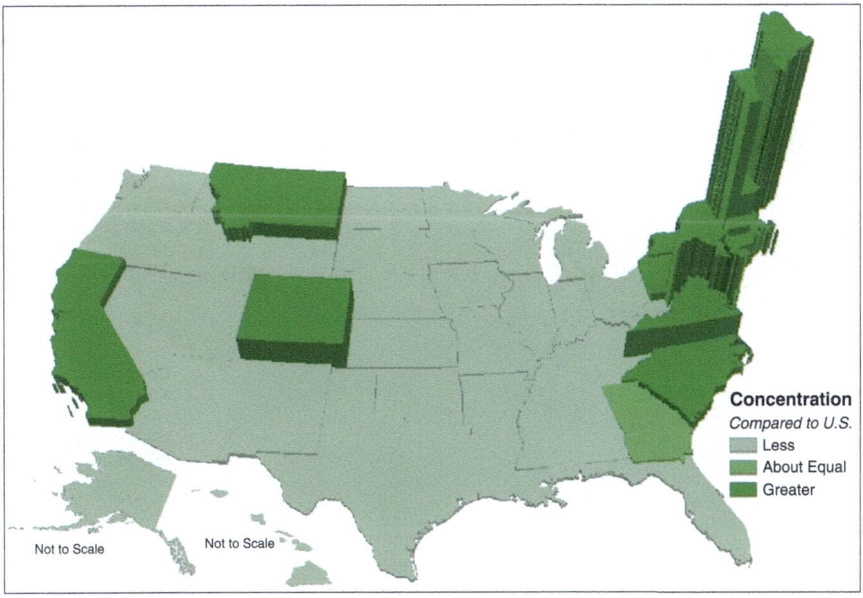

Fig. 8.11 Concentration of acres under conservation easements by land trusts. Concentration determined using location quotients. Location quotients calculated as ratio of easement acres per state to area of state divided by the ratio of easement acres per U.S. to area U.S. Classes manually defined: LQ<0.81 (concentration < U.S.), LQ 0.81–1.20 (concentration ≈ U.S.), LQ>1.20 (concentration > U.S.). State heights extruded proportional to actual LQ value. (*Source:* 2005 National Land Trust Census, Land Trust Alliance (2009))

acres owned, New Mexico should be in this highest class, as acres owned by land trusts in New Mexico are the highest in the U.S. exceeding California, the state that ranks second with a difference of almost 100,000 acres.

It is also r'ecognized that choropleth maps, and all maps in general, have their limitations, and one has to be careful about the judgments or conclusions arrived solely from interpretations from the maps, which is why there was a reluctance to point out specific causal factors in the absence of overwhelming evidence. Location quotients also have problems in that values below the equally concentrated level, that is, values less than one, are compressed, and values above one can go to infinity (Wikle 1998).

8.6 Summary and Conclusion

This study utilized a variety of techniques to analyze the geographic patterns of land trust activities across the U.S. These techniques included classifying and mapping aggregate totals for each of the land trust activities considered, making some

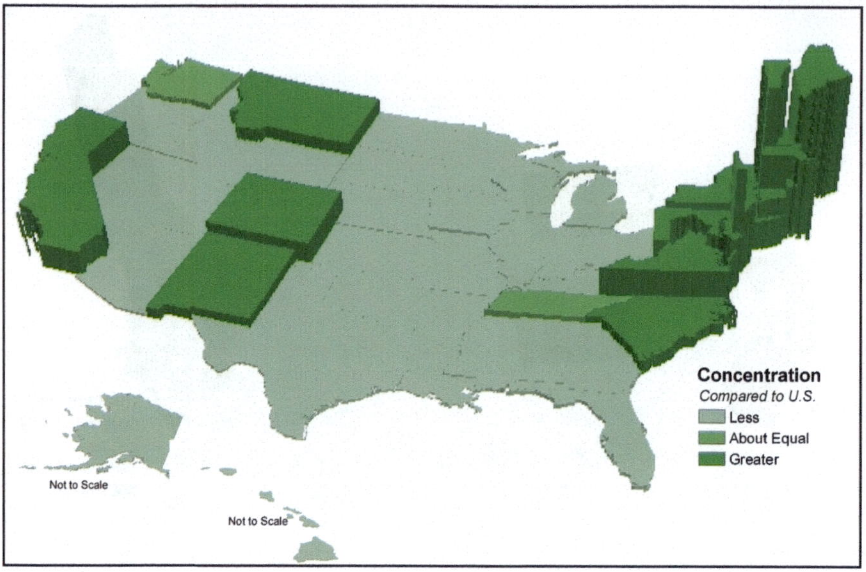

Fig. 8.12 Concentration of total acres protected by land trusts. Concentration determined using location quotients. Location quotients calculated as ratio of total acres protected per state to area of state divided by the ratio of total acres protected per U.S. to area U.S. Classes manually defined: LQ < 0.81 (concentration < U.S.), LQ 0.81–1.20 (concentration ≈ U.S.), LQ > 1.20 (concentration > U.S.). State heights extruded proportional to actual LQ value (*Source*: 2005 National Land Trust Census, Land Trust Alliance (2009))

comparisons in search of causal factors for such patterns, using location quotients to determine geographic patterns of relative concentration, and further, utilizing a variety of two-dimensional and three-dimensional mapping techniques. Analyses included computing location quotients, which were mapped and used to show the spatial patterns of relative concentrations of each of these land trust activities within the U.S. Comparative analysis and even limited statistical calculations were employed to find possible causal factors such as population and per capita income that might be associated with the spatial patterns found. However, none stood out distinctly. A multiple regression analysis in combination with similar techniques utilized in this study may find such predictive factors. Studies utilizing a combination of multiple regression analysis, correlations, location quotients, and mapping may reveal specific causal factors not uncovered in this study and would provide an avenue for future research.

Table 8.4 Location quotients: acres under easement by land trusts in state per area of state divided by acres owned by land trusts in the United States per area of U.S., ranked in three reverse-order quantiles

Highest quantile			Middle quantile			Lowest quantile		
State	Location quotient	Rank	State	Location quotient	Rank	State	Location quotient	Rank
Maine	26.6409	51	New Mexico	0.6698	34	Arizona	0.1799	17
Vermont	23.8922	50	Delaware	0.6689	33	Minnesota	0.1664	16
Maryland	11.2768	49	Mississippi	0.5837	32	Nebraska	0.0959	15
New Hampshire	8.2973	48	Michigan	0.5433	31	Indiana	0.0891	14
Virginia	5.2670	47	Alabama	0.5375	30	South Dakota	0.0801	13
Colorado	4.6862	46	Ohio	0.4914	29	Missouri	0.0778	12
Massachusetts	4.3247	45	Florida	0.3852	28	Illinois	0.0768	11
Rhode Island	4.3199	44	West Virginia	0.3828	27	Kentucky	0.0716	10
Connecticut	2.8570	43	Washington	0.3728	26	Iowa	0.0612	9
Montana	2.7876	42	Wisconsin	0.3470	25	District of Columbia	0.0521	8
New York	2.2590	41	Louisiana	0.3011	24	Kansas	0.0348	7
South Carolina	1.8290	40	Oregon	0.2994	23	Oklahoma	0.0288	6
Pennsylvania	1.7631	39	Wyoming	0.2897	22	Arkansas	0.0252	5
California	1.5560	38	Tennessee	0.2874	21	Hawaii	0.0191	4
North Carolina	1.3211	37	Texas	0.2855	20	Alaska	0.0048	3
New Jersey	0.9047	36	Utah	0.2328	19	North Dakota	0.0000	2
Georgia	0.8581	35	Idaho	0.2065	18	Nevada	0.0000	1

Source: National Land Trust Census (2005)

Table 8.5 Location quotients: acres protected by land trusts in state per area of state divided by acres protected by land trusts in the United States per area of U.S., ranked in three reverse-order quantiles

Highest quantile			Middle quantile			Lowest quantile		
State	Location quotient	Rank	State	Location quotient	Rank	State	Location quotient	Rank
Vermont	18.1127	51	North Carolina	1.4052	34	Oregon	0.2318	17
Maine	16.1097	50	Tennessee	1.1879	33	Idaho	0.2132	16
Delaware	15.1424	49	Washington	0.9860	32	Indiana	0.1706	15
Massachusetts	10.5018	48	Alabama	0.5650	31	Louisiana	0.1635	14
New Hampshire	8.5867	47	Michigan	0.5509	30	Nebraska	0.1352	13
New Jersey	8.5588	46	Georgia	0.5299	29	Minnesota	0.1205	12
Rhode Island	7.1302	45	Iowa	0.4699	28	Missouri	0.0969	11
Maryland	6.6203	44	Florida	0.4686	27	Kentucky	0.0855	10
Connecticut	5.8306	43	Wisconsin	0.4192	26	South Dakota	0.0802	9
Virginia	5.0161	42	Ohio	0.3712	25	Nevada	0.0399	8
New York	3.8768	41	Illinois	0.3446	24	Oklahoma	0.0394	7
California	3.3121	40	Mississippi	0.3445	23	District of Columbia	0.0320	6
Pennsylvania	2.9299	39	Wyoming	0.3261	22	Hawaii	0.0257	5
Colorado	2.7792	38	West Virginia	0.3147	21	Arkansas	0.0241	4
Montana	1.9422	37	Arizona	0.2826	20	Alaska	0.0226	3
South Carolina	1.7241	36	Texas	0.2776	19	Kansas	0.0212	2
New Mexico	1.4230	35	Utah	0.2724	18	North Dakota	0.0041	1

Source: National Land Trust Census (2005)

References

Brewer R (2003) Conservancy: the land trust movement in America. Dartmouth, Lebanon, 364 p

Daniels T (2004) Conservancy: the land trust movement in America/land conservation financing. J Am Plan Assoc 70(4):494–496

Ducks Unlimited [homepage on the Internet] (2009) Memphis: [Cited 20 Dec. 2009]. Available from: http://www.ducks.org/

Gustanski JA, Squires RH (2000) Protecting the land: conservation easements past, present, and future. Island Press, Washington, DC, p 567

Kiesecker JM, Shaw MR et al (2007) Conservation easements in context: a quantitative analysis of their use by The Nature Conservancy. Front Ecol Environ 5(3):125–130

Land Trust Alliance [homepage on the Internet] (2009) [Cited 18 Dec. 2009]. Available from: http://www.landtrustalliance.org/

Lieberknecht K (2009) Public access to U.S. conservation land trust properties. J Am Plan Assoc 75(4):479–491

National Land Trust Census (2005) Land Trust Alliance survey of land trusts. Data provided courtesy Ron Aldrich, Land Trust Alliance

The Conservation Fund [homepage on the Internet] (2009) Arlington (VA). [Cited 20 Dec. 2009]. Available from: http://www.conservationfund.org/

The Nature Conservancy [homepage on the Internet] (2009) Arlington (VA). [Cited 20 Dec. 2009]. Available from: http://www.nature.org/

The Trust For Public Lands [homepage on the Internet]. 2009 Nov. [Cited 20 Dec. 2009]. Available from: http://www.nature.org/

Trustees History [home page on the Internet] (2009) [Cited 19 Dec. 2009]. Available from: http://www.thetrustees.org/about-us/history/

U.S. Census Bureau (2009) [Cited 20 Dec. 2009]. Available from: http://quickfacts.census.gov/qfd/states/

Wikle TA (1998) A comparison of geographic membership patterns in three national environmental organizations. J Environ Educ 29(3):39–49

Yuan-Farrell C, Marvier M, Press D, Kareiva P (2005) Conservation easements as a conservation strategy: is there a sense to the spatial distribution of easements? Nat Areas J 25:282–289

Chapter 9
Poverty Reduction and Social Development in Bangladesh

Adil Mohammed Khan and Ishrat Islam

Abstract This study is an attempt to analyze the spatial variations of development in Bangladesh in terms of poverty reduction and social development at regional levels. The overarching goal is to determine the relationship among poverty levels and various indicators of development. The analysis reveals that although Bangladesh has been successful in reducing poverty level at the national scale, the poverty level is not homogeneous at the subnational level. The western part of the nation is lagging relative to the eastern part, and the northwestern part is lagging relative to the central part, with respect to poverty reduction. This chapter examines several factors of development such as electrification, urbanization, cooperatives, industrial establishments, and overseas employment. The study suggests that a well-functioning institutional mechanism will aid in poverty reduction, and that a committee be set up to address spatial disparity in development for proposing policies to ameliorate balanced regional development.

Keywords Poverty reduction • Social development • Education • Health • Investment • Electricity • Industry • Spatial disparity • Bangladesh

9.1 Introduction

Regional difference in economic development is a common phenomenon in the developed as well as developing countries because economic development never progresses at the same pace in all the regions of a country. However, the vision of balanced regional growth must be addressed by any government as it ensures equity and fairness among citizens. Bangladesh is a developing country of the Third World that achieved its independence in 1971. Article 19(2) of the constitution of the

A.M. Khan (✉)
Department of Urban and Regional Planning, Jahangirnagar University, Dhaka, Bangladesh
e-mail: adilmdkhan@yahoo.com

I. Islam
Department of Urban and Regional Planning, Bangladesh University of Engineering and Technology, Dhaka, Bangladesh
e-mail: ishratislam@urp.buet.ac.bd

A.K. Dutt et al. (eds.), *Spatial Diversity and Dynamics in Resources and Urban Development*, DOI 10.1007/978-94-017-9771-9_9

Peoples' republic of Bangladesh states: "*The state shall adopt effective measures to remove social and economic inequality between man and man and to ensure the equitable distribution of wealth among citizens, and of opportunities in order to attain a uniform level of economic development throughout the republic.*" In spite of the constitutional obligations, regional variations have taken place in terms of development of Bangladesh since the independence. Development plans and programs of the country have not been adopted with a regional approach for ensuring balanced development in the country (Jahir 2008). As a result, variations in socioeconomic development among regions are presently quite evident for the country (Ali and Sen 2003; Mcleod 2007).

Bangladesh has made strong progress toward reducing poverty in the past decade at the aggregate level (World Bank 1998, 2002, 2007). Poverty incidence reduced from 51.0 % in 1995–1996 to 40.0 % in 2005 (HIES 2005). The poverty gap also declined from 17.2 in 1992 to 9.0 in 2005, and the squared poverty gap declined from 6.8 to 2.9 during the same period. But at the regional level, the pace of poverty reduction is not equal, and some regions are lagging behind while others are performing well. This phenomenon is not uncommon in many developing and developed countries.

Bangladesh is divided into 6 administrative *Divisions* and 64 administrative *Districts* (before the 1990s, Bangladesh was divided into 20 administrative units termed *Greater Districts*). The incidence of poverty is disproportionately higher in the Divisions of Barisal, Khulna, and Rajshahi on the basis of the preliminary report of the Household Income and Expenditure Survey (HIES) 2005. This survey report shows that poverty is still higher in the western region of the country while lower in the eastern region. The poverty incidence in the three northwestern Divisions, namely, Khulna, Barisal, and Rajshahi, was 45.7 %, 52.0 %, and 51.2 %, respectively, whereas the incidence of poverty in the southeastern Divisions, namely, Dhaka, Chittagong, and Sylhet, was 32.0 %, 34.0 %, and 33.8 %, respectively.

9.2 Poverty Reduction in Bangladesh

Poverty reduction is one of the major targets of the government to achieve the Millennium Development Goals (MDG). The vision of poverty reduction has been encompassed in the poverty reduction strategy papers (CPD 2003). Although Bangladesh has been successful in reducing poverty at the national level, still there exists an acute spatial disparity at different regional levels (Quadir and Islam 1999). Using the HIES data for 2005, it is observed that the Kushtia and Dhaka regions were most successful in reducing poverty, at the rate of 27.83 % and 26.1 %, respectively, from the poverty level of 1995 (Table 9.1). Comilla, Tangail, Sylhet, and Chittagong are other regions that have reduced poverty significantly from 1995 to 2005 (Fig. 9.1) (BBS 1994, 2002, 2004a, b, 2005a, b, 2006). In Dinajpur, Khulna, Jamalpur, and Patuakali, the lagging regions of the country, poverty has increased from the baseline year of 1995 (BBS 1999, 2001, 2002). Between these periods, the national level of poverty has been reduced at the rate of 12.67 % from 1995 to 2005.

Table 9.1 Poverty reduction rate for greater districts, 1995–2005

Category	Poverty reduction (%)	Name of districts
Extreme poor	Poverty increased by 1–10 %	Dinajpur, Khulna, Jamalpur, Patuakali
Moderate poor	Poverty reduced by 1–5 %	Rangpur, Pabna, Jessore, Bogra
Average	Poverty reduced by 5–15 %	Faridpur, Noakhali, Rajshahi, Barisal, Mymensingh
Relatively better	Poverty reduced by 15–20 %	Tangail, Sylhet, Chittagong
Better	Poverty reduced by 20–30 %	Kushtia, Dhaka, Comilla

Source: Bangladesh Bureau of Statistics (BBS)

From the choropleth map of the changes in poverty incidence, it is found that the eastern part of the country has been successful in reducing poverty relative to the western part of the country (Fig. 9.1). The eastern region, including Dhaka, Comilla, Sylhet, and Chittagong, has tremendously reduced the poverty level whereas some parts of the northwestern region such as Dinajpur, Jamalpur, southwestern regions of Khulna, and Patuakhali have failed to improve relative poverty levels; rather, their poverty level has increased to some extent. This trend clearly depicts an east–west divide in the development pattern of the nation during 1995–2005.

9.2.1 Gini Coefficient of Poverty Reduction (in Terms of Poor Population)

Poverty reduction was not uniform at regional levels in Bangladesh during 1995–2005. The Gini coefficient value of poverty reduction during this period was 0.46 (Table 9.2); this signifies severe disparity existed in development in Bangladesh during that period, which resulted in disparity in the reduction of poverty in Bangladesh at regional levels.

9.2.2 Trend Analysis of Poverty by Divisions

Despite the reduction in poverty by nearly 9 % during 2000–2005, incidence of poverty is still high, as much as 40 % according to HIES 2005 based on the Cost of Basic Needs (CBN) method using the upper poverty line. The number of poor people was substantially larger during 2005 (about 5.6 crore; 1 crore is equivalent to 10 million). Also, the return caused by decline in poverty incidence at the national level was not equitably distributed in all the divisions of the country. In three divisions, namely, Rajshahi, Barisal, and Khulna, the incidence remained higher than those of the national and other divisional levels (Tables 9.2 and 9.3; Fig. 9.2).

The regional distribution of poverty levels illustrates that its incidence is higher in the rural than the urban areas (Bangladesh Institute of Development Studies and Chronic Poverty Research Centre 2006). The pattern is equally true for both the national and the division levels, except for Sylhet in 2000. Also, with the declining

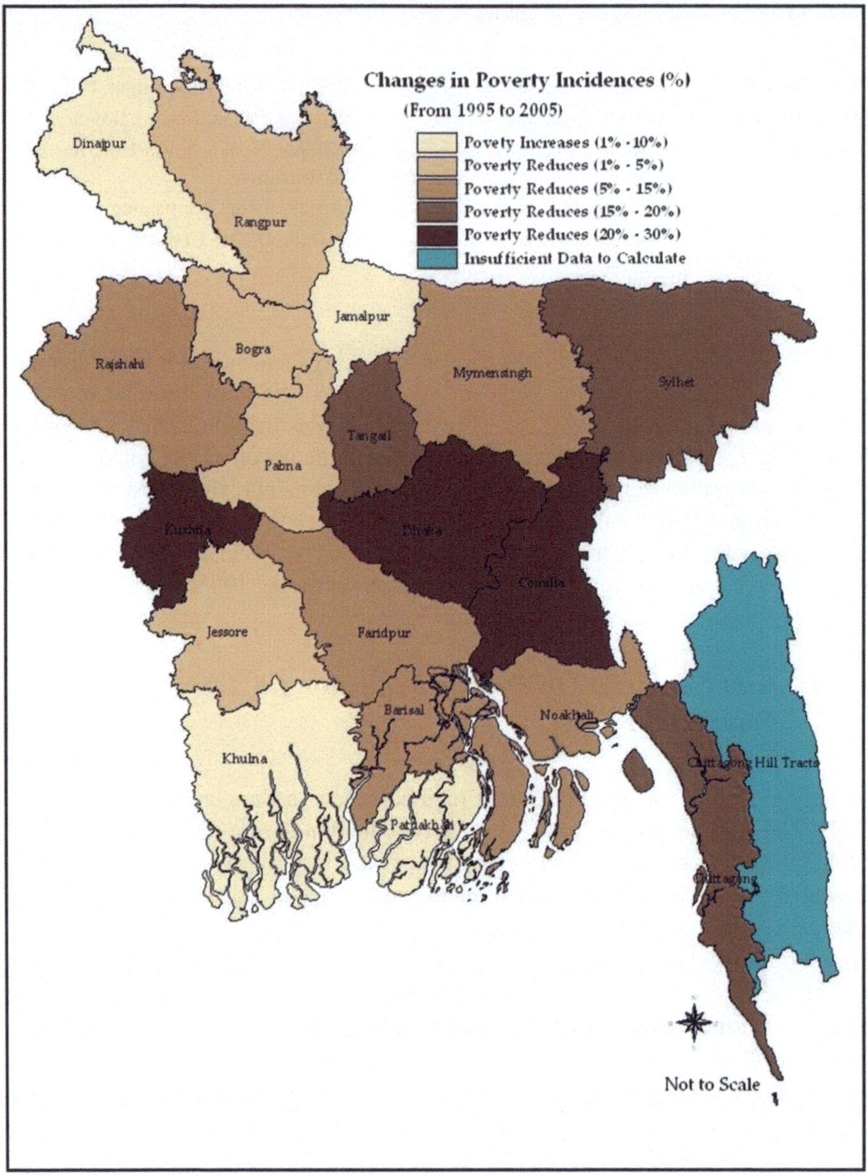

Fig. 9.1 Spatial variation in changes in poverty incidences by greater districts from 1995 to 2005 (From Bangladesh Bureau of Statistics; map prepared by author)

Table 9.2 Gini Index values for poverty in different years

Year	Variables	Gini Index value
1995	Poverty vs. population	0.05
1995	Poverty vs. area	0.243
2005	Poverty vs. population	0.184
2005	Poverty vs. area	0.21
2003	Poverty vs. population	0.10
2003	Poverty vs. area	0.28
1995–2005	Poverty reduction vs. population	0.46

Source: Calculated by the authors

Table 9.3 Incidence of poverty by divisions (percentage of population below/upper poverty line)

	1995			2000			2005		
Division	Total	Rural	Urban	Total	Rural	Urban	Total	Rural	Urban
Barisal	49.9	50.2	44.4	53.1	55.1	32.0	52.0	54.1	40.4
Chittagong	52.4	54.0	40.8	45.7	46.3	44.2	34.0	36.0	27.8
Dhaka	40.2	48.5	18.4	46.7	55.9	28.2	32.0	39.0	20.2
Khulna	55.0	56.0	48.7	45.1	46.4	38.5	45.7	46.5	43.2
Rajshahi	61.8	65.0	36.8	56.7	58.5	44.5	51.2	52.3	45.2
Sylhet				42.4	41.9	49.6	33.8	36.1	18.6
National	51.0	55.3	29.5	48.9	52.3	35.2	40.0	43.8	28.4

Source: BBS; Reports of HIES 2000 and HIES 2005

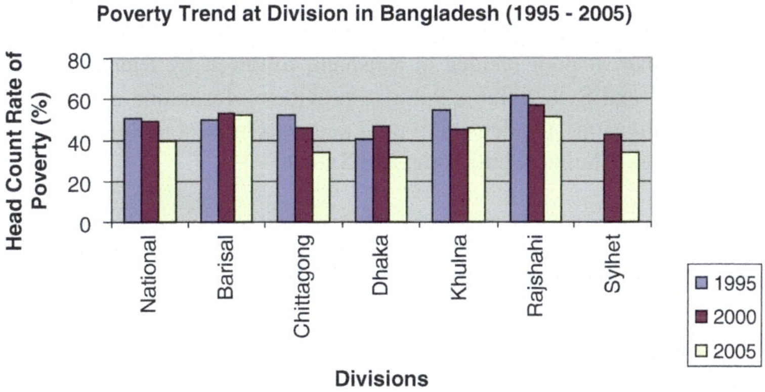

Fig. 9.2 Trend of poverty incidences by divisions (1995–2000) (From BBS; Reports of HIES 2000 and HIES 2005)

Table 9.4 Number and density of poor population by region, 2005

Region	Area (square kilometers)	Population			Poverty incidence	Poor population		Population density (per square kilometer)	
		%	Crore	%	(%)	Crore	%	Poor	All
Barisal	13,297	9.01	0.89	6.42	52	0.46	8.3	346	669
Chittagong	33,771	22.88	2.67	19.25	34	0.91	16.4	270	790
Dhaka	31,120	21.09	4.47	32.23	32	1.43	25.8	460	1,436
Khlna	22,273	15.09	1.62	11.68	45.7	0.74	13.3	332	727
Rajshahi	34,514	23.39	3.34	24.08	51.2	1.71	30.8	495	967
Sylhet	12,596	8.54	0.88	6.34	33.8	0.3	5.4	238	699
Bangladesh	147,571	100	13.87	100	40	5.55	100	376	940

Source: A strategy for poverty reduction in the lagging regions in Bangladesh Planning Commission 2008

trend in poverty at the national level, the rural areas of all the divisions have experienced the same movement. However, alarmingly poverty incidence in the urban areas of Barisal, Khulna, and Rajshahi shows an opposite tendency. Not only are the incidences in these areas very high, they have been on a rising trend since 2000.

9.2.3 *Spatial Distribution of the Poor Population*

Given the size of the population, the poverty incidence of a division shows the number of poor in that particular area. Counting the number of poor in this way, in turn, yields a scenario of poverty distribution across the divisions. Table 9.4 shows that the largest number of poor resides in Rajshahi, followed by Dhaka, Chittagong, Khulna, Barisal, and Sylhet. Table 9.4 also provides information about density of poor people by divisions: the density of poor people is highest in Rajshahi, followed by Dhaka, Barisal, Khulna, Chittagong, and Sylhet.

9.3 Social Development in Bangladesh

Social development is a process that results in the transformation of social structures in a manner which improves the capacity of the society to fulfill its aspirations (ADB 2004). Inequality in social development can also reflect the ill-being of people across regions. According to the Poverty Reduction Strategy Paper (PRSP 2005), education and health are major determinants of social development. As a result, levels of educational attainment and health indicators represent the variation in some non-income aspects of poverty (Planning Commission 2008).

Basic education for all children is clearly stated in the constitution: "*The state shall adopt effective measures for the purpose of establishing a uniform, mass oriented and universal system of education and extending free and compulsory education to all children to such stage as may be determined by law"*. Acknowledging education as a national responsibility of the State and recognizing the fundamental rights of the people to education ushered in a new era in Bangladesh.

The country has undertaken a number of measures to improve education since its independence. Commendable growth in access and achieving gender equity are the major achievements of those efforts. However, the rate of growth in the education sector is not equal in terms of either educational attainment or educational facilities at regional levels in Bangladesh.

9.3.1 Regional Variation in Educational Attainment

9.3.1.1 Literacy Rate

At the divisional level, the literacy rate is improving, but regional inequality exists temporally and spatially. In 2005, the highest literacy rate was observed in Barisal (62.1 %), followed by Khulna (53.9 %), Dhaka (53.1 %), Chittagong (51.3 %), Rajshahi (49.7 %), and Sylhet (42.1 %), according to the Household Income and Expenditure Survey (HIES 2005). The national average literacy rate during 7 years and more was 51.9%, which was 46.7% in urban areas and 67.6% in rural areas during 2005. It is noteworthy that the rural areas are more advanced in terms of literacy rate than the urban areas. However, the national average literacy rate was 44.9% in 2000, 40.9% in urban areas and 60.2% in rural areas (Table 9.4).

9.3.1.2 Primary and Secondary Enrollment Rate

At the national level, the school enrollment rate in the age group 6–10 years was 80.42 %, 79.47 % in the rural areas and 83.98 % in the urban areas (HIES 2005). The school enrollment rate in the age group 11–15 years was lower than in the age group 6–10 years: 69.97 % at the national level, 69.75 % in the rural areas, and 70.72 % in the urban areas (Table 9.5).

Table 9.5 Literacy rate (%) (7 years and above) by divisions, 2005

Divisions	National	Urban	Rural
Both sexes	51.9	46.7	67.6
Barisal	62.1	59.6	76.2
Chittagong	51.3	46.7	65
Dhaka	53.1	42.9	69.9
Khulna	53.9	50.8	65.6
Rajshahi	49.7	47.3	62.3
Sylhet	42.1	37.8	68.6

Source: HIES (2005)

Regional variations are observed in the school enrollment rate for the age groups of both 6–10 years and 11–15 years (DPE 2007). It is evident from Table 9.5, at the aggregate level, that in the age group of 6–10 years the highest enrollment rate was observed in Khulna Division (87.15 %) and the lowest in Chittagong (72.29 %). In the age group 11–15 years, the highest enrollment rate was found in Khulna Division (76.75 %) and the lowest in Sylhet (59.14 %) (HIES 2005).

9.3.2 Analysis of Regional Variations in Educational Attainment

Factors explaining regional variations in educational attainment are important in policy analyses. It is quite noteworthy that Khulna and Barisal Divisions, in spite of being the poorest, have higher primary enrollment rates among both boys and girls than Dhaka, Chittagong, and Sylhet. Khulna also had the highest enrollment rates at both primary and secondary levels during 2005 (BANBEIS 2009).

In Sylhet, the situation is alarming, as it is 8 % less likely for girls to attend school compared to Dhaka Division during 2005. Gender gaps in enrollment rates have historically been high in Sylhet, with the male primary gross enrollment rate at 15 percentage points higher than the female rate of 76 % in 2000 (countrywide female primary gross enrollment was 93 % in the same year). Gender gaps in primary and secondary education have narrowed substantially in Sylhet since 2000, to the extent that gross primary enrollment among females surpassed that among males in 2005 (Table 9.4) (HIES 2005) (Fig. 9.3).

In terms of divisional scenario, the fact that economic opportunities in a region do not cast a positive influence on education outcomes is somewhat of a puzzle, and especially so given that returns to endowments including education are lower in lagging regions, which should dampen the demand for education. More complex phenomena may however be at work: for example, the ability to migrate from a lagging

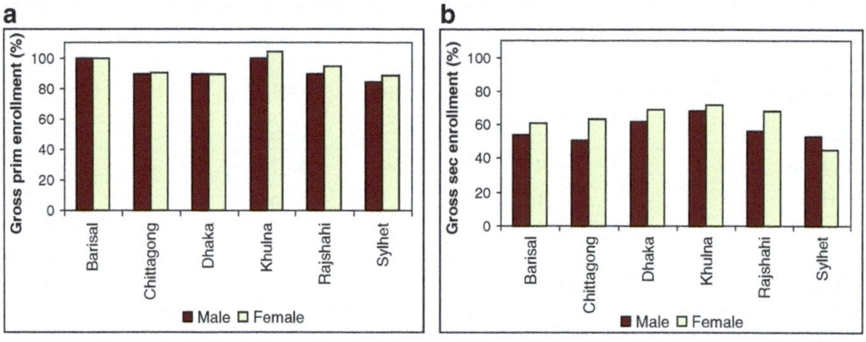

Fig. 9.3 Gross primary (**a**) and secondary (**b**) enrollment by gender in divisions, 2005 (From HIES 2005)

region is linked to better human (and physical) endowments, which can serve as an incentive for staying in school. Conversely, the greater labor market opportunities in the more vibrant Dhaka and Chittagong Divisions may translate into a greater demand for child labor, which would raise the opportunity cost of attending school, particularly for poorer households.

Other factors that could explain the paradox include the greater concentration and longer presence of NGOs (non-governmental organizations) in the economically lagging regions and the impact of their awareness-raising activities on social outcomes, positive spillover effects on non-NGO members in the same community, and differences in historical social norms, particularly as relates to the empowerment and mobility of women. The last factor is likely to be particularly relevant for Sylhet. Given that the HIES data do not allow a close examination of these questions, more detailed analysis using alternate data sources will be necessary to explain this apparent paradox among the spatial patterns of income poverty and human development (World Bank 2008).

A district-wide analysis of education and the incidence of poverty gives a more clear and realistic picture regarding this relationship. The correlation coefficient between literacy rate (7 years and above in 2001) and poverty incidences (2003) has been found to be −0.40, which implies that a higher level of educational attainment certainly reduces the poverty incidence in a region. As a result, it can be said that education has a higher impact on reducing poverty at the regional level in Bangladesh; nevertheless, there are other contributing factors besides poverty that affect the level of regional education.

9.3.3 Comments on the Education Sector of Bangladesh

To improve the quality of education and to initiate a set of reforms to develop the education sector, the government undertook several actions, of which the formation of the education commission named the Mohammad Moniruzzaman Mia Commission in January 2003 was one of the major initiatives. The Commission submitted its report to the government in March 2004. The Commission Report suggests 880 recommendations on all aspects of educational development. One of its recommendations is "*The Policy of increasing access to education in the rural areas through establishment of new educational institutions with GOB financing in the underserved areas*." Spatial concentration of developed regions in educational attainment was found in the central region, southern coastal region, and southeastern region of the nation during 2001. The northern part of the country is quite underdeveloped in comparison to the southern part of the country and this division is quite apparent. The lagging regions are the northeastern region, central-north region, and central-western regions of the country. These lagging regions should be identified properly and adequate measures should be taken to improve educational attainments in these regions. Moreover, qualitative analysis of education in terms of both educational attainment and educational facilities across regions are quite important as this may reveal spatial variation in the education sector at the various geographic scales in Bangladesh.

9.4 Spatial Variation in Health Outcomes and Health Facilities in Bangladesh

The Government of Bangladesh is committed to ensure good status of the health, nutrition, and livelihood of its citizens (DGHS 2007). To facilitate implementation of this commitment, it needs faster implementation of programs and policies. The Government's Ministry of Health and Family Welfare has undertaken a sector-wide approach for program planning and operation for such an environment (DGHS 2008). However, regional considerations in policy planning for the health sector were not given proper importance in the past. As a result, health facilities are not evenly distributed, thereby leading to variations in health outcomes across the country.

9.4.1 Regional Variations in Health by Divisions

Significant variations among divisions exist in terms of health outcomes in the country. Divisional variations in health in Bangladesh in terms of various health indicators such as the infant mortality rate (IMR), under-age-5 mortality rate (U5MR), maternal mortality rate (MMR), underweight children, and life expectancy are discussed next.

9.4.1.1 Infant Mortality Rate

According to SVRS 2006, at the national level IMR per 1,000 live births has decreased from 111 in 1981 to 45 in 2006. IMR was highest in Rajshahi Division (52) and was the lowest in Khulna Division (34) for both sexes in 2006. For males, IMR was the highest in Sylhet Division (56) and lowest in Khulna Division (37). For females, it was highest in Barisal Division (52) and lowest in Khulna Division (30). In the rural areas, IMR was the highest in Barisal Division (54) and the lowest in Khulna Division (36), but for urban areas, it was highest in Dhaka Division (42) and lowest in Khulna Division (25).

IMR had decreased substantially during 1993–2006 in all urban and rural areas of all the divisions for both males and females. According to Table 9.6, it has decreased by 35 % in Barisal Division, by 50 % in Chittagong Division, by 49 % in Dhaka Division, by 57 % in Khulna Division, by 42 % in Rajshahi Division, and by 24 % in Sylhet Division (1996–2006).

9.4.1.2 Under-Five-Age Mortality Rate

Under-five mortality rate (U5MR) is defined as the number of deaths in children under 5 years of age per 1,000 live births in a given year. According to SVRS 2006, the U5MR rate was 62, 65 for males and 59 for females. In the rural area, these rates

Table 9.6 Primary and secondary enrolment rate (%) for 2005

	Children aged 6–10 years			Children aged 11–15 years		
Division	Total	Rural	Urban	Total	Rural	Urban
National	80.42	79.47	83.98	69.97	69.75	70.72
Barisal	83.16	82.37	87.50	72.75	71.78	78.85
Chittagong	72.29	69.73	82.45	68.77	68.48	69.72
Dhaka	81.70	81.03	83.05	69.96	70.19	69.55
Khulna	87.15	87.23	86.85	76.75	77.47	73.54
Rajshahi	83.48	83.22	85.21	69.79	69.56	71.07
Sylhet	76.40	75.08	87.10	59.14	57.20	73.03

Source: HIES 2005

Table 9.7 Infant mortality rate (IMR) (%) by divisions (per 1,000 live births), 2005–2006

		2001			2005			2006			Decrease
Divisions	1993	Total	Rural	Urban	Total	Rural	Urban	Total	Rural	Urban	(1993–2006)
Barisal	86	51	53	44	49	50	42	51	54	28	35
Chittagong	90	58	60	45	48	46	54	40	42	35	50
Dhaka	94	56	61	43	52	56	44	45	46	42	49
Khulna	91	51	54	38	49	54	25	34	36	25	57
Rajshahi	94	59	62	43	49	50	41	52	54	40	42
Sylhet	74	61	62	54	50	49	66	50	51	40	24

Source: SVRS (2004, 2006)

were 64, 69, and 60, respectively and in the urban area 53, 49. and 57. respectively. In 2006, U5MR was highest in Barisal Division (75.9) and lowest in Khulna Division (47.1). In females, it was highest in Chittagong Division (69.9) and lowest in Khulna Division (38.8). In rural areas, it was highest in Barisal Division (78.6) and lowest in Khulna Division (45.4), but in urban areas the highest rate was in Dhaka Division (58.9) and lowest was in Khulna (33.5) (Table 9.7).

9.4.1.3 Maternal Mortality Rate

The maternal mortality rate (MMR) is a very important mortality index of mothers, who are exposed to the risk of death during childbirth. It is generally expressed as the ratio of maternal deaths in a period to live births during the same period, expressed per thousand live births. According to SVRS 2006, MMR was slightly reduced by 0.39 points at the national level, by 0.27 points at the rural level, and by 0.74 points at the urban level during 2003–2006.

The maternal mortality rate (MMR) still continues to be very high in Bangladesh. However, inequity in the rate across the Divisions has lessened over the years. In 2004, the highest rate, per thousand live births, was observed in Sylhet (4.50), followed by Barisal (4.38), Chittagong (4.14), Rajshahi (3.96), Khulna (3.95), and

Dhaka (3.51). During 2003 and 2004, the rate has increased in Dhaka, Khulna, and Rajshahi. During 2003–2006, MMR was lowest in Dhaka Division (2.93) and highest in Sylhet Division (5.19). Barisal, Chittagong, and Sylhet are relatively poor regions with a high MMR and Dhaka, Khulna, and Rajshahi are developed regions with a lower MMR.

9.4.1.4 Underweight Children

According to the Child Nutrition Survey 2000, the national average of underweight children is 51.1 %, which is higher in rural areas (52.8 %) and relatively lower in urban areas. The rate of underweight children was lowest is Khulna Division (38.6) and highest in Barisal Division (57.4). Dhaka Division (50), Chittagong, and Sylhet Division (51.8) have an average level of underweight children in terms of the whole country in 2000. According to the Directorate of Health Services (DHS), as in 2000, underweight children were highest in Barisal Division (46) and lowest in Khulna Division (34) in 2007 (Fig. 9.4).

9.4.2 Role of NGO Programs in Health Outcomes

The extent of NGO program coverage could be an explanation of the spatial divide in health indicators. Lagging regions of Bangladesh ("lagging" defined in income-poverty terms) actually have a much higher concentration of NGO activities than income-affluent regions. From Directorate of Health Services (DHS) data (2004), the proportion of rural respondents covered by NGOs ranges from 18 % in Sylhet and 23 % in Chittagong division to 34–35 % in Rajshahi and Khulna. Analysis using DHS health data shows the relative advantage of NGO membership across poverty

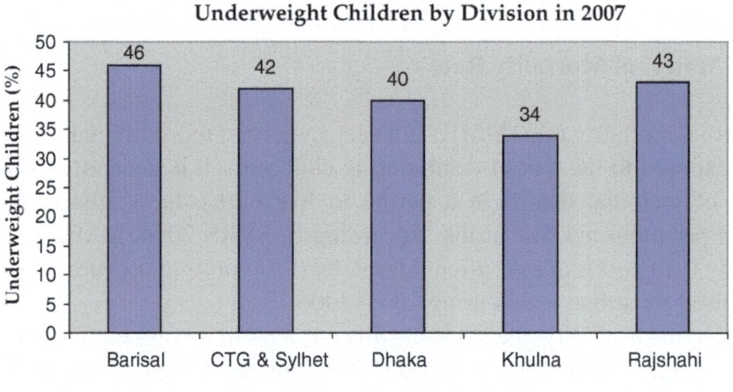

Fig. 9.4 Variation in underweight children (in percent) by divisions, 2007 (From Directorate of Health Services (DHS) 2007)

categories and higher marginal effects of NGO membership on health outcomes after taking into account household and community-level controls (World Bank 2008). There are a number of plausible explanations for this "NGO effect", such as higher awareness among NGO member households and higher expenditure by NGOs on services for member households and spillover effects on non-NGO members residing in the same community.

However, NGO presence is clearly not the only explanation of these somewhat counterintuitive trends. Another possible explanation for the slower progress in health performance in Chittagong and Sylhet is that these regions have a greater historical backlog of relatively conservative social norms, as expressed in larger desired family size, more restrictive attitudes on women's physical mobility, and related indicators of female empowerment. These issues merit further explorations in future research.

9.4.3 Spatial Variation in Health Facilities

Health facilities in Bangladesh are not equitably distributed; rather, there are regional disparities in distribution. Health facilities in Bangladesh have been concentrated in the central regions, especially in the capital city of Dhaka. In 2007, health facilities in Dhaka numbered 318 whereas the next in the hierarchy was Chittagong district with 130 hospitals (BBS 2007a, b). These data clearly show the level of concentration of health facilities in the central part of the country (Fig. 9.5).

9.4.3.1 Regional Disparity in Health Facilities

The Gini coefficient for health facilities during 1991 was 0.23, and it was 0.22 during 2007, which implies that no significant change has been observed in spatial distribution of health facilities during the intervening period. However, the Gini value of government health facilities in 2007 was 0.20 and the Gini value of private health facilities during 2007 was 0.57; this implies that although small variations exist in the distribution of government hospitals, the existing variation in the distribution of private hospitals in the country is phenomenal. Overconcentration of private hospitals in some of the divisional headquarters and major districts has contributed largely to this regional inequality.

9.4.4 Comments on Health Issues in Bangladesh

Spatial concentrations of developed regions in health outcomes are clustered in the central, northeast, central north, and southwest regions. The western part of the country is a comparatively better situation in health than the eastern part of the

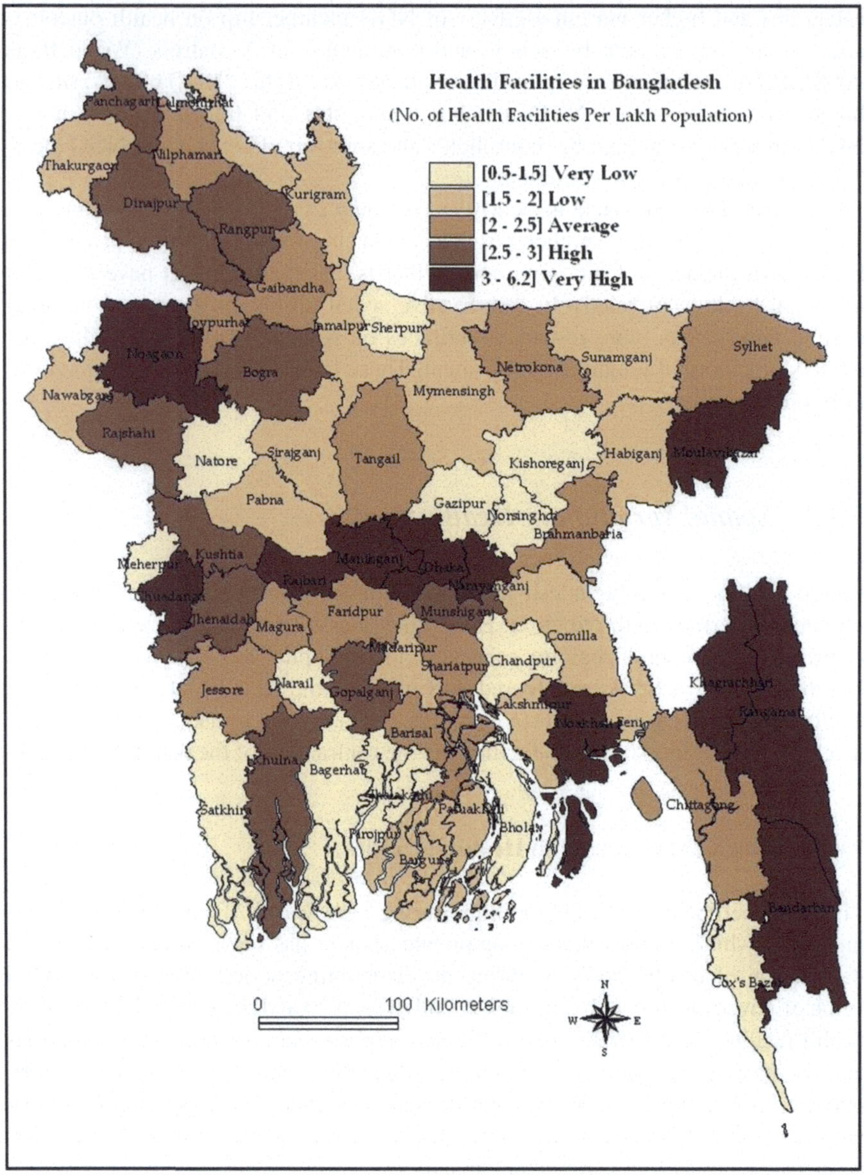

Fig. 9.5 Spatial variation in health facilities in Bangladesh (number of facilities per Lakh population), 2007 (From DHS 2007; map prepared by authors)

country. The lagging regions in terms of composite score of health indicators are northwest, central north, central east, southern coastal, and southeast hilly regions. These impoverished regions should be identified properly and adequate measures should be taken to improve the health situation of the inhabitants. The Government should undertake necessary policies to encourage health facilities in these lagging regions, thus diffusing health facilities from the central region especially away from Dhaka City.

9.5 Development Factors in Bangladesh

Spatial variations exist in various factors affecting regional development in the country. Major factors that significantly contribute to the economic development of Bangladesh are public and private investment, industrialization, roadway development, number of establishments, agricultural growth, electrification, overseas employment, micro-credit disbursement, land ownership, population growth, and urbanization.

9.5.1 Public Investment

Since the First Five-Year Plan (1973–1978) to the Fourth Five-Year Plan (1990–1995), the issue of regional inequality has not been addressed, whereas it remains a major objective in all the five-year plans of India. Bangladesh being a mono-ethnic society, such an objective was not important in the past. After a paradigm shift from five-year plans to poverty reduction strategy, the I-PRSP and the PRSP of Bangladesh have also not addressed the issue of regional gaps in poverty (Planning Commission 2008). In terms of Annual Development Program (ADP) allocation within 2003–2006, Sylhet Division has the highest allocation, with LQ of 1.40, and Dhaka was the next to receive ADP, with a location quotient (LQ) of 1.12. The other divisions are less than the national average of unity LQ. This pattern of distribution of ADP shows some hints of east–west division of distribution of development budgets of the nation. The most effective government measure for poverty reduction is the Social Safety Net Programme (SSNP). The allocation under different SSNP should be devoted to the lagging divisions with a higher proportion of poor, but this is not considered properly. Sylhet and Dhaka Division have received more than the national average of social safety net benefits. Especially, Sylhet Division was allocated remarkable SSNP benefits with LQ of 1.73. Conversely, the lagging divisions of Rajshahi and Khulna had received SSNP benefits less than the national average. Khulna Division received only 8.91 % with a population of 12.07 %. This picture of distribution of development funds within regions in Bangladesh is particularly gloomy in terms of poverty reduction efforts.

9.5.2 *Industrialization*

Manufacturing industries create many employment opportunities, thus reducing poverty. The correlation coefficient between manufacturing industrialization and poverty incidences for districts is −0.46 for Bangladesh. Spatial concentration of districts with a high level of population engaged in manufacturing industries has been found in central districts of Bangladesh, namely Dhaka, Narayangong, Manikaganj, and Gazipur. The lagging regions are concentrated in the southern coastal and northern regions of the country. The Gini coefficient for the number of manufacturing industries per 100 square kilometers (km^2) is 0.61, which clearly shows the wide level of divergences in the spatial distribution of manufacturing establishments in the country.

9.5.3 *Private Investment*

In terms of private investment, the east–west divide of the nation is quite noteworthy. Private investment in the eastern region is quite high in comparison to the western region of Bangladesh. The central districts of Dhaka, Gazipur, Naryangonj, Munshigonj, and Norshingdi have enjoyed the locational advantage of being close to a big capital market. Being a port city, private investment in Chittagong District is also quite higher than the national average. In contrast, the lagging regions are concentrated in the northern, northwestern, and southern coastal areas. The correlation coefficient among private investment (BOI 2003) and poverty incidences (PKSF 2003) is calculated as −0.31; this explains the level of poverty incidence is lower where private investments are high. Further, extreme regional imbalance can be observed at the local level with respect to investments in Bangladesh. The Gini coefficient value of private investment was 0.87, which shows that private investment has been concentrated in parts of Bangladesh.

9.5.4 *Agricultural Growth*

Bangladesh is a country where agriculture predominates as the major source of livelihood. Almost 50 % of the labor force is employed in the agricultural sector. As a result, the poverty levels in different regions, especially in rural areas, are largely affected by the vagaries in agricultural sector performance. In terms of annual compound growth rate, the eastern districts of the country observed significant growth over the years. Greater districts of Khustia, Jessore, Dinajpur, Bogra, and Rajshahi have a high growth rate in agriculture whereas the central districts of Dhaka, Tangail, and the southern coastal districts of Barisal, Patuakhali, and Noakhali had a lower annual compound growth rate in agriculture within the period of 1980 to 2000.

The correlation coefficient between agricultural growth rate and rural poverty is –0.37, which indicates that a higher growth rate in agriculture simultaneously reduces poverty. However, in the broader regional consideration it is observed that agricultural production has a miniscule relationship with poverty incidence. The estimated correlation coefficient between value addition by agriculture and regional poverty incidence was –0.06, which implies that, although a high level of agricultural production slightly reduces poverty, other factors exist that strongly influence the regional poverty levels. Level of urbanization and industrialization of a particular region in addition to agricultural production may have agglomerative effects on overall regional poverty incidence.

9.5.4.1 Electrification

According to the Census of 2001 for Bangladesh, 30 % of households had access to electricity, which is comparatively high in urban areas. Spatial concentrations of developed regions in terms of electricity coverage are found in the central region (Dhaka, Narayangonj, Gazipur, Munshigonj, Norshingdhi) and southeastern region (Comilla, Feni, Chittagong). The northern part of the country is relatively underdeveloped in comparison to the southern part of the country. Divisional headquarters are quite developed in electricity coverage compared to other districts of the country. Regions underdeveloped in terms of electricity coverage are the northwest, central north, northeastern (except Sylhet), central north, southern coastal, and southeastern hilly regions. The correlation coefficient between electricity coverage to households and regional poverty incidence was –0.67, which shows a high negative relationship. Thus, the implications if that provision of electrification can reduce regional poverty incidence to a significant extent.

9.5.4.2 Micro-credit Disbursement

Bangladesh has made revolutionary progress in disbursing micro-credit to the poor people of the country. It is observed that micro-credit to poor people has significantly contributed to the improvements in their standard of living. However, there are some divergences in the distribution of micro-credit in different regions of the country. According to PKSF, coverage of micro-credit was 60.5 % in terms of poor households in 2003 but 47 % in terms of total households (PKSF 2003). The western region of the country where poverty incidence is relatively higher has a higher percentage of micro-credit coverage than the eastern part of the country (Fig. 9.6). The correlation coefficient between micro-credit coverage to poor households (2003) and poverty incidence (2003) in district has been estimated to be –0.24, which implies that disbursement of micro-credit to poor people has a positive impact on them, reducing the level of poverty. Poor households with greater coverage of micro-credit have lower incidences of poverty. As a result, to improve the poverty situation in the lagging regions, micro-credit coverage should be increased for poor people.

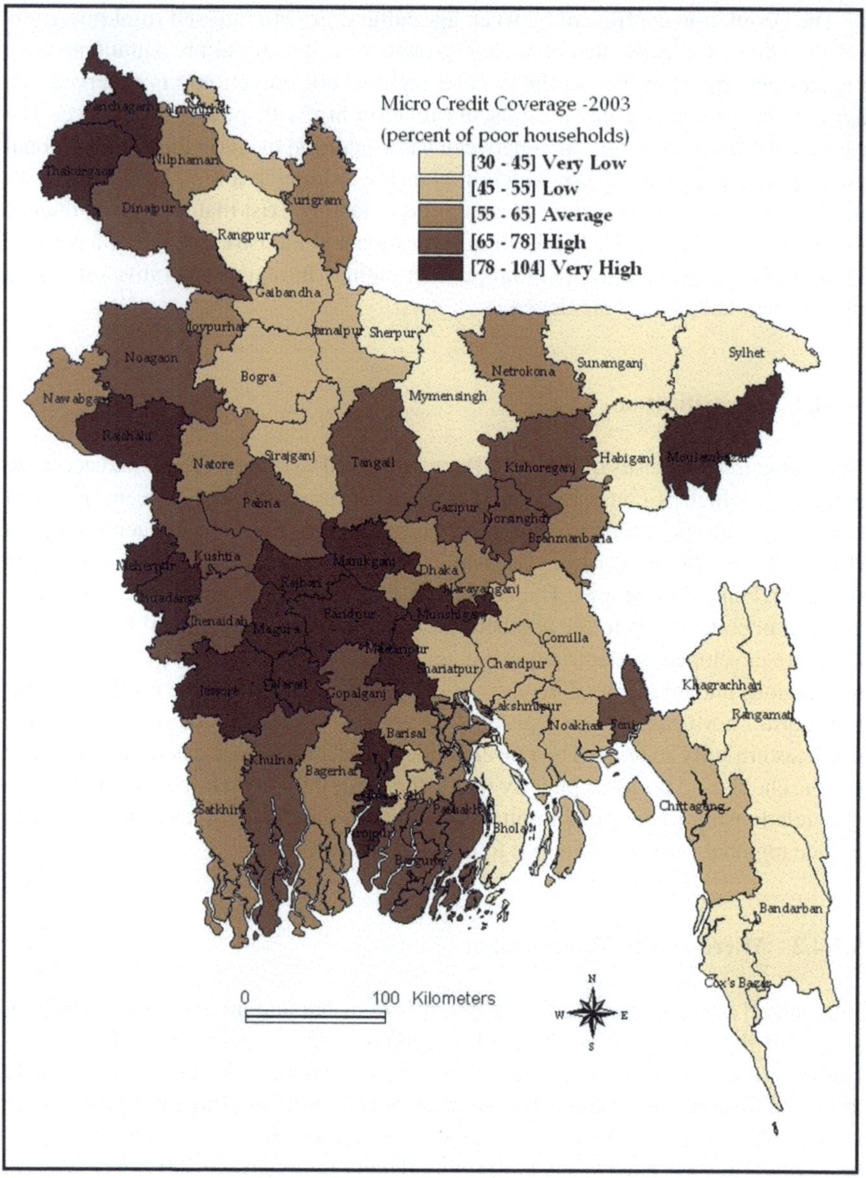

Fig. 9.6 Spatial variation of micro-credit program (% coverage of poor households), 2003 (From Report on 'Maps on Micro-credit Coverage in Bangladesh' PKSF 2003; map prepared by authors)

9.6 Development in Bangladesh: Major Findings and Remarks

Experiences from Bangladesh reveal that there are many interrelated factors constructing the complex idea of development closely intertwined with each other. Table 9.8 shows correlations among various factors of development related to poverty and social development.

According to Table 9.8, the prominent factors in Bangladesh that significantly influence the incidence of poverty of a region include electrification, urbanization, number of primary cooperatives, number of manufacturing industries and establishments, and overseas employment.

9.6.1 Spatial Disparity in Development Using the Gini Index Value

Disparities in terms of various indicators of development are quite apparent for Bangladesh. Table 9.9 shows Gini coefficients for various factors of development in various different years for Bangladesh.

According to Table 9.9, spatial disparity was severe in terms of poverty reduction among regions during 1995–2005. Moreover, spatial disparity is extreme among regions in terms of manufacturing industrialization, investment from the Bangladesh Small and Cottage Industries Corporation (BSCIC), private investment, private health facilities, overseas employment, and urbanization (Table 9.10).

9.6.2 Poverty Reduction Programs and Strategies of Government

From the First Five-Year Plan (1973–1978) to the Fourth Five-Year Plan (1990–1995), the issue of regional equality has not been addressed. As Bangladesh is basically a mono-ethnic society, such an objective was not important in the past. The Fifth Five-Year Plan (1995–2000) recognized the importance of mitigating regional disparity in development as "development of hitherto neglected areas like the north-western region, Chittagong Hill Tracts and Coastal areas …" It also addressed the balanced regional development and recognized significant differences in regional development. It directed each ministry for taking projects related to development of various regions from their sectoral allocation under the ADPs. The maintenance of the projects was then the responsibility of the relevant local governments. After a paradigm shift from 5-year plans to the poverty reduction strategy paper, the I-PRSP and PRSP of Bangladesh have also not addressed the issue of regional gaps in poverty (Planning Commission 2008).

Table 9.8 Mortality rate (%) under 5 years of age by divisions (per 1,000 live births), 2006

Division	Total	Female	Male	Urban	Rural
Barisal	75.9	64.4	85.1	55.3	78.6
Chittagong	67.9	69.9	65.0	49.8	72.8
Dhaka	59.1	54.8	62.3	58.9	59.1
Khulna	43.1	38.8	47.1	33.5	45.4
Rajshahi	65.7	62.7	68.5	53	67.6
Sylhet	64.8	63.5	66.0	51.4	66.3
Total	62.0	59.0	64.8	53.2	64.3

Source: SVRS 2006

Table 9.9 Interrelationship between poverty and associated factors of development

Investment type	Correlation coefficient	Year
Electrification	–0.67	2001
Urbanization	–0.57	2001
Primary cooperatives (per thousand poor population)	–0.5	2005
Manufacturing industries	–0.46	2003
Establishment	–0.46	2003
Overseas employment	–0.40	1976–2007
Micro-credit by Grameen Bank	–0.38	2006
Road development (1992–2005)	–0.33	1992–2005
Local investment (board of investment)	–0.31	2003
Industrial plot of BSCIC	–0.26	2008
NGO coverage of poor households	–0.24	2003
Credit by BRDB	–0.18	2005
Roadway (per km^2)	–0.14	2005
Agricultural growth (with rural poverty)	–0.37	1980–2000
Agricultural production	–0.06	2000
Economically active population (labor force)	0.04	2003
Landless people	0.12	1996
Agricultural credit by Bangladesh Krishi Bank	–0.54	2005
Population density	–0.42	2001
Population growth change (1991–2001)	–0.41	1991–2001

9.6.3 *Vision of the Government in Poverty Reduction*

The Economic Relations Divisions of the Government of Bangladesh observed: "To acquire the Millennium Development Goals which were declared by United Nations, in September, 2000, the Government of Bangladesh has been giving the highest priority to alleviation of poverty and in the light of the MDGs, the government has compiled Poverty Reduction Strategy Paper (PRSP)." The vision of this

Table 9.10 Spatial disparity in development among regions using Gini Index value

Factors of development	Year	Gini value
Poverty (with respect to population)	1995	0.05
Poverty (with respect to area)	1995	0.243
Poverty (with respect to population)	2005	0.184
Poverty (with respect to area)	2005	0.21
Poverty (with respect to population)	2003	0.10
Poverty (with respect to area)	2003	0.28
Poverty reduction (with respect to population)	1995–2005	0.46
Establishment	2003	0.35
Manufacturing industries	2003	0.61
Roadway	1992	0.22
Roadway	2005	0.19
Roadway development (1992–2005)	1992–2005	0.23
Government-allocated industrial plot	2008	0.42
Local investment	2003	0.87
Health facility (total)	1991	0.23
Health facility (total)	2007	0.22
Government health facility	2007	0.20
Private health facility	2007	0.57
College (total)	2005	0.21
College (total)	2001	0.17
Government college	2001	0.32
Primary school (total)	2001	0.15
Government primary school	2001	0.13
Secondary school (total)	2005	0.21
Secondary school (total)	2001	0.23
Government secondary school	2001	0.18
Micro-credit coverage for poor households	2003	0.16
Primary co-operatives	2005	0.23
Electrification	2001	0.33
Overseas employment	1976–2007	0.52
Landless rural households	1996	0.25
Population growth change (1991–2001)	1991–2001	0.20
Urbanization	2001	0.39

Calculated by authors

strategy is to accelerate the economic growth and poverty reduction efforts by proper utilization of economic and social energies of the nation (I-PRSP 2003). To achieve the desired goal, the Government of Bangladesh has prepared an Eight Point Strategic Agenda that includes Employment, Nutrition, Quality Education (particularly at primary, secondary, and vocational levels with strong emphasis on girls' education), Local Governance, Maternal Health, Sanitation, and Safe Water,

Criminal Justice, and Monitoring. Although it is a long-term vision, the main targets of these strategies are to build a poverty-free society. In the meantime Bangladesh has been able to achieve considerable success in many sectors that are distinctive as important indicators to education, health, and family welfare, population control, the development of women and children, social welfare activities, etc. Besides these, Bangladesh has also made notable progress in the fight against poverty. Between 1991–1992 and 2000, the incidence of national poverty declined from 58.8 to 49.8 %, indicating a modest reduction rate of 1 percentage point per year (Ministry of Finance 2003).

The goal of poverty reduction is based on a vision formed on the basis of understanding of key issues of the present state of the economy. To fulfill the vision of poverty reduction, four strategic blocks are identified: enhancing pro-poor growth, boosting critical sectors for pro-poor economic growth, devising effective safety nets and targeted programs, and finally ensuring social development. The framework also identifies four supporting strategies or issues: (1) ensuring participation, social inclusion, and empowerment of all sections, groups, and classes of people; (2) promoting good governance by ensuring transparency, accountability, and rule of law; (3) providing service delivery efficiently and effectively, particularly to the poor; and (4) caring for the environment and sustainable development on a long-term basis. Identification of problems and recommendations of actions to be taken in four strategic blocks and four supporting strategies are based on nationwide consultation with stakeholders at various levels as well as the Thematic Group Reports (PRSP 2005). However, all the program and strategies for poverty reduction were based on sectorial approaches of planning, ignoring the spatial approaches of development of lagging and underdeveloped poor regions of the country.

9.7 Conclusion

Balanced and homogeneous growth of the country could only be achieved by governmental intervention through uplifting the backward regions of the country. The regions that are 'backward' in terms of poverty reduction and social development should be identified properly to take immediate measures for development in those regions. According to the experiences of Bangladesh, spatial disparity in development is extreme among different regions of the country in terms of manufacturing industrialization, investment from BSCIC, private investment, private health facilities, overseas employment, urbanization, etc. Significant factors that considerably influence the poverty incidence of any region in Bangladesh are electrification, urbanization, number of cooperatives, industrialization, establishments, and overseas employment. A well-functioning institutional mechanism and a high-level committee should be established within the government of Bangladesh to address all issues of spatial inequality in Bangladesh that may help in framing a suitable development policy at a national level to pursue balanced development at regional strata.

References

Ali Z, Sen B (2003) Spatial inequality in social progress in Bangladesh. Bangladesh Institute of Development Studies, Dhaka

Asian Development Bank (ADB) (2004) Bangladesh: gender, poverty and the millennium development goals: country gender strategy. Asian Development Bank, Manila

BANBEIS www.banbeis.com. Retrieved on 20 Apr 2009

Bangladesh Bureau of Statistics (1993) Statistical year book of Bangladesh 1992. Bangladesh Bureau of Statistics, Ministry of Planning, Government of Bangladesh, Dhaka

Bangladesh Bureau of Statistics (1994) Bangladesh Population Census 1991, Vol. 3. Urban Area Report, Bangladesh Bureau of Statistics, Statistics Division, Ministry of Planning, Government of People's Republic of Bangladesh, Dhaka

Bangladesh Bureau of Statistics (1999) Poverty monitoring survey 1999. Ministry of Planning, Dhaka

Bangladesh Bureau of Statistics (BBS) (2000) Report of the household income and expenditure survey 2005. Bangladesh Bureau of Statistics, Ministry of Planning, Government of Bangladesh, Dhaka

Bangladesh Bureau of Statistics (2001) Report on Child Nutrition Survey 2000. Bangladesh Bureau of Statistics, Ministry of Planning, Government of Bangladesh, Dhaka

Bangladesh Bureau of Statistics (2002) Report on Poverty Monitoring Survey 1999. Ministry of Planning, Dhaka

Bangladesh Bureau of Statistics (2004a) Bangladesh population census 2001. Preliminary report, Bangladesh Bureau of Statistics, Statistics Division, Ministry of Planning, Government of People's Republic of Bangladesh, Dhaka

Bangladesh Bureau of Statistics (2004b) Statistical year book of Bangladesh 2002. Bangladesh Bureau of Statistics, Ministry of Planning, Government of Bangladesh, Dhaka

Bangladesh Bureau of Statistics (2005a) Analysis of poverty on basic needs dimension, vol 1. Ministry of Planning, Dhaka

Bangladesh Bureau of Statistics (2005b) Household Income and Expenditure Survey 2005. Ministry of Planning, Dhaka

Bangladesh Bureau of Statistics (2006) Preliminary report on household income and expenditure survey 2005. Ministry of Planning, Dhaka

Bangladesh Bureau of Statistics (2007a) Population census 2001, vol I, Analytical series. Bangladesh Bureau of Statistics, Ministry of Planning, Government of Bangladesh, Dhaka

Bangladesh Bureau of Statistics (2007b) Report on sample vital registration system (SVRS) 2005–2006. Bangladesh Bureau of Statistics, Ministry of Planning, Government of Bangladesh, Dhaka

Bangladesh Institute of Development Studies and Chronic Poverty Research Centre (2006) Chronic poverty in Bangladesh: tales of ascent, descent, marginality and persistence. Bangladesh Institute of Development Studies, Dhaka; Chronic Poverty Research Centre, Manchester

BOI (2003) http://www.boi.gov.bd/ Retrieved on 12 October 2005

Center for Policy Dialogue (2003) Poverty reduction strategy for Bangladesh: a CPD review of the I-PRSP draft, Second Interim Report. I-PRSP Review, Centre for Policy Dialogue, Dhaka

Directorate General of Health Services (DGHS) (2007) DGHS yearbook, 2007. Published in May 2008. Ministry of Health and Family Welfare, Dhaka

Directorate General of Health Services (DGHS) (2008) Health bulletin 2008. Management information system. Directorate General of Health Services, Ministry of Health and Family Welfare, Government of Bangladesh, Dhaka

Directorate of Primary Education (DPE) (2007) School survey report 2007 of Second Primary Education Development Programme (PEDP-II). Directorate of Primary Education, Ministry of Primary and Mass Education, Government of Bangladesh, Dhaka

Household Income and Expenditure Survey (HIES) (2005) Report of the Household Income and Expenditure Survey 2005. Bangladesh Bureau of Statistics, Ministry of Planning

Interim Poverty Reduction Strategy Paper (I-PRSP) (2003) A national strategy for economic growth, poverty reduction and social development. Ministry of Finance, Government of Bangladesh, Dhaka

Jahir J (2008) Spatial modeling of regional divergences of urban development in Bangladesh. Unpublished BURP thesis, Jahangirnagar University, Dhaka

Mcleod D (2007) Is poverty increasing in Bangladesh: reconciling national and global monitoring estimates. New York

Ministry of Finance (2003) Bangladesh economic review 2003. Ministry of Finance, Government of Bangladesh, Dhaka

PKSF (2003) Maps on micro-credit coverage in Bangladesh. Palli Karma Shahayak Foundation, Dhaka

Planning Commission (2008) A strategy for poverty reduction in the lagging regions in Bangladesh. Ministry of Planning, Government of Bangladesh, Dhaka

PRSP (2005) Poverty reduction strategy paper. National Strategy for Accelerated Poverty Reduction, General Economics Division, Planning Commission, Government of Bangladesh

Quadir DA, Islam AZMZ (1999) Monitoring poverty in the spatial domain in Bangladesh. In: Geo-informatics Conference, Ann Arbor, MI, 19–21 June

SVRS (2004) Report on Sample Vital Registration System (SVRS) 2002–03. Bangladesh Bureau of Statistics, Ministry of Planning, Government of Bangladesh, Dhaka

SVRS (2006) Report on Sample Vital Registration System (SVRS) 2004–05. Bangladesh Bureau of Statistics, Ministry of Planning, Government of Bangladesh, Dhaka

World Bank (1998) Bangladesh from counting the poor to making the poor count. Poverty Reduction and Economic Management Sector Unit, South Asia Region

World Bank (2002) Poverty in Bangladesh: building on progress. Poverty Reduction and Economic Management Sector Unit, South Asia Region

World Bank (2007) Bangladesh: strategy for sustained growth. Bangladesh Development Series, Paper No. 18. World Bank Office, Dhaka

World Bank (2008) Poverty assessment for Bangladesh: creating opportunities and bridging the East–West Divide. Bangladesh Development Series, Paper No. 26. World Bank Office, Dhaka

Chapter 10
Mineral Resource Potential and Prospects in Chotanagpur Region

Rameshwar Thakur and Swati Thakur

Abstract The potential and prospects of natural resources are the subject matter of this paper. The study analyzed land, forest, water, and mineral resources data. The emphasis is on distribution, production, problems, and prospects of mineral resources in Chotanagpur Region. The study suggests that after facing various problems, such as resource depletion, environmental pollution, and ecological and social concerns, the region has a very high industrial potential because of the rich mineral resource base. There is an urgent need to protect and utilize all resources at microlevel units for regional development.

Keywords Mineral resources • Natural resources • Distribution • Geology • Regional development • Chotanagpur Region (India)

10.1 Introduction

Resources provide the basic foundation for economic development. Their importance is particularly heightened in developing countries where indigenous natural resources are an important means of capital formation and accelerated economic growth. However, the value of such resources depends on scientific and technological knowledge, cultural attitudes, and the feasibility of their economic utilization. Natural resources include land, water, forest, and mineral resources. Chotanagpur is relatively rich in all these resources and yet, paradoxically enough, has remained a poor region of the country. Overexploitation of resources with the thrust of rapid economic development in the recent past has posed many themes of research in the

R. Thakur (✉)
Department of Geography, B.R. Ambedkar College, University of Delhi, New Delhi, Delhi, India
e-mail: rthakur.du@gmail.com

S. Thakur
Department of Geography, Dyal Singh College, University of Delhi, New Delhi, Delhi, India
e-mail: sthakur.dsc@gmail.com

A.K. Dutt et al. (eds.), *Spatial Diversity and Dynamics in Resources and Urban Development*, DOI 10.1007/978-94-017-9771-9_10

Chotanagpur Region (Thakur 2000). This region has been the nucleus of sociological, anthropological, and geological studies because of its tribal culture and complex geology. The geologic variation accounts for a great diversity in landforms as well as natural resources, and has attracted the attention of geographers since independence. Thus, geographers have paid attention to a variety of themes, viz., rural and urban settlement, economic region, transport network, and the geologic and geomorphologic history of Chotanagpur Region. Unfortunately, no attempt has been made to study the mineral resource development and its management in the Chotanagpur Region. This chapter attempts to fills the void in this topic of research.

The Chotanagpur Region is selected as a study area because it is one of the most fragile and sensitive ecosystems. Rapid changes in land use and an emergent economy coexisting with the traditional economy have affected the region. Similarly, the association between mineral resources management and economic development shows a visible linkage that provides an opportunity to understand the development process. The other reason to choose Chotanagpur Region is the lack of socioeconomic development, which is correlated with poor utilization of mineral resources. The region is characterized by a fragile highland ecosystem with a tremendous potential of mineral resource development. Given this overview, the purpose of this chapter is to analyze the natural resource potential and interpret the distribution, production, problems, and prospects of mineral resource in the Chotanagpur Region.

10.2 Study Area

The Chotanagpur plateau, surrounded by hills, intermountain valleys, rivers, streams, and forest, occupies a portion of the area of Jharkhand State in India. It is the easternmost continuation of the Indian peninsular plateau. It is bounded by the fertile alluvial Middle Ganga plain in the north (or the 150-m contour line), Orissa in the south, West Bengal in the east, and Uttar Pradesh and Chhattisgarh in the west. The region has an amorphous shape with a narrow base in the south and a broader extent in the northern portion. It comprises 24 districts (2007) under the administrative boundary of the Jharkhand State and covers an area of about 77,096 km^2, which constitutes 2.34 % of the total area of the country. Chotanagpur Region is the home of 26,909,428 people, with a population density of 338 persons/km^2 (Census of India 2001) The principal tribes of this region are Munda, Oraon, Santhal, Bhumij, Bhuiya, and Kharia.

Chotanagpur is a landlocked area, with a heterogeneous landscape, endowed with natural resources, dominated by the aboriginal habitat. It is a unique geographic unit. The resources in this area have given rise to a large number of industries and have led to the growth of industrial cities such as Ranchi, Bokaro, and Jamshedpur (Ram 1968). It is the homeland of tribal populations and the economic hub of mineral-based industrial activities. Nature has carved out its form, but man has defined its significance. Structurally, the region consists of Dharwar rocks, which include Archaean granites with some sedimentary and metamorphic deposits of iron ore. This region has considerable mineral deposits in a broad belt that runs roughly east–west in the Singhbhum copper belt, and the Damodar Valley coalfield belt, where the chief mining centers and major industrial centers of the region are concentrated. The Chotanagpur Region produces 37.3 % of the mineral production of India. The Dharwarian rock found predominantly in the region is the source of many minerals and is designated as the 'Mineral Monarch' of the country, which seems to be true in both variety and richness. With its noteworthy deposits of coal and iron ore it is often labeled as the 'Rhur of India.' The area has huge reserves of good-quality coal containing 92 % of India's known deposits of coal and nearly 100 % of its cooking coal. As a result of the concentration of mining and industrial activities in the region, about one fourth of the total workers (Census of India 2001) are engaged in these economic activities. The distribution of these activities, however, is not uniform; rather, they occur in patches along the main transport routes and in different resource centers. It is thus possible to distinguish various resources with distinct characteristics.

10.3 Natural Resource Potential

Chotanagpur is blessed with relatively diverse natural resources, fertile soils, and forests rich in floristic composition. Water resources can be a boon to the economy along with mineralized rock beds with a rich quality and quantity of minerals. But these resources are not evenly distributed, and the plateau is marked by an uneven distribution of natural resources and variation in their development. There are few geographic studies of the relationship between the natural resources and economic development (Ginsburg 1957; Tosi and Voertmau 1964; White 1965; Hodder 1971). Much emphasis has been given to the mechanism of economic development rather than explaining the perspective of the local resource base (Thakur 2003). In the context of Chotanagpur, an attempt is made here to understand the disparities in levels of development.

It can be asserted that levels of development are directly related to development potentials. Exploitation of natural resources plays a significant role in setting the development process in industrialized and developed countries (Smith 1977, 1979). There is disparity in the spatial distribution of resources in Chotanagpur region, and increases in the exploitation of natural resources have contributed least to the economic growth of the area; indeed, an abundance of resources is coupled with economic backwardness in this region. Thus, the distribution of natural resources must be analyzed to assess their role in the local development of the Chotanagpur Region.

10.3.1 Land Resource

Land is one of the most important resources in an area. It encompasses soil, geology, hydrology, and plants over and above any specific area of earth surface. The quality of land and its proper use is significant in the development of the region. It is the basic resource that determines the economy of the region. The development of the transport network, agricultural infrastructure, restriction of forest depletion, and educational development have motivated people to use the land. Despite these criteria, people in Chotanagpur do not use the land rationally and efficiently. It is, therefore, essential to know the various land uses. Of the total area, 7,971 ha, the major land uses and changes in these during 1990–1991 and 2001–2002 appear in Table 10.1.

The decrease in forest cover is caused by expansion of cultivable land and non-agricultural use of land. About 48 % of the reporting area is under cultivation, of which 22 % is the net sown area, and the balance, 26 %, comprises fallow land (Fig. 10.1). The net sown area varies in different districts. It is highest in Ranchi (37.8%) and Santhal Pargana (37.0 %), and lowest in Hazaribag (20.6 %); Dhanbad (33.0 %), Singhbhum (25.5 %), and Palamau (24.4 %) are at an intermediate level in net sown area.

Table 10.1 Land use change in Jharkhand: 1990–1991 to 2001–2002 (in thousand hectares)

Land use	1990–1991	Percent	2001–2002	Percent
Forest	2,332	29.25	2,333	29.26
Wasteland (1, put to non-agriculture use; 2, barren land)	1,231	15.45	1,366	17.14
Cultivatable wasteland (1, permanent pastures and other grazing land; 2, cultivable waste)	498	6.25	475	5.96
Fallow land (1, current fallow; 2, other fallow)	1,904	23.89	2,028	25.44
Net sown area	2,006	25.16	1,769	22.20
Total	**7,971**	**100.00**	**7,971**	**100.00**

Source: Bihar through figures, 1991–1999, directorate of statistics, Bihar, Patna and Chotanagpur Survey Report

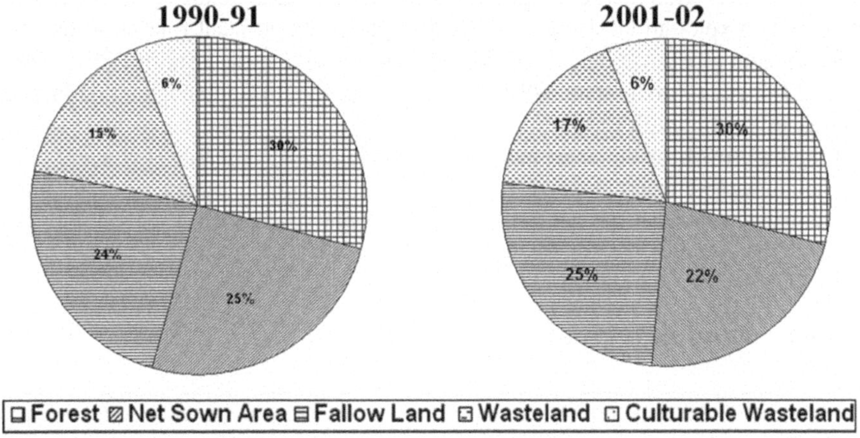

Fig. 10.1 Jharkhand: land use change

The percentage of net sown area is intimately related to the degree of flat land available in different districts. The areas affected by gully erosion or consisting of steep slopes are either covered by forest or fall into the category of culturable wasteland. The percentage of culturable waste is quite high all over Chotanagpur, being closely comparable to the net sown area: 16.8 % in Singhbhum, 18.4 % in Palamau, 20.35 % in Hazaribag, 24.4 % in Dhanbad, 26.8 % in Ranchi, and 30.0 % in Santhal Pargana. Because of limited irrigation facilities, thin soil cover, excess drainage, and vagaries of the monsoons, the upland covered by thin forests is left as fallow land.

Wasteland and land devoted to settlements and other nonagricultural uses comprises about 12 % and forests cover about 29 % of the classified area during 2001–2002. The remaining 11 % is made up of village pastures and groves, including tree crops and bamboo, and cultivable wasteland. Much of this submarginal and wooded area requires heavy capital outlay to bring under cultivation. Because of regional variations, not all the land is of the same quality. Consequently, the land of a region remains suitable for particular purposes in the service of mankind.

The significant features of land utilization in the plateau are (1) a high percentage of area in forest-covered land, (2) the very small areas under pastures for grazing and culturable waste, despite a large bovine population, and (3) scope for further extension of cultivation. Fertile soils coupled with the monsoonal rhythm of the climate have long been valuable and the most exploited natural resources. The region contains pockets of rich soil in valleys that respond well to irrigation and provide sustenance to the majority of the rural population. Considerable scope exists for expanding land under cultivation as the percentage of cultivatable waste is high.

10.3.2 Forest

The forests are one of the most abundant and valuable resources. They covered 2,333,000 ha of land during 2001–2002, which is about 30 % of the land area of the region. Variation in rainfall and altitude causes Chotanagpur forests to contain a variety of natural vegetation, ranging from tropical moist deciduous to dry deciduous forests. A large part of the region receiving more than 120 cm of rainfall is occupied by forest and woodland, ranging from stunted open growth on slopes, barren soil, or the southern drier slopes facing the sun, to fair-sized dense jungle in more favourable situations. It is one of the most heavily wooded tracts of India. Now only two categories of forest, the reserve forest and protected forest, are found in the region. Thr reserve forest is most thickly planted and consists of the best timber reserves (9 % of the total forest land) in the southern part of the plateau that consists of Porahat and Kolha Forest. Other areas of reserve forest are in the Pat region, Kodarma plateau, and the Rajmahal hills.

Forest occupies a considerable proportion of land in different districts, varying between 25 and 50 %, except for a few. The highest percentage of land under forest cover lies in Palamau (50 %), followed by Hazaribag, Singhbhum, Ranchi, and Santhal Pargana, at 48.2 %, 47.0 %, 25.0 %, and 23.75 %, respectively. Dhanbad and Bokaro have less than 15 % under forest. The forest of Chotanagpur, especially that of Hazaribag and Ranchi, being the most accessible, has received the full brunt of the fury of cutting. Besides the indiscreet acts of the people, natural processes such as fire and gully erosion reduce the forest cover.

Forest is more luxuriant and a valuable timber supply over extensive areas in the districts of Palamau, Garhwa, Latehar, Chatra, Lohardaga, Ranchi, Hazaribag, Bokaro, and Singhbhum. The Singhbhum Valley possesses considerable reserves of good-quality timber. An important raw material for paper and pulp is obtained from the Palamau forest. *Sabai* grass, used for manufacturing rope and paper, is very common in Singhbhum. The forest in Chotanagpur contributes nearly 40 % of lac production in the country. In addition, forest yields a number of minor products, such as tendu leaves for *biri* making, mahua, and fruits such as *myrabolaur kath*. The problem of deforestation has received the maximum attention from the Government and the general public, but ironically, Government policies on afforestation have attracted public criticism.

10.3.3 Water Resource

Water is an important gift of nature. Fresh water is vital for irrigation, domestic uses, industrial uses, inland fisheries, forest growth, public health, generation of hydro electricity, recreation, and numerous other uses. Human beings depend on sufficient water supply for their sustenance and progress. The supply of water is abundant, including both surface water and underground water. Rainfall is an important source of water supply to plants, surface runoff, and groundwater. Chotanagpur is one of the rainiest parts of eastern India. It is drained by numerous rivers, of which the South Koel, North Koel, Subarnarekha, Damodar, Ajay, and Barakar Rivers are prominent. The distribution of water resources is variable and the contribution of different river basins also varies with the basin (National Commission of Agriculture 1976). Influenced by the precipitation pattern and orography, water resources in Chotanagpur show high spatial variability. Because overall the climate is hot, runoff in Chotanagpur is highly influenced by rugged topography and very high evaporation. High summer temperature and other meteorological factors evaporate the precipitation so that only a small percent of the precipitation can replenish the groundwater. Both the supply and demand of water vary locally and regionally and so also does the contrasting pattern of utilization. The rural water supply has been quite neglected until recently. Both surface water and groundwater are pumped out after purification and supplied to households and industries for consumption.

A major focus of water development was to expand irrigation facilities by utilizing surface water resources through construction of dams, reservoirs, barrages, weirs, and irrigation canals. The region's rapidly growing urban population and income, and improved lifestyles of the beneficiaries of the rapid economic growth, expanding industry, and development of rural centers, have all triggered high water and energy demands. Further, the regional imbalance in water availability, problems emerging from the legacy of past water development policy, and fast economic growth has caused regional water development to face critical challenges.

10.3.4 Minerals and Power

Among natural resources, minerals occupy a significant position because they provide raw materials as well as power for industrialization, but they are nonrenewable resources, and thus these are called *wasting assets*. Once they are extracted, they are lost forever. The guiding principles of mineral use policy, therefore, are social development and conservation (Thakur 1998).

The efficiency of their utilization can be increased with improving technology, but their potentiality cannot be changed because it depends on geologic processes. Therefore, minerals should be conserved for a longer time and more social benefits. Growing population pressure, increasing per capita consumption, and demand have thus caused the exploitation of minerals to increase.

Chotanagpur mineral and power resources are abundant in supply. A survey of recent studies has improved our knowledge of the quality and quantity of the reserves of the more important minerals and their utilization. The plateau's endowment for heavy metallurgy is particularly rich in reserves of high-grade iron ore, manganese, copper, and chromite, and limestone and refractory hold a strong position. Coal reserves are adequate. In petroleum and nonferrous metallic minerals, the position is very weak except for a good deposit of bauxite, but in mica and in some strategic minerals such as limonite, china clay, and thorium the plateau holds a strong position (Indian Bureau of Mines 2002).

The largest concentration of minerals occurs in the southeastern section of the plateau and the adjoining areas in Orissa, West Bengal, and Madhya Pradesh. Chotanagpur leads the others in variety and value of minerals produced. The extensive mineral deposits located in the plateau can be classified into three groups: metallic, nonmetallic, and fuel minerals. Metallic minerals are further subdivided into ferrous and nonferrous minerals. Ferrous minerals are those that contain iron in substantial quality. These minerals provide a strong base for the development of metallurgical industries, particularly iron, steel, and alloys. Nonferrous minerals are those that do not contain iron and are highly important in day-to-day life.

10.3.5 Geology

Geology forms an important basis of the regional divisions and the extent and alignment of the mining regions. Geologic age has a close relationship with the formation of minerals, which acts as a transport corridor leading to transparent vaults in Chotanagpur. The structural trends of the rocks from east to west have influenced the localization of minerals. The mineral deposits are, therefore, found in east–west elongated zones in either down-faulted or depositional basins or in fissures, cracks, or veins. The structural frame of the region was laid down by the Archaean movements that folded the Dharwarian sediments and caused by batholithic intrusions into the folded Dharwarians. The Archaean folds were subjected to further movements, which caused the great thrust through which welled up the ultrabasic magma preserved in the Dalma and Dhanjri Ranges. Each age is the replica of the formation aggregation of minerals in their richness, extent, and uses.

Most of the rocks of Chotanagpur belong to three important geologic formations: Archaean, Gondwana, and Dharwarian. The Archaean are the most important, covering more than 90 % of the plateau surface. The sedimentary rocks, grouped under the Dharwarian system, are found in the extreme southern and northern portion of the plateau. They are best developed in the Singhbhum District. Here the quartzite, interblended with a thick mass of iron ore and called banded hematite quartzite, is the most important source of iron ore deposits. For this reason, the Dharwarian formation in this part has been named an iron ore series. Gondwana deposits in the trough basin of the Damodar Valley facilitated the formation of thick coal deposits and seams of bituminous and other types of coal. The formation of Vidhyan deposits in the region has provided the occurrence of limestone. Lava capping during the Cretaceous period enriched this region with laterites containing bauxite. Besides these important minerals, igneous intrusive eruptions have also played an important role in the formation of a number of other minerals.

10.3.6 Distribution of Minerals

According to the Ninth Plan, the metallurgical and mineral industries constitute the bedrock of the industrial sector as they provide the basic raw materials for most of the industries (Tiqqa Hema Malini 2004). Coal and iron, for instance, are the basic minerals needed for the growth of the iron and steel industry, which in turn is vitally necessary for the region's development. Similarly, other minerals such as mica and manganese, copper, lead, and zinc are of economic significance. Also, there are mineral fuels such as coal, petroleum, thorium, and uranium that are of national importance. Thorium and uranium, the atomic energy minerals, are a tremendous source of power. There are also a number of minor minerals with varying degrees of utility to the region. The reserves of Chotanagpur minerals for basic industries such as coal and iron are ample, but there is a fairly long list of vital minerals such as tin, lead, zinc, nickel, cobalt, and sulfur (Table 10.2). Most of all is petroleum, in which Chotanagpur is deficient. In the Eighth and Ninth Plans the Government laid great emphasis on mineral exploitation by adoption of improved technologies such as remote sensing and geotechnics, particularly for those minerals in which the resources of the country are poor.

Table 10.2 Jharkhand: reserves of important minerals, 2001–2002

Minerals	Proved	Probable	Possible	Total
Asbestos	6,756	24,078	124,059	154,893
Bauxite	34,048.83	28,178	55,124.93	117,352.59
China clay	40,935.33	20,230	149,321.88	210,487.32
Chromite	15	97.53	35.3	465.53
Copper	68,966	74,243	37,953	181,162
Dolomite	20,323.32	19,262.15	17,597.57	57,183.04
Fire clay	2,554.34	1,024.52	64,614.20	68,193.06
Feldspar	216,011	552,792	730,812	1,499,615
Graphite	271,142.70	3,986,827.45	1,371,132	5,629,102
Gold	8,010	–	–	8,010
Iron ore (hematite)	2,560,232.67	334,830.30	386,019.29	3,281,082.26
Iron ore (magnetite)	1,343	3,846	2,358	7,547
Kyanite	943,183	108,735	114,805	1,166,723
Limestone	238,214	56,913	367,351	662,478
Mica (kg)	–	–	1,494,430	1,494,430
Manganese ore	2,930	23,632	3,394.02	29,956.02
Coal	22,168.29	–	35,402.61	57,570.90
Total				**14,626,250.72**

Source: Jharkhand; Indian Bureau of Mines 2002, pp. 76/31/32

Minerals occurring in the region are asbestos (Singhbhum District); bauxite (Dumka, Gumla, Lohardaga, and Palamau Districts); china clay (Dumka, Hazaribag, Lohardaga, Singhbhum, Sahibganj, and Ranchi Districts); chromite (Singhbhum District); copper (Hazaribag and Singhbhum Districts); dolomite (Palamau District); feldspar (Deoghar, Dhanbad, Giridih, Hazaribag, Kodarma, and Palamau Districts); fire clay (Dhanbad, Dumka, Giridih, Hazaribag, Palamau, Singhbhum, and Ranchi Districts); gold (Singhbhum District); graphite (Palamau District); hematite (Singhbhum District); magnetite (Palamau District); kyanite (Singhbhum District); limestone (Bokaro, Garhwa, Giridih, Hazaribag, Pakur, Ranchi, and Singhbhum Districts); manganese ore (Singhbhum District); mica (Giridih, Kodarma, and Hazaribag Districts); and coal (Dhanbad, Giridih, Bokaro, Hazaribag, Daltonganj, Palamau, Garhwa, and Godda Districts) (Fig. 10.2).

10.3.7 Metallic Minerals

10.3.7.1 Iron Ore

The Dharwar rocks of Chotanagpur contain one of the world's largest and richest deposits of iron ore, mainly hematite and magnetite. The percentage of iron in the ores varies from 60 to 80 %. It contributes to about 40 % of the production of iron

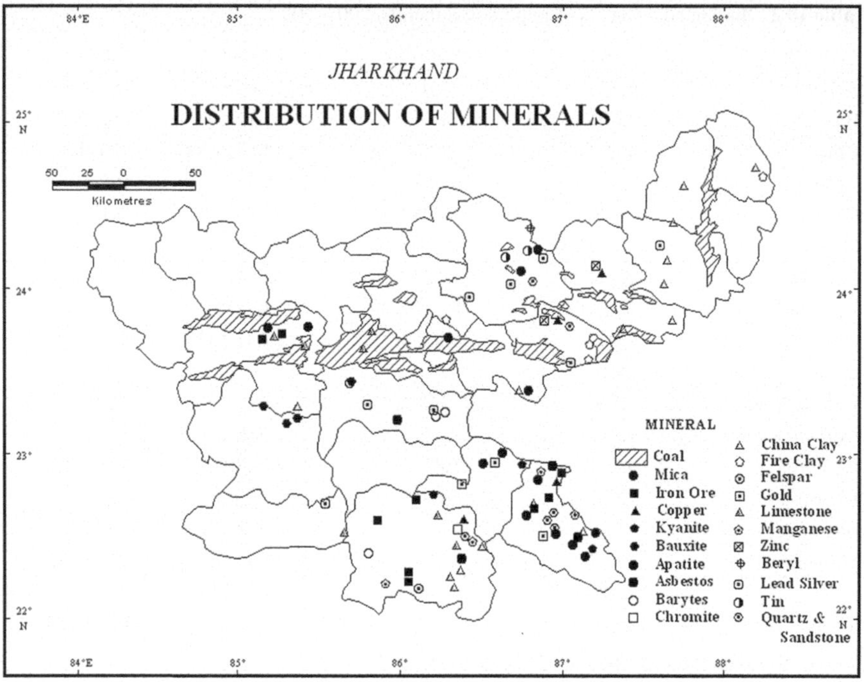

Fig. 10.2 Distribution of minerals. Source: State Atlas of Jharkhand, (NATMO) 2007, p. 26

ore in India. Magnetite contains iron with a dark brown to blackish color and is referred to as black ore. Vast deposits of iron ore lie in South Singhbhum in the old Pre-Cambrian rocks popularly known as the iron ore series and Kolhan series. The iron ore consists mainly of banded hematite, quartzite, phyllite, tuffs, and lavas, and forms a horseshoe shape closed on the south in Orissa and open in the north in Singhbhum. Important mines are located as Noamundi, Paursia Baru, Bada Baru, Gua, Jamda, Jhihing Baru, Lukud Baru, and North Baru. This belt provides iron ore to the steel plants of Jamshedpur, Bokaro, Raurkela, Durgapur, and Bhilai. Low-grade iron ore is found in some areas of the Damodar basin. A small amount of magnetite of iron ore is found about 11 km southwest of Daltonganj. The total estimated reserves or iron ore in the region during 2001–2002 are placed at 3,288 million tons, of which 3,281 million tons are hematite and only seven million tons are magnetite, about one fifth of the country's reserves. Most of these iron mines have been connected with electrified rail routes and other transport, which have enabled the development of mining sites. Rich iron content is a great asset to Chotanagpur. Its increasing use in India's industrialization has made the iron belt significant.

10.3.8 Manganese

Manganese is an essential ingredient in the manufacture of iron and steel metals; its production is closely linked with the demand for steel factories. It is used in the making of dry batteries and in the chemical, glass, and match industries. Manganese is a multipurpose metallic mineral found in ancient crystalline Dharwarian rocks in Singhbhum District and is used in the making of a special type of steel known as ferro-manganese. It is also found in association with iron ore deposits of the Keonjhar-Bonai tract that stretches south from Gua to Limtu between Chaibasa and Jamda and between Jamda and Noamundi. The proximity of manganese mines to steel plants is a great advantage from the production aspect. The region has an estimated reserve of 2.3 million tons, and the annual production was about 3,000 tons in 2002. Besides Singhbhum, it is also found in Dhanbad, Giridih, and Hazaribag Districts. These districts have medium-quality manganese, as well as meager amounts, but its significance lies in the fact that it is found in locations close to the areas of iron ore and coal. It has attracted the establishment of the steel industry, increasing the demand for manganese.

10.3.9 Chromite

Chromite is both an alloy and a refractory of which India has an adequate supply. It is an ore of chromium metal. Reserves of chromite are reported in Singhbhum District and also around Jojahatu and Keraikela near Saraikela. The region has superior quality chromite as the ore contains 48 % chromium. The area possessing chromite ore is estimated to be 2.4 lakh tons, of which 2,250 tons are known and the remaining amount is probable. The annual production of chromite is about 170 tons. Some smaller deposits are also traceable at Gurgaon, Kusmita, Janoa-Ranjra-Kocha, and Tonto. Increasing use of chromite in the steel and chemical industries has led to decrease in its export. Nearly one half of the annual production is exported to Japan and Europe for use as an alloy.

10.3.10 Copper

Among nonferrous minerals, Chotanagpur has adequate reserves of copper and bauxite, but is deficient in tin, nickel, silver, gold, lead, and zinc. The only area where copper ore is being successfully mined in India is Singhbhum. A copper-bearing belt extends in a 130-km-long stretch from Chakradharpur to Singuri. Copper is mined at Ghatshila, Mosaboni, Turandih, Tamapanar, Durda, Dhobani, and Rajdah. These mines produce more than 1 million tons of copper ore annually. The copper ores occur in veins in the granitic rocks and in the adjoining mica-schist, quartz-schist, and epidiorites. Although the soda granite outcrops are missing for some distance, the copper belt persists through Rekha Mines, Mosaboni, and eventually ends at Baharegora. Besides, there are some important mines at Pathargora and Kenduadih. There is also an occurrence of copper in the Rajmahal highland and Hazaribag plateau, particularly at Barganda, Taridih, Dandlo, Bagadar, and Parasnath. The total reserves are placed at 154 million tons, of which 58 million are known and 96 million tons are unknown. Its production in 2002 was 280,020 tons. It is mined at Rakha Mines and processed at Man Bhander.

10.3.11 Bauxite

Bauxite is a nonferrous metallic mineral, located in and around the "Pat region" on the west side of Ranchi District and adjoining highlands in Palamau District. It is concentrated in Sarendag, Bagru, Dudha, Kotcha, Khamar, and Banjara; there are other mines at Pakhar, Chapuduadhia, Pakri, Oranga, Banda, and Bar. Abundant resources of high-grade bauxite, which amount to more than 12 million tons, provide bright prospects for the growth of the aluminum industry. The continuously rising annual production reached 8 lakh tons during 2002, or 37 % of the country's total production. More than 90 % of the annual production is smelted to meet the growing domestic requirements. It is mined near Lohardaga and transported by railed to Muri for processing into alumina. The mining of bauxite is facing the problem of poor transport in the Pat region. Bauxite is used in making airplane parts, electrical appliances and goods, and household fittings and utensils. It is also used for manufacturing of white-colored cement and certain chemicals.

In addition to the aforementioned nonferrous metallic minerals, other minerals such as gold, silver, lead, tin, zinc, and nickel are found in negligible amounts. Gold is confined to the Subarnarekha valley where the sand contains a few small particles. Silver is found in association with sulfur, lead, and copper; it is found in Hazaribag, Chatra, Palamau, Ranchi, and Singhbhum Districts but its concentration has not been commercialized. Lead is found in scattered form in Hazaribag, Palamau, Ranchi, and Singhbhum Districts. Tin, being obtained from cassiterite ore, is found in the districts of Hazaribag, Giridih, Ranchi, and Lohardaga.

10.4 Nonmetallic Minerals

Significant nonmetallic minerals are mica, limestone, dolomite, kyanite, graphite, fireclay, china clay, asbestos, feldspar, sulfur, gypsum, and phosphate. These minerals are used in a variety of industries such as cement, fertilizers, refractories, and electrical goods in Chotanagpur. Mica was one of the indispensable minerals used in the electrical and electronic industries until recently, but the synthetic substitute has considerably reduced our production as well as exports: it is in high demand in the world market because of its better quality and its ruby color. The principal mica belt,

128 km long and 32 km wide, lies along the northern forested fringe of the plateau in Kodarma, where the world's best muscovite is mined. The premier mica-producing areas are Kodarma Reserve Forest, Domchanch, Dhorhakola, Masnodih, Dhab, Gawan, and Tisri. In addition to these mica mines, some mica is also found in Singhbhum, Palamau, and Hazaribag District. The region supplies more than 80 % of India's output. The average annual production is about 2,527 metric tons, which is 53 % of the country's reserves. About 90 % of its production is supplied to world markets. The mica belt of Chotanagpur deserves special status because it has huge occurrences of ruby mica, which fetches a good price in the world market. Kodarma and Jhumri Tilaya are the main mica industrial centers where different types of mica are processed for market. Its importance is increasing day by day because of its varied uses in industrialization and electrification.

Among building materials the most important is limestone. Chotanagpur has small reserves, which are almost inexhaustible, in the Vidhayan limestone in Son Valley, but there is some reserve in Palamau, Ranchi, Hazaribag, and Singhbhum Districts. These areas have small reserves of limestone. The nodular limestone called *kankar* or *ghuting* occurs in many places in the Rajmahal hills and adjoining areas. It is the chief raw material of cement and is also used in the making of lime, in the iron and steel industry as a flux, in sugar refining, in chemicals, and in the paper, glass, textile, and ceramic industries. Limestone with a high silica content is used in various cement factories located nearby.

Two refractory minerals connected with heavy metallurgy of which India has adequate reserves are magnetite and fireclay. Singhbhum produces magnetite as well as quartzite. Good-quality fireclay is found in the Damodar valley coalfields and in the neighboring Rajmahal hills. Chotanagpur has a good reserve of kyanite at Lapsa-Buru near Rajkharsawan in Singhbhum along the northern flank of the copper belt; it is used in the iron and steel industries and in the lining of furnaces. Most of the kyanite mined is exported. Dolomite has numerous uses, the more important being in the paper, glass, rubber, and several other industries. It is mainly produced in Palamau and Garhwa Districts, and also occurs in the Koel basin near Daltonganj and Chaibasa in Singhbhum District. The reserve of dolomite is about 34 million tons and production is 1.66 lakh tons. China clay is a white powdery clay derived from the decomposition of feldspar in gneisses, schists, granites, pegmatites, shales, and sandstone. It is used in the textile industry, paper industry, and insulator factories as a bleaching material and as pot clay. It is found in Singhbhum, Ranchi, Santhal Parganas, and Dhanbad Districts. About 41 million tons is in reserve, of which 4 million tons is known. The annual production is 60,000 tons. Fire clay is used as a refractory material as well as in ignition of furnaces in steel plants and in the manufacture of cement, firebricks, and in paper and textile industries. The chief producing areas in the Damodar Basin are associated with the coal seams of Jharia, Raniganj, and Daltonganj. It also occurs widely in Palamau, Ranchi, and Singhbhum Districts. Its annual production is about 261 tons.

In addition to the aforementioned nonmetallic minerals, some other minerals such as apatite, beryl and sulphur are also found in Singhbhum and Hazaribag Districts, where the Archaean and Dharwarian rocks are suitable for the production of these minerals. Apatite is used in chemical fertilizers.

10.5 Fuel Minerals

Fuel minerals include coal, petroleum, natural gas. and atomic or radioactive minerals. The Chotanagpur plateau has been fortunate enough to have abundant reserves of coal. The occurrence of coal coincides with the rocks of the Damodar Valley, which contains nearly half of the coal reserves of the country. Petroleum similar to coal is a fossil fuel with varied uses, but Chotanagpur has no petroleum reserve. There are indications of the presence of sediments favorable for the occurrence of oil in Singhbhum and Hazaribag Districts. Juduguda and surrounding areas near Rakha mines contain uranium deposits, but there is no nuclear power plant in the State.

Coal is one of the primary sources of energy, accounting for about 69 % of total energy consumption in the country. It is used as fuel in industries, in thermal power stations, and for domestic purposes. It is also used as a raw material in chemical and fertilizer industries and in the production of thousands of items of daily use. Coal is principally found in two geologic formations in India, Lower Gondwana and Tertiary (Dayal 1972). The Lower Gondwana is preserved in a longitudinal trough in the Damodar Valley in Chotanagpur. As per the assessment of the Geological Survey of India, (2005–2006) the total coal reserves of Chotanagpur are around 76 million tons spread over 207 coal mines. It ranks highest in production as well as reserves of coal in India. The total reserves of coal in the region are estimated as 57,570.90 million tons; that is, 45 % of the total Indian reserves and about two thirds of the current national output. It provides 92 % of the total value of mineral production in the region. The major coalfields in the valley are Jharia, Bokaro, Ramgarh, and Karnpura, which extend in an east–west direction conforming approximately to the alignment of the Auranga and Damodar Rivers from the Hutar fields in the west to Jharia field in the east. A narrow belt is also confined to the Rajmahal area running from north to south. No coalfields are located in the southern and western part of the plateau region. Hence, long and costly haulage is necessary if coal to be supplied to these regions to develop industries, as well as power from the main producing areas (Dayal 1977). Production has increased rapidly in 50 years, from 33 million tons in 1950–1951 to 333 million tons during 2000–2001. The phenomenal growth in production was the result of nationalization of coalfields in 1973, which paved the way for the introduction of newer technology, standardization of equipment, and creation of infrastructural facilities (Misra and Puri 1996, p. 111). This growth was also the result of a change in the government policy concerning energy development. The increasing demand for electricity and the limited possibilities of increasing power generation from hydro and nuclear stations led to increasing demand for coal (Dayal 1977). As per the Geological Survey of India, this region has three to six coal seams having a thickness from 8.3 to 62.06 m. The State Mining Development Corporation (MDC) has estimated a reserve of 8 million tons of coal in the forest area of Palamau near Duga region. Other important areas in Dhanbad District near Nirsa, Lakhima, and Kapasara; in Dumka District at Hura and Chuparibita; in Godda District at Chitra. They have estimated a reserve of 164.12 million tons in west Bokaro coal field (at Loiya and Gose), 30 million tons at the Medanirai of Hutar coalfield, and 362.45 million tons from the southern Karanpura coalfield at Maurya and Lapanga.

The coalfields of Chotanagpur valley are divided into the following five regions. (1) The Damodar valley coalfield includes Jharia, Bokaro, Chandrapara, Ramgarh, North Karanpura, and South Karanpura; (2) the Giridih and Hazaribag coalfield includes Itkhori, Chop, and Giridih; (3) the Ajay basin coalfield contains Jayanti, Sahjori, and Kunditkaria; (4) the Rajmahal Hills coalfield includes Brahmani, Pachwara, Chaparbhila, Jilbari, and Hura; and (5) the Koel valley coalfield contains Daltenganj, Hutar, and Auranga. Table 10.3 and Fig. 10.3 gives the total known and estimated coal reserves in the region.

Table 10.3 Jharkhand: coalfield reserves, 2001–2002 (million tons)

	Coalfield region	Known	Estimated	Total	Percentage
1	Damodar valley coalfields	20,713.78	24,919.77	45,633.55	59.00
2	Giridih and Hazaribag coalfield	54.06	345.78	399.84	01.00
3	Ajay basin coalfield	6,002.09	13,234.01	19,236.10	25.00
4	Rajmahal Hills coalfield	1,090.60	7,982.70	9,073.30	12.00
5	Koel valley coalfield	309.78	2,154.35	2,464.13	03.00
	Total	**28,170.31**	**48,636.61**	**76,806.92**	**100.00**

Source: Indian Mineral Yearbook 2002

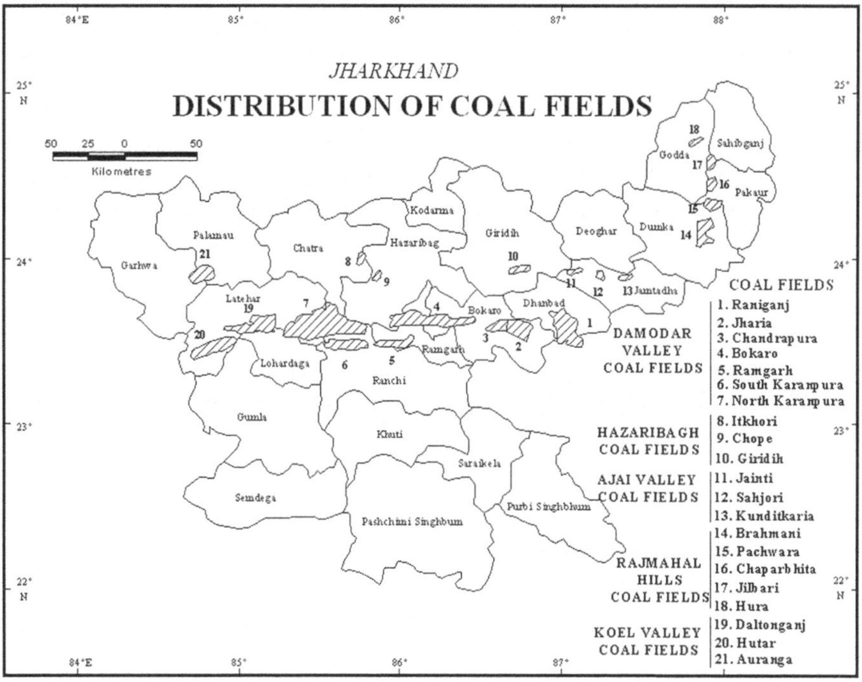

Fig. 10.3 Coalfields in Jharkhand

Recently, coal consumption has increased in the generation of electricity (67 %), production of steel (13 %), cement manufacturing (4 %), and domestic and other purposes. At current rates of production, existing coal reserves will last about 100 years. The production of coalfields in Chotanagpur has been constantly increasing, although its relative share in the country is decreasing because of the increasing exploitation of inferior-grade coal in other states. Further, coal reserves lie too deep and their extraction is not always profitable under current technological knowledge. Reorganization of old mines, introduction of new methods and techniques in the development of new mines, and standardization of plants and equipment are some of the steps adopted by the government to increase coal production. However, the coal industry has not been geared to meet the challenge of the increased demand for coal.

10.6 Atomic Minerals

The state of Jharkhand is well placed with respect to atomic mineral reserves. This region has atomic mineral deposits containing enormous quartzites of uranium, graphite, limonite, and thorium. Uranium is found in association with sulfide copper and oxidized iron ore. It occurs mainly in the Archaean crystalline schist and pre-Cambrian metamorphosed slates and phyllites. Uranium is mined in Singhbhum. The important mines are located in Dhalbhum near Sungri, Jaduguda, Narwa, Jurmadih, Bagjat, Kanyaluka, and Karuadungre, but its production comes mainly from the Jaduguda mine in Singhbhum. Graphite, a form of carbon known as bleach lead or plumbgo, is found in North Koel basin. The region has a total estimated reserve of 5.6 million tons with a value of 18.45 lakh during 2001–2002. It is used in the refractory and foundry industries, although the quality is poor. Limonite is found mainly in Ranchi district and has an estimated reserve of 0.75 million tons. It occurs as small veins and aggregates in mica-bearing pegmatites. It is used as the raw material for titanium metal in spacecraft (airplanes, rockets) and nuclear-powered submarines. Thorium is found in the rugged and dissected upper surface of Ranchi plateau and Panch Pargana in Jharkhand. This area comprises an estimated reserve of 2 lakh tons. In addition to the aforementioned atomic minerals, beryllium is found in Hazaribag, and monazite and zircon are found in Ranchi and Hazaribag Districts, respectively. It is assumed that this region has the potentiality of generating the power equivalent of 59 million tons of coal.

10.7 Production of Minerals

The richness of mineral and fuel resources is evident from the fact that this region contributed more than 50 % of the total value of all minerals produced in the State during 2000–2001. About 92 % of the estimated coal resource reserves of India is localized in this region, but it provides more than three fourths of the total coal production in Jharkhand. The region is a leading producer of minerals, such as mica, copper, appetite, bauxite, and china clay and about 40 % of iron ore. Thus, the region could emerge as a leading mining area, especially of coal, which provides power and energy to many areas. The relative position of production of various minerals and their values is presented in Table 10.4 and Fig. 10.4. The value of

Table 10.4 Jharkhand: output and value of minerals, 2001–2002

Minerals	Production	Value (thousand rupees)
Coal	76,807	46,168,688
Bauxite	1,215,301	276,287
Copper (ore + cone)	290,484	473,754
Gold	337	127,246
Iron ore	13,053	2,547,171
Manganese ore	1,161	442
Silver	15,973	116,787
Dolomite	94,048	56,053
Feldspar	10,852	1,305
Fire clay	5,163	621
Graphite	11,417	1,845
Kainite	3,620	2,816
Kaolin	39,252	22,079
Limestone	694	153,413
Mica (crude)	140	2,712
All minerals	49,963,868	49,966,297

Source: Indian Bureau of Mines 2002, p. 76

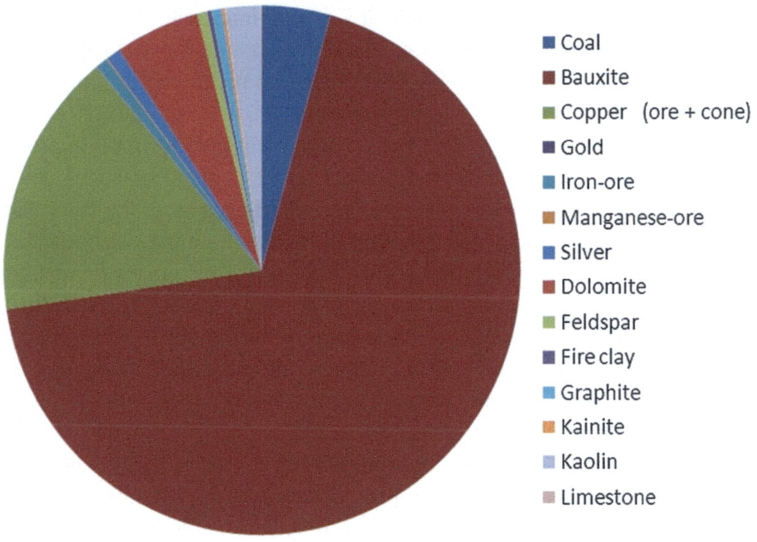

Fig. 10.4 Minerals production in Jharkhand

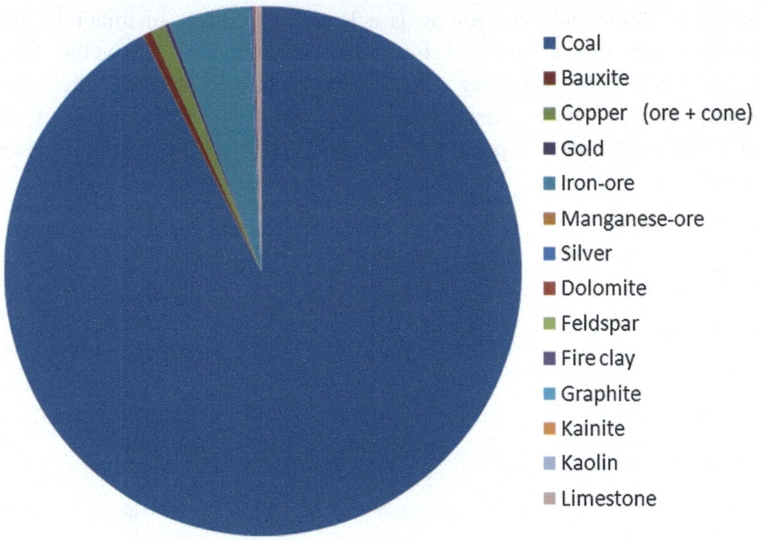

Fig. 10.5 Production value in Jharkhand (Rs. '000)

mineral production (excluding minor minerals) in Chotanagpur during 2001–2002 was estimated at rupees (Rs.) 4,997 crores, which marginally increased by about 1 % during 2000–2001 (Fig. 10.5). The region claims first position and accounts for a little more than 8 % of the total value of mineral production in the country during 2001–2002 (Indian Bureau of Mines: State Reviews, 2002, p. 76.34). The state has achieved its position mainly because of the contribution of coal. Coal alone contributed 92 % to the total value of minerals. Other principal minerals produced are bauxite, copper, iron ore, gold, dolomite, silver, and limestone.

Among the important minerals, coal, iron ore, silver, and quartz reported higher production, whereas decrease in production was noticed for dolomite (67 %), manganese ore (55 %), kaolin (21 %), kyanite (11 %), mica (crude) (22 %), copper cone (20 %), and bauxite (8 %.). The basic cause leading to decrease in production of kyanite and kaolin is labor shortage and lack of demand, particularly during 2001–2002. Production of mica (crude) also decreased because of the labor problem. Breakdown of machinery and abandonment of the Rakha mine of Hindustan Coal Limited (HCL) from February 2002 because of uneconomic operation resulted in a fall in production of copper concentrate. There was an increase of Rs. 736 crore in the total value of mineral production in the country during 2001–2002 compared to the 2000–2001 period. The principal states that reported increase in value of mineral production at national levels were Jharkhand, Madhya Pradesh, Gujarat, Chhattisgarh, Andhra Pradesh, Orissa, Maharashtra, and West Bengal. In Chotanagpur there is an excellent combination of various minerals with adequate power resources. The Damodar valley provides unique opportunities for development of a diversified industrial base, particularly ferrous and nonferrous metallurgical industries. Besides having enormous reserves of good-quality coal, the Damodar valley is the only source of cooking coal in the country. Bokaro, Chandrapura, and Patratu are important thermal power plants in the region, built after the completion of the Damodar Valley Project.

10.8 Problems

Various problems have resulted from exploitation of mineral resources in Jharkhand State. The major problems follow.

10.8.1 Depletion of Mineral Resources

Because of excessive exploitation in Jharkhand, many minerals will be depleted in the near future, which calls for conservation and judicious utilization.

10.8.1.1 Ecological Problems

Mineral extraction has led the area to suffer serious environmental problems such as land degradation, particularly in open-cast mining, land subsidence in underground mining, deforestation, atmospheric pollution, pollution of rivers and streams, and disposal of solid wastes. Open-cast mining in areas with forest cover leads to deforestation. Rapidly growing mining activity has rendered large agricultural tracts almost useless. Natural vegetation has been removed from vast tracts from Dhanbad, Bokaro, and Hazaribag Districts. Many parts of southern districts, particularly Palamau, Ranchi, and Singhbhum are worst affected. Such areas suffer from a water problem, particularly lack of proper drainage, and ultimately they become breeding grounds for mosquitoes spreading malaria with a vengeance. In hilly mining areas, landslides are a common phenomenon taking a toll on life, cattle, and property. In many mines, miners have to work under most hazardous situations. Hundreds of lives are lost each year by fire in coal mines and occasional flooding. Occurrence of poisonous gas in mining pockets is a great hazard for miners.

10.8.1.2 Pollution

Many mineral-producing areas lead to air and water pollution in the surrounding regions, which in turn leads to various health hazards (particularly for children and women), and dust-related occupational diseases such as asbestosis and silicosis. In coal-mining areas the increasing use of coal, particularly, of ash content, has adverse environmental impacts arising from emissions because coal will continue to be the main source of commercial energy of Jharkhand.

10.8.1.3 Social Problems

New discoveries of minerals often lead to displacement of people. In Jharkhand State many tribal areas are rich in minerals but the tribal population are most affected. The process of industrialization has badly shattered Jharkhand State's economy, values, and lifestyle.

10.9 Prospects

In this world of diminishing resources, it becomes essential that the mineral resources should be judiciously used by the present generation to ensure a resource base for future generations. Efforts should be made to reclaim various minerals as much as possible, which can be done by using the latest technology. Remote sensing satellites have rendered great help in identifying mineral resources. Mineral resources management requires consideration of the geologic characteristics of the area in question, its topography, and its hydrology. There is a need to consult local communities at the time of planning and to enforce any waste disposal mechanisms at the outset of the operations and even afterward (Thakur 2007). Reuse and recycling of waste materials to handle with the waste problem, which is taking a heavy toll on our topsoil, needs to be worked out for specific solutions. The necessity of having the mining and processing industries in proximity is very important (Prasad 2008).

Since the formation of the new State of Jharkhand in 2002, the government has demonstrated a vision for development in different resource and economic sectors. Resource management is dependent on the appraisal of the social, cultural, economic, and political environment and the technology available for the existing resource. The control and distribution of resources would determine their use and sustainability. Jharkhand can be rated as one of the richest states of the country as far as potential and production of mineral resources is concerned. If produced minerals had been put to industrial uses they would have given a stable industrial base, but the local situation is quite different. In spite of emphasis on regional development during the past five decades and rapidly increasing mineral output, no manufacturing industry, except for the two biggest steel plants, Bokaro and Tata Iron Steel Company (TISCO), was initiated in this State.

Among major minerals, some quantities of coal and iron ore are used for industrial purposes within the State. Several established factories such as mica, cement (Dalmianagar), copper, and aluminum remained shut down. At the same time, it can also be argued that development of such a poor state, which lacks all other factors of production other than natural resources such as minerals and forest, cannot be initiated without efficient resource use.

Simply stated, collection of revenue from resources will not create a congenial structure for development. In fact, acts and regulations of mines and minerals get in the way of the establishment of mineral-based industries. All major minerals are under the direct control of the Central Government, and location of industries based on them is not possible unless sanctioned by the government (Government of India 2002). Under these circumstances, mining areas have come up, while a major part of the state could not get the benefit of being part and parcel of the mineralized region.

In the manufacturing sector, industries with the brightest future are those related to iron and steel: metal fabrication, production of wires, nails, galvanized sheet-working hardware, aluminum hardware, and hollow-ware. Among mineral industries, much scope exists for cement, refractories, potteries and sanitary ware, abrasives, paints, and mineral wood. Among forest-based industries, lac, tasar, biri, furniture, and paper and animal products such as bone meal, glues, and leather tanning can be developed. The government has proposed setting up a number of new industries in the region, but the program is still too materialistic.

Immense potential exists for developing a systematic tourism infrastructure that will bring financial revenue. The greatest potentiality for expansion of industries in proximity to the cities of Ranchi, Bokaro, and Jamshedpur is the result of (1) locational advantages and market, and (2) in part, the incentives given by the government in terms of provision of developed land, electricity, water, and other faculties. For the region as a whole, the industrial potential is very high with regard to the rich resources because of the rapidly growing demand and increasing urbanization.

10.10 Conclusion

The landscapes of Chotanagpur, as a result of mining and industrial development, have been modified and are continually changing. The development of coal mining in Damodar valley, iron mining in Kolhan, copper mining in eastern Singhbhum, mica mining in Kodarma, and quarrying of bauxite in Ranchi have brought changes in the use of agricultural land in the surrounding regions. Accordingly, the heavy industries at Jamshedpur, the copper smelting industry in the Moshabani-Ghatshila area, and the varied industrial development in the Damodar Valley have their impact upon the economy, land use, and population of the immediate environment. The mining and industrial centers have modified not only the physical landscape but also the old patterns of settlement and human occupation. There has been a shift of population and labor force from rural areas to urban centers and the consequent decrease in the presence of the population on the land.

The rural landscape of Chotanagpur has been superimposed by a modern landscape in the chimneys and head frames of collieries, blast furnaces and coke ovens, pit heaps, quarries, modern factories, engineering works, railroads, and sprawling towns. These new features of the geographic landscape, resulting from mining and from industrial and urban developments, have also modified and are continually changing the face of the plateau. The character of farming has changed. Most of the crops and farm produce are now raised for nearby mining and industrial centers. The development of railways and roadways has also changed the economy of the hinterland trade. Transport is welding all these minerals, and the agricultural and industrial resource regions of Chotanagpur, into one major region of India. If the development of the entire area is well planned and efficiently carried out and managed, this will produce an economic or resource region of world importance.

The preceding analysis leads to the conclusion that Jharkhand State is rich in mineral resources, particularly in power resources. The situation is like sailing in a sea where there is "water, water everywhere but not a drop to drink." Minerals are there but they cannot be utilized for the development of immediate areas. The State has the responsibility of safeguarding these resources but not the right of their efficient utilization. Under these conditions, tremendous growth in mineral output simply tells a story of expanding export of mineral resources (Sharma 1992). Thus, expectations of regional development have not been fulfilled. There is urgent need to develop mineral industries and power resources at the microlevel (Thakur 2004).

To establish a sound resource base, it is necessary that the state government take immediate steps to declare a land use policy after carrying out land capability classification. The land use policy should assign and demarcate zones for each type of activities in a village or group of villages, and specifically mark areas for rehabilitation and for the four bioproductive systems (agriculture, grasslands, forests, and water bodies) for major watersheds. The inventory of resources for each watershed and land use allocation plan should be prepared at the microlevel for each village so that it forms the basis for rural development. In addition to this, there is urgent need to prevent indiscriminate conversion of agriculture and vegetation/fruit-bearing lands for the construction of buildings. There should be a strict policy for colonization in the cities and semi-urban areas, and no construction should be allowed unless all the infrastructure (adequate water, sewer system, rainwater harvesting, electricity, waste disposal, etc.) is available.

References

Dayal P (1972) Agricultural and industrial potentialities. In: Dutt AK (ed) India: resource potentiality and planning. Kendall/Hunt, Dubuque, pp 38–71

Dayal P (1977) The energy resource of India: presidential address, geology and geography section. Indian Science Congress Association, Calcutta

Gensburg N (1957) Natural resources and economic development. Ann Assoc Am Geogr 47:196–212

Government of India (2002) Ministry of coal and mines. Indian Bureau of Mines, December

Hodder BW (1971) Natural resources and tropical development. In: Mukherjee AB (ed) Developing the under-developing countries. Macmillan, London, pp 69–75

Indian Bureau of Mines (2002) Indian minerals year book, vol 1–3, Indira Bhawan, Civil Lines, Nagpur

Misra SK, Puri VK (1996) Indian economy: its development experience. Himalayan Publishing House, Bombay

Prasad M (2008) Impact of new mining technology on land degradation in coalfields of Chotanagpur. In: Thakur B (ed) Perspectives in resource management in developing countries, vol 3, Ecological degradation of land. Concept, New Delhi, pp 212–231

Ram LN (1968) Industrial belts in south-east Chotanagpur. Transact Indian Counc Geogr, Special IGU 5:79–90

Sharma SK (1992) Resource utilisation and development. Northern Book Centre, New Delhi, pp 103–111

Smith DM (1977) Human geography: a welfare approach. Edward Arnold, London

Thakur B (1998) The energy resources of India: problems and planning. Reg Symbiosis 6:75–91

Thakur R (2000) Development of natural resources in Chotanagpur Regions. In: Kumar A, Khushwaha RS, Thakur B (eds) Earth system sciences, vol 2. Concept, New Delhi, pp 415–437

Thakur B (2003) Introduction. In: Thakur B (ed) Perspectives in resource management in developing countries, vol 1, Resource management: theory and techniques. Concept, New Delhi, pp 1–45

Thakur R (2004) Utilisation, problems and management of coal energy in Jharkhand. Nat Geogr J India 50(3-4):67–78

Thakur B (2007) Emerging trends in India: resource development implications for sustainable management in developing countries. In: Thakur B (ed) Perspectives in resource management in developing countries, vol 2, Population, resource and development. Concept, New Delhi, pp 3–87

Tiqqa A, Hema Malini B (2004) Resource development vs. urbanization in Jharkhand State. Ann Nat Assoc Geogr India 24(1):48–55

Tosi JA, Voertman RF (1964) Some environmental factors in the economic development of the tropics. Econ Geogr 40(3):189

White GF (1965) The special problem of the underdeveloped countries with respect to natural resources. In: Burton I, Kates RW (eds) Reading in resource management and conservation. University of Chicago Press, Chicago, pp 422–426

Part V
Ecological Perspectives

Chapter 11
Land, Life, and Environmental Change in the Himalaya

P.P. Karan

Abstract The long-time pristine characteristics of the Himalayas have been threatened by environmental degradation. Destruction of forests and vegetation, the shrinking of the glaciers, and population pressure are many examples of the regional environmental problems faced by the Himalayan region. The region has developed distinctive patterns of land use and land ownership in response to environmental changes. The region can be subdivided into three geologic zones: the Outer Himalayas, the Middle Himalayas, and the Great Himalayas. Regional variations in environmental degradation can be observed. Conditions range from the extremely severe situation in Nepal to a moderately serious situation in the Indian Himalayas and somewhat less serious conditions in Bhutan. A geographic analysis of environmental degradation in the region posits several responses to environmental stress, such as 'planning for conserving soil and water in various subregions,' 'preservation of Himalayan environment,' 'a holistic as opposed to piecemeal approach seeking solutions to environment problems,' 'examining uncontrolled growth with conservation,' and 'addressing the role of women in Himalayan society.'

Keywords Environmental change • Himalayas • Geo-ecological zones • Regional variations in environmental degradation • Environmental conservation

11.1 Introduction

"In a thousand ages of the gods, I could not tell you of the glories of the Himalaya." So wrote a Sanskrit poet who found himself at a loss for words to describe the beauty and magnificence of the Himalaya. The beauty and magnificence remain, but there is another element that is beginning to threaten them: environmental degradation. The snow-covered peaks glisten in the sun, but barrenness marks the mountains below. The Himalayan farmer is no expert in ecology, but during the past three

P.P. Karan (✉)
Department of Geography, University of Kentucky, Lexington, KY 40506-0027, USA
e-mail: ppkaran@uky.edu

A.K. Dutt et al. (eds.), *Spatial Diversity and Dynamics in Resources and Urban Development*, DOI 10.1007/978-94-017-9771-9_11

decades he has seen his land gradually yield less and has watched with anguish the loss of his fields and cattle to landslides during the wet monsoon. The desolation of the mountains is a problem created by humans. The destruction of forests and vegetation is not an isolated instance in the Himalaya; it is only one example of regional environmental problems. On a global scale, a noxious cocktail of soot, smog, and toxic chemicals is blotting out the sun and altering weather patterns in large parts of Asia. Rain can cleanse the skies, but some of the black grime that falls to earth ends up on the surface of the Himalayan glaciers that are the source of water for billions of people in India, Pakistan, and China. As a result, the glaciers that feed into the Ganges, Indus, Yangtse, and Yellow Rivers are absorbing more sunlight and are melting more rapidly. According to the United Nations report, these glaciers have shrunk by 5 % since the 1950s and, at the current rate of retreat, could shrink by an additional 75 % by 2050. Although the overall impact of the clouds is not entirely understood, what is happening in the Himalaya may be affecting precipitation in parts of India and Southeast Asia, where monsoon rainfall has been decreasing in recent decades, and central China, where devastating floods have become more frequent.

The Himalaya extends 2,720 km from the southern edge of Central Asia to the borders of Burma and Yunan Province in China. More than 50 million people live in this region. Increasing at a rate of nearly 2 % annually, the population is being forced ever upward on the slopes of the Himalaya. The slopes are too steep and the soils too thin for intensive cultivation, even with the aid of terracing. Nevertheless, with a rapidly growing population, each hectare of arable land on the slopes must support larger numbers of people. Despite hard work, many Himalayan farmers get poor yields from their land, for it lacks the physical characteristics to give more than a meager level of sustenance. The fragile mountain environments are subjected to uses that cannot be sustained, and the middle Himalayan slopes are becoming unproductive. The farmers become ecological refugees who move to the foothills and the piedmont plain.

The rapidly increasing interregional migration of people from the mountains to the piedmont plain in Nepal, Bhutan, and the Indian Himalaya is more than a demographic phenomenon. High land–man ratios, limited employment opportunities, low income, food shortages, and steadily deteriorating economic conditions, combined with natural hazards such as soil erosion, floods, and landslides, contribute to the migration of people from the mountains. This migration is causing environmental modifications that affect long-term productivity.

The slopes of the Himalaya contain tightly compressed vegetative belts that range from warm valley floors through deciduous and coniferous forests and alpine grasslands to permanent snow cover. Human survival in the mountainous environment is dependent on efficient exploitation of these narrow zones, and Himalayan societies have developed several strategies to cope with the limitations of vertical terrain where a single productive zone cannot meet the demands of a population. Some groups specialize in herding instead of cultivation and meet agricultural needs through trade with farmers at lower altitudes. This type of exchange allows exploitation of several available eco-zones. Some communities combine farming and herding at different altitudes with a seasonal trek of livestock and herders between valley floors and high pastures. In glacial valleys such as Langtang and Khumbu in Nepal, this kind of mixed agriculture or agropastoral transhumance is especially important because each altitudinal zone has a different function in the economy. The system requires a complex, cyclical movement of men and animals to suitable climates: an upslope trek in summer in pursuit of sun and forage, and a fall retreat in search of protection from the harshness of winter. This pendulum-like rhythm unites alpine pastures and valley floors to give mountain agriculture its characteristic ecological design.

The highlands of the Himalaya have developed distinctive patterns of land use and landownership to facilitate these intricate movements. Community law typically allows private ownership of small, cultivated fields and hay meadows located near the main village, whereas upland forests and alpine pastures are held in common. Garden plots and hay meadows require intensive care, and they can be exploited by private households. However, successful grazing on high pastures and use of forests require coordinated efforts. In recent years uncontrolled cutting of trees in some parts of the Himalaya to meet the demands of trekkers and tourists have left many highland communities without essential firewood or building materials, and overgrazing on the high pastures have led to irreversible erosion.

Traditional communities and economies in the Himalaya are also being transformed by influences from outside the region. Population pressures on the limited amount of land have sent highlanders upslope as well as to areas far from their home valleys to supplement incomes. For example, wage labor and military service in India and Nepal have been outlets for some groups for a long time.

During the past half century, tourism has played an important role in the Himalayan economy. The influx of mountain enthusiasts has created new employment opportunities for the highlanders, but tourism has also accelerated environmental deterioration. Tourism is only one threat to Himalayan ecology. Construction of roads, bridges, and hydroelectric dams has negative ecological effects on the finely tuned mountainous habitats. As economic development and population growth continue to threaten environmental stability in the Himalaya, the problem of preventing further environmental deterioration becomes urgent.

11.2 Geoecological Zones

The Himalaya may be divided into three geoecological zones: the Outer Himalaya, the Middle Himalaya, and the Great Himalaya. The Outer Himalaya is the lowest zone, with an average elevation between 1,000 and 2,000 m. The zone is contiguous with piedmont plains in Pakistan, India, and Nepal. The zone varies in width from approximately 48 km in the west to a narrow strip in the east. Except for a gap south of Bhutan and Darjeeling, the Outer Himalaya zone is continuous from the Indus River to the Brahmaputra. The zone has numerous longitudinal flat-bottomed valleys, called duns, which are usually spindle shaped and filled with gravelly alluvium.

One malarial dun, the Rapti Valley in southern Nepal, was transformed by an 80-km road built with United States aid in the late 1950s. The new road and a DDT-spraying program opened the valley, which previously had been a hunting preserve for the Rana rulers of Nepal, for settlement by more than 30,000 homeless, landless farmers from the Middle Himalaya. The success of the malaria eradication program led to other resettlement and land distribution efforts in southern Nepal. The resettlement programs in the Outer Himalaya have resulted in widespread clearing of forests to make land available for newcomers. Forests of the zone in most of Nepal and India are now restricted to small, scattered patches. Reforestation efforts in the past three decades have been successful in some areas.

The Middle Himalaya zone is approximately 80 km wide and abuts the flank of the Great Himalaya. This zone consists principally of branches running obliquely from the Great Himalayan Range and other disconnected units. The chief oblique branches are the well-known Mahabharat Range, which stretches on an east–west axis across Nepal, the Dhauladhar Range, and the Pir Panjal Range. The Middle Himalaya has a remarkable uniformity of elevation, between 3,000 and 5,000 m.

About 60 % of the Himalayan population lives in the valleys and on the slopes of this geoecological zone. Sizable concentrations of population exist in some large valleys such as the Kathmandu and Pokhara Valleys in Nepal, the Paro Valley in Bhutan, and the Vale of Kashmir. The density of population in the Kathmandu Valley is more than 1,200 persons per square kilometer, in the Pokhara Valley more than 500, and in the Vale of Kashmir more than 400. The small valleys and basins have fewer people, but even in the steepest and most rugged areas every patch of arable land has been settled, so that the most thinly populated regions still have densities of approximately 25 persons/km^2.

Most cultivators in the Middle Himalaya are sustained by less than a quarter hectare of land and supplement their incomes by raising livestock and participating in cottage manufactures. The forests of this zone are receding further under the combined pressures of clearing for cultivation, uncontrolled grazing, and wood gathering for fuel. The removal of forests along the construction routes for roads also contributes to loss of topsoil, landslides, and excessive water runoff. The population of the Middle Himalaya is precariously dependent on the natural environment: any negative change in the physical environment adversely affects their survival. Environmental stress is generating a rural-to-rural migration flow within this geoecological zone and from it to the piedmont plain and the Outer Himalaya.

The highest geoecological zone is the Great Himalaya, which consists of a single range with 50 peaks of elevations more than 7,000 m. The peaks include Mount Everest (8,848 m), Kanchenjunga (8,578 m), Nanga Parbat (8,126 m), Makalu (8,470 m), Dhaulagiri (8,172 m), and Nanda Devi (7,817 m). The zone has an average width of 24 km, but southward-projecting spurs extend for a distance of 16 or more km. The Tibetan Himalaya comprises subordinate ranges on the northern flank of the Great Himalaya.

The Great Himalaya is sparsely populated; settlements are restricted to forest clearings in the high mountain valleys. An increased demand for firewood and overgrazing by livestock has initiated destruction of forests and alpine pastures in many areas. Most destructive than these demands of local origin is the use of firewood by campers and hikers who have become numerous during the past four decades. Tourism has created demands for services and materials that are slowly changing the ecology, the environment, and the economy of the Great Himalayan zone. For example, as firewood becomes increasingly scarce, cattle and yak dung is collected and burned instead of being used for fertilizer.

The net effect of overcropping, overgrazing, and overcutting in each geoecological zone is accelerated erosion of fertile topsoil. Brown-colored, silt-laden rivers–the Indus and its tributaries, the Ganges, the Yamuna, the Gandak, the Tista, and the Brahmaputra–carry away the soil that forms the basis of life for the Himalayan people. An estimated 280 million cubic meters of topsoil is washed away annually from Nepal Himalaya to the Gangetic Plain in India and Bangladesh. Himalayan silt stains the Bay of Bengal as far as 700 km from the shore, and during the devastating annual floods the debris spreads over the delta. A vast portion of the population of the Indian subcontinent lives on plains of streams that flow from the Himalaya and is thus vulnerable to consequences of ecological misuses upstream. Of major concern is the impact of environmental stress in the Himalaya on the expensive and ambitious irrigation projects that have sustained the green revolution in northern India.

11.3 Regional Variations in Environmental Degradation

Regional variations in environmental degradation exist in the Himalaya. Conditions range from an extremely crucial situation in Nepal to moderately serious in Indian Himalaya and a somewhat better situation in Bhutan. However, if rapid development continues in Bhutan without due regard for conservation, the problem may assume crucial proportions in the coming decade.

11.3.1 Western Himalaya

Environmental degradation in the western Himalaya includes widespread deforestation. The per capita forested area has continued to decline, no forests remain below 2,000 m, and the forested area in a middle belt rising to an average height of 3,000 m has been reduced substantially. In addition to increased demand for firewood and extensive lopping of trees to feed livestock, construction of a network of roads during the past five decades for strategic and developmental purposes in the border regions has been a major factor in destruction of forests, environmental degradation, and an increased number of landslides.

The rapidly increasing population has accelerated manmade pollution. Streams that formerly were clear are now polluted with refuse and domestic effluents. Hill people who use the water for drinking suffer from diarrhea: cholera and typhoid epidemics are frequent during pilgrimages to sacred places in the Himalaya. Large lakes such as Dal in Kashmir or Nainital in Uttarakhand have become polluted after an influx of tourists and pilgrims. The demand for firewood from these groups further depletes the forests. The effects of this type of influx on ecologically sensitive areas such as the Valley of Flowers between 3,658 and 3,962 m near Joshimath can be readily seen in the landscape.

Activities of the local forest departments intensify with the advent of roads in mountain areas. The main objective is to increase the yield of commercial timber and firewood. Contractors once cut only valuable timber and left secondary products such as roots and vegetal cover to stabilize the soil. Now indiscriminate felling that divests forests of their vital undergrowth prevails in many areas. Additionally, building sites are cleared to accommodate the hordes of road construction workers in temporary encampments. Replanting the cleared slopes is obviously essential to reduce further run.

The Gulaba Pass (3,600 m) was covered by dense forests only a few years ago, but it is now bare and barren. The Kulu Valley, formerly a picturesque scene of deodar trees, some 45 m high and 3 or more m in girth, is now almost barren. The felling of the tall coniferous trees has been followed by increase in the number of avalanches. Private landowners have cleared mountain slopes to plant apple orchards. The enterprise is profitable, and Himachal Pradesh has become a leading producer of apples in India. The need for packing cases to ship the apples is an additional burden on the limited forest resources. An apple tree requires 5 to 7 years of growth to bear fruit; meanwhile, farmers cultivate potatoes between the rows of apple trees and thereby promote soil erosion from the slopes.

In Kashmir, logging operations in remote forests roll cut logs down the hillsides, an activity that leads to soil erosion and landslides. The rich chir-pine forests in Kashmir have been damaged by excessive tapping of resin, which weakens the trees by sapping them dry. The weakened trees then break easily during storms. Fast-growing willow trees are traditionally a source of firewood for the people of Kashmir. Because a large portion of the harvest is used in the sporting goods industry, the population now experiences a shortage of firewood.

Livestock far outnumber people in western Himalaya. Nomadic graziers were once allowed to cross the international boundaries freely so that they had access to pastures in the Tibetan Himalaya. After the Sino-Indian hostilities, border crossing was stopped in 1962. Overgrazing and depletion of vegetal cover resulted.

11.3.2 Nepal Himalaya

Nepal, with an estimated population of nearly 30 million in 2009 living within an area of 140,797 km^2, is one of the densely settled mountain countries and one of the most populated areas of the Himalaya. With increased population, forests have been depleted, and the consequent runoff produces erosion and loss of cultivable land. As much as 40 % of former farmland in eastern Nepal has been abandoned because it is no longer fertile enough to produce crops. One fourth of the forests in the country have been cut during the past four decades. These statistics are disturbing for a predominantly agricultural economy. About 95 % of the Nepalese population lives directly off the land. Some grow rice on the irrigated piedmont plain, but most subsist on maize, millet, and potatoes grown on narrow tracts of cultivated land along mountain slopes.

The lack of land leads poor people to extend the frontiers of agriculture into marginally productive areas of the mountains. Currently nearly 60 % of the Nepalese lives in villages in the Middle Himalayan valleys at altitudes between 450 and 3,600 m. Agricultural land is scarce in the mountains, but the population is dense. Since 2001 agricultural production has contracted significantly because of pressures by Maoist insurgents on farm owners and farm workers. The impact of more than a decade of Maoist insurgency in Nepal on the environment and the economy has been significant, and research is needed for a firm assessment.

As a result of the erosion of topsoil, the bed level of many rivers on the piedmont of Nepal and adjacent India is rising. The stream courses meander, and arable land is lost. Time is running out for efforts to cope with the consequences of environmental degradation in Nepal. An effective environmental preservation program must be accompanied by efforts to improve the general living conditions of most of Nepal's poor and dispossessed people who subsist in the mountains. Many rural development projects with assistance from foreign countries or international agencies are trying to improve the environmental situation in Nepal. However, the political instability in the country during the past two decades has made it difficult to offset the damage to the environment.

11.3.3 Sikkim Himalaya

Population growth and demand for new farmland during the past four decades have pushed forest clearing higher and higher up the mountain slopes in Sikkim. The newly cleared land was first planted with maize or rice, but these crops gave little protection to the soil on the slopes. Rains washed away some of it and leached minerals on which soil fertility is dependent. The rivers of Sikkim become heavily silted during the rainy season. Dried gullies on the mountainsides dot the landscape during the dry season as visible evidence of the destruction of soil resources and environmental problems. Checking the loss of agricultural land through erosion control is a principal problem in Sikkim and Darjeeling, where hope for increased agricultural productivity is being eroded along with the soils that are necessary to support it.

Steepness of terrain, tectonic instability, heavy monsoonal rainfall, and rapid population growth are the factors that adversely affect the environment in Sikkim Himalaya. Sikkim has one of the highest rates of population growth in the Himalaya. Excluding areas above 3,000 m, cultivated land constitutes approximately 35 % of Sikkim. Since the integration of Sikkim into India in 1975, immigration has worsened population pressures on the land.

Blasting during construction of new roads has loosened the mountain slopes in many parts of the state. Excessive landslides are a consequence, and numerous landslip zones have been identified. Some slips have been forested. Stabilization of these areas requires major forestation and engineering work. Streams are polluted by debris from road construction, and the inadequate maintenance of roads and mountain slopes leads to gullying. As in the western Himalaya, opening the interior to motorized transportation hastens the destruction of otherwise inaccessible virgin forests. A developmental program consistent with environmental protection and improvement is clearly needed for Sikkim. The designation of an 850-km^2 national park near Kanchenjunga in 1977 was a step in that direction.

11.3.4 Bhutan Himalaya

Bhutan has a small population (682,000) in relationship to its area (46,500 km^2). The country has avoided many severe environmental problems that affect other parts of the Himalaya. Almost half the country remains forested. King Wangchuk has transformed Bhutan from one of the world's most poor countries to one of its more enlightened in terms of environmental protection and cultural preservation. The economy has grown at an average annual rate of 7 % during the past 25 years, largely the result of exports of hydroelectricity to India. At the same time, the king maintained strict control over the country through decrees preserving the environment and the Buddhist culture of the majority of the Bhutanese. In a society rich in sacred streams and mountains, this approach is popular. It also ensures the protection of a precious ecosystem; under Bhutan's constitution, at least 60 % of the country must be forested. The king's actions are formulated under the overarching policy of "Gross National Happiness" as opposed to economic growth at any cost. Stringent checks on the exploitation of timber have created relatively few jobs. The government will have to create jobs for the growing number of educated youths. In the short term, ecotourism is the best hope. A plan to double the number of tourists (20,000 in 2007) will create a total of 100,000 jobs in the sector. This plan will require a new international airport as the current one, at Paro, is not adequate to handle the traffic. To improve the lot of poor peasants, the government will have to lead them out of subsistence farming. To bring their produce to market, many more roads must be built through pristine forests.

11.3.5 Eastern Himalaya

The Eastern Himalaya in Arunachal Pradesh until recently contained one of the largest reserves of subtropical forests in the Himalaya. The reserves have a wide range of flora and fauna, including the rare, one-horned rhinoceros. However, pressure on the forests is mounting as a result of growing population, road building, and improved communications. The forest products are the most significant sector of Arunachal Pradesh economy. To preserve the forest, Arunachal Pradesh has stopped additional sawmills and plywood production. However, large-scale felling in recent decades has already extended into some areas of virgin forest.

Large projects as part of the flood control program on Subansiri and Dibang Rivers involve deforestation and population resettlement. A negative impact on the ecosystem can be expected. The dams would submerge large tribal habitations, including rice terraces. The tribe's people who adopted sedentary, terraced rice cultivation on the hill slopes during the past several decades may be forced to return to marginal slash-and-burn farming in the hills.

11.4 Responses to Environmental Stress

The governments of India, Nepal, and Bhutan are aware of the dangers from environmental degradation in the Himalaya. India has established one of the largest forest protection funds and plans to set up a regulatory body modeled on the U.S. Environmental Protection Agency in an effort to improve its dismal environmental track record. The $2.5 billion fund will be earmarked for the regeneration and management of forests, which have been identified as an important means of reducing carbon emissions according to a 2009 report from the Indian Ministry of Environment and Forests. An additional $1 billion in public funds will be allocated for forestry-related activities. These measures should help preserve the forests and protect the environment in the Himalaya. Various state governments have banned commercial felling in areas above 1,000 m. All forest areas in Himachal Pradesh have been nationalized to prevent overexploitation. Environmental preservation is one of the principal goals of the development plan in Bhutan. Various international agencies aid the Nepalese government in creating programs for the integrated development of mountain areas to arrest further ecological damage.

The people of the Himalaya are becoming increasingly aware of the serious ecological and environmental problems in the region. The chipko movement, launched in 1972 in Kumaon to protest the sale of ash trees to a sporting goods manufacturer by the state forestry department, has done much to create an awareness of environmental degradation in the Himalaya. The movement began when villagers, protesting the felling by contractors, literally hugged the trees. The villagers were successful, and the forestry department allotted another area to the sporting goods firm. The chipko movement has broadened from a local protest to a principal activity for environmental conservation. Its leaders have traversed the Himalaya from Kashmir to Arunachal Pradesh on foot to assess environmental damage and to mobilize public opinion in support of conservation of the soil, water, and forest resources.

Environmental degradation involves several factors, and solutions must be sought by accounting for these variations. Different methods include education, integrated corrective activities, institutional change, and encouragement of decision making at the local level. A solution to the environmental crisis cannot be general because the causes lie in varied local and regional conditions, but I outline here five policies to handle the problems.

Because environmental problems in the Himalaya have arisen from the processes of rapid development and population growth as well as physical processes, one of the challenges facing scholars of the mountain environment is to separate environmental changes caused by human activities from changes that would have occurred without human interference. Research should stress identification of the role of human and physical factors in the nature and extent of environmental degradation in Himalayan subregions. A land classification survey, accounting for soil types, erodibility, and rockiness as well as slopes and drainage, and a land use study would be necessary to development a plan for conservation of soil and water in various subregions. With data on land capacity, farmers in the Himalaya could be encouraged to put land to its best use, and a set of conservation measures could be devised to maintain the land in optimal condition.

Concern for preservation of the fragile Himalayan environment should be at the core of future developmental plans. Resources such as forests, hydroelectricity, and scenic beauty must be considered in the context of preserving the quality of mountain landscapes and environments. Overall developmental plans and specific projects must contain measures to protect the environment.

The complex environmental problems of the Himalaya require an integrated approach to their solution. Currently projects for environmental protection are spread throughout numerous governmental agencies. It is important to have an administrative structure that facilitates coordination among agencies that implement plans for resource exploitation and environmental protection.

Himalayan governments must continue to weigh carefully the relative desirability of uncontrolled growth or guided growth with conservation, of sophisticated mechanized techniques or appropriate technology in resource development, and of urban growth or rural development. It is hoped that the development in the Himalaya will stress both economic and environmental goals.

Plans for development and environmental preservation must consider the important role of women in Himalayan society. In Himalayan villages, women perform almost half the agricultural labor and most of the domestic chores. Their workload increases when environmental degradation forces them to journey ever farther in search of firewood or fodder for livestock and water. Women not only do most of the labor in obtaining foodstuffs and fuel and water, but also make many decisions about these tasks, but they have limited access to information about conservation and techniques of resource management. Women should participate in formal and informal nontechnical training programs to be effectively involved in conservation of resources and environmental management in the Himalaya.

Selected Bibliography

Barnard PL, Owen LA, Sharma MC, Finkel RC (2001) Natural and human induced landsliding in the Garhwal Himalaya of northern India. Geomorphology 40(1-2):21–35

Chadha SK (ed) (1989) Himalayan ecology. Ashish Publishing House, New Delhi

Government of India (2005) Himachal Pradesh development report. Planning Commission, New Delhi

ICIMOD (2006) Environmental assessment of Nepal: emerging issues and challenges. International Centre for Integrated Mountain Development and Asian Development Bank, Kathmandu

Jodha NS (2005) Adaptation strategies against growing environmental and social vulnerabilities in mountain areas. Himal J Sci 3(5):33–42

Karan PP (1984) Sikkim Himalaya: development in mountain environment. Institute for the Study of Languages and Cultures of Asia and Africa, Tokyo

Karan PP (1987) Bhutan: development amid environmental and cultural preservation. Tokyo University of Foreign Studies, Tokyo

Karan PP (1990) Bhutan: environment, culture and development strategy. Intellectual Publishing House, New Delhi

Karan PP (1994) Nepal: development and change in a landlocked Himalayan kingdom. Tokyo University of Foreign Studies, Tokyo

Karan PP (1996) Nepal: a Himalayan kingdom in transition. United Nations University Press, Tokyo

Khawas V (2005) Conservation and management of water resources in Sikkim Himalaya; some suggestions. Spatio-Econ Dev Record 12(4):27–31

Zurick D, Karan PP (1999) Himalaya: life on the edge of the world. Johns Hopkins University Press, Baltimore

Zurick D, Pacheco J (2006) Illustrated atlas of the Himalaya. The University Press of Kentucky, Lexington

Chapter 12
Adaptive Management of India's Wildlife Sanctuaries

Paul Robbins and Anil K. Chhangani

Abstract This chapter examines India's wildlife sanctuaries as laboratories for understanding the nuanced relationship between science, democracy, and conservation. India's 523 wildlife sanctuaries have been set aside by the Indian Wildlife Act for the purpose of conserving biodiversity and the natural heritage of the county. Amidst social and economic change, these sanctuaries face the threats of urban growth, agricultural encroachment, and exploitation by mining and timber extraction. Even so, these wildlife sanctuaries cannot be considered pristine or characterized by wilderness because they have been recently created and have long been influenced by people. The purpose of this essay is to investigate the Kumbhalgarh Wildlife Sanctuary in Rajasthan in detail to understand the intertwining relationship between new institutions and complex landscapes, thereby providing a scientific opportunity to enhance our understanding of conservation science. A close examination of this sanctuary suggests: (1) ecological transitions in the sanctuary are multidirectional, (2) extensive conservation resources reside outside the sanctuary, and (3) ecological knowledge resources are extensive and untapped. These results suggest an adaptive management approach is paramount, insofar as it would take advantage of the complex dynamics of reserves, as well as the inevitable human impact on the landscape and the considerable ecological knowledge possessed by local communities.

Keywords Wildlife reserve • Human Impact • Ecological knowledge • Kumbhalgarh • Political ecology

P. Robbins (✉)
Nelson Institute of Environmental Studies, University of Wisconsin, Madison, WI, USA
e-mail: director@nelson.wisc.edu

A.K. Chhangani
Department of Environmental Science, Maharaja Ganga Singh University, Bikaner 334001, Rajasthan, India
e-mail: chhanganiak@yahoo.com

A.K. Dutt et al. (eds.), *Spatial Diversity and Dynamics in Resources and Urban Development*, DOI 10.1007/978-94-017-9771-9_12

12.1 Introduction

This chapter argues that India's wildlife sanctuaries—distinct from national parks or other kinds of designated forest areas—present unique opportunities for science, democracy, and conservation to be better intertwined. Based on observations made by Indian colleagues and naturalists in the field of conservation, and especially upon data from the Kumbhalgarh Wildlife Sanctuary in Rajasthan over the past decade, it is clear that these opportunities are predicated on the fact that most sanctuaries are not necessarily pristine areas, but are instead *recently created and long influenced by people*. The opportunities that sanctuaries present, therefore, can only be realized by admitting the following: (1) although they are wild, they are not wildernesses; (2) they can be managed to nurture a range of species, but they cannot be "restored" to an imaginary pristine condition; and (3) they will continue to be influenced by local people and some human activities, no matter what rules and restrictions are implemented for their care. Once the distracting ideals of wilderness, pristinity, and non-humanity are abandoned, possibilities for experimentation, observation, and cooperation abound. Specifically, sanctuaries open the door to (1) the implementation of adaptive management regimes where ongoing earth observation and surveying can be used to set goals and evaluate outcomes in real time; (2) the improved reliance on local, lowest-level foresters and forest guards to help evaluate, monitor, and nurture conservation; and (3) the use of citizen science data collection among local resource users. To make the most of India's sanctuaries, we must treat them as ongoing, in situ socioecological experiments, not miniature national parks.

To begin, it is essential to assert and recognize that *India's wildlife sanctuaries are a vast socioecological science experiment*. India is home to 523 wildlife sanctuaries, areas set aside by the Indian Wildlife Act to conserve the biodiversity and natural heritage of the country, which is faced with challenges of urban growth, agricultural encroachment, and reckless exploitation from industries ranging from mining to timber extraction. Although both parks and sanctuaries are critical parts of the government's extensive efforts to protect wildlife, they differ in many important respects. The characteristics of these conservation spaces, as a result, provide a remarkable opportunity to conduct a national-scale science experiment through the implementation of adaptive management. India's 99 national parks cover 39,000 km^2, but the country's wildlife sanctuaries cover nearly 119,000 km^2 (Table 12.1). At first blush, therefore, it is possible to think of sanctuaries as the leading edge of the country's efforts. However, looks can be deceiving.

Table 12.1 Number and area of India's Wildlife Sanctuaries relative to its National Parks

	Total area (km^2)	Number	Average area (km^2)
National Parks	39,155	99	396
Wildlife Sanctuaries	118,417	523	226

Source: State of the Environment Report 2009. Government of India, Delhi

Significantly, sanctuaries, although more than five times as numerous as parks, are almost half as large on average. This is no accident. Parks have been organized around single key endemic at-risk species with large home ranges and requiring large buffers (e.g., tigers and elephants), but sanctuaries have been adopted opportunistically by carving out available Reserve Forest land that provides habitat to an enormous range of miscellaneous fauna. So too, many of these sanctuary areas are small precisely because they are surrounded by, and embedded within, landscapes that have had long historical human usage as well as intensive development and habitation. If national parks represent Indian conservation's "low-hanging fruit," that is, large areas of land under mostly state control that could be more easily converted into coherent polygons for protection of keystone species, Wildlife Sanctuaries frequently represent ad hoc and interstitial spaces, lands carved on the margins and fringes of human life and industries.

In light of this, it is also notable that many sanctuaries (as well as some parks) are, on many occasions, human in genesis, their ecologies the product of the actions of rulers, farmers, and grazers, at least in part. Some of these cases are obvious ones and are internationally famous as a result. The area around the Bharatpur Bird Sanctuary (only recently converted to Keoladeo National Park) was originally flooded by Maharaja Suraj Mal around 1760 to produce hunting habitat. With more than 300 species of birds permanently resident and seasonally migrant to the site, the reserve is unquestionably an anthropogenic wilderness. Other protected areas may not be quite so dramatically artificial, but most of them, especially smaller sanctuaries, bear the marks of human action, species selection, and landscape modification. In sum, it can be observed in general that India's Wildlife Sanctuaries:

1. Opportunistically conserve areas that have a diversity of species
2. Incorporate small, fringe parcels of marginal and interstitial land
3. Have ecological parameters that often emerge directly from human actions on the landscape

These characteristics are not unique to Indian conservation areas; rather, they are part of a global trend. The number of protected areas around the world recognized by the International Union for the Conservation of Nature (IUCN) increased from 27,794 to 102,102 between 1982 and 2003. During the same period, however, the total area under such reserves increased from 8.8 million km^2 to only 18.8 million km^2. Although the number of protected areas has tripled, the average size of a protected area is almost half what it once was (Robbins et al. 2007). These new conservation territories are not only smaller, they are increasingly close to human populations, and are carved from historically settled and heavily used lands (Parks and Harcourt 2002; Zimmerer 2000; Zimmerer and Young 1998). Couple this with the fact that the establishment of a protected area tends to increase human population density in proximity (Wittemyer et al. 2008) and that proximate populations have measurable effects on land cover within conservation sites (Karanth et al. 2006). The problem of India's proliferating sanctuaries can be seen as merely part of a global trend: an exploding conservation mandate facing the diminishing margins of frontier space.

What is unclear, however, is how the large-scale, dramatic imposition of entirely new systems of (frequently draconian and restrictive) rules on human land use will affect areas long influenced by people, close to human habitation, and historically linked to human activity. India's proliferation of protected areas amidst human activity represents, therefore, an *enormous scientific opportunity* to observe the effects of new institutions on complex landscapes. By experimenting with differing rules of use, rigorously monitoring their effects on land cover and on species diversity and structure, India stands to become a world leader for in situ conservation science, for a twenty-first century in which the luxury of pristine wild spaces will be rare around the world. Beyond this, moreover, observation of current sanctuary conditions dictates that the alternative to this sort of intentional experimentation—the naïve expectation that restrictions will produce predictable human behaviors and primitive ecological outcomes—is not substantiated by the facts. Research from Kumbhalgarh Wildlife Sanctuary underlines this basic truth.

12.2 The "Bad News" from Kumbhalgarh Wildlife Sanctuary

The Kumbhalgarh Wildlife Sanctuary is a large protected area well above the national average in size at 610 km^2. Similar to almost all other sanctuaries, however, it is *interstitial*, the product of remnant Forest Department Reserve Forest land in highly marginal uplands in the Aravalli hills. As such, it spans a corridor-shaped diagonal from 73°15′E, 25°00′N to 73°45′E, 25°30′N, flanked on its northwest side by dozens of densely populated small settlements and towns. Although the reserve is lengthy, some 50 km in length, it is as narrow as 8 km in places. The morphology of the reserve is, in this sense, a classic fragment of ad hoc conservation history.

Characterized by vegetation with deciduous forest patches, the reserve is dominated by *Anogeissus pendula* (local: Dhaw), *Boswellia serrata* (local: Salar), *Acacia senegal* (local: Kumbhat), and *Butea monosperma* (local: Palas). This patchy forest cover and steep relief provide habitat for endemic wildlife species including leopard (*Panthera pardus*), hyena (*Hyaena hyaena*), Indian wolf (*Canis lupus*), Hanuman langur (*Semnopithecus entellus*), and nilgai or blue bull (*Boselaphus tragocamelus*) (Chhangani 2000). A large number of migratory waterfowl also annually visit the reserve, bolstering the diversity of its already high avian diversity.

12.2.1 *Lesson 1: Kumbhalgarh Is Wild, but It Is Not a Wilderness*

As a sanctuary, the reserve has obvious merits. It is biodiverse. It is a breeding site for rare species (e.g., the wolf). Despite some historical inholdings, it is under the historical control of state authorities. But its credentials are also highly suspect, at least in terms of its status as a wilderness area. The area has been under the extractive control of the Forest Department since the colonial period. The earliest boundary pillars and administrative maps of the forest were probably first created in the wake of the 1887 Forest Survey of Marwar, after which explicit restrictions were phased in over time, subsistence practices were criminalized within forest boundaries, and the cutting of trees and the collection of non-timber forest products was forbidden (Chief Wildlife Warden Kumbhalgarh Wildlife Sanctuary 1996). But just as quickly as such local uses of the forest were banned, large-scale extraction began in earnest. Between 1900 and 1950 industrial forestry was instituted for the extraction of timber for railroad sleepers and the harvesting of *Acacia catechu*, a species now rare in the forest (Robbins 2000). This phase was followed by a period of intensive contract forestry between 1950 and 1972, when private companies leased timber extraction rights across the Reserve. The Reserve was also heavily hunted by colonial officers and local ruling elites over this period, with birds, wild boar, and top predators eliminated in numbers that are hard to estimate. The last tiger was likely shot dead before the early 1960s.

For three quarters of a century, therefore, the ecological structure, vegetation profile and density, and diversity profile of the forest were undoubtedly permanently altered. The establishment of the Wildlife Sanctuary in 1972 did not occur in a wilderness, or a place "untrammeled by man." The contemporary legacy of this extractive activity is difficult to evaluate without baseline data, which exist for very few if any of India's hundreds of sanctuaries, but Kumbhalgarh is not a wilderness. Similarly, we suggest, most of India's sanctuaries lack conditions close to anything that can be described as wilderness. These sanctuaries, therefore, must be managed without the luxury of any such assumption.

12.2.2 Lesson 2: Pristinity Is Impossible

Ideally, a wilderness management regime would seek to remove human impacts from the forest in their entirety and so allow the restoration of the landscape to a "pre-human" state, whatever that might be. There are several aspects about the current ecosystem status of Kumbhalgarh, however, that make that extremely unlikely. First among these is the absence of the historic top predator of the system. The absence of tigers from the forest means that, as a result of concomitant changes in top controls in the trophic system, nothing resembling pre-human conditions is likely to be achieved even if all human impacts were removed from the forest. By removing the "top cat," many prey species have come to thrive, possibly in populations that exceed in density those that existed before the elimination of tigers. Populations of langur monkeys, for example, and of Indian blue bulls are large and on the rise. More interesting, the absence of the apex predator has opened the habitat to the successful rise and persistence of second-tier predators, most notably the leopard and the Indian wolf, which have become keystone species for the reserve and central conservation concerns in the Sanctuary's management plan (Robbins et al. 2009).

Second, the likelihood of simple return to a prior state is drawn more dramatically into question by the ecological restructuring of the forest system resulting from invasive plant species. At Kumbhalgarh, the two central invasive species problems are *Prosopis juliflora* (known locally as Angrezi babul, Vilayati babul, and Sarkari babul) and *Lantana camera*. Both are found dominating many sections of the reserve and have expanded in coverage, especially in the past 15 years. Although the introduction of the species was anthropogenic (both were brought into the region by direct and intentional introductions by the Rajasthan Forest Department), it is extremely unlikely that a reduction of human activity in the sanctuary will lessen the rate and trajectory of their increase.

There are multiple vectors for seed dispersal among wild animal species present in the reserve, most notably with langur scat showing significant numbers of *Lantana* seeds and nilgai scat being a major source of *P. juliflora* seed distribution and reproduction. It is possible that other disturbance forces might lessen or curb the expansion of these invasives. It is not clear, for example, how the forest responds to fire events, and whether these would favor native species recovery or rather exacerbate invasives. It is also unclear the degree to which direct human removal of invasives from blocks or sections of the forest might retard their continued expansion. Each such intervention, however, would have to be considered an anthropogenic disturbance and highly experimental.

These changes and outcomes that make a restoration of pristinity impossible are not necessarily "bad" ones, especially if the conservation of wolves, panthers, nilgai, and langur monkeys, for example, is an important management goal. They merely draw into question the concept that removing human influences will lead to a creation of pristine environmental conditions at Kumbhalgarh. We suggest that for many. if not all, of India's sanctuaries, comparable arguments could be sustained.

12.2.3 Lesson 3: Human Impacts Are Ongoing and Difficult to Curb

Any plan to manage India's sanctuaries as wildernesses also confronts the very real fact that removing human influences and land uses from the forest has proven extremely difficult. Forest officers, in cooperation with state authorities and Central Government agents, have done a fairly good job of retarding the most destructive, widespread, large-scale activities in sanctuaries, including large-scale commercial forestry, most notably, as well as mining. More intractable have been the daily household extraction practices including grass and leaf collection, fuelwood harvesting, and grazing.

Kumbhalgarh is emblematic in this regard. A household survey of forest uses showed enormous and wide-scale use of forest products by residents of villages and towns adjacent to the reserve, *despite a total ban* on all such extractive activities, as of 2002. These are summarized, stratified by caste, in Table 12.2.

Table 12.2 Proportion of members of caste groups participating in specific forest uses (by percent)

	Ever collect firewood in forest	Collect firewood only in forest	Ever collect timber in forest	Ever graze CB in forest	Ever graze SG in forest	Ever collect palas in forest	Ever collect Dhav in Forest
Total (*n*=708 households)	70.1 %	48.4 %	52.0 %	48.7 %	40.3 %	25.0 %	19.0 %
Rajput (*n*=71, 10 % of all households)	70.4 %	53.5 %	29.6 %	70.4 %	7.0 %	15.5 %	2.8 %
Scheduled tribes (*n*=86, 12.1 %)	72.1 %	53.4 %	67.4 %	46.5 %	48.8 %	24.4 %	30.2 %
Scheduled caste (*n*=141, 19.9 %)	64.5 %	40.4 %	41.8 %	46.1 %	26.2 %	9.9 %	7.8 %
Jat (*n*=54, 7.6 %)	61.1 %	48.1 %	63.0 %	74.1 %	0.0 %	70.4 %	5.6 %
Rabari (n = 230, 32.5 %)	71.7 %	50.7 %	70.0 %	40.4 %	81.3 %	32.6 %	40.0 %
Brahman (*n*=31, 4.4 %)	64.5 %	38.7 %	16.1 %	48.4 %	9.7 %	22.6 %	3.2 %
Mali (n=23, 3.2 %)	47.8 %	39.1 %	34.8 %	47.8 %	13.0 %	30.4 %	0.0 %
Vaishya (*n*=21, 3.0 %)	50.0 %	9.5 %	0.0 %	4.8 %	0.0 %	0.0 %	0.0 %
Muslim (*n*=16, 2.3 %)	62.5 %	25.0 %	25.0 %	25.0 %	18.8 %	0.0 %	0.0 %
Kumhar (*n*=35, 4.9 %)	42.9 %	25.7 %	34.3 %	71.4 %	14.3 %	11.4 %	0.0 %

Source: Robbins et al. (2007)
Respondents were allowed to provide more than one type of forest use; percentages may add to more than 100 % across caste rows. CB = Cattle/Buffalo; SG = Sheep/Goats

Notably, a significant proportion of households rely on the forest as an important basis for household maintenance and many use the forest exclusively for sources such as firewood and grazing. This study further demonstrates that the presence or absence of offsetting resources (village common grazing land, most notably) had no influence on whether households utilized the forest. The clear conclusion is that small-scale but widespread use of the reserve is pervasive and persistent, despite national and regional efforts to halt such activities. It is likely that such activities have had a negative effect on some forest canopy. Time-series analysis of satellite images of the forest suggest some loss in forest canopy, as summarized in Fig. 12.1.

What we do *not* know is the degree to which such activities and land cover changes are damaging to wildlife habitat and, if so, for which species. Given the wide range of adaptations and landscape ecologies of the diverse animal and avian species of the forest, this remains a point only for speculation. We further suggest that this situation, of ongoing human forest use in sanctuaries, is ubiquitous in the sanctuary system, and that it has proven difficult to control. We also suggest that the actual effects of this human activity remain poorly understood.

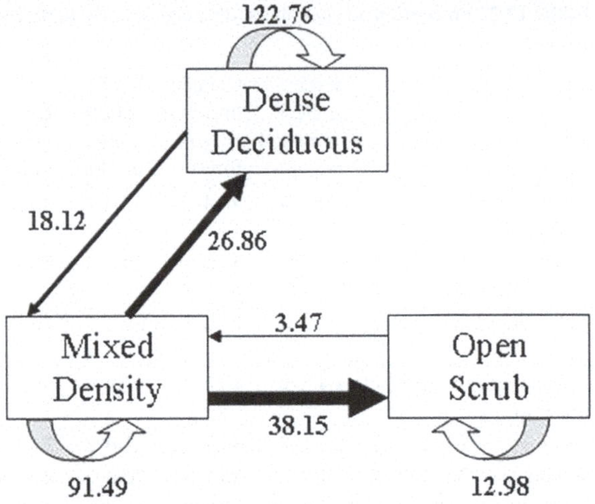

Fig. 12.1 Change in forest cover at Kumbhalgarh Wildlife Sanctuary between 1986 and 1999. (From Robbins 2000)

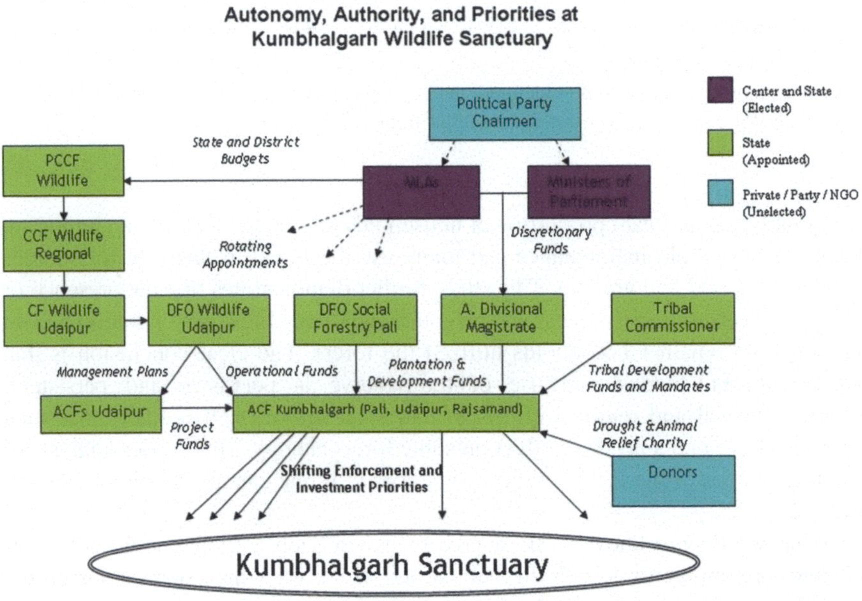

Fig. 12.2 Simplified diagram of enforcement authority at Kumbhalgarh Wildlife Sanctuary, stressing the lack of autonomy of lowest level personnel and the most local managers and observers

12.2.4 *Lesson 4: Little Room to Adapt or Change Rules and Few Resources to Monitor Impacts*

Given the ongoing changes in sanctuary conditions (e.g., invasive species, human uses, canopy transformations) and the complexity of its management regime, observation of the situation at Kumbhalgarh has underlined one further problem in management: the degree to which sanctuary managers have discretionary authority to adapt to changes, write new rules, and experiment with new techniques or activities (i.e., plantation or controlled use of fire). Between 1986 and 2009, the only major policy/management changes that have occurred at Kumbhalgarh have been increasing stringency in the rules against human forest use. These rules have largely been mandated from the district and divisional levels, and also from the Central Empowered Committee of the supreme court, rather from the level of the Wildlife Wardens, range officer, or foresters observing changes in habitat, cover, and animal populations locally. Figure 12.2, which summarizes the structure of decision making and flow of authority over Kumbhalgarh, in a highly simplified way, stresses the lack of autonomy and authority possessed by the most local level of wildlife managers.

This lack of autonomy means (1) that the ability to adapt or change rules or strategies in the face of ongoing change is limited, and (2) that the possibility to transmit learning or observation up the chain of control is impaired. We would further suggest that such a situation is in no way unique to Kumbhalgarh, but instead is typical of the structure surrounding all sanctuaries.

To summarize, observations of the situation at Kumbhalgarh stress that sanctuaries are not wildernesses in any real sense, that their restoration to pristine conditions is unlikely, that human use is ongoing, and that adaptive autonomy of foresters is limited. These points would all seem like "bad news" from the point of view of wildlife management.

12.3 The *Good News* from Kumbhalgarh Wildlife Sanctuary

Yet the situation at Kumbhalgarh shows remarkable and surprising conservation successes, and the potential for many more such successes through experimental and open-ended and adaptive management. Specifically, under close observation: (1) ecological change at the reserve has proven to be multidirectional, (2) the knowledge resources of local people and foresters have proven to be extremely rich, if untapped, (3) many of the conservation resources for wildlife have been demonstrated to be extensive, but that these frequently lie outside the sanctuary boundaries, and therefore success is possible.

12.3.1 Ecologies Are Multidirectional

Although it is true that heavy extractive uses of the forest at Kumbhalgarh have resulted in deleterious land cover change as described previously, it is also the case that surprising increases in forest cover can be observed. Figure 12.1 shows, for example, that change in forest density is bidirectional. Although there has been forest clearance, a significant proportion of forest cover has actually increased, especially succession of mixed-density forest to closed canopy. Further examination of the spatiality of the change suggests that both sites of forest increase and decrease occur in areas proximate to human populations. In a period of ongoing and heavy forest use, this raises a number of questions. Where precisely is forest cover increasing? What are the landscape-scale changes in forest pattern and structure? What impact does forest cover increase have on the conservation mission? The key general insight, however, is that sanctuaries can change in multiple directions. They may not be liable to recovering pristinity, therefore, but may be able to transform, even under human use, into viable and sustainable habitat.

Similarly, although decline of key species has been noted at Kumbhalgarh, many fauna have thrived under these recent transformations. Table 12.3 summarizes wildlife population change data for the sanctuary between 1991 and 2005. Some species of concern have clearly experienced precipitous declines, with implications for the sustainability of the sanctuary as a whole, especially including wolves, jungle fowl, and wildcats.

Remarkably, however, a number of other species are thriving in precisely this turbulent and dynamic environment, especially including the sloth bear and leopards, but also the elusive sambar deer. It is certainly the case that interpreting so short a data time series for a complex ecology is difficult. Species experiencing short-term declines may recover dramatically and vice versa, following classic

Table 12.3 Selected wildlife population change trends between 1991 and 2005 at Kumbhalgarh Wildlife Sanctuary

Species		1991	2005	Δ
Langur	*Presbytis entellus*	3,071	4,894	59 %
Sloth bear	*Melwasus ursinus*	105	162	54 %
Blue bull	*Boselaphus tragocamelus*	604	931	54 %
Leopard	*Panthera pardus*	54	82	52 %
Sambar	*Cervas unicolor*	88	122	39 %
Jackal	*Canis aureas*	312	300	–4 %
Hyena	*Hyaena hyeana*	125	119	–5 %
Mongoose	*Herpestes smithi*	162	149	–8 %
Wildcat	*Felis chaus*	76	65	–14 %
Grey jungle fowl	*Gallus sonneralii*	629	430	–32 %
Wolf	*Canis lupus*	85	47	–45 %

Δ refers to change (in %) in the table between 1991 and 2005

Lotka–Volterra long-wave interactions between predators and prey (Lotka 1925). Preliminary evidence suggests, however, that many of these trends are real, and the result of adaptation of species to changing conditions. Langur monkeys (which provide a key prey base for leopards), for example, have adapted to invasive species, especially *Lantana camera*, for new and expanding forage resources. Leopards have come to prey extensively on local livestock so as to maintain strong reproductive populations. These results suggest much the same implications as those for land cover. The ecology of Kumbhalgarh is going through an ongoing transformation that makes it unlikely to ever recover to pristine wilderness conditions. This caveat does not rule out in any way, however, the possibility of species-specific successes and successful conservation. What we do not know, however, is which, if any, of the anthropogenic impacts on the forest (e.g., grazing, tree clearance, grass harvesting) have come to favor which species, and why. In the absence of experimental monitoring of these impacts and direct intervention into the sanctuaries rules of access and use, these questions will remain unanswered.

12.3.2 Extensive Conservation Resources Lie Outside the Sanctuary

It is also increasingly clear that the conservation successes observed at Kumbhalgarh are not independent from human land uses and actions outside the reserve. Specifically, cropping and livestock raising unquestionably provide subsidies for conservation efforts, supporting both predatory and grazing species in the reserve.

Results from Chhangani's recent village survey (2008) reveal that approximately 75 % of households report experiencing crop raids by wild animals from the reserve, 50 % of these reporting nightly raids during the cropping season. Of these, 98.5 % report blue bull raids, 47.4 % wild boar raids, and 14.9 % raids by langur monkeys. Notably, at least two of these species are species thriving in the reserve. Although it is true that increased wildlife populations unquestionably lead to crop raiding, the reverse is also clear: agricultural resources are important to the conservation success of these species.

Much the same can be said of predatory animal activities. Of the households, 37 % report livestock loss to predators, with averages as high as 3.9 animals reported lost per month per household with large herds (local herd sizes can exceed 300 sheep and goats). Aggregating this effect over the study region, this represents losses of dozens or hundreds of animals per month, and must therefore be considered a substantial proportion of food resources for the current 129 wild predators in the reserve. Notably, 79 % of these losses were reported by herders admitting that their herds experienced this predation while grazing (illegally) within the sanctuary. Because these illegal losses cannot be remunerated, herders simply consider this an overhead operating expense, a kind of tax paid to conservation activities in exchange for grazing resources in the reserve.

Table 12.4 Species conservation success and nuisance status

Mammals of conservation concern	Households reporting nuisance (%)	Species population change 1991–2005	Species population trend 1991–2005 (% of 1991)
Blue bull	73.34	327	54.14 %
Wild boar	35.96	–408	–64.66 %
Leopard	26.79	28	51.85 %
Hyena	26.37	–6	–4.80 %
Sloth bear	23.69	57	54.29 %
Jackal	20.73	–12	–3.85 %
Langur	18.19	1,823	59.36 %
Wolf	08.46	–38	–44.71 %
Four-horned antelope	00.00	–105	–49.76 %
Chinkara gazelle	00.00	–27	–72.97 %

Source: Chhangani et al. (2008)

Table 12.4 summarizes the relationship between conservation successes and nuisance status for key species of concern. Clearly there is a positive relationship overall between the ability of animals to access important resources from outside the reserve and their ability to thrive and so meet the conservation goals of the reserve.

Nor is any of this unique to Kumbhalgarh. The great success of the aforementioned Bharatpur Reserve, which notably attracts hundreds of species of migrant birds annually, is precisely a result of the fact that its wetland marshes are surrounded by grain fields rich in forage resources and nest thatching. The success of many such reserve "islands" is likely linked to their surroundings. This likelihood is reinforced by the findings of Jai Ranganathan and colleagues (2008), who have established that avian diversity in production-oriented forests associated with betel nut production is equal to or higher than that of nearby conservation forests. That there is some interaction between these "natural" and "artificial" landscapes in the conservation of species in difficult to doubt. Thus, the good news from Kumbhalgarh, applicable to other sanctuaries in India, is that conservation success can be as much *aided* as thwarted by humanized resources and activities in the areas surrounding reserves.

12.3.3 Ecological Knowledge Resources Are Extensive but Untapped

Equally good news from Kumbhalgarh is the increasing evidence that local producers and local foresters have extensive ecological knowledge of forest conditions. A great deal of attention is paid (around the world and in the United States as much as in India, it should be noted) to the ecological knowledge of local people. Surveys of local people suggest that producers who use forest resources can identify dozens of species, can enumerate the interactions between human and climate impacts and species distribution (Robbins 2000), and can identify the range and habits of most local fauna. They know where panthers, sloth bear, and wolves predominate and can track their observations and encounters over time. Although this knowledge is occasionally brought to bear on individual management challenges (for example, finding a nuisance predator preying on village sheep), there is currently no formal way for local information about plants and animals to be used in the design of rules or the execution of conservation activities.

Much the same can be said of forester knowledge. Most of the foresters at Kumbhalgarh have been in their posts for more than 20 years. Their amassed collective knowledge of plant and animals species has, unsurprisingly, proven enormous. And yet, as for that of local people, the foresters' knowledge goes largely untapped. When researchers from the School of Desert Sciences conducted a training of foresters to collect botanical information at sites throughout the forest in 2006, foresters demonstrated knowledge of almost all tree, shrub, and grass species in the forest and the favorability of specific floral configurations for habitats of birds and mammals. At the conclusion of the exercise, however, foresters noted that this was the first "actual forestry" they had been called upon to do for many years. As the previous generation of foresters continues to age, moreover, it is entirely unclear that the next generation of practitioners, although well trained in geographic information science and other critical skills, will possess anything close to the contextual ecological knowledge required for success of the Sanctuary's mission. Nor is it clear how current institutional configuration might take advantage of that knowledge, assuming they had it.

In sum, Kumbhalgarh is a reserve that is deeply humanized and impossible to restore to a pristine state, but it is also one where conservation successes are ongoing, where resources exist both inside and outside the forest to conserve critical species, and where local people and foresters possess extensive knowledge. What form of management might take advantage of the inevitable dynamics of the reserve, the extensive reality of human impact, and the intellectual and knowledge base possessed within local communities and the forest department itself?

12.4 Adaptive Management: A Modest Proposal

The answer must be adaptive management. Adaptive management (following Holling 1978) is here understood as a management regime that proceeds based on the assumption that the way the managed system operates is *poorly understood and unpredictable*. Such an assumption, as we have demonstrated here, definitely applies to Kumbhalgarh and probably applies to almost all reserve areas. With that in mind, management decisions are made to intentionally provoke experimental conditions. By simultaneously implementing varied policy treatments and then comparing their results, it is possible to test hypotheses about the behavior of complex systems, even while managing them for desired outcomes.

> Experimentation in this sense goes beyond management through trial and error and casual observation; it is structured and theoretically driven, designed to elicit specific responses from systems under study such that new knowledge can be incorporated systematically into future treatments.... The approach also recognizes that managed systems present moving targets influenced largely by human drivers and, therefore, explicitly incorporate these human factors into management experiments. (Arvai et al. 2006, p. 218)

When applied to Kumbhalgarh, and by implication to all Sanctuaries in India, what this would entail is (1) the *freedom* to adapt and invent new conservation rules, (2) the *opportunity* to directly intervene in environmental systems, (3) the *necessity* of democratic and scientific monitoring, and (4) the *obligation* to change rules to create new outcomes.

12.4.1 The Freedom to Adapt and Invent New Rules

Because India's wildlife sanctuaries are by their nature experimental, there is no reason to believe that any given, centrally determined rule system (e.g., one that disallows grazing, no matter what, everywhere, all the time) will always help to achieve conservation goals, under all circumstances. As a result, it is essential to deliberately design experimental interventions in reserve rules, *based on what we want to know* (e.g., how do grazed areas actually differ from ungrazed ones?). The development of these rules would necessitate allowing responsible regional and local authorities, in consultation with appropriate experts from the Indian Wildlife Institute, for example, the freedom to invent management rules that may differ from those of other reserves, and that may be uneven across a single reserve, as where part of a reserve may be subjected to a treatment, such as allowing grazing, deliberate burning, or plantation of a specific species, and other parts of the reserve kept as untreated controls.

12.4.2 The Opportunity to Intervene

These rules must then lead to deliberate human actions that follow from the designed rules. For conservation, the list of interventions is lengthy, but some areas or subareas would have to have grazers removed under full enclosure but others would allow them access, for example. Some areas might be subject to plantation while others to deliberate species removal (of exotics, for example). Fire might be applied in some areas and withheld elsewhere. Culling of species might take place, or reintroduction. By admitting that human actions in part craft the landscapes of conservation, no single management technique should, a priori, be eliminated from consideration as a tool for any given reserve.

12.4.2.1 The Necessity of Democratic and Scientific Monitoring

None of these interventions makes any sense if the outcomes are not monitored. To know whether a specific practice has a significant result requires conditions to be observed before, during, and after new rules are implemented. Such monitoring, across more than 500 reserves, would be too resource demanding. Fortunately, available resources are more extensive than previously recognized.

In the first place, the power of remotely sensed data, both from India's own impressive satellite systems as well as those of other international consortia (e.g., SPOT Image), allows regular, controlled, monitoring opportunities, at least of land cover change in Reserves. By better coordinating the monitoring activities of the Indian Space Agency with local reserve manager needs, more deliberate efforts can be made to test specific hypotheses. Remotely sensed data are, by their nature, limited in resolution and their ability to discern relevant objects. Wildlife themselves, along with specific distributions of floral resources and habitat conditions, will necessarily require monitoring at a finer scale. Here, the vast and totally untapped knowledge of local people and foresters might be brought to bear.

In terms of forestry, the role of local foresters and cattle guards in recording and monitoring environmental conditions has been extremely limited, and on the decrease in the period since forests set aside for economic production have been transitioned to sanctuary status. For adaptive management to work, the lowest level foresters would need to be empowered to collect data (e.g., scat, pugmarks, etc.) not only during the periodic census that now constitute the forest mandate, but on a rolling and deliberate experimental basis, precisely to sample the forest to discern meaningful differences or changes where new rules have been implemented.

But even full forester participation would likely be unequal to the data challenges of an adaptive regime. Citizen science, where monitoring, reporting, and accounting for local species is performed by the people themselves (Irwin 1995), is the only viable model for decentralized experimental sanctuary design and management. The incentive for participation in such activities, especially on the part of local producers, is necessarily their inclusion in the decision-making process and rule crafting, the forwarding of actual researchable questions, the allowance of their uses of sanctuary areas under certain regime conditions, and/or their share in any remunerative benefits of conservation. More than participatory management, therefore, there is a need for *participatory science*.

12.4.3 The Obligation to Adapt

Finally, any effort at adaptive sanctuary management requires that once an outcome has been observed for better or for worse or, whenever a change in conditions occurs for reasons exogenous to management plans (as a result of climate stress, for example, or catastrophic events), it is necessary to change the management rules. The challenge that such an approach represents is profound, because it requires a reorientation of thinking within management bureaucracies. Managers must be allowed to consider new conditions and imagine different outcomes. In so doing, they can then begin the consultative process of designing changes to existing rules, precisely to test new ideas about how the managed system works. By allowing change, new techniques must be allowed for conservation on a continuous basis, so that managers can be made ready to prepare for contingency and empowered to make substantive and desirable changes.

12.5 Conclusion

The barriers to such a change in mental outlook within sanctuary management in India are precisely those that opened this essay, however. Assuming that wildlife sanctuaries are wildernesses, that they can be conserved back to a pristine state, and that human impacts can be fully disallowed in conservation areas, makes it literally impossible to even imagine adaptive governance. It further imposes an undemocratic and necessarily unscientific attitude toward wildlife governance, excluding the possibility that wildlife scientists, foresters, and local people might pursue answerable questions to preserve wildlife, precisely by sharing the forest. As we have tried to demonstrate here, these assumptions are empirically unsustainable and politically dangerous. They are bad for panthers and for people. By jettisoning these impediments for India's sanctuaries, the country can lead the way internationally, through its ongoing experiment to save the country's treasured wild heritage.

References

Arvai J, Bridge G, Dolsak N, Franzese R, Koontz T, Luginbuhl A, Robbins P, Richards K, Korfmacher KS, Sohngen B, Tansey J, Thompson A (2006) Adaptive management of the global climate problem: bridging the gap between climate research and climate policy. Clim Change 78(1):217–225

Chhangani AK (2000) Ecobehavioural diversity of langurs (*Presbytes entellus*) living in different ecosystems. Unpublished Ph.D. thesis. J. N.V. University, Jodhpur

Chhangani AK, Robbins P, Mohnot SM (2008) Crop raiding and livestock predation at Kumbhalgarh Wildlife Sanctuary, Rajasthan, India. Hum Dimens Wildl 13(5):305–316

Chief Wildlife Warden Kumbhalgarh Wildlife Sanctuary (1996) Management plan: Kumbhalgarh Wildlife Sanctuary 1996–1997 to 2000–2001. Rajasthan Forest Service, Udaipur

Holling CS (1978) Adaptive environmental assessment and management. Wiley, New York

Irwin A (1995) Citizen science: a study of people, expertise, and sustainable development. Routledge, New York

Karanth KK, Curran LM, Reuning-Scherer JD (2006) Village size and forest disturbance in Bhadra Wildlife Sanctuary, Western Ghats, India. Biol Conserv 128:147–157

Lotka AJ (1925) Elements of physical biology. Williams & Wilkins, Baltimore

Parks SA, Harcourt AH (2002) Reserve size, local human density, and mammalian extinctions in U.S. protected areas. Conserv Biol 16:800–808

Ranganathan J, Daniels RJR, Chandran MDS, Ehrlich PR, Daily GC (2008) Sustaining biodiversity in ancient tropical countryside. Proc Natl Acad Sci U S A 105(46):17852–17854

Robbins P (2000) The practical politics of knowing: state environmental knowledge and local political economy. Econ Geogr 76(2):126–144

Robbins P, Chhangani AK, Rice J, Trigosa E, Mohnot SM (2007) Enforcement authority and vegetation change at Kumbhalgarh Wildlife Sanctuary, Rajasthan, India. Environ Manag 40(3):365–378

Robbins P, McSweeney K, Chhangani AK, Rice JL (2009) Conservation as it is: illicit resource use in a wildlife reserve in India. Hum Ecol 37(5):559–575

Wittemyer G, Elsen P, Bean WTA, Burton CO, Brashares JS (2008) Accelerated human population growth at protected area edges. Science 321(5885):123–126

Zimmerer KS (2000) The reworking of conservation geographies: nonequilibrium landscapes and nature-society hybrids. Ann Assoc Am Geogr 90(2):356–369

Zimmerer KS, Young KR (1998) Nature's geography: new lessons for conservation in developing countries. University of Wisconsin Press, Madison

Chapter 13
Nutrition and Nutritional Deficiency Diseases in Sonbhadra District, Uttar Pradesh (India)

Mahendra Bahadur Singh and Vimalesh Kumar Singh

Abstract The present study points out substantial variation in food consumption across communities. Hindu respondents have a better dietary intake than the Muslim counterparts. This statement is not in conformity with the NFHS-2 India survey report (1998–1999, p. 161). With regard to composite nutritional status index, 96.9 % of Muslim respondents fall in the category of poor nutritional status whereas in Hindus this proportion is 66.8 %. Religion, education, occupation, and income are important predictors for better nutritional status. Of the total 420 respondents, about 63.3 % are suffering from some kind of deficiency disease. This proportion is slightly higher (68.8 %) in Muslims. Among Hindus, the condition of scheduled caste/scheduled tribe (SC/ST)[1,2] respondents is poorer (68.6 %) compared to Other backward classes (OBC)[3], and General caste (general caste). Among SC/ST respondents, asthma is most prevalent followed by scurvy, anemia, and dental decay. Among OBC, scurvy is the most prevalent followed by anemia and dental decay. In contrast, general caste respondents suffer more from diabetes, followed by leprosy and asthma.

Keywords Poverty • Nutritional food intake • Composite nutritional status index • Multivariate analysis • Nutritional deficiency diseases

---

[1] Scheduled Caste (SC)- is a constitutionally declared collection of castes, which suffered from the practice of untouchability.

[2] Scheduled Tribe (ST)- is a tribe listed in the scheduled list of tribes by Government of India and is identified on the basis of certain criteria such as primitive traits, distinct culture, geographical isolation and general backwardness.

[3] Other Backward classes (OBC)- are castes which have been identified as economically/socially backward by the Central and State governments in India.

M.B. Singh (✉) • V.K. Singh
Department of Geography, Banaras Hindu University, Varanasi, India
e-mail: singhmb@rediffmail.com; vksgeography@rediffmail.com

A.K. Dutt et al. (eds.), *Spatial Diversity and Dynamics in Resources and Urban Development*, DOI 10.1007/978-94-017-9771-9_13

13.1 Introduction

To achieve sustainable human development, it seems imperative to determine nutritional food intake and nutritional deficiency across the society so that nutritional deficiency diseases could be brought to the fore, thus enabling planners to suggest a sound and purposeful plan pertaining to future well-being and health. An assessment of caloric intake has long emerged as a significant geographic problem, but no satisfactory and complete answer has so far been proposed (Singh 1972).

Poverty is an enemy of society, and its elimination requires global attention through a well-formulated strategy. Poverty is the root cause for poor nutritional intake and low nutritional level in any society. Malnutrition adversely affects mental development, physical growth, productivity, and the span of working hours, and as such significantly influences the economic potential of men and women, thereby hindering the economic progress of the country (Chitralekha 1982).

A nutritional availability study becomes essential for several reasons. On the one hand, it provides reliable information for food planning. On the other hand, nutritional deficiency affects the quality of the population and its mobility behavior and as such it may be employed as a yardstick for measuring the level of socioeconomic development (Singh et al. 1997). Nutritional availability reveals the actual food available for human consumption in an area at a given point of time. The measurement of nutritional availability rests on the determination of the quality of food that may affect human consumption locally or otherwise (Dubey and Mishra 1984). After gaining some idea of nutritional deficiency, the diseases related to it can be ascertained.

13.2 Methodology and Study Area

The present study was conducted in Sonbhadra District, which is located between latitudes 23°51′22″N and 24°53′16″N and longitudes 82°31′55″E and 83°33′45″E covering an area of 6,788 km^2. The Sonbhadra District is divided into eight community development blocks: Robertsganj, Ghorawal, Chatara, Nagwa, Chopan, Meyorpur, Dudhi, and Babhani (Fig. 13.1). Sonbhadra District may be classified as a largely rural district with merely 19–20 % of the population living in urban areas (Obra, Robertsganj, Churk, Duddhi, Chopan, Ghorawal, Renukoot, and Pipri towns). According to the 2001 census, the study area was inhabited by 1,463,519 people, and during the past decade (1991–2001), the population in each block has increased tremendously with an average increase of 36.49 %.

13.2.1 Data

The present work is an outcome of intensive fieldwork. This study covers almost all salient features of nutrition and nutritional deficiency diseases in the people of Sonbhadra District. The primary information related to nutrition and nutritional

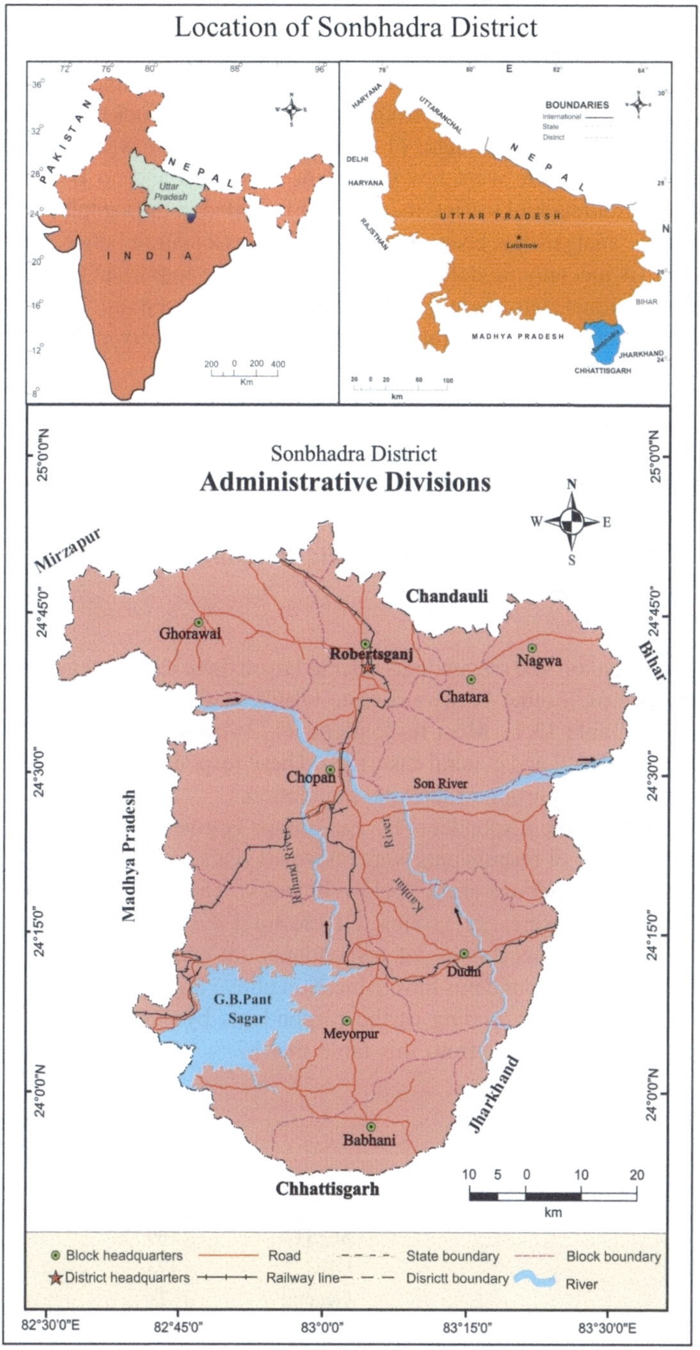

Fig. 13.1 Location map of Sonbhadra district

deficiency diseases has been generated through a questionnaire-based survey of 420 respondents. These respondents were taken from 28 villages (15 respondents from each village) based on purposive random sample. In fact, samples had to be collected from 32 villages (4 villages from each development block) but as a result of guerrilla groups affecting 4 villages of the Nagwa block these were left in the sample survey. Because there were few Christians and Sikhs and no Baudh and Jain populations in Sonbhadra District, they are not included in the sample survey.

Multivariate analysis has been used to show the association between nutritional diseases and socioeconomic determinants. To know the important predictors for a composite nutritional status index, logistic regression has been run. For multivariate and composite index analysis, Strata SE 9.0 and SPSS 16.0 software were used, and Sigma plot 8.0 was used for graph preparation.

13.3 Results and Discussion

13.3.1 Socioeconomic Profile of the Respondents in Sonbhadra District

This study is based on 420 respondents selected from three caste categories: General caste (general caste), other backward caste OBC), and scheduled caste/scheduled tribe (SC/ST) (Table 13.1). Most respondents (173) belong to the OBC caste followed by SC/ST (169) and general caste (78). These respondents have been selected according to their proportion of the total population.

Classification of the respondents by age groups is given in Table 13.2, showed that more than 49 % of respondents were above 40 years of age. Only 11.2 % of the respondents are of lower age group (below 30 years). The number of respondents is increased with increasing age groups. Of the total 420 respondents, 43.8 % of respondents are illiterate, 28.8 % of respondents are educated up to high school level, and 27.4 % of respondents have education above high school (Table 13.3).

Table 13.4 indicates grouping of the respondents by caste according to their educational level. It is clear from this table that 87 % of SC/ST respondents are

Table 13.1 Caste-wise survey information in Sonbhadra district, 2008

	Respondents	
Castes	Number	Percent
SC/ST	169	40.2
OBC	173	41.2
General	78	18.6
Total	420	100.0

Source: Personal survey (2008)
SC/ST scheduled caste/scheduled tribe, *OBC* other backward caste, *General* general caste

Table 13.2 Age structure, 2008

Age group (years)	Respondents Number	Percent (%)
Less than 30	47	11.2
30–35	58	13.8
35–40	108	25.7
More than 40	207	49.3
Total	420	100.0

Source: Personal survey (2008)

Table 13.3 Level of literacy, 2008

Education level	Respondents Number	Percent (%)
Illiterate	184	43.8
Up to high school	121	28.8
Above high school	115	27.4
Total	420	100.0

Source: Personal survey (2008)

Table 13.4 Caste-based literacy, 2008

	Education							
	Illiterate		Up to high school		Above high school		Total	
Castes	Number	%	Number	%	Number	%	Number	%
SC/ST	147	87.0	16	9.5	6	3.6	169	100.0
OBC	37	21.4	75	43.4	61	35.3	173	100.0
General	–	–	30	38.5	48	61.5	78	100.0
Total	184	43.8	121	28.8	115	27.4	420	100.0

Source: Personal survey (2008)

illiterate, with only 9.5 % literate up to high school and 3.5 % above high school. About 21.4 % of OBC respondents are illiterate; 43.4 % are literate up to high school level and 35.3 % above high school. In general caste, all respondents are literate. Table 13.5 reveals that 44.3 % of respondents belong to the lower income group (less than Rs 2,000 per month), 39.5 % belong to the income category of Rs 2,000–3,500 per month, 9 % to the income category of Rs 3,500–5,000 per month, and 7.1 % respondents have an income above Rs 5,000 per month.

13.3.2 Analysis of Nutritional Food Intake

The consumption of a variety of nutritious food is essential for maintaining good health. A well-balanced diet contains adequate amount of protein, fat, carbohydrates, vitamins, and minerals. Meat, fish, eggs, milk, and pulses are rich in

Table 13.5 Income group and number of respondents, 2008

Income [in rupees (Rs)/month]	Respondents	
	Number	Percent (%)
Less than 2,000	186	44.3
2,000–3,500	166	39.5
3,500–5,000	38	9.0
More than 5,000	30	7.1
Total	420	100.0

Source: Personal survey (2008)

Table 13.6 Nutritional food intake, 2008

Food items	Condition of intake				Total
	Daily (%)	Weekly (%)	Occasional (%)	Never use (%)	
Milk/curd	53.3	6.0	40.7	–	420
Fruits	–	29.0	71.0	–	420
Vegetables	70.2	21.2	8.6	–	420
Eggs	2.4	29.0	56.9	11.7	420
Meat/fish	–	3.6	82.4	14.0	420
Sugar	95.0	–	5.0	–	420
Ghee/oil	100.0	–	–	–	420
Pulses	61.4	31.7	6.9	–	420

Source: Personal survey (2008)

protein. Green vegetables are rich sources of iron, folic acid, vitamin C, carotene, riboflavin, and calcium. Vitamin C is also obtained from many fruits. Bananas are rich in carbohydrates. Papayas, mangoes, and other yellow fruits contain carotene that is converted into vitamin A. Vitamin A is also present in milk and milk products as well as in egg yolks (Gopalan et al. 1996).

To obtain information about nutritional food intake, the people were questioned about how often they consume various types of food (daily, weekly, occasionally, or never use). In the study area respondents consume pulses on a daily basis and vegetables frequently; 53.3 % of respondents use milk and curd each day while 40.7 % of respondents use milk occasionally. A variety of fruits is not eaten every day. This food item is consumed on a weekly (29 %) and occasional (71 %) basis. The majority of respondents (70.2 %) consume vegetables at least each day. Akin to fruits, in context of the consumption of eggs, fish, meat, and chicken, the condition of the respondents is pitiable (Table 13.6). About 56.9 % of respondents eat eggs occasionally, 82.40 % use meat and fish occasionally, about 11.70 % of respondents say that they never eat eggs, and 14 % of respondents never eat meat or fish. However, the never use category for such items in the Indian dietary system does not

Table 13.7 Religion and nutritional food intake, 2008

	Muslims				Hindus				
	Daily	Weekly	Occasional	Never use	Daily	Weekly	Occasional	Never use	
Food items	%	%	%	%	%	%	%	%	Total
Milk/curd	25.0	21.9	53.1	–	55.7	4.6	39.7	–	420
Fruits	–	3.1	96.9	–	–	31.2	68.8	–	420
Vegetables	37.5	62.5	–	–	72.9	17.8	9.3	–	420
Egg	3.1	25.0	68.8	3.1	2.3	29.4	55.9	12.4	420
Meat/fish	–	–	96.9	3.1	–	3.9	81.2	14.9	420
Sugar	96.9	–	3.1	–	94.8	–	5.2	–	420
Ghee/oil	100.0	–	–	–	100.0	–	–	–	420
Pulses	59.4	40.6	–	–	61.6	30.9	7.5	–	420

Source: Personal survey (2008)

reveal the true picture of nonavailability/affordability because a sizeable proportion of Hindu community respondents refrain from consuming these food items for religious taboo reasons.

The present survey points out a substantial variation in food consumption of respondents across the communities. Hindu community respondents have a better dietary intake than Muslims (Table 13.7). This statement does not conform to the NFHS-2 India survey report (1998–1999), which states that Muslims consume every food item except dairy products more often than Hindus. The most interesting fact of this survey is that only in the Muslim community is there not a single respondent in the "never consume" category. A large number of the respondents of the SC community consume fruits, meat/fish, eggs, and milk/curd occasionally (Table 13.8). In context of dietary intake, OBC respondents come next to SC respondents. Thus, it can be inferred that respondents of SC and OBC communities have a relatively poor diet that is particularly deficient in milk and curd, fruits, eggs, and fish, meat, and chicken.

13.3.3 Composite Nutritional Status Index (CNSI)

To avoid difficulty in interpretation of the dietary system (daily, weekly, occasional, and never use), as well as the variety of food items, composite nutritional status index has been computed by assigning a weight to each item. This measure helps to provide a rationalized means for discussion and multivariate analysis.

Table 13.9 reveals the composite nutritional status index in the study area by religion. Among the total sampled respondents, 7.62 % of respondents belong to the Muslim religion and the majority, 92.38 % of respondents, are of the Hindu religion. Of the total Muslim respondents, 96.9 % of respondents are characterized by poor

Table 13.8 Caste and nutritional food intake, 2008

	SC/ST				OBC				General				
	Daily	Weekly	Occasional	Never use	Daily	Weekly	Occasional	Never use	Daily	Weekly	Occasional	Never use	
Food items	%	%	%	%	%	%	%	%	%	%	%	%	Total
Milk/curd	45.0	4.1	50.9	–	46.2	10.4	43.4	–	87.2	–	12.8	–	420
Fruits	–	11.2	88.8	–	–	21.4	78.6	–	–	84.6	15.4	–	420
Vegetables	46.2	32.5	21.3	–	80.3	19.7	–	–	100.0	–	–	–	420
Egg	–	24.9	70.4	4.7	2.9	36.4	59.0	1.7	6.4	21.8	23.1	48.7	420
Meat/fish	–	–	95.3	4.7	–	7.5	90.8	1.7	–	2.6	35.9	61.5	420
Sugar	95.9	–	4.1	–	96.5	–	3.5	–	89.7	–	10.3	–	420
Ghee/oil	100.0	–	–	–	100.0	–	–	–	100.0	–	–	–	420
Pulses	30.8	54.4	14.8	–	76.3	21.4	2.3	–	94.9	5.1	–	–	420

Source: Personal survey (2008)

Table 13.9 Religion and composite nutritional status index

	Composite nutritional status index				
	Poor nutritional status		Better nutritional status		
Religion	Number	%	Number	%	Total
Muslim	31	96.9	1	3.1	32
Hindu	259	66.8	129	33.2	388
Total	290	69.0	130	31.0	420

Source: Personal survey (2008)

Fig. 13.2 Nutritional status index by region

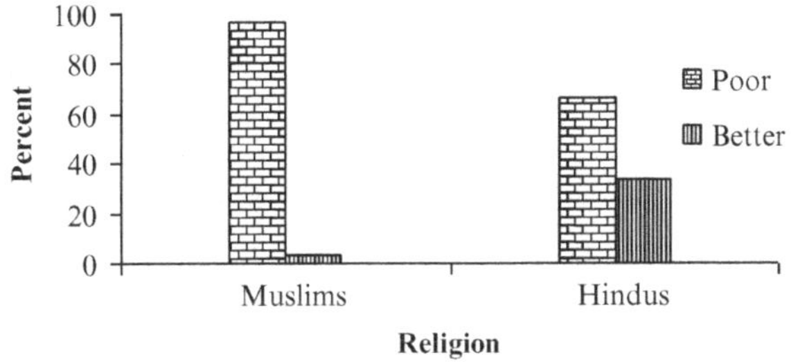

Table 13.10 Caste and composite nutritional status index (CNSI)

	Composite nutritional status index				
	Poor nutritional status		Better nutritional status		
Caste	Number	%	Number	%	Total
SC/ST	142	84.0	27	16.0	169
Non-SC/ST	148	59.0	103	41.0	251
Total	290	69.0	130	31.0	420

Source: Personal survey (2008)

nutritional status (Fig. 13.2); only 3.1 % of Muslim respondents were found to have a better nutritional status. In the Hindu religion, 66.8 % of respondents record poor nutritional status and 33.2 % of respondents have better nutritional status. This table exhibits a notable variation in the nutritional intake of Hindus and Muslims.

Table 13.10 indicates composite nutritional status index by caste in the Sonbhadra District. Among SC/ST, 84 % of respondents revealed poor nutritional status and only 16 % of respondents showed better nutritional status. Compared to SC/ST, good nutritional status can be seen in non-SC/ST castes as only 59 % of respondents of non-SC/ST (general and OBC castes) experience poor nutritional status and 41 % of respondents have a better nutritional status (Fig. 13.3). On average in the study area, 69 % of respondents are characterized by poor nutritional status and only 31 % have shown better nutritional status (Fig. 6.1).

Table 13.11 indicates the relationship between educational status and composite nutritional status index in the study area. Among the illiterate, 84.8 % of respon-

Table 13.11 Education and composite nutritional status index

Education	Composite nutritional status index				Total
	Poor nutritional status		Better nutritional status		
	Number	%	Number	%	
Illiterate	156	84.8	28	15.2	184
Up to high school	72	59.5	49	40.5	121
Above high school	62	53.9	53	46.1	115
Total	290	69.0	130	31.0	420

Source: Personal survey (2008)

Fig. 13.3 Composite nutritional status index (CNSI) by caste

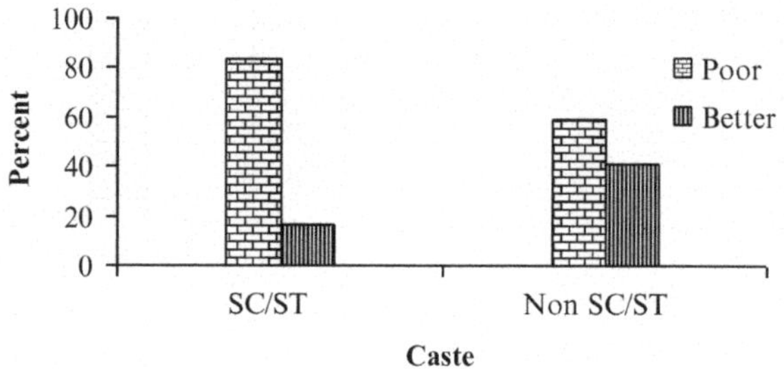

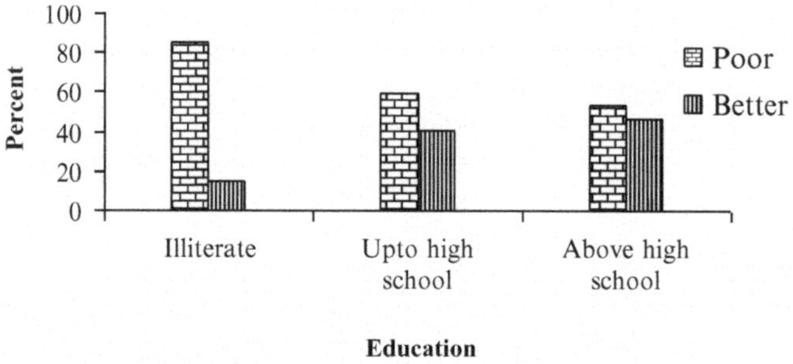

Fig. 13.4 Education-based composite nutritional status index (CNSI)

dents reveal poor nutritional status and only 15.2 % of illiterate respondents record a better condition of nutritional status (Fig. 13.4). Of respondents having education up to high school, 59.5 % have poor nutritional status and 40.5 % are characterized by better nutritional status. In above high school educational status, 53.9 % of respondents have been found with poor nutritional status and 46.1 % with better condition of nutritional status. Thus, it can be inferred that the higher the educational level, the better is the nutritional intake. The analysis of composite nutritional status index and its association with family type is given in Table 13.12. It is clear from this table that in a nuclear family 71.3 % of respondents exhibit poor nutritional intake whereas this proportion is lower in a joint family (Fig. 13.5). Better

Table 13.12 Family type and composite nutritional status index

Family type	Composite nutritional status index				Total
	Poor nutritional status		Better nutritional status		
	Number	%	Number	%	
Nuclear	239	71.3	96	28.7	335
Joint	51	60.0	34	40.0	85
Total	290	69.0	130	31.0	420

Source: Personal survey (2008)

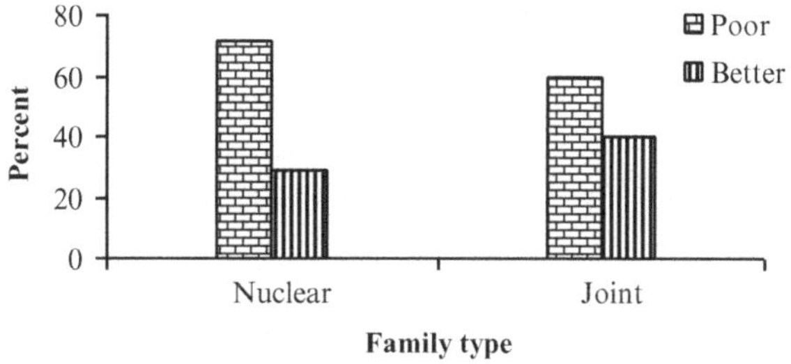

Fig. 13.5 Family-based composite nutritional status index (CNSI)

Table 13.13 Occupation and composite nutritional status index

Occupation	Composite nutritional status index				Total
	Poor nutritional status		Better nutritional status		
	Number	%	Number	%	
Labor	133	88.7	17	11.3	150
Farmer	114	56.4	88	43.6	202
Government job/business	43	63.2	25	36.8	68
Total	290	69.0	130	31.0	420

Source: Personal survey (2008)

nutritional status has been found in the joint family (40 % respondents). In the nuclear family system, the proportion of respondents having better nutritional status is much lower than that of the joint family.

Table 13.13 portrays composite nutritional status index according to occupation categories. Remarkable variation is seen in composite nutritional status index (CNSI) performance among the respondents deriving their livelihood through various occupations. The most important and striking feature of this analysis is that 63.2 % of respondents with a government job reveal poor CNSI, even poorer than respondents engaged in farming. The highest proportion of better CNSI (43.6 %) is found in the farming occupation because the farmer is a self-producer of milk and milk products, vegetables, pulses, and chickens, etc., and also eggs by keeping hens in his household (Fig. 13.6).

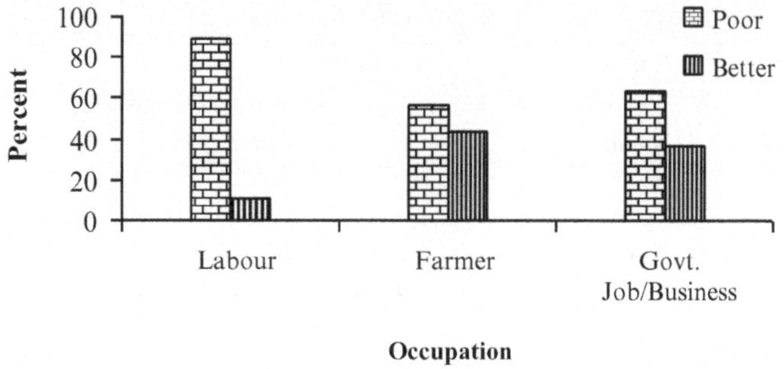

Fig. 13.6 Composite nutritional status index (CNSI) by occupation

Table 13.14 Income and composite nutritional status index

	Composite nutritional status index				
	Poor nutritional status		Better nutritional status		
Income group (in Rs/month)	Number	%	Number	%	Total
Below 2,000	163	87.6	23	12.4	186
2,000–3,500	88	53.0	78	47.0	166
3,500–5,000	27	71.1	11	28.9	38
Above 5,000	12	40.0	18	60.0	30
Total	290	69.0	130	31.0	420

Source: Personal survey (2008)

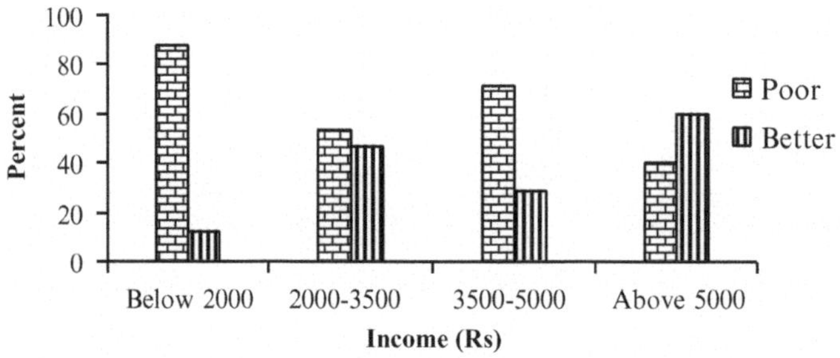

Fig. 13.7 Composite nutritional status index (CNSI) by income

Table 13.14 shows the CNSI by income, which exhibits significant variation across various income groups. However, these variations are in the positive direction. For instance, the magnitude of better nutritional status is increased in respondents having higher income. About 60 % better nutritional status has been reported in the highest income group (Rs 5,000 per month) compared to the lower income groups (Fig. 13.7).

Table 13.15 reveals the CNSI by age group. Respondents having a lower share of better nutritional status have been found in the lower age groups such as groups less

Table 13.15 Age group and composite nutritional status index

Age group (in years)	Composite nutritional status index				Total
	Poor nutritional status		Better nutritional status		
	Number	%	Number	%	
Below 30	43	91.5	4	8.5	47
30–35	46	79.3	12	20.7	58
35–40	81	75.0	27	25.0	108
Above 40	120	58.0	87	42.0	207
Total	290	69.0	130	31.0	420

Source: Personal survey (2008)

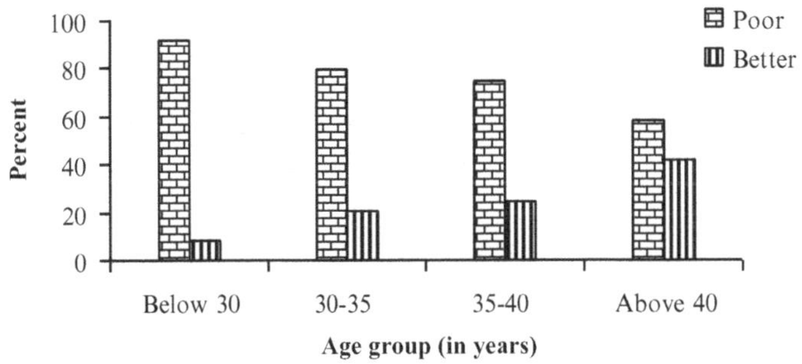

Fig. 13.8 Age group and composite nutritional status index (CNSI)

than 30, 30–35, and 35–40 years of age. It is highest (42 %) in respondents aged above 40 years. Thus, it can be inferred that the proportion of better nutritional status is higher in respondents above 40 years of age (Fig. 13.8). The reason for this may be the higher level of awareness about intake of safe food and less use of tobacco and alcohol and drugs in aged respondents compared to the awareness of younger age group respondents about the risk of nutrition deficiency.

13.3.4 Multivariate Analysis for Composite Nutritional Status Index (CNSI)

Binary logistic regressions to obtain the adjusted effect of the predictor variables on the dependent variables have been applied here. Table 13.16. presents the results of logistic regression assessing the association between composite nutritional status and the explanatory variables. Because of the smaller sample size, the categories of religion and caste have been merged into only two categories. The 95 % confidence intervals are also presented in the table. The results show that a respondent's religion, education, income,

Table 13.16 Logistic regression results predicting the odds of composite nutritional status index (CNSI) according to selected socioeconomic and demographic characteristics

Covariates and categories	Odds ratio exponent (β)	95 % confidence interval	
Religion***			
Muslim®	1.00		
Hindu	30.77	3.60	262.52
Caste*			
SC/ST®	1.00		
Non-SC/ST	2.19	0.96	4.97
Education***			
Illiterate®	1.00		
Up to high school	3.46	1.96	6.10
Above high school	4.63	2.63	8.15
Income groups***			
Below 2,000®	1.00		
2,000–3,500	9.68	3.84	24.42
3,500–5,000	1.63	0.46	5.82
Above 5,000	16.73	3.68	75.99
Occupation***			
Labor®	1.00		
Farmer	7.12	2.36	21.53
Government job and business	2.55	0.57	11.37
Family type			
Joint®	1.00		
Nuclear	1.38	0.68	2.82
Age group			
Below 30®	1.00		
30–35	0.74	0.18	2.98
35–40	1.20	0.33	4.36
Above 40	3.82	1.11	13.12

Source: Personal survey, 2008
Dependent variable: better nutritional status (1); poor nutritional status (0); ® reference category
Comment: Religion, education, income, and occupation are important predictors of having better nutritional status
$^{*}p<0.10$; $^{**}p<0.05$; $^{***}p<0.01$

and occupation are highly significantly associated ($p<0.01$) with better CNSI and caste is significantly associated ($p<0.05$) with better CNSI. The probability for better CNSI is much higher in Hindus (odds ratio, 30.77 times better condition) than Muslims. Similarly, in comparison to SC/ST, the probability for better CNSI is higher in non-SC/ST castes (odds ratio, 2.19). For a comparison of literate and illiterate respondents for having better CNSI, illiterate has been taken as the reference category.

Respondents belonging to the educated up to high school category (odds ratio, 3.46) and above high school category (odds ratio, 4.63) are more likely to have a better CNSI compared to those who belong to the illiterate (no formal schooling) category. Income is the most important factor for better CNSI. The analysis for this aspect indicates that for the below Rs 2,000 per month income group, probability is 1 (one) compared to the Rs 2,000–3,500 income group (odds ratio, 9.68), Rs 3,500–5,000 income group (odds ratio 1.63), and above Rs 5,000 group (odds ratio, 16.73). Similarly, the person's occupation is also positively associated with better CNSI. In this analysis the labor category has been referenced as one. The farmer category (odds ratio, 7.12) and government job and business category (odds ratio, 2.55) respondents have a better chance for good nutritional status. Age group and family type have not appeared to be statistically significant in this analysis.

13.3.5 Nutritional Deficiency Diseases

13.3.5.1 Scurvy

Scurvy is a nutritional deficiency disease directly associated with the lack of vitamin C. The signs of this condition are swollen and bleeding gums and bleeding into the skin or joints.

13.3.5.2 Anemia

Anemia usually accompanies scurvy; it is partially caused by hemorrhagic blood loss, but also by the faulty metabolic interrelationship of vitamin C with folic acid and with iron. Concurrent deficiency of other nutrients also contributes to the anemia. Normal percentages are highest in the youngest individuals and decline as people age.

13.3.5.3 Rickets

Rickets, a disease directly related to impaired metabolism of calcium and phosphorus, is manifested in defective bone growth and changes in the body musculature. The impaired mineral metabolism in rickets may have many causes, but by far the commonest cause is a deficiency of vitamin D. Vitamin D may be found in food or formed in the body through the action of short ultraviolet radiations such as those in sunlight. Vitamin D is necessary for the absorption of calcium and phosphorus and for their deposit in bone tissue. A dietary lack of vitamin D may occasionally occur in people on a vegetarian diet who do not consume milk products or in people who have trouble in digesting milk products. A dietary lack of calcium and phosphorus may also have a part in the nutritional causes of rickets.

13.3.5.4 Dental Decay

Dental caries, or cavities, are very common. They are caused by acid on the teeth. The acid is made by the bacteria in dental plaque. The plaque bacteria feed on sugars and starches from the diet and change them into acid. This acid eats into tooth enamel, or the outer layer of the tooth, and dentin, the major part or core of the tooth. The tooth then gradually dissolves. Tooth decay (also known as dental decay) results from a bacterial infection of the teeth.

13.3.5.5 Diabetes

Diabetes mellitus is a clinical syndrome characterized by hyperglycemia and disturbances of carbohydrate, fat, and protein metabolism that are associated with absolute or relative deficiencies in insulin action and/or insulin secretion.

13.3.5.6 Leprosy

Leprosy is a disease that affects primarily the skin and nerves. The disease was thought to have started somewhere in India and then was passed on to Africa and Europe. In the late 1800s leprosy was very common throughout Europe. Today the disease is not very common. Worldwide it is estimated that only 5 % of the population is susceptible.

13.3.6 Distribution of Nutritional Deficiency Diseases

In the study area, scurvy, anemia, asthma, dental decay, diabetes, and leprosy have been found to be the major deficiency diseases (Table 13.17). Of the total 420 respondents, about 63.3 % are suffering with some kind of deficiency diseases. The percentage of respondents having no deficiency diseases is highest in OBC (42.19 %), followed by general caste (35.9 %) and SC/ST (31.36 %); thus, the highest percentage of respondents suffering with deficiency diseases is recorded in SC/ST. Among SC/ST respondents, asthma is more prevalent, followed by scurvy, anemia, and dental decay. Among OBC, scurvy is at the top, followed by anemia and dental decay; among respondents of general caste, diabetes is more common, followed by leprosy and asthma. The reason behind the higher frequency of diabetes in the general caste may be that they do less physical work than OBC and SC/ST respondents.

It is clear from Table 13.18 that in Muslims the occurrence of deficiency diseases is slightly higher (68.8 %) than in the Hindu counterpart (Fig. 13.9). Among Hindus, the highest percentage of nutritional deficiency diseases has been found in SC/ST respondents (68.6 %), followed by general caste respondents (64.1 %). The OBC castes report the lowest percentage of nutritional diseases (Fig. 13.10). Table 13.19 shows the higher prevalence of nutritional deficiency diseases among illiterate

Table 13.17 Nutritional deficiency diseases by major caste groups

Nutritional deficiency diseases	Castes						Total
	SC/ST		OBC		General		
	Number	%	Number	%	Number	%	
No deficiency	53	31.36	73	42.19	28	35.90	154
Anemia	17	10.08	20	11.56	4	5.13	41
Dental decay	15	8.88	19	10.98	4	5.13	38
Leprosy	10	5.92	9	5.20	12	15.38	31
Asthma	26	15.38	8	4.63	6	7.69	40
Diabetes	13	7.69	11	6.36	14	17.95	38
Scurvy	22	13.02	28	16.18	0	0	50
Others	13	7.69	5	2.90	10	12.82	28
Total	169	100.0	173	100.0	78	100.0	420

Source: Personal survey (2008)

Table 13.18 Nutritional deficiency diseases by caste and religion

Religion	No nutritional disease		Having nutritional disease		Total
	Number	%	Number	%	
Muslim	10	31.2	22	68.8	32
Hindu	144	37.1	244	62.9	388
Castes					
SC/ST	53	31.4	116	68.6	169
OBC	73	42.2	100	57.8	173
General	28	35.9	50	64.1	78
Total	154	36.7	266	63.3	420

Source: Personal survey (2008)

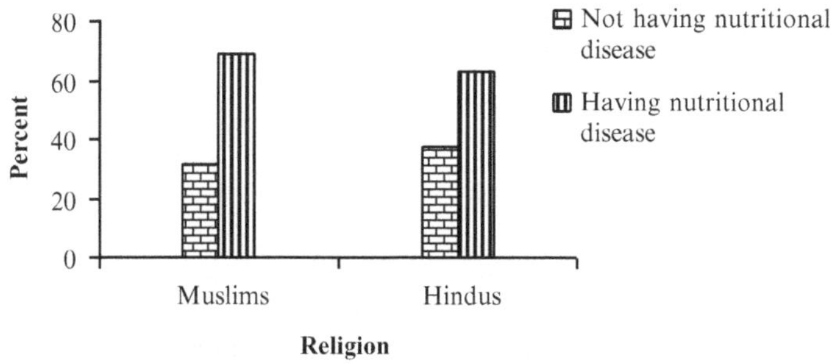

Fig. 13.9 Religion and nutritional deficiency diseases

respondents (70.1 %), followed by respondents having education above high school (65.2 %) and up to high school (51.2 %) (Fig. 13.11). There is no clear-cut positive relationship between education and nutritional deficiency diseases.

Table 13.20 highlights that the highest (above Rs 5,000) and lowest (below Rs 2,000) income groups have a high percentage of respondents suffering from nutri-

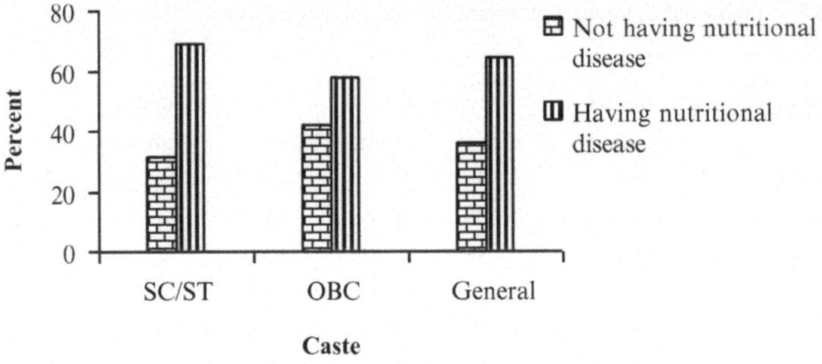

Fig. 13.10 Caste and nutritional deficiency diseases

Table 13.19 Education and nutritional deficiency diseases

	No nutritional disease		Having nutritional disease		
Education	Number	%	Number	%	Total
Illiterate	55	29.9	129	70.1	184
Up to high school	59	48.8	62	51.2	121
Above high school	40	34.8	75	65.2	115
Total	154	36.7	266	63.3	420

Source: Personal survey (2008)

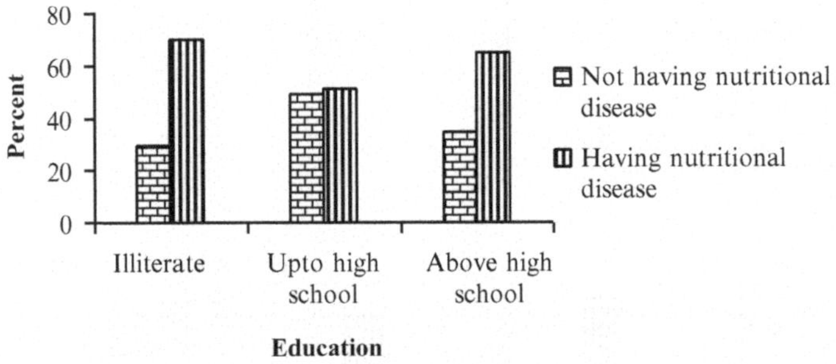

Fig. 13.11 Education and nutritional deficiency diseases

tional deficiency diseases (Fig. 13.12). A lower percentage of respondents having such diseases has been found in the middle income group. Thus, on the basis of these data no inference can be derived showing an association between income and nutritional diseases. Table 13.21 evinces that the respondents having higher percentage of nutritional deficiency diseases belong to a nuclear family, whereas the respondents of a joint family witness a lower percentage of nutritional deficiency diseases (Fig. 13.13). Table 13.22 revealed that 80.9 % respondents of the age group below 30 years have nutritional deficiency diseases. With increase in age, the percentage of respondents having nutritional deficiency diseases declines. Thus, the inference 'the higher the age the lower are deficiency diseases' can be drawn. Relationships between occupation and nutritional deficiency diseases are shown in Table 13.23. It is interest-

Table 13.20 Income and nutritional deficiency diseases

Income group (Rs/month)	No nutritional disease		Having nutritional disease		Total
	Number	%	Number	%	
Below 2,000	52	28.0	134	72.0	186
2,000–3,500	83	50.0	83	50.0	166
3,500–5,000	14	36.8	24	63.2	38
Above 5,000	5	16.7	25	83.3	30
Total	154	36.7	266	63.3	420

Source: Personal survey (2008)

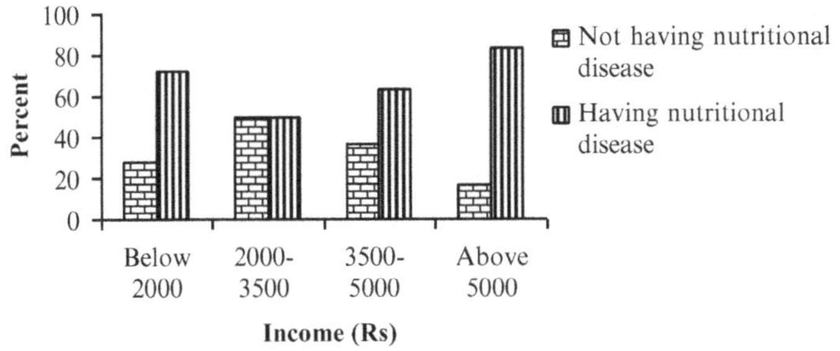

Fig. 13.12 Income and nutritional deficiency diseases

Table 13.21 Family type and nutritional deficiency diseases

Family type	No nutritional disease		Having nutritional disease		Total
	Number	%	Number	%	
Nuclear	120	35.8	215	64.2	335
Joint	34	40.0	51	60.0	85
Total	154	36.7	266	63.3	420

Source: Personal survey (2008)

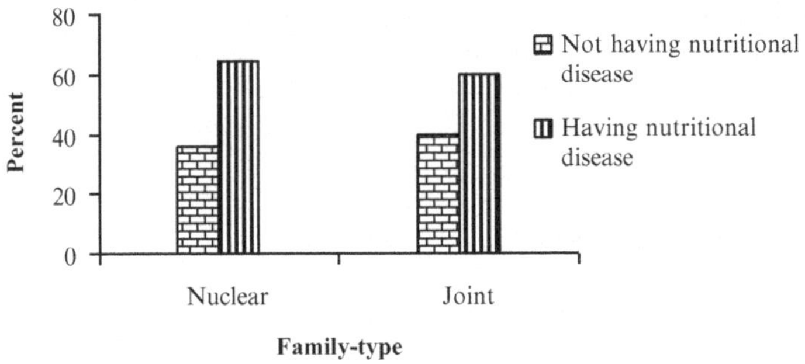

Fig. 13.13 Family and status of nutritional deficiency diseases

Table 13.22 Age group and nutritional deficiency diseases

Age group (years)	No nutritional disease		Having nutritional disease		Total
	Number	%	Number	%	
Below 30	9	19.1	38	80.9	47
30–35	22	37.9	36	62.1	58
35–40	39	36.1	69	63.9	108
Above 40	84	40.6	123	59.4	207
Total	154	36.7	266	63.3	420

Source: Personal survey (2008)

Table 13.23 Occupation and status of nutritional deficiency diseases

Occupation	No nutritional disease		Having nutritional disease		Total
	Number	%	Number	%	
Labor	40	26.7	110	73.3	150
Farmer	93	46.0	109	54.0	202
Job/business	21	30.9	47	69.1	68
Total	154	36.7	266	63.3	420

Source: Personal survey (2008)

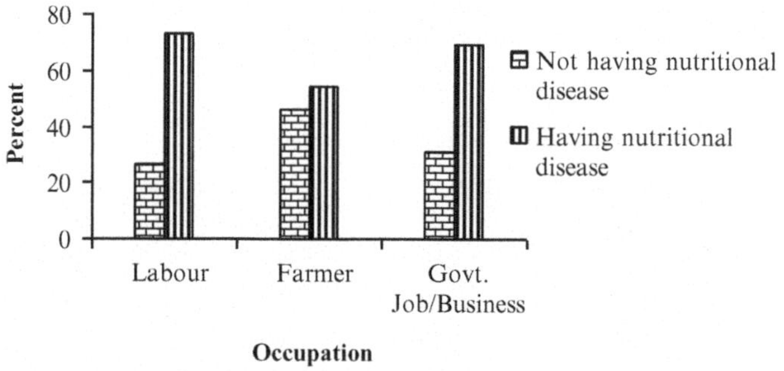

Fig. 13.14 Occupation and status of nutritional deficiency diseases

ing to observe that 73.3 % of respondents having nutritional deficiency diseases belong to the labor group, followed by 69.1 % of respondents having nutritional deficiency diseases in the job/business group (Fig. 13.14). The lowest nutritional deficiency disease occurrences were found in the farmer group.

13.3.7 Multivariate Analysis

Binary logistic regressions to obtain the adjusted effect of the predictor variables on the dependent variables have been applied here. The results of the logistic regressions are presented in Table 13.24, which shows the results of logistic regression assessing the association between nutritional deficiency diseases and other

Table 13.24 Logistic regression results predicting the odds of having nutritional diseases according to selected socioeconomic and demographic characteristics

Covariates and categories	Odds ratio exponent (β)	95 % confidence interval	
Religion			
Muslim®	1.00		
Hindu	0.77	0.31	1.91
Caste**			
SC/ST®	1.00		
OBC	2.59	1.20	5.59
General	1.71	0.60	4.86
Education***			
Illiterate®	1.00		
Up to high school	0.51	0.31	0.82
Above high school	0.51	0.31	0.83
Income groups***			
Below 2,000®	1.00		
2,000–3,500	0.31	0.15	0.65
3,500–5,000	0.25	0.07	0.85
Above 5,000	0.23	0.05	1.18
Occupation***			
Labor®	1.00		
Farmer	0.42	0.18	0.96
Government job and business	0.17	0.05	0.59
Family type*			
Nuclear®	1.00		
Joint	1.84	0.99	3.42
Age group			
Below 30®	1.00		
30–35	1.11	0.46	2.71
35–40	0.70	0.32	1.53
Above 40	0.71	0.34	1.47

Source: Personal survey, 2008
Dependent variable: having nutritional disease (1); no nutritional disease (0); ® reference category
Comment: Caste, education, income, occupation, and family type are important predictors of having nutritional diseases
*$p<0.10$, **$p<0.05$, ***$p<0.01$

explanatory variables. The relationship between religion and nutritional deficiency diseases is insignificant. The OBC category (2.59) and general category respondents have higher odds ratio (1.71) compared to the SC/ST reference category (1.0). Thus, caste has been found to be statistically significant ($p<0.05$). If the educational category is considered, a very significant association is observed ($p<0.01$) with up to high school (odds ratio, 0.51) and above high school (0.51) education compared to the illiterate reference category (1.0). Thus, the higher the education, the lesser is the chance of nutritional deficiency diseases. Similarly, the income groups appeared statistically significant ($p < 0.01$) because they revealed that income groups second (odds ratio, 0.31), third (odds ratio, 0.25), and fourth (odds ratio, 0.23) possess a lesser odds ratio for having nutritional deficiency diseases compared to the reference category (1.0). Table 13.24 indicates that occupations are also significantly associated ($p < 0.01$): farmers (odds ratio, 0.42) and government job and business respondents (odds ratio, 0.17) have a lesser odds ratio than the reference category labor (1.0), with a higher percentage of nutritional deficiency disease. Family type is not as significant ($p < 0.10$) as education, income, and occupation. Age groups are also not statistically significant for the occurrence of nutritional deficiency diseases.

13.4 Summary and Conclusion

To understand nutritional food intake, nutritional status, and nutritional deficiency diseases across various communities, this study has been carried out in the Sonbhadra District of U.P. Substantial variation in nutritional food intake and nutritional status has been found across the communities, as well as in terms of education, family type, occupation, income, and age group. Religion, occupation, income, and education have been found to be important predictors for having better nutritional status. Scurvy, anemia, asthma, leprosy, dental decay, and diabetes are major nutritional deficiency diseases occurring in the area under study. Similarly, education, income, occupation, and caste appear important predictors for nutritional deficiency diseases. The results of this study suggest improving the educational and economic condition of the rural masses of Sonbhadra District in particular and of India in general. A strong campaign for awareness toward nutritional food intake and nutritional deficiency diseases is necessary for reducing the suffering of the people from nutritional deficiency diseases.

References

Chitralekha R (1982) Poverty and levels of nutrition: some more evidences. IGJ Madras 57(1, 2):58–61

Dubey RS, Mishra RP (1984) Nutritional availability patterns in Madhya Pradesh. Trans Inst Indian Geogr 6(2):79–89

Gopalan C et al (1996) Nutritive value of Indian foods. National Institute of Nutrition, Indian Council of Medical Research, Hyderabad

NFHS-2 (1998–1999) National family health survey, India. International Institute of Population Science, Mumbai

Singh B (1972) A geographical appraisal of the caloric intake in Shahganj Tehsil. NGJI Varanasi XVIII(2):102–113

Singh MB et al (1997) Poverty, nutrition and nutrition deficiency diseases in Kailhat village, Mirzapur district (U.P.). NGJI 43(4):307–315

Part VI
Water Management

Chapter 14
Measurement of Water Scarcity

Rajesh Kumar Abhay

Abstract When the balance between availability and demand of water is disturbed, problems of water scarcity begin. In this chapter an attempt is made to explain the methods of measuring water scarcity based on secondary and primary data for which Thornthwaite's water balance method and Sullivan's water scarcity index method have been taken, respectively. The paper explains how water scarcity can be calculated by using different data sets by different methods, and their positive and negative aspects have also been evaluated with respect to the outcome.

Keywords Scarcity • Water balance • Deficit • Water scarcity index

14.1 Introduction

Water scarcity is a most serious hindrance to agricultural development and a major threat to the environment in dry areas (Oweis 2005, p. 192). Once considered an abundant resource, now water is increasingly seen as a 'scarce' resource, one that needs to be managed judiciously. Before going into the details of this problem, it is necessary to understand what water scarcity really is. The etymological roots of the word 'scarcity' go back to the Old Northern French word '*escartre*,' which meant insufficiency of supply (Mehta 2003, p. 5067). Actually, the term scarcity of water refers to situations in which water resources available for producing output are insufficient to satisfy human wants. According to the European Environment Agency, "*water scarcity occurs where there are insufficient water resources to satisfy long-term average requirements. It refers to long-term water imbalances, combining low water availability with a level of water demand exceeding the supply capacity of the natural system.*" As used by water engineers, if the annual

R.K. Abhay (✉)
Department of Geography, Delhi School of Economics, University of Delhi, New Delhi, Delhi 11007, India
e-mail: abhay.dse@gmail.com

A.K. Dutt et al. (eds.), *Spatial Diversity and Dynamics in Resources and Urban Development*, DOI 10.1007/978-94-017-9771-9_14

availability of renewable freshwater is 1,000 m^3 or less per person in the population, this is a situation of water scarcity.

Water scarcity is, thus, inevitably and also incorrectly assumed to be caused only by deviations from normal in rainfall. However, the most alarming and exponential increases in water scarcity during the past few years cannot be linked exclusively to the abnormalities in rainfall because there is no long-term change in rainfall, although there have been some variations in annual rainfall. Further, water scarcity and drought are no longer restricted to arid regions with scanty rainfall. Areas receiving high rainfall, such as Kerala and Goa, are also experiencing acute water scarcity and have been demanding drought relief. There are, however, many other intangible and ambiguous aspects of the problem, leading to different types of scarcities experienced by a wide range of actors. Hence, the responses to 'scarcity' are also varied, and there is a need to understand their relational aspects.

Water scarcity is closely linked with the availability and demands of water in a particular area. Availability of water depends upon the surface water and groundwater resources. The water requirement now and in the future will also increase, not only because of increasing population but also because of changing lifestyles in both rural and urban areas. The utilization patterns of water are also changing. On the basis of present-day utilization patterns, future water requirements can be projected. The water requirement would be higher in urban areas because of the high concentration of population. Of the water required in urban areas, the maximum would be utilized to maintain cleanliness in public and private places. Hence, the water requirement will increase on a large scale. On the other hand, water available per capita decreases very sharply with an increase in population density.

When the balance between availability and demand of water is disrupted, problems emerge. Water scarcity is also the result of this imbalance. Not only does this imbalance create water scarcity, but mismanagement of available water resources also exacerbates the situation. For example, in the northeastern states of India, where there is enough rainfall and water is available in the rainy season, the physiography of the area as well as lack of proper management techniques creates a water scarcity situation during the summer season.

In popular usage, "scarcity" is a situation in which there is insufficient water to satisfy normal requirements. There are degrees of scarcity—absolute, life threatening, seasonal, temporary, and cyclical. Populations with normally high levels of consumption may experience temporary scarcity more keenly than other societies who are accustomed to using much less water. Scarcity often arises because of socioeconomic trends that have little to do with basic needs. Defining scarcity for policy-making purposes is very difficult. Terms such as water scarcity, shortage, and stress are commonly used interchangeably, although each term has different specific meanings. *Water shortage* is a dearth, or absolute shortage; low levels of water supply relative to minimum levels necessary for basic needs can be measured by annual renewable flows (in cubic meters) per head of population, or its reciprocal, that is, the number of people dependent on each unit of water, such as millions of persons per cubic kilometer. *Water scarcity* is an imbalance of supply and demand under prevailing institutional arrangements and demand in excess of available supply; a high rate of utilization compared to available supply, especially if the remaining supply potentials are difficult or costly to tap. Because water scarcity is a relative concept, it is difficult to capture in a single index.

The symptoms of water scarcity or shortage are growing conflict between users and competition for water, declining standards of reliability and service, harvest failures, and food insecurity. The term water stress was coined in 1992 by a Swedish hydrologist, Malin Falkenmark. To understand 'water scarcity' and 'water stress,' Falkenmark concluded that countries are characterized as water stressed and water scarce depending on the amount of renewable water available. Water stress is a situation when a region faces acute water crises to such an extent that the basic water requirement is not met (Falkenmark 1983). The circumstances may range from supply hindrances to distance factors or economic constraints. Water stress, thus, represents widespread and chronic lack of access to safe and affordable water. Water scarcity, on the other hand, indicates that there is an insufficiency of the resource to meet the demands. Water-scarce regions are those where water resource availability is less; hence, the means of water availability and accessibility to the concerned population become restrained (Sahay 2003).

The discussion concludes that water scarcity is a relative concept, depends on local conditions, and is difficult to measure. Thus, in this study an attempt is made to explain the methods for measuring water scarcity in arid and semiarid parts of Rajasthan, that is, Sikar and its five surrounding districts: Jaipur, Churu, Pilani, Nagaur, and Ajmer.

14.2 Review of the Literature

In recent years natural resources have been studied from various aspects such as physical, economic, political, and social. Water scarcity as a major threat to water resources has been studied by many researchers. They have used different methods to measure the intensity and magnitude of the problem, but neglected the suitability of the technique without knowing its merits and limitations. Some of the important studies have been reviewed in this context.

Subramaniam and Vinayak (1982) stated that water balance studies in relationship to climatic studies are one of the important disciplines of applied climatology. They have studied the climatic features of Trivandrum and its vicinity. The water balance methods of Thornthwaite (1948) and Thornthwaite and Mather (1955) were adopted. On the basis of humidity, aridity, moisture indices, and moisture adequacy calculation, they found that the study area reveals a stable humid climate, and currently a shift toward dry climate was evidenced by the researchers in the study area. Joshi (2004) analyzed the relationship between requirements and availability of water in Port Blair town. The availability of groundwater and its changing nature with seasons was discussed with the variations in the consumption pattern of water for domestic use depending upon season and the beliefs, customs, and habits of the users. It is concluded that water crisis is a permanent feature in Port Blair town.

Soundaravalli (1992) identified surplus and deficit areas in Dindigul city of Anna district, Tamil Nadu and explained the causes of nonavailability of water over the years. Bandyopadhyay (1987) suggested analysis of interaction among environmental and political forces in finding a solution to the problem of water scarcity. Drought and water scarcity are positively related to each other. The political ecology approach has been emphasized to cope with water scarcity. Oweis (2005) stated that in dry areas, water is most scarce, land is fragile, and drought inflicts severe hardship on already poor populations. The efficient use of water can help to alleviate the problems of water scarcity and drought.

Among the numerous techniques for improving water use efficiency, the most effective are water harvesting and supplemental irrigation. Water harvesting stores the water by different methods and effectively combats desertification and enhances the resilience of the communities and ecosystem under drought. Khurana (2003) found that the problem of scarcity of water can be solved by adopting the rainwater harvesting system. Rainwater harvesting is not just the starting point for meeting drinking water needs, but also the starting point of an effort to eradicate rural poverty itself. It can generate massive rural employment and reduces distress-based migration from rural to urban areas. The harvesting system also generates community spirit within the village, and builds up what economists call 'the social capital.'

The foregoing discussion explains that the problem has been discussed and analyzed from various points of view. Thornthwaite's water balance method remained quite popular to identify the water-scarce areas. However, the method is quite old, using natural climatic conditions (mainly temperature and rainfall) to determine the deficit conditions in an area, but now the modern conditions have become so complex that water scarcity is no more a natural phenomenon. It depends more on human aspects such as accessibility, availability, and the quality of water as deteriorated by humans. At present, it is a man-made problem. Water has been made scarce by humans misusing and mismanaging it. This study highlights the importance of measuring water scarcity in arid areas of India in particular and in the world in general.

14.3 The Study Area

Rajasthan State is the largest in India, at present, in terms of areal coverage. The study area lies in the northeastern part of the State (Fig. 14.1). Mainly six districts, Sikar, Jaipur, Churu, Ajmer, Nagaur, and Pilani, have been selected to study Thornthwaite's method of water balance, and only Sikar district was selected to test Sullivan's water scarcity index. The Sikar district is located in the Shekhawati region of the State, the second most developed district after Jaipur (capital of Rajasthan). This administrative headquarters of Sikar District is located in the northeastern part of Rajasthan between 27°21′ and 28°12′N and 74°44′ and 75°25′E at an average elevation of 432 m above mean sea level, with an area of 7,732 km^2. Sikar is the most important town in the Shekhawati region. Sikar, besides being the

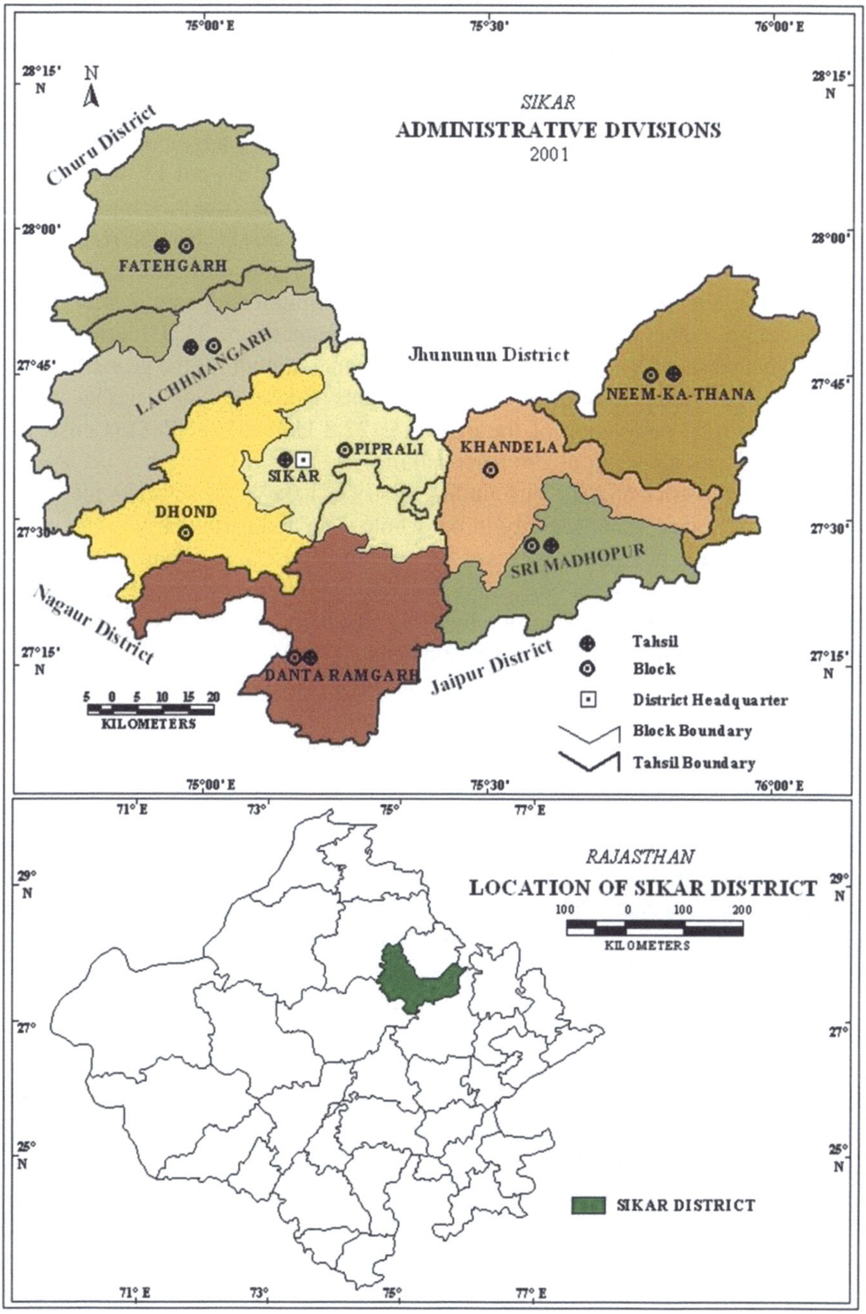

Fig. 14.1 Rajasthan: location of study area, 2001

district headquarters administratively, constitutes six tahsils, namely, Fatehpur, Lachhmangarh, Sikar, Danta Ramgarh, Neem Ka Thana, and Sri Madhopur, and eight blocks, namely Fatehpur, Lachhmangarh, Dhond, Piprali, Danta Ramgarh, Khandela, Neem Ka Thana, and Sri Madhopur.

The Aravalli ranges divide the district into two main topographic areas. The western region is characterized by sand dunes and the eastern by hill ranges. There are no perennial rivers in the district. Sikar district occupies a part of the eastern limit of the desert tract. The climate of the study area is characterized by hot summers, scanty rainfall, a chilly winter season, and general dryness of the air except during the monsoon season. The average maximum temperature was 46 °C, the minimum temperature 0.0 °C, and the mean temperature 23 °C during 2001. Normal annual rainfall is 466 mm; the average annual rainfall for the year 2001 was recorded as 250 mm. Rainfall is very low, highly indefinite, and variable in the district. The total population of the district is 22.8 lakhs (2,287,788) (Census of India 2001), with a density of 296 persons/km^2.

The Sikar district and its surrounding districts have been selected for study because of the acute water scarcity in the whole state and particularly in the study area. The climatic conditions increase the magnitude of the problem in the study area. The gap between water requirements and the total utilizable supply is increasing day by day. Sikar and its surrounding districts, therefore, have been selected as a case for the measurement of water scarcity.

14.4 Data and Methods

The study is based on both primary and secondary sources of data. The secondary data have been collected from the Public Health and Engineering Department (PHED), Census of India Publications, and data related to temperature and rainfall to calculate water balance were collected from the Indian Meteorological Department (IMD). In addition, books, theses, journal articles, newspapers, and Internet websites were also consulted.

The secondary data for temperature and rainfall were available at the district level; water balance was calculated at district level. Water balance has been calculated for Sikar and its five surrounding districts in the State. To calculate water deficit, the Thornthwaite (1948) method and the Thornthwaite and Mather (1955) water balance method have been used. Here, it has been assumed that high deficit means higher water scarcity and lower deficit means a lower level of water scarcity.

The magnitude of water scarcity, the Water Scarcity Index (WSI) developed by Caroline Sullivan (2002a), has been used, and primary data were collected and analyzed to determine the WSI scores. In this method, a higher WSI score means a lower magnitude of water scarcity and a lower WSI score means a higher scarcity. On the basis of WSI scores, *iso-scarcity* lines have been drawn and spatial variations in the water scarcity have been determined.

The *stratified proportionate random sampling* technique was adopted for selecting the village and household for primary data collection. Villages were selected proportionately from deficit zones, that is, 2.5 % of villages from each deficit zone. The deficit zones were identified on the basis of Thornthwaite's (1948) and Thornthwaite and Mather's (1955) water balance method. Based on the level of deficit, water-scarce areas have been delimited. Of 992 villages, 25 villages in eight blocks and four towns (total, nine towns in the district), 1 from each deficit zone, were selected for primary survey. The sampling frame was chosen in a manner that would provide an even spatially distributed sample. A total of 261 households (225 rural and 36 urban) were covered, 9 households from each selected unit of analysis.

14.5 Discussion

14.5.1 Thornthwaite's Water Balance

It is assumed that high deficit means a higher magnitude of water scarcity and vice versa. To identify the deficit in the district, the water balance method has been used. For the climatological water balance for any period and place, Thornthwaite's formula as a useful tool is best known. He had actually arrived at a complex exponential formula for evaluating daily or monthly potential evapotranspiration (PET) from more generally available data of mean air temperature, rainfall, length of day, and latitude. The whole computational procedure carried out on the water balance of a place is expressed by the equation given below:

$$\mathrm{PET} = 1.6\left(10 \times \frac{\mathrm{Ti}}{\mathrm{I}}\right)\mathrm{a}$$

where

PET = potential evapotranspiration
Ti = mean monthly or daily temperature
I = monthly heat index
and **a** is a constant (*where* $\mathbf{a} = 6.75\mathrm{E}^{-7}\,\mathrm{I}^3 - 7.71\mathrm{E}^{-5}\,\mathrm{I}^2 + 1.79\mathrm{E}^{-2}\,\mathrm{I} + 0.49239$).

The spatial distribution of annual estimated PET reveals an interesting pattern (Table 14.1). Because data were available to the district level, six districts were taken into consideration to determine the pattern of deficit.

Figure 14.2 exhibits a decreasing tendency of PET values toward the center, that is, Sikar District, reflecting the temperature of the study area, which is comparatively low. The annual PET calculated for Sikar is 1,398 mm. It increases as we move in all directions from the center, although its increasing trend is different. It increases rapidly in the north and northeastern directions from the center, but in the

Table 14.1 Mean monthly potential evapotranspiration by the Thornthwaite method, 2000

Station	Jan	Feb	Mar	Apr	May	Jun	Jul	Aug	Sep	Oct	Nov	Dec
1. Sikar	$I=119.11$		$a=2.6712$									
T (°C)	13.10	14.65	22.05	27.85	30.10	33.45	27.55	27.80	29.10	10.55	17.80	10.05
PET (mm)	20.63	27.81	82.91	154.70	190.38	252.37	150.29	153.96	173.95	11.57	46.79	10.16
C.F.	0.92	0.88	1.03	1.07	1.16	1.16	1.18	1.13	1.02	0.99	0.90	0.90
Adj. PET	18.88	24.48	85.39	165.53	220.84	**291.48**	177.34	173.97	177.43	11.40	42.12	9.15
2. Jaipur	$I=150.89$		$a=3.75709$									
T (°C)	16.30	16.95	23.75	32.35	33.95	34.10	29.5	29.30	29.25	28.65	22.70	17.40
PET (mm)	21.38	24.76	87.95	280.84	336.68	342.31	210.22	193.59	192.35	177.94	74.20	27.33
C.F.	0.92	0.88	1.03	1.07	1.16	1.15	1.18	1.13	1.02	0.99	0.90	0.90
Adj. PET	19.67	21.79	90.58	300.50	390.55	393.65	248.06	218.75	192.35	177.94	74.20	27.33
3. Pilani	$I=145.55$		$a=3.5458$									
T (°C)	13.60	15.25	21.45	30.45	34.35	33.80	31.20	30.85	29.35	28.0	21.40	15.70
PET (mm)	12.58	18.88	63.28	219.18	336.03	317.34	238.92	229.56	192.37	162.79	62.76	20.93
C.F.	0.91	0.88	1.03	1.07	1.16	1.16	1.18	1.13	1.02	0.98	0.90	0.90
Adj. PET	11.45	16.61	65.18	234.52	389.79	368.11	281.93	259.40	196.22	159.53	56.48	18.83
4. Churu	$I=152.82$		$a=3.8364$									
T (°C)	14.60	16.0	22.7	31.7	35.7	35.35	32.35	32.05	30.25	27.85	21.05	15.70
PET (mm)	13.43	19.08	73.01	262.89	414.73	399.34	284.18	274.20	219.67	159.97	54.66	17.74
C.F.	0.91	0.88	1.03	1.07	1.17	1.16	1.19	1.13	1.03	0.98	0.90	0.90
Adj. PET	12.22	16.79	75.20	281.29	485.23	463.24	338.17	309.85	226.26	156.77	49.19	15.97

5. Nagaur	*I* = 156.84		*a* = 4.008									
T (°C)	15.00	18.50	24.50	30.00	34.50	36.00	33.00	32.00	31.50	28.00	22.50	16.50
PET (mm)	13.38	31.01	95.59	215.22	376.82	446.89	315.33	278.74	261.70	163.23	67.95	19.61
C.F.	0.92	0 88	1.03	1.07	1.16	1.15	1.18	1.13	1.02	0.99	0.90	0.90
Adj. PET	12.31	27.29	98.45	230.28	437.11	**513.93**	372.09	314.98	266.93	161.60	61.15	17.64
6. Ajmer	*I* = 158.49		*a* = 4.0800									
T (°C)	18.45	18.60	25.00	32.65	33.35	33.40	29.60	29.55	30.00	30.75	24.85	20.45
PET (mm)	29.74	30.74	102.73	305.31	332.91	334.95	204.63	203.22	216.15	239.06	100.24	45.26
C.F.	0.92	0.88	1.03	1.07	1.16	1.15	1.18	1.13	1.02	0.99	0.91	0.91
Adj. PET	27.36	27.05	105.81	326.68	386.18	385.20	241.46	229.64	220.47	236.67	91.22	41.19

Source: Calculated by the author
T temperature, *PET* potential evapotranspiration, *Adj. PET*, adjusted PET, *C.F.* correction factor

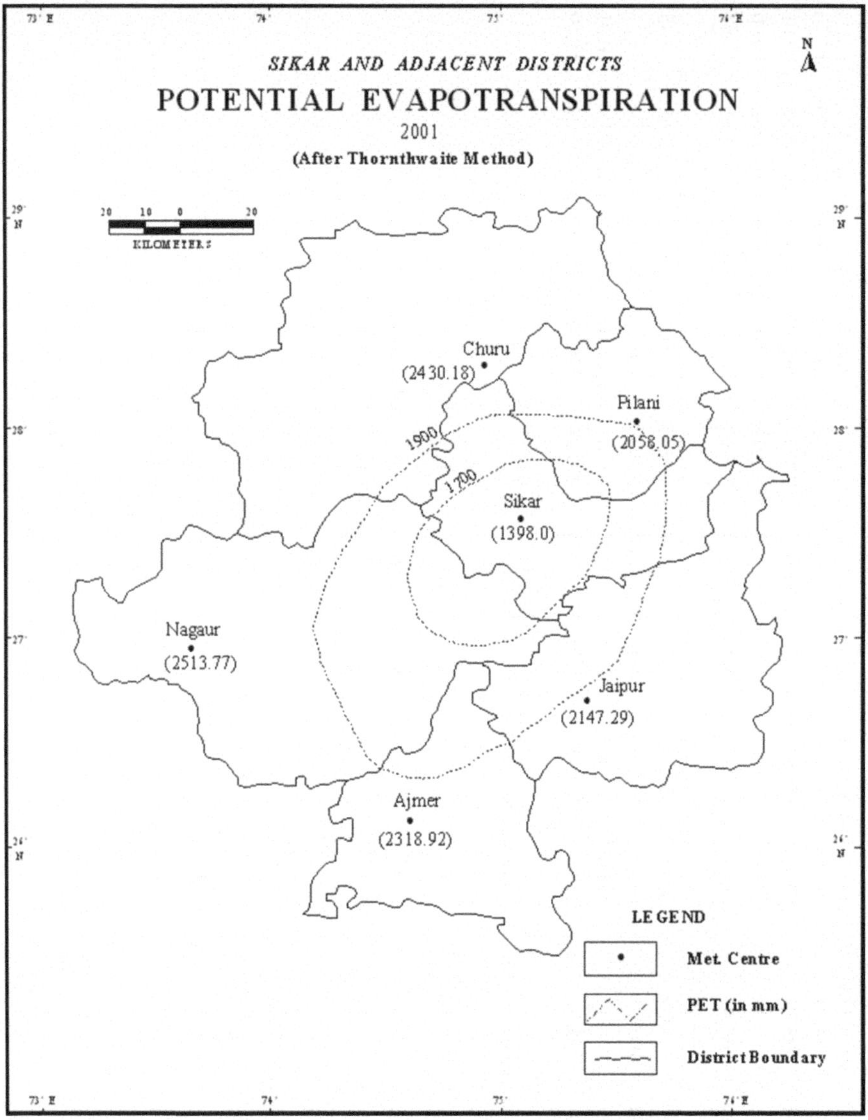

Fig. 14.2 Sikar and adjacent districts: potential evapotranspiration, 2001

southwest, south, and southeast, PET increases slowly because of distance between district headquarters, the point where temperature is recorded, and thus the iso-PET lines become wider. The highest PET value, 2,513.77 mm, was obtained at Nagaur, the southwestern margin, followed by Ajmer (2,318.92 mm) in the south and Jaipur (2,147.29 mm) in the southeast.

14.5.1.1 Surplus and Deficit

The amount of water that cannot be stored is termed the soil moisture (SM) surplus and is calculated by subtracting recharge from excess precipitation during wet months:

$$\text{Thus, surplus} = (\text{P} - \text{PET})\ \ (+\Delta\text{SM}) \qquad \text{when P} > \text{PET}$$

Because all months have no soil moisture, there is no surplus of water at all the stations. Thus, all the stations face deficits that vary in magnitude. The amount by which actual evapotranspiration (AET) and potential evapotranspiration (PET) rates differ in any month is called the soil moisture deficit:

$$\text{Deficit} = \text{PET} - \text{AET}$$

Deficit is the difference between PET and AET, respectively. Once a soil moisture deficit develops, it can be reduced when excess precipitation is stored in the soil at the beginning of the wet season. Thus, on the basis of *iso-deficit* lines, spatial variations in deficit are identified. The water balance is been shown (Fig. 14.3a–f) for all six stations.

Deficit varies from 1,380 mm at Sikar to 2,325.7 mm at Nagaur (Fig. 14.4). Deficit increases in all directions from the center at Sikar. In the northwestern direction, *iso-deficit* lines are closer to each other, signifying the rapid increase in the deficit level. It can also be calculated that scarcity of water increases toward the northwestern side. Finally, the deficit area has been extracted for the Sikar district and the deficit zones are demarcated.

14.5.1.2 Deficit Zones

On the basis of intensity of deficit, the district is divided into four deficits zones: very high, high, medium, and low (Fig. 14.5). Temperature and rainfall were low in the central part of the study area. A low difference between PET and AET has resulted in a low deficit in the center in comparison to surrounding areas; this is a low-deficit zone, and the deficit range is less than 1,500 mm. As we move toward the northwest and eastern directions, the deficit increases. Very high deficit zones have a deficit level of more than 1,900 mm, located in the northwestern part of the district. Thus, the Thornthwaite's water balance method provides deep insight into water scarcity based on collected meteorological data. The higher deficit indicates a higher level of water scarcity and vice versa. The figure also explains increase in level of water scarcity from the center. Thus, in the study area deficit and water scarcity are positively correlated. The method provides a regional pattern of water scarcity at the district level or the point where the climate data is collected.

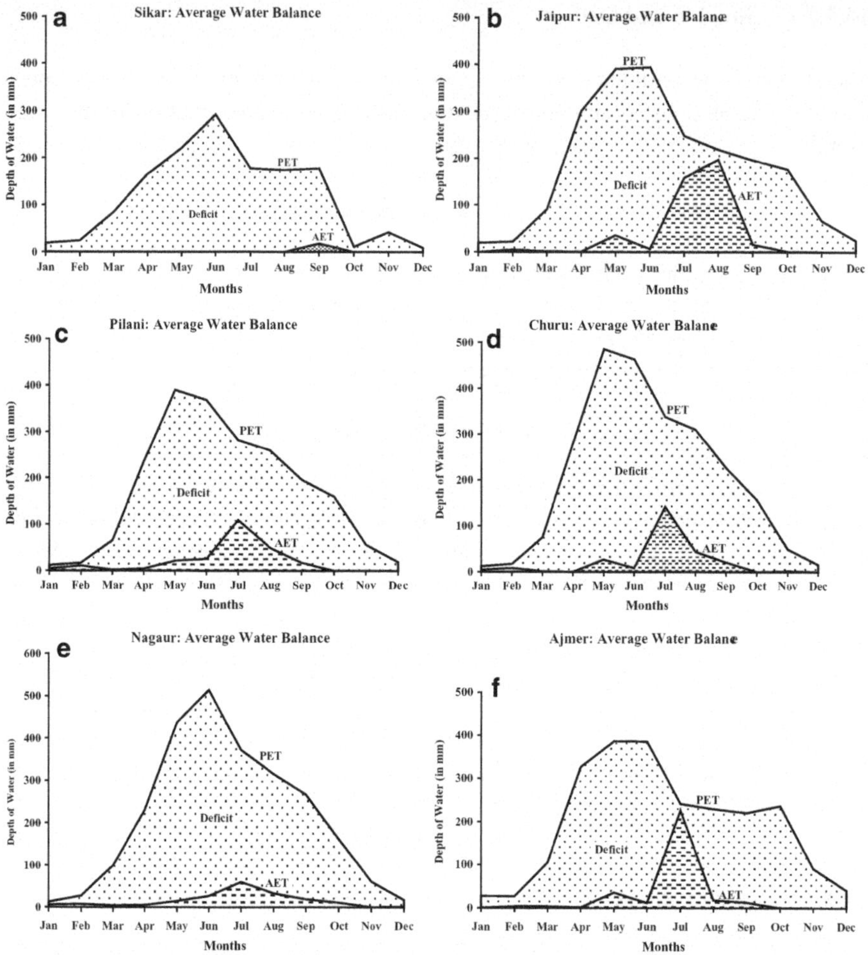

Fig. 14.3 Water balance, 2001 (Based on Thornthwaite method): (**a**) Sikar, average water balance; (**b**) Jaipur, average water balance; (**c**) Pilani, average water balance; (**d**) Churu, average water balance; (**e**) Nagaur, average water balance; (**f**) Ajmer, average water balance

14.5.2 Water Scarcity Index

Another widely known assessment of global water resource scarcity is the work published by Caroline Sullivan (2002a). Sullivan is an environmental economist specializing in water management and policy. She and her team of researchers of the Centre for Ecology and Hydrology (CEH), Wallingford, discussed various indicators that could be used in the water scarcity index (WSI). The water scarcity index is designed as a composite, interdisciplinary tool, linking indicators of water and human welfare to indicate the degree to which water scarcity impacts on human populations.

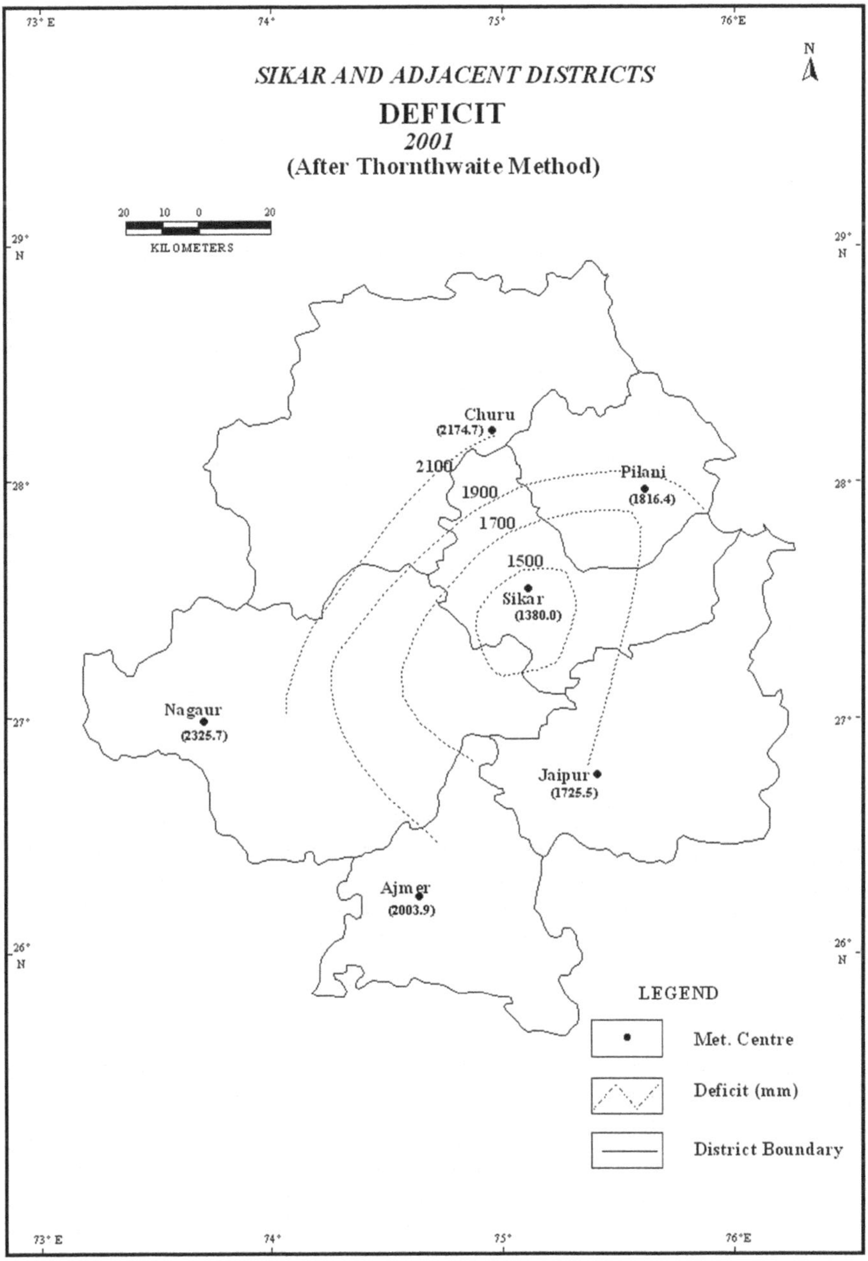

Fig. 14.4 Sikar and adjacent districts: deficit, 2001

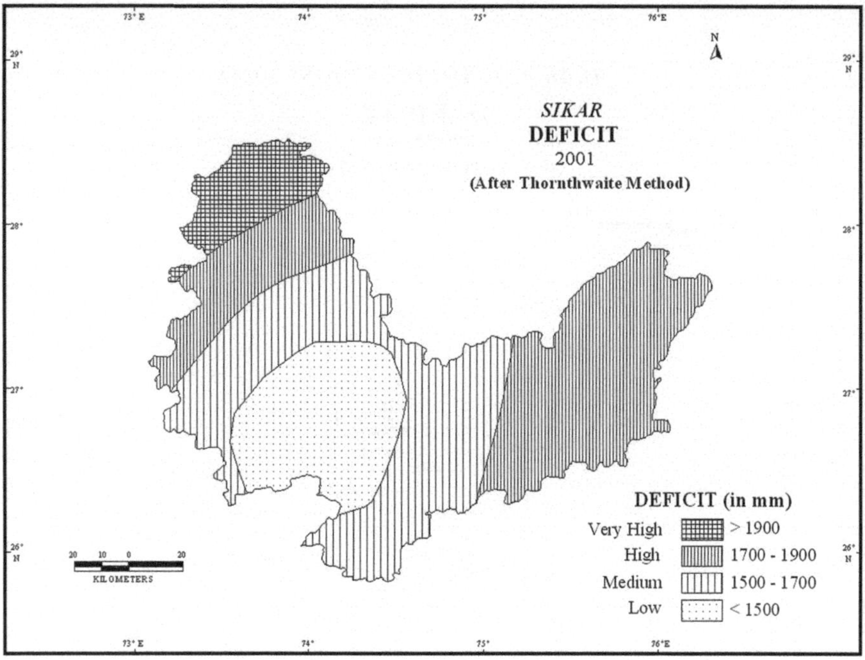

Fig. 14.5 Sikar: deficit, 2001

The primary focus of the index is on poor people, who suffer most from inadequate access to water. The WSI combines physical, social, economic, and environmental information associated with water scarcity, access to water, and ability to use water for productive purposes. The identified components are, first, water availability, which refers to water resources. The actual physical availability of surface water and groundwater make up this factor. Various aspects of this component, such as types of water demand—domestic, agricultural, and time taken in collecting water—are included in assessing this element. In the present case, the resource component was represented by availability of water in number of hours. Because this single aspect inherently includes seasonal variability and collection time, duration of water availability in hours was considered for the water scarcity index.

The chief component is water quality. Safe water quality is considered as a true representative of community environmental concern: it is taken as an environmental indicator. The population with access to safe water has been taken into consideration. Here, it is considered only from the respondent standpoint of water's fitness for drinking, cooking, washing, and bathing, and fitness for other purposes is not included. The last component is time taken to collect domestic water. With variations in rainfall and groundwater availability, difference in access to water is inevitable. Access to water for human use, including time and effort required to collect water for the household, is another major component of the water scarcity index.

The heterogeneity of water's physical availability is generally compounded by unequal access to water resources in a district or even within a community. This variability in one's access to water, spread over a region or a community, is perhaps the essence of water scarcity. In the proposed index, access to the resource is represented by ownership of a water source. Private ownership of a water source (as taps, hand pumps, wells, or tube wells) converts into rights of access to water, whereas non-ownership or use of a common water source denotes difficulty in accessing water.

14.5.2.1 Estimation of WSI

In this method, the index is constructed from a series of variables that capture the essence of being used to measure scarcity, which is done using national level data, or at a local level using locally determined values and parameters. Using the composite index approach, the following elements are included in calculating the water scarcity index:

1. Water availability
2. Access to safe water
3. Time taken to collect domestic water

These data result in the WSI formula as follows:

$$\mathrm{WSI} = w_{\mathrm{a}} A + w_{\mathrm{s}} S + w_{\mathrm{t}} \left(100 - T\right)\left(\text{Sullivan 2002b, p.1203}\right)$$

where

- A is adjusted water availability (AWA) assessment as percentage. Calculated on the basis of groundwater and surface water availability related to ecological water requirement, plus all other domestic demands, as well as the demand from agriculture and industry. The resource component is represented by availability of water in number of hours, which includes seasonal variability and collection time for water.
- S is a population with access to safe water and sanitation in percentage.
- T is an index to represent time and effort taken to collect water for the households, for example, from proportion of population having access in or near the home (common or private ownership), etc. $100 - T$ is the structure used to take account of the negative relationship between the time taken to get water and the final level of the WSI.
- w_{a}, w_{s}, and w_{t} are the weights given to each indicator of the index, and their total should be 1 to produce a WSI between 0 and 100.
- Because A, S, and T are all defined to be between 0 and 100, and w_{a}, w_{s}, and w_{t} are between 0 and 1, to develop a water scarcity index value between 0 and 100, the formula needs to be modified as follows:

$$\mathrm{WSI} = 1/3\left(w_{\mathrm{a}} A + w_{\mathrm{a}} S + w_{\mathrm{t}} \left(100 - T\right)\right)\left(\text{Sullivan 2002b, p.1203}\right)$$

The method is important to identify those areas where water scarcity conditions are severe. With the help of water scarcity index values, information about the magnitude of scarcity in the area can be derived. Figure 14.6 shows the spatial variation in the magnitude of water scarcity in the Sikar district. On the map, *iso-scarcity lines* are plotted that join equal scarcity level based on WSI values. The study has identified very high, high, and medium levels of water-scarce regions; their WSI values are given here:

1. Very high (less than 20): *Villages*: Ramgarh (6.5), Hathideh (8.3), Dhandhela (9.3), Piprali (13.9), Kuli (14.8), Gurara (16.7), Kalyanpura (16.7), Gaonri (18.5), Hirna (18.5), Bheekamsara (18.5), Mundru (19.5)
Towns: Neem Ka Thana (11.1), Khandela (13.0), Sikar (13.9), Fatehpur (16.7)

2. High (20–30): *Villages*: Dhod (20.4), Hetamsar (22.2), Doodwa (24.1), Garh Bhopji (24.1), Kachhwa (25.9), Palsana (25.9), Sanwali (26.9), Lapuwa (26.9), Khanri (27.8), Hukumpura, Maganpura (28.7)

3. Medium (more than 30): *Villages*: Peethampuri (31.5), Napawali (32.4), Goriyan (33.3)

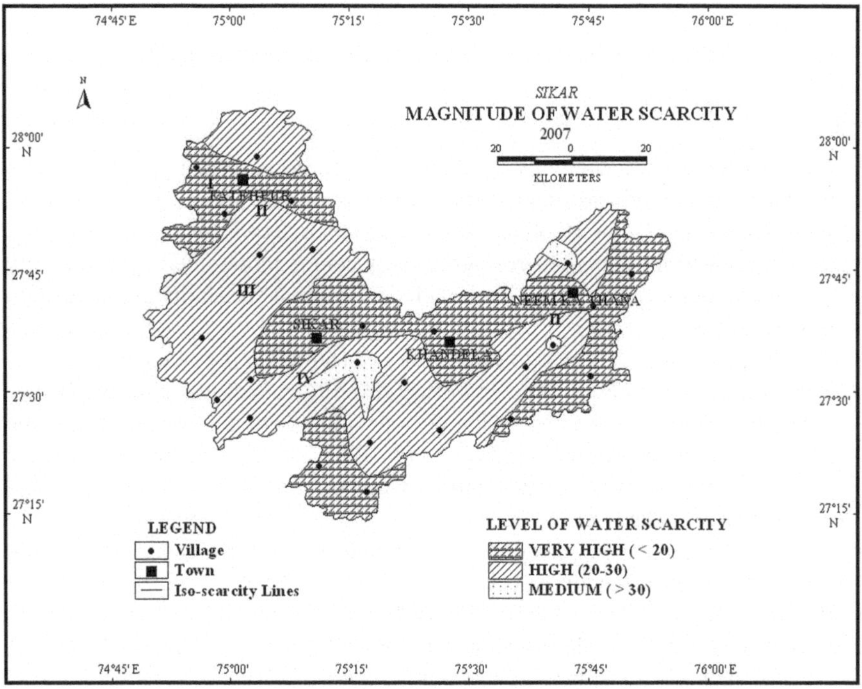

Fig. 14.6 Sikar: magnitude of water scarcity, 2007

The WSI scores of all 25 sampled villages and four towns indicate that 11 villages and four towns fall in the first category, that is, very high scarcity category with less than 20 in WSI values (Table 14.2). Eleven villages are found in the second category, that is, high scarcity of water (20–30), and 3 villages are moderately scarce with WSI scores of 30 and above.

Table 14.2 reveals that the WSI scores of rural and urban areas differ. As mentioned earlier, the higher the WSI score, the lower the degree of water scarcity, whereas a lower WSI score means high. Thus, the highest water scarcity and the lowest rural WSI score of Ramgarh village is 6.5, followed by Hathideh (8.3) and Dhandhela (9.3), while Neem Ka Thana town (11.1) is highly scarce, first in all urban examples. Surprisingly, all towns fall in the very high scarcity category with scores less than 20. With the maximum score, Fatehpur (16.7) is comparatively less scarce. On the other hand, in rural areas the maximum WSI scores are for Goriyan village, displaying a 33.3 score, which means moderately scarce.

14.6 Comparison of the Methods

Based on the foregoing analysis, it is pertinent to evaluate and compare the characteristics of both methods. Both methods are equally important to judge the availability and amount of water available to a particular area. The different types of data used by both secondary and primary methods by Thornthwaite and Sullivan give different patterns of water scarcity at different hierarchical levels. The availability of data for the Thornthwaite method is easier than for the Sullivan method, which uses primary data, often hard to collect. Secondary data (temperature and rainfall) were available up to district level, whereas primary data were collected up to individual level. Thus, the results of secondary data will provide district level patterns; conversely, the local level pattern can be extracted with the help of primary data. Secondary data are considered authentic and reliable because they are collected and published by the concerned government departments from time to time. Thus, the temporal analysis becomes easy in this method. On the other hand, primary data are raw data collected by an individual surveyor and may have errors because of manipulation. Thus, the actual picture may remain out of sight, although not entirely.

The water balance method requires data such as temperature, rainfall computational tables, and the moisture-holding capacity of the soil to calculate the water balance of an area. This method has difficult concepts with difficult calculations that require efficiency to calculate water balance. The WSI method requires first-hand information regarding accessibility, availability, and quality of water, which is easily available by talking to the people and spending time with them. Thus, the data cost is less, but it is a time-consuming process. Therefore, temporal analysis based on primary data becomes very difficult to use. This method has simple concepts that can be handled.

Table 14.2 Sikar: water scarcity index scores for villages

Rank	Village (block)	Availability	Safe water	Time taken	Wgt I	Wgt II	Wgt III	wA	wS	100–T	wt (100–T)	Total	WSI
1	Ramgarh (DRG)	33.3	0	88.9	0.5	0.25	0.25	16.7	0.0	11.1	2.8	19.4	6.5
2	Hathideh (KHN)	11.1	44.4	66.7	0.5	0.25	0.25	5.6	11.1	33.3	8.3	25.0	8.3
3	Dhandhela (NKT)	0	100	88.9	0.5	0.25	0.25	0.0	25.0	11.1	2.8	27.8	9.3
4	Piprali (PPR)	33.4	100	100	0.5	0.25	0.25	16.7	25.0	0.0	0.0	41.7	13.9
5	Kuli (DRG)	55.6	66.7	100	0.5	0.25	0.25	27.8	16.7	0.0	0.0	44.5	14.8
6	Gurara (KHN)	44.4	88.9	77.8	0.5	0.25	0.25	22.2	22.2	22.2	5.6	50.0	16.7
7	Kalyanpura (FPR)	100	0	100	0.5	0.25	0.25	50.0	0.0	0.0	0.0	50.0	16.7
10	Gaonri (NKT)	22.2	77.8	0	0.5	0.25	0.25	11.1	19.5	100.0	25.0	55.6	18.5
8	Hirna (FPR)	100	0	77.8	0.5	0.25	0.25	50.0	0.0	22.2	5.6	55.6	18.5
9	Bheekamsara (FPR)	100	0	77.8	0.5	0.25	0.25	50.0	0.0	22.2	5.6	55.6	18.5
11	Mundru (SMP)	88.9	55.6	100	0.5	0.25	0.25	44.5	13.9	0.0	0.0	58.4	19.5
12	Dhod (DHD)	66.7	55.6	44.4	0.5	0.25	0.25	33.4	13.9	55.6	13.9	61.2	20.4
13	Hetamsar (FPR)	100	55.6	88.9	0.5	0.25	0.25	50.0	13.9	11.1	2.8	66.7	22.2
14	Doodwa (LMG)	100	77.8	88.9	0.5	0.25	0.25	50.0	19.5	11.1	2.8	72.2	24.1
15	Garh Bhopji (KHN)	100	77.8	88.9	0.5	0.25	0.25	50.0	19.5	11.1	2.8	72.2	24.1
16	Kachhwa (LMG)	100	77.8	66.7	0.5	0.25	0.25	50.0	19.5	33.3	8.3	77.8	25.9
17	Palsana (PPR)	77.8	100	44.4	0.5	0.25	0.25	38.9	25.0	55.6	13.9	77.8	25.9
18	Sanwali (LMG)	100	100	77.8	0.5	0.25	0.25	50.0	25.0	22.2	5.6	80.6	26.9
19	Lapuwa (SMP)	88.9	77.8	33.3	0.5	0.25	0.25	44.5	19.5	66.7	16.7	80.6	26.9
20	Khanri (DRG)	77.8	88.9	11.1	0.5	0.25	0.25	38.9	22.2	88.9	22.2	83.4	27.8
21	Hukumpura (DHD)	88.9	100	33.3	0.5	0.25	0.25	44.5	25.0	66.7	16.7	86.1	28.7
22	Maganpura (DRG)	100	66.7	22.2	0.5	0.25	0.25	50.0	16.7	77.8	19.5	86.1	28.7
23	Peethampuri (NKT)	100	100	22.2	0.5	0.25	0.25	50.0	25.0	77.8	19.5	94.5	31.5
24	Napawali (NKT)	100	100	11.1	0.5	0.25	0.25	50.0	25.0	88.9	22.2	97.2	32.4
25	Goriya (PPR)	100	100	0	0.5	0.25	0.25	50.0	25.0	100.0	25.0	100.0	33.3

Source: Calculated by the author
DRG Danta Ramgarh, *FPR* Fatehpur, *DHD* Dhond, *NKT* Neem Ka Thana, *KHN* Khandela, *PPR* Piprali, *SMP* Sri Madhopur, *LMG* Lachhmangarh

Thornthwaite's method is a nonempirical estimation of water balance because it uses secondary data. The results do not represent ground reality because of the large areal coverage and one point of data collection, as up to district level. On the other hand, Sullivan's water scarcity index method is more practical, contains easy concepts, and represents ground reality as a result of the units of data collection. In both methods, graphical representation is done by plotting *iso-scarcity lines* on the map. The Thornthwaite method depends on secondary data, and the Sullivan method, which is quite new, represents more ground reality. Thus, it is necessary that researchers should deeply evaluate the nature of the problem and the availability of data for the study area. Both methods have their own implications and limitations depending upon the local conditions.

14.7 Findings

The results of both methods are quite noticeable from Figs. 14.5 and 14.6. The most important finding, after discussing the methods and their outcome, is that the results of Thornthwaite's water balance method are important if the study is to be conducted for a large region. The reason can be the unit of analysis, which can be a district or group of districts for which secondary data can be easily collected and analyzed. In this method, the variation will not be clearly visible beyond the district level because of data limitation. But if the study is conducted at the local or village level, then, being primary data based, a water scarcity index method will be very effective to obtain the local level spatial variations in water scarcity. In the WSI method, the variations are more evidently detected than the water balance method. As we know, a microregion has also local-level variations, thus, it is difficult to suggest any water management strategies for a wide area that has been identified as water deficit based on the water balance method. On the other hand, WSI is mostly based on face-to-face interviews and data collection, and it is quite easy to suggest management strategies that could be taken up for water management at the microlevel.

14.8 Conclusions

The study has prepared the ground for a systematic evaluation of water scarcity with the help of Thornthwaite's water balance and Sullivan's water scarcity index method. The results by both methods cast useful light on the problem of water scarcity measurement. Both methods present their own merits and demerits related to data availability, processing and analysis, and provided results. It can be concluded that both methods provide good insight into the patterns of water scarcity in a region at different regional levels according to the nature of the data used.

References

Bandyopadhyay J (1987) Political ecology of drought and water scarcity: need for an ecological water resources policy. Econ Polit Wkly 12:2159–2169
Census of India (2001) Provisional population totals. Paper I. Rajasthan
Falkenmark M (1983) Main problems of the water use and transfer of technology. In: Glassner MI (ed) Global resources. Praeger Publishers Inc., New York, pp 468–478
Joshi KC (2004) Water scarcity in Port Blair: problems and prospects. Geogr Rev India 66(4):391–397
Khurana I (2003) Water scarcity? Try capturing the rain. In: Prasad K (ed) Water resource and sustainable development: challenges for the 21st century. Shipra Publication, New Delhi, pp 137–147
Mehta L (2003) Contexts and constructions of water scarcity. Econ Polit Wkly 38(48):5067–72
Oweis TY (2005) The role of water harvesting and supplemental irrigation in coping with water scarcity and drought in the dry areas. In: Wilhite DA (ed) Drought and water crises. Taylor & Francis, London, pp 191–213
Sahay R (2003) Water stress and coping strategies in the Mehrauli Block of Metropolitan Delhi. M. Phil. dissertation, Department of Geography, University of Delhi
Soundaravalli G (1992) Water scarcity in Dindigul city. In: Vidyanath V, Rao RR (eds) Development of India's resource base: patterns, problems and prospects. Gian Publishing House, New Delhi, pp 270–279
Subramaniam AR, Vinayak VSSK (1982) Water balance studies of Trivandrum and its vicinity. Trans Inst Indian Geogr 4(1):39–42
Sullivan C (2002a) Calculating water poverty index. Centre for Ecology and Hydrology, Wallingford
Sullivan C (2002b) Calculating water poverty index. World Dev 30(7):1195–1210
Thornthwaite CW (1948) An approach towards a rational classification of climate. Geogr Rev 38(1):56–94
Thornthwaite CW, Mather JR (1955) The water balance. Publ Climatol 8(1):87–94

Chapter 15
Nonlinear Groundwater and Agricultural Land Use Change in Rajasthan, India

Trevor Birkenholtz

Abstract Since the 1950s, the rapid expansion of groundwater irrigation globally has led to dramatic shifts in land use. Nowhere is this more true than in India where, since the 1960s, groundwater irrigation has expanded to 34.5 million hectares, 70 % of the country's total. Yet we do not know the character of this landscape nor the degree to which changes in land use are the result of multiple ecological and social drivers. Therefore, this article asks: (1) what is the relationship between groundwater decline and agricultural land use change in India, and what does it mean for the future of agricultural intensification; and (2) in what ways do social institutions produce and adapt to this change, while leading to yet further shifts in land use? This chapter draws on government statistics and from household surveys and interviews from a case study in the semiarid state of Rajasthan, India, to examine these questions.

Findings suggest that the relationship between the expansion of groundwater-irrigated area and land use change is nonlinear, in that the expansion of irrigated area initially led to the expansion of market-oriented crops but rapid groundwater decline is demanding a return to local cropping varieties, particularly among the most marginal producers. This return is being facilitated through the creation of new adaptive social institutions, such as tube well irrigation partnerships, under conditions of dynamic ecological change, including synergistic groundwater and land use change. The conclusion offers suggestions toward a second Green Revolution in agriculture via the strengthening of local social institutions.

Keywords Adaptation • Groundwater • India • Irrigation • Second Green Revolution • Social institutions

T. Birkenholtz (✉)
Department of Geography and Geographic Information Science,
University of Illinois at Urbana-Champaign, Champaign, IL, USA
e-mail: trevbirk@illinois.edu

A.K. Dutt et al. (eds.), *Spatial Diversity and Dynamics in Resources and Urban Development*, DOI 10.1007/978-94-017-9771-9_15

15.1 Introduction

More than 300 million hectares (ha) of agricultural land are irrigated globally (Shah 2005). Historically supplied by surface water, this pattern continues into the present with more than 66 % being surface water dependent (Shah 2005). But this is changing. In the United States, for example, between 1950 and 2000 groundwater withdrawals, as a percentage of total water use for irrigation, increased from 23 to 42 % (Hutson et al. 2005). In India, groundwater-based irrigated area rose from 7.4 million hectares in 1962 to 34.5 in 2003 (Narayanamoorthy 2006), currently accounting for 70 % of total irrigated area (World Bank 2005). The rise in groundwater-based irrigation in India therefore a significant and rapid land use transformation, therefore.

Although irrigated area has expanded significantly, intensification has also occurred through the adoption of further Green Revolution technologies, such as chemical fertilizers and high-yielding seed varieties (HYVs). The extensification and intensification of agriculture, made possible through the spread of groundwater irrigation systems, has resulted in vastly increased agricultural yields and in increased capital accumulation among farmers. Taken together, these shifts support an increasingly globalized production regime. Wheat exports in India, for example, rose from 632,468 million tons (mt) in 1995 to 2,007,947 in 2004 (FAO 2008). It has also exacerbated disparities between classes and castes of farmers (Jeffrey 2001; Birkenholtz 2008a). So too, the rapid increase in groundwater-irrigated area in India is dependent not only on the availability of groundwater and the spread of new groundwater-lifting technologies but also on farmers' abilities to create adaptive social institutions to access them, while mediating ecological and global political-economic change.

However, demand for groundwater in India is expected to exceed supply by 2020, which may lead to new forms of conflict or cooperation as resource scarcity increases (World Bank 2005). Consequently, the future viability of groundwater-led intensified agriculture is of significant concern. It is not clear, however, whether groundwater decline will result in a linear decline in either groundwater-irrigated area or in HYVs, or what this means for land use change and farmers' abilities to adapt to these dynamic political-economic and ecological shifts. The questions become: What is the relationship between groundwater decline and agricultural land use change in India, and what does it mean for the future of agricultural intensification?; and second, in what ways do social institutions produce and adapt to this change, while leading to yet further shifts in land use?

In human–environment research, increasing attention is focused on the degree to which land use transformations are the product of multiple social and ecological processes (Mertz et al. 2005; Chowdhury and Turner 2006; Caldas et al. 2007). Hazell et al., for example, identified a number of drivers of global agricultural change that are informative in thinking about land use change specifically (Hazell and Wood 2008): these include global-scale drivers such as globalization of markets, OECD agricultural support and privatization of agricultural science;

country-scale drivers such as agricultural supports, energy policy, and water scarcity; and local-scale drivers such as local natural resource availability, property rights regimes, and non-farm employment opportunities (Hazell and Wood 2008, p. 502). To these local-scale drivers I would add local social institutional change or adaptation that may result from but also feed back into these multi-scalar drivers. Adaptation in this sense is the shifting practices that farmers (i.e., land managers) employ to meet ecological or social challenges to continue or expand production. For example, farmers may apply gypsum to the soil to reduce its alkalinity, or they may form partnerships for tube well irrigation systems, which are too expensive (and/or risky) to adopt on their own, to intensify production.

Much related work to date, however, examines the political and ecological drivers of changing cultivation and fallow practices under globalization pressures around deforestation (Vasquez-Leon and Liverman 2004; Zimmerer 2006; Lawrence et al. 2008), particularly in the Amazon (Walker and Homma 1996; Caldas et al. 2007; Hecht and Saatchi 2007), rather than the conversion of dry-land agriculture or grazing areas to irrigated agriculture (but see Wadley et al. 2006). Work in India, for example, has shown that increases in groundwater use via tube well adoption led to radical increases in winter cropping and decreases in fallow grazing area (Robbins 2001). Yet looking at the shifts in agricultural area does not detail the actual and shifting character of the agricultural landscape. There is a need to link the positive, but nonlinear, social and ecological feedbacks that result from agricultural intensification (Peters et al. 2007). And, although some rightly point to the need to examine the effects of land use change on hydrological processes (DeFries and Eshleman 2004), few have attempted to examine the positive feedbacks between land use and hydrological processes and social decision making (Alauddin and Quiggin 2008; Gaur et al. 2008). Therefore, the relationship between groundwater decline, agricultural land use change, and social decision making in India is in need of explication.

This chapter examines these relationships through a case study from the semiarid northwestern Indian state of Rajasthan, where 71 % of irrigated agriculture and 80 % of the state's drinking water needs rely on groundwater (Directorate of Economics and Statistics 2003; World Bank 2005). In Rajasthan, as with much of India, groundwater withdrawal and tube well construction are not regulated. This lack of regulation, along with State efforts to expand rural electrification and to provide low-interest loans to farmers for agricultural intensification since the late 1960s, has been a dramatic driver of country-scale agricultural change (Birkenholtz 2008a). This scenario has resulted in rapid technological diffusion of 1.4 million agricultural tube wells across the state, with a 33 % increase between 1991 and 2001 alone (GORGB 2003). Groundwater irrigated area also increased dramatically, by 52 %, to more than four million hectares between 1994 and 2001.

Given the state's (and India's) heavy reliance on this water source, understanding the relationship between social, groundwater, and land use change is very important. Moreover, it informs our theoretical understanding of land use change under multiple and shifting driving forces. This research, carried out in Rajasthan's Jaipur District (Fig. 15.1), suggests that the relationship between the expansion of

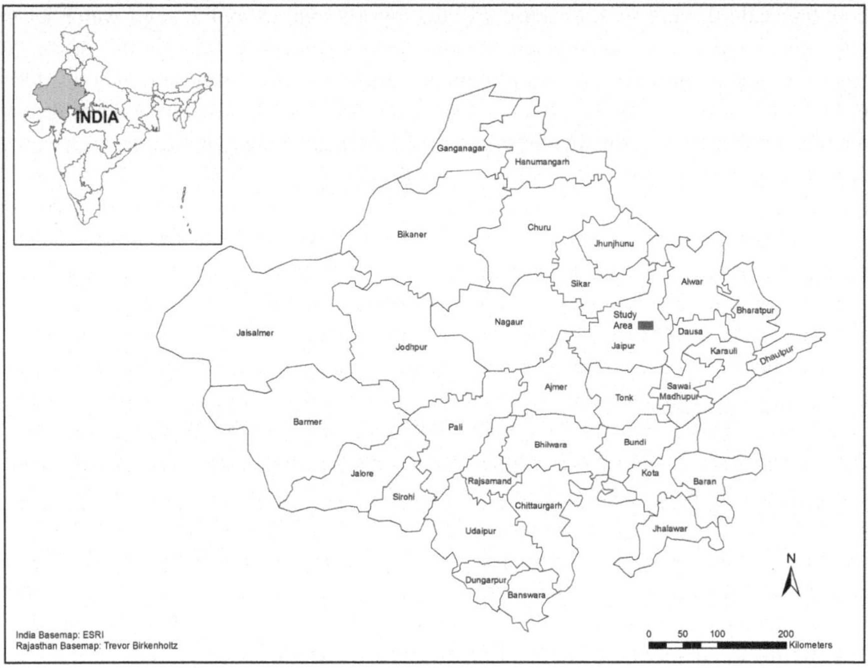

Fig. 15.1 Study area in Jaipur District, Rajasthan, India

groundwater-irrigated area and land use change is nonlinear in that the expansion of irrigated area initially led to the expansion of market-oriented crops but that rapid groundwater decline is demanding a return to local cropping varieties, particularly among the most marginal producers. This return is being facilitated through the creation of new adaptive social institutions, under conditions of dynamic political-economic and ecological change, including synergistic groundwater and land use change. These ecological constraints are felt by farmers differentially and may undermine some farmers' abilities to adapt to broader political, economic and ecological shifts (such as climate change) in the future.

15.2 Study Area and Methods

15.2.1 Study Area: Jaipur District, Rajasthan

Jaipur District is a semiarid region of moderately productive, yet spatially uneven, agricultural land, with nitrogen-poor alluvial soils and reasonable groundwater recharge (Singhania and Somani 1992). The area is entirely reliant on groundwater for domestic and irrigation needs. There is no surface water and no government water supply. Summer temperatures commonly reach 44 °C. Average annual rainfall is typically between 500 and 600 mm, occurring mostly between July and September, but can be highly variable; for example, rainfall in 2002 was 207 mm. There are two cropping seasons: the *khariph* (summer) crop, which is cultivated from July to October (i.e., during the rainy season), and the *rabi* (winter) crop, which is cultivated between October and March. The main *khariph* crops are millet, peanuts, sesame, legumes, spices, vegetables (for those who can grow them), and fodder crops; the main *rabi* crops are wheat, barley, fodder crops, and a limited number of vegetables. The *rabi* crop is fully dependent on (mostly flood) irrigation whereas the *khariph* crop relies on monsoon rains. Those with the capability irrigate the *khariph* as well. The *rabi* crop, therefore, can only be produced by those households with access to irrigation.

The social environment of the study area is highly stratified, composed of low- and high-caste Hindus and small, medium, and large landholders. There is a significant and interdependent relationship between caste and landholdings, with χ^2 (4, $n=151$) = 17.556, $p=.001$, where marginal castes own the least amount of land. Access to resources, including groundwater and irrigation, is mediated through these relationships, which impacts land use. Farmers have formed partnerships for the construction, use, and maintenance of tube wells for irrigation. The size of these partnerships is stratified by landholdings (class), which is also related to caste (see previous). Farmers form these partnerships, not because one tube well can irrigate more area than some farmers own, but because of the high cost of the tube well and electrical connections and the high rates of tube well failure as groundwater declines (Birkenholtz 2009). Agriculture is the main occupation, with high levels of off-farm employment, particularly among the most marginal households (Tables 15.1 and 15.2).

15.2.2 Methods

Research for the present study was carried out between 2002 and 2007. First, it draws on district-level statistical data published by the Department of Economics and Statistics of the State of Rajasthan. It analyzes cropping pattern change over the period between 1993–1994 and 2001–2002, the longest period for which comparable district-wide data were available. Second, in 2005 a household survey, utilizing an every-third-household selection technique, was conducted of 151 farmers in six villages of Bassi Tehsil, about 60 km east of Rajasthan's capitol city, Jaipur. The surveys detailed basic household production information (including cropping and income), groundwater use, and irrigation, along with more open-ended questions regarding the groundwater situation. Third, the surveys were followed up with repeated in-depth interviews with 78 farmers and with government engineers and tube well-drilling firms in 2005 and 2007. These interviews provide detailed accounts of adaptive cultivation strategies under dynamically changing conditions of groundwater quality and quantity.

Table 15.1 Percentage change in area and production of principal crops in Jaipur District, Rajasthan, between 1993–1994 and 2001–2002

Crop	Area change	Production change	Subsistence (S) or commercial (C)	Irrigation requirement
Linseed	−96.55	−90.91	C	High
Rape/mustard	−60.12	−66.97	C	High
Red chili	−36.82	−36.61	C	High
Total oilseed	−38.97	−25.75	C	High
Maize	−6.42	132.91	S/C	High
Pulses	−1.04	−6.41	S	None
Barley	6.69	71.17	S	High
Total food grain	10.17	42.61	C	–
Wheat	12.30	56.23	S/C	High
Pearl millet	13.73	37.15	S	Low
Sorghum	23.39	−60.97	S	Low
Total condiments and spices	32.10	−41.02	C	–
Sesame	48.96	27.14	C	Low
Groundnut	52.93	171.94	C	Low
Total vegetables	53.58	11.01	C	High

Table data arranged by "area change" from most negative to most positive

Table 15.2 Income and land use indicators from survey ($n = 151$)

Land category (hectares, ha)					
	0.25–0.5	0.51–1.0	1.10–2.5	2.51–3.9	4.00–7.5
No. households	6	28	81	24	12
Tube well partners	7.1	3.11	2.96	2.68	2.15
Average income[a]	46,591	82,176	104,457	137,477	205,766
Average crop income	9,425	14,212	40,716	74,636	127,600
Average crop income/ha	27,013	20,759	28,935	24,695	24,451
Average	1.33	0.96	1.91	2.33	2.00
Crops					
Average number of summer crops	1.16	1.57	2.17	2.54	3.00
Average percent irrigated, winter	94	63	83	80	69
Average percent irrigated, summer	67	63	74	76	62
Average total hectares	2.15	19.11	116.08	72.66	65.82
Total irrigated area, winter	2.03	12.28	93.75	57.47	45.44
Total irrigated area, summer	1.39	12.05	85.32	54.43	41.27

[a]Average income in rupees; at time of research, 45 rupees = $1

15.3 Groundwater Decline, Agricultural Land Use Change, and the Future of Intensification

Between 1994 and 2002, total net irrigated area in Jaipur District rose by more than 9 % from 302,428 to 330,569 ha. Groundwater irrigation became more prominent over this period with an increase in net tube well irrigated by 11 % to 329,000 ha. However, the composition and productivity of groundwater-irrigated areas changed dramatically. Table 15.1 illustrates the change in area and productivity of subsistence and commercial crops in the district. Irrigation-intensive oilseeds have declined significantly in both categories. Commercial spices have increased in area but have declined in production. Overall, there is an area and productivity increase in low irrigation-demanding mixed subsistence and commercial crops such as pearl millet, sorghum, sesame, and groundnut. And finally, the production of wheat, barley, and maize, which have high irrigation requirements, has gained in productivity with relatively little change in the amount of land devoted to them.

These district-level patterns, although more precise that most measures of land use, say little of the processes producing them. One possibility is a change in market prices, encouraging farmers to switch production to more lucrative crops, but commodity prices overall during this period have moved downward (Barker and Molle 2005; FAO 2006). Therefore, this does not explain the decline in oilseed production or the rise in sorghum, sesame, or groundnuts. The change in land use over the period could also be caused by a lack of available inputs, such as seeds, fertilizer, or pesticides. According to farmers and local government officials, however, this is not the issue either.

This significant transformation in land use is actually the result of rapid groundwater decline. Rapid utilization of groundwater for irrigation throughout the 1990s has resulted in falling water tables of as much as 60 m throughout the study area (GORGB 2003). There is also a yearly groundwater overdraft of 410 million m^3. This considerable groundwater decline in the area has led to increases in concentrations of naturally occurring minerals, such as sodium and calcium, and to the creation of saline water and sodic soils (Jacks et al. 2005). This change has encouraged a return to local crop varieties that are more tolerant to these soil and water conditions. Quoting one farmer:

> Formerly we could grow tomatoes, okra, chili peppers and eggplant, but now we do not like to because we cannot produce much of it due to the salty water. So now we grow mostly sesame, groundnuts, fodder crops and some lentils [which require less irrigation].

Groundwater salinity is not constant spatially or temporally. One hundred percent of 151 farmers surveyed indicated that they had at least seasonal groundwater salinity and/or hardness (*talia*). Following another farmer:

> The groundwater becomes more saline throughout the summer. It did not used to be like this; it [the salinity of the water and the calcification/sodic soil] happened with the irrigation.

Therefore, to answer the first question—*what is the relationship between groundwater decline and agricultural land use change and what does it mean for the future of agricultural intensification*—is that groundwater decline and, in particular, the changing character of groundwater, is undermining the continued production of water-demanding, mostly market (e.g., oilseeds, spices, vegetables), varieties, which without future adaptive management action (such as investment into less water demanding varieties and irrigation efficiency-enhancing technologies) will likely undermine continued intensification. Looking to the household, however, the role of social institutions in producing and adapting to this change becomes apparent. It also illustrates the ways that these regional land use shifts emerge from the disaggregated decisions of individual farmers (and farmer partnerships).

15.4 Social Institutions and Shifting Land Use

Stratifying the survey into five landholding categories, the relationship between the number of tube well irrigation partners, income, and cropping variety becomes apparent. Of the sample, 83 % of all tubewells were owned in partnerships and 76.6 % of the sample coordinated irrigation timing with their partners or neighbors. This practice allows the smallest farmers to irrigate large proportions of their land, 94 % in the winter and 67 % in the summer (see Table 15.2), and to maintain high income to crop area ratios relative to larger landholding classes. Partnerships for the construction and use of tube wells initially underwrote the expansion of groundwater-irrigated area and the cultivation of water-demanding crops, but the size of the partnership impacts the variety of crops and the quality of irrigation.

There is an inverse relationship between the number of irrigation partners and cropping variety in both winter and summer (see Table 15.2). Smaller landholders have larger numbers of partners, which reduces the availability of irrigation per partner. This arrangement also produces a higher spatial concentration of wells, which places heavier demands on groundwater, further exacerbating localized groundwater drawdown and mineralization (as well as mutual interference between wells in close proximity). These factors limit the variety of crops that can be grown to low irrigation-demanding crops, such as fodder crops and crops for subsistence use.

Indeed, in land category one, 67 % of the sample produced sorghum and 83 % produced wheat during the winter for home consumption. In the summer, 33 % produced both pearl millet and sorghum for home consumption. Only one farmer in this category produced the primary winter crops of sorghum and wheat and two farmers produced groundnut and pearl millet in the summer for the market. None of the farmers produced vegetables, which are primarily market crops. These crops are also very water demanding and are sensitive to irrigation timing. Farmers are adapting to this change by returning to local crop varieties (see previous farmer quote). Thus, tube well adoption, rather than leading to further capitalization and to market integration, has had the opposite effect of disengagement from the market and an increase in subsistence cultivation as a consequence of groundwater quality constraints.

Small farmers adapt to these conditions, therefore, by returning to local, more resilient cropping varieties such as pearl millet, sorghum, sesame, and groundnut. They are also abandoning monocultural cropping in favor of intercropping. Figures 15.2 and 15.3 illustrate the intercultivation of pearl millet, sorghum, sesame, and watermelon in the same field. Two years earlier, this field was divided into HYV millet and spices. Intercropping, along with prevalent home use, also makes it difficult to obtain accurate data for planted area or crop production (hence their absence from this study). Moreover, typical LULC classifications are too coarse to register this shift, even though the social and ecological impacts are quite diverse. These patterns of production and adaptation strategies are in contrast to those of larger farmers.

In land category four, 50 % produced sorghum, 100 % produced wheat, 50 % produced barley, and 17 % produced alfalfa for home consumption in the winter. In the summer, 88 % produced pearl millet and 38 % produced sorghum, in addition to six other crops, for home consumption. But in the winter, 95 % produced wheat, 4 % produced pearl millet, gram, and chickpeas, 67 % produced sorghum and barley, and 8 % produced sesame for the market. Finally, 25 % of the farmers in this category produce lucrative but water-demanding vegetables (during the winter), indicating their greater access to groundwater of acceptable quality. Thus, for these larger farmers, who have 2.68 tube well partners on average, tube well adoption allows them to integrate with the market and to expand their cropping variety. Table 15.2 shows the average number of crops being grown during the winter and

Fig. 15.2 Image of study area indicating multicropping

Fig. 15.3 Detailed image of study area indicating multicropping

summer seasons for each of the five land categories. Clearly, larger farmers produce a more diverse variety of crops, but this is not solely the result of larger landholdings. It is the result of fewer irrigation partners, which provides them a more reliable supply of irrigation, enabling them to increase their cropping variety and integrate more fully (and selectively) into commodity markets.

Larger farmers are also able to grow more varieties because their beneficial economic position enables them to engage in further adaptive strategies, such as the addition of gypsum to the soil. Adding gypsum to soil, then flushing it with irrigation (or carefully timed precipitation), can reduce alkalinity and sodicity (Ramesam and Barua 1973; Fullen and Catt 2004). Of 151 farmers, 100% indicated that they had at some time added gypsum to their soil to "loosen" it up, but this is a very expensive adaptation. One (large) farmer in the survey, with 6.33 ha, adds $275 worth of gypsum yearly to the soil. This adaptation allows some farmers to grow lucrative, but less salt tolerant, varieties such as vegetables.

Therefore, the answer to the second question—*what ways do social institutions produce and adapt to this change, while leading to yet further shifts in land use?*—is that tube well irrigation partnerships initially led to the expansion of irrigation-demanding, market-oriented crops. But the spread of groundwater irrigation resulted in groundwater drawdown, diminishing the quality and availability of groundwater for irrigation, particularly among the smallest producers, who have to form larger partnerships. Therefore, the cause and consequences of ecological feedbacks are socially stratified, which has also resulted in socially differentiated adaptation. Small producers have returned to intercropping local varieties that are less water demanding and more subsistence oriented. Larger farmers generally have better access to high-quality groundwater, but have also turned to mitigation techniques, such as adding gypsum as a soil beneficiation practice. These divergences in adaptive ability have led to yet further differentiation, which is visible in the landscape via its impacts on land use diversity, in income disparities, and in degree of market integration.

15.5 Discussion: Groundwater Decline, Social Institutions, and Land Use

The rapid diffusion of groundwater irrigation throughout the study area since the Green Revolution initially led to the expansion of water-demanding and market-oriented crop production, a significant transformation in land use. Because of the success of this process of state-supported, but farmer-led, agricultural intensification and local institutional adaptation, however, significant groundwater decline has also occurred, undermining the original ecological conditions under which this revolution occurred. This contretemps is leading to yet another significant shift in land use and to further local adaptations, a shift that is not caught by typical coarse categories of land use classification.

Returning to the typology of drivers of change in agriculture set out by (Hazell and Wood 2008), the authors point to global-, country-, and local-scale drivers. Although global-scale political economic drivers, such as OECD agricultural subsidies that make it difficult for farmers in less developed countries (LDC) to compete on global markets, are important, particularly in driving the initial rise in agricultural intensification in the study area, the most significant drivers of changing land use in this chapter are more localized, at least for the time being. Of course the future potential of global climate change (GCC) to increase climate variability and to produce more extreme events, particularly in the tropics, is an issue (IPCC 2007). So too, future trends in global agricultural policy could cause shifts as well.

Therefore, at different moments in the economic and ecological process, the drivers of land use change are divergent. Initially, the innovation and diffusion of Green Revolution technologies faced few ecological hurdles. The barriers were rural electrification and sufficient capital investment and adaptation (i.e., tube well adoption via collective investment), which the state and local farmers addressed, respectively. But currently, local-scale social and ecological change is driving these shifts in land use. Hazell and Wood (2008) point out that "water shortages within river basins and aquifers will curtail irrigated agriculture in many countries if not addressed soon" (p. 502). The present research suggests, however, that groundwater decline does not curtail irrigated agriculture in a simple linear fashion. Instead, this is a dynamic process wherein the groundwater-irrigated area is still expanding but the character of the crops and the adaptive diversification strategies of the farmers are changing, both producing and mediating these ecological shifts.

Rapid groundwater decline, however, has begun, and without further country-scale technical intervention, such as the expansion of efficiency-enhancing drip irrigation systems, irrigated area and market crop production will continue to decline. Rather than promoting the spread of these technologies with state investment, as with the Green Revolution, the state is proposing a series of state- and local-level institutional reforms, including the privatization of water resources, the liberalization of markets, and new nondemocratic forms of decentralized governance, which will undermine local institutions (Birkenholtz 2008a, b). These proposed interventions are being influenced by a current political-economic climate that promotes neoliberal, market-led approaches to property institutions and natural resource management.

These series of proposed interventions represent another particular economic and ecological moment, which are unlikely to produce the conditions necessary for farmers to continue to evolve local adaptive management strategies. It is clear that farmers need to develop land use strategies to help them mitigate unpredictable economic and ecological conditions (Hanson et al. 2007). But the ecological conditions described here, along with the currently proposed state interventions, reduce farmer flexibility and the dynamism in decision making that is needed to mediate the unpredictability of globalized markets and nonlinear ecologies.

15.6 Conclusions

The case study presented here illustrates that the global expansion of groundwater-irrigated area is the result of positive feedbacks between social and ecological drivers. These drivers, however, vary nonlinearly from moment to moment within political, economic, and ecological processes, rendering land use a complex process. Although there continues to be vast potential for further groundwater irrigation globally (Faurès et al. 2007), local groundwater decline and social adaptive potential are highly uneven. This caveat will continue to impact land use and social well-being, including food security. However, the continued expansion of irrigated area and crop production need not face future limits, currently imposed proximately by groundwater decline but ultimately by political and economic will.

Clear scope exists to enhance the efficiency of groundwater irrigation systems. Research in Southern India has recognized, for example, up to 60 % efficiency gains with drip irrigation (Narayanamoorthy and Deshpande 2005). Thus, there is potential and need for a second Green Revolution in India and throughout the Global South (Postel et al. 2001; Wollenweber et al. 2005). The potential for this second Green Revolution is dependent on strengthening local social institutions and adaptive capacity with focused public and private support.

References

Alauddin M, Quiggin J (2008) Agricultural intensification, irrigation and the environment in South Asia: issues and policy options. Ecol Econ 65:111–124

Barker R, Molle F (2005) Perspectives on Asian irrigation. In: Shivakoti G, Vermillion D, Lam WF, Ostrom E, Pradhan U, Yoder R (eds) Asian irrigation in transition: responding to challenges. Sage, New Delhi/London

Birkenholtz T (2008a) Contesting expertise: the politics of environmental knowledge in Northern Indian groundwater practices. Geoforum 39:466–482

Birkenholtz T (2008b) 'Environmentality' in Rajasthan's groundwater sector: divergent environmental knowledges and subjectivities. In: Goodman M, Boykoff M, Evered KT (eds) Contentious geographies: environment, meaning, scale. Ashgate Press, Aldershot

Birkenholtz T (2009) Irrigated landscapes, produced scarcity, and adaptive social institutions in Rajasthan, India. Ann Assoc Am Geogr 99:118–137

Caldas M, Walker R, Arima E, Perz S, Aldrich S, Simmons C (2007) Theorizing land cover and land use change: the peasant economy of Amazonian deforestation. Ann Assoc Am Geogr 97:86–110

Chowdhury RR, Turner BL (2006) Reconciling agency and structure in empirical analysis: smallholder land use in the southern Yucatan, Mexico. Ann Assoc Am Geogr 96:302–322

DeFries R, Eshleman NK (2004) Land-use change and hydrologic processes: a major focus for the future. Hydrol Process 18:2183–2186

Directorate of Economics and Statistics (2003) Statistical abstract: Rajasthan 2001. Directorate of Economics and Statistics, Jaipur

FAO (2006) FAOSTAT core prices data. Food and Agricultural Organization of the United Nations, Geneva

FAO (2008) Key statistics of food and agriculture external trade. Geneva: United Nations Development Program

Faurès J-M, Hoogeveen J, Bruinsma J (2007) The FAO irrigated area forecast for 2030. Available at: http://www.fao.org/nr/aboutnr/nrl/en/

Fullen MA, Catt JA (2004) Chapter 3: Desertification, salinization and amelioration of arid soils. In: Soil management problems and solutions. Oxford University Press, Oxford

Gaur A, Biggs TW, Gumma MK, Parthasaradhi G, Turra H (2008) Water scarcity effects on equitable water distribution and land use in a major irrigation project: case study in India. J Irrig Drain Eng ASCE 134:26–35

GORGB (2003) Government of Rajasthan Groundwater Board. Personal communication. In: Birkenholtz T (ed). Rajasthan Groundwater Board, Jaipur

Hanson JD, Liebig MA, Merrill SD, Tanaka DL, Krupinsky JM, Stott DE (2007) Dynamic cropping systems: increasing adaptability amid an uncertain future. Agron J 99:939–943

Hazell P, Wood S (2008) Drivers of change in global agriculture. Philos Trans R Soc B Biol Sci 363:495–515

Hecht SB, Saatchi SS (2007) Globalization and forest resurgence: changes in forest cover in El Salvador. Bioscience 57:663–672

Hutson SS, Barber NL, Kenny JF, Linsey KS, Lumia DS, Maupin MA (2005) Estimated use of water in the United States in 2000. United States Geological Survey, Washington, DC

IPCC (2007) Contribution of working group I to the fourth assessment report of the Intergovernmental Panel on Climate Change. In: Solomon S, Qin D, Manning M, Chen Z, Marquis M, Averyt KB, Tignor M, Miller HL (eds) Climate change 2007: the physical science basis. Cambridge University Press, Cambridge/New York

Jacks G, Bhattacharya P, Chaudhary V, Singh KP (2005) Controls on the genesis of some high-fluoride groundwaters in India. Appl Geochem 20:221–228

Jeffrey C (2001) 'A fist is stronger than five fingers': caste and dominance in rural north India. Trans Inst Br Geogr 26:217–236

Lawrence D, D'Odorico P, Diekmann L, DeLonge M, Das R, Eaton J (2008) Land change science special feature: ecological feedbacks following deforestation create the potential for a catastrophic ecosystem shift in tropical dry forest (vol 104, pg 20696, 2007). Proc Natl Acad Sci U S A 105:3169–3169

Mertz O, Wadley RL, Christensen AE (2005) Local land use strategies in a globalizing world: subsistence farming, cash crops and income diversification. Agric Syst 85:209–215

Narayanamoorthy A (2006) Trends in irrigated area in India: 1950–1951 to 2002–2003

Narayanamoorthy A, Deshpande RS (2005) Where water seeps! Towards a new phase in India's irrigation reforms. Academic Foundation, New Delhi

Peters DPC, Sala OE, Allen CD, Covich A, Brunson M (2007) Cascading events in linked ecological and socioeconomic systems. Front Ecol Environ 5:221–224

Postel S, Polak P, Gonzales F, Keller J (2001) Drip irrigation for small farmers: a new initiative to alleviate hunger and poverty. Water Int 26:3–13

Ramesam V, Barua S (1973) Preliminary studies on the mechanisms controlling the salinity in northwestern arid region of India. Indian Geohydrol 9:10–18

Robbins P (2001) Tracking invasives in India, or why our landscapes have never been modern. Ann Assoc Am Geogr 91:637–659

Shah T (2005) Groundwater and human development: challenges and opportunities in livelihoods and environment. Water Sci Technol 51:27–37

Singhania RA, Somani LL (1992) Soils of Rajasthan. In: Sharma HS, Sharma ML (eds) Geographical facets of Rajasthan. Kuldeep Publications, Ajmer

Vasquez-Leon M, Liverman D (2004) The political ecology of land-use change: affluent ranchers and destitute farmers in the Mexican municipio of Alamos. Hum Organ 63:21–33

Wadley RL, Mertz O, Christensen AE (2006) Local land use strategies in a globalizing world: managing social and environmental dynamics. Land Degrad Dev 17:117–121

Walker R, Homma AKO (1996) Land use and land cover dynamics in the Brazilian Amazon: an overview. Ecol Econ 18:67–80

Wollenweber B, Porter JR, Lubberstedt T (2005) Need for multidisciplinary research towards a second green revolution: commentary. Curr Opin Plant Biol 8:337–341

World Bank (2005) India: bracing for a turbulent water future. Retrieved January 3, 2005, from http://www.worldbank.org.in/WBSITE/EXTERNAL/COUNTRIES/SOUTHASIAEXT/INDIAEXTN/0,contentMDK:20668501~pagePK:141137~piPK:141127~theSitePK:295584,00.html

Zimmerer KS (ed) (2006) Globalization & new geographies of conservation. The University of Chicago Press, Chicago

Chapter 16
Political Ecology of Groundwater Depletion in Northwestern India

Inder Jeet

Abstract Reduction in the stock of groundwater in terms of quantity and quality by utilization of the groundwater resource by farming communities has been recognized as one of the major environmental problems in northwestern parts of India. The study area as a part of the Indus basin shows clear signs of groundwater depletion in many districts. Pressure on the groundwater resource has become linked to a few major developments in the cropping systems of Punjab, Haryana, and western Uttar Pradesh in recent years. This change in cropping pattern and certain governmental policies has tremendous negative effects on groundwater resources. In the present research, it is inquired how the patterns of groundwater utilization are influenced by the social structure of the farmers, government policies, penetration of market forces in agriculture systems, and surplus extraction of groundwater for irrigation. The chapter argues that the decreasing size of land holdings as a result of disintegration of the joint family system in farming communities under the law of inheritance and governmental policies are the main causative factors of groundwater exploitation in northwestern parts of India. Consequently, a *political ecology approach* has been used in the present research.

Keywords Ground water depletion • North-Western India

16.1 Introduction

Water is a scarce resource for agriculture in India, with a distinctly seasonal climatic regime and spatial variations in groundwater utilization. The study of natural resource utilization can be successfully achieved via the combined efforts of physical and human geographers, each with their established skills (Bell and Roberts 1986, 1991). One of the most robust and profound approaches to analyze

I. Jeet (✉)
Department of Geography, Maharshi Dayanand University, Rohtak, Haryana 124001, India
e-mail: ij_ghangas@yahoo.com

A.K. Dutt et al. (eds.), *Spatial Diversity and Dynamics in Resources and Urban Development*, DOI 10.1007/978-94-017-9771-9_16

groundwater utilization in the context of human–environment interaction is political ecology. Political ecology in geography has increasingly sought the incorporation of ecological analysis. This approach provides a framework for human ecologists interested in examining the interrelationships between local patterns of resource and larger political economy (Bassett 1988). Basic to this approach is to understand how political interest, social structure, social institutions, and human–environment interactions act together to cause groundwater depletion (Dhawan 1995; Bhullar and Sidhu 2006).[1] Groundwater depletion ecology is inherently interactive and illustrates the potential for geography to unite the physical (environment), social, and institutional (man-made) worlds. The political ecology concept is employed here in a comparative analysis of regional variations in the nature and intensity of groundwater exploitation in Northwestern India.

Agriculture in India has made amazing progress since the introduction and widespread use of high-yielding varieties of seeds, particularly in areas endowed with assured irrigation. It is beyond doubt that groundwater irrigation, being the stable source of irrigation, has made possible changes in cropping pattern that can contribute to greater increases in productivity (Bharadwaj 1990). The new agriculture technology program, which has been vigorously pushed by government and industry, demands much more water than traditional practices. Strikingly, it is pointed out that the need for agricultural water is going to increase every year as we need more and more food grains to sustain our rapidly growing population, increasing by about 10.7 million per year. Farmers nowadays have easy access to modern water-lifting technology that is capable of rapid groundwater withdrawal.

The area under tube well irrigation, which accounted for less than 1 % of net irrigated area up to 1960, increased to about 37 % by 2006–2007. The development of the rural electrification program and the availability of credit at highly subsidized rates have helped the farmers to increase the area under groundwater irrigation significantly (Shah 1985; Vaidyanathan 1994). The rapid development of groundwater irrigation helped not only the well-owning farmers but also non-well-owning farmers through groundwater markets. The Punjab State Environment Report (2007) indicates that these conditions, which determine access to groundwater, hold the responsibility of overextraction and, hence, decline in the water table (The Tribune 2007).[2] Further, there is debate regarding the efficacy of government subsidies in agricultural production, especially in areas especially where groundwater has been overexploited and agriculture production has become a nonprofitable proposition.[3]

[1] In technical terms the groundwater resource can be deemed to be in depleted state when the groundwater stock of a region diminishes in volume (Dhawan 1995). Generally, such a situation arises if groundwater use or withdrawal in a year exceeds groundwater recharge over the year.

[2] The State of Environment Report (2007) prepared by Punjab State Council for Science and Technology (The Tribune, September, 6, 2007) presents that 103 of 137 blocks in the State are overexploited by tube wells drawing groundwater for paddy cultivation.

[3] The decline in water productivity growth in the food grains sector of Indian agriculture in the past decade may be attributed partly to the loss of momentum in the development of yield-increasing technologies and partly to the political economy of irrigation from groundwater. The existing regulatory measures in India are totally ineffective or not implemented (Janakarajan 1996). This also

Given this background this chapter focuses upon groundwater irrigation and portrays that the small size of land holdings (social structure and economic conditions) and government policies are responsible for resource degradation. The analysis employs the *regional political ecology* perspective. Through the lenses of this framework, the following analysis provides added insights on how the economic behavior of small farmers is inextricably bound to change the ecology of groundwater depletion, and the politics within which farmers are embedded. The discussion indicates that agricultural practices of small 'progressive' farmers in the study region are accountable for depletion of the water table because of changing cropping patterns.

Small farmers have relatively less access to their own groundwater-lifting technologies such as bore wells, tube wells, and submersible pumps. Therefore, they depend on the farmers having large holdings with irrigational facilities. Consequently, there is a positive correlation between farm size and ownership of irrigation assets; the probability of groundwater purchase can be expected to be inversely related to farm size (Shah 1985). However, even a farmer with a large but highly fragmented farm may go for groundwater purchase to irrigate some of his plots that are located farther from the main plot having the irrigation facilities (Kolavalli and Chicoin 1989; Haque 1996; Saleth 1996). Irrigated land capable of producing two or more crops in a year, using the new agricultural technology, may sustain small and marginal farmers on a viable basis. Small farmers obtain higher yields per hectare because of account of careful culture, intensive cultivation, and easily available irrigation from groundwater from nearby farmers of large holdings. However, the variability and sustainability of farmers should not be considered as a static concept.[4] Moreover, what is found viable today may or may not remain so in future, depending on the population pressure on land. In fact, given the present trend of population pressure on land, most of the small and marginal farmers may join the rank of landless laborers in the future.

16.2 Purpose

The current research investigates the relationship among patterns of groundwater utilization and the influence of social structure of farmers, government policies, and penetration of market forces in the agriculture system and the surplus extraction of groundwater for irrigation. Various dimensions of the problems, such as the economic, institutional, environmental, legal, and political aspects, are addressed to diagnose

raises a fundamental question: what went wrong with the existing regulatory measure? This warrants a thorough and systematic examination of the socioeconomic and political environment in which they are operating. The logic of groundwater-led agrarian change proceeds in three stages, as in the case of Gujarat State particularly, and is applicable to the entire country by and large; the conditions of access to groundwater; pattern of groundwater use; and conjunction of the two.

[4] It varies because of change in technology, cropping pattern, farming system, population growth, infrastructural facilities, and possibilities of integrations between farm and non-farm activities.

the problem of groundwater depletion. Further, the analysis delves into examining the influence between patterns of groundwater exploitation and social structure of the farming communities. It is hypothesized that this connection has an effect on markets and cropping patterns, leading to surplus extraction of groundwater for irrigation.

16.3 The Study Area

The study area is a part of the Indo-Gangetic plains and the densely populated region of India. It comprises the States of Punjab, Haryana, and Western Uttar Pradesh (the districts of Uttar Pradesh situated adjoining Haryana). This region is one of the largest groundwater repositories of the world. However, rising population and unsustainable agricultural practices have created greater stress on the groundwater resource. The adoption and diffusion of high-yielding varieties of seeds, pesticides, irrigation, and fertilizer intake has changed the entire scenario of this region in achieving food sufficiency. After 1965, emphasis was given to two major crops, wheat and rice. With the increase of irrigation, fertilizers, and high-yielding seeds, the best pulse-growing areas were converted to produce these two crops. Currently, these two crops occupy more than 80 % of gross cultivated area. Paddy cultivation, which requires a large amount of water, has changed the groundwater ecology of this area. Farmers are exploiting groundwater in this area unsystematically and injudiciously. Today, the farming community of this area owns larger land holdings, with their own tractors, combine harvesters, pumping sets, and tube wells. Consequently, an interesting picture has emerged. Small landholders are hiring these facilities from the large farmers to plough, irrigate, and harvest crops. Subsequently, an unhealthy competition has taken place among the small and large farmers to increase agricultural productivity by applying more water and more fertilizers. The sustainability of the groundwater has, therefore, been questioned in this entire tract.

16.4 Methodology

This study draws its inspiration from Blakie and Brookfield's (1987) interpretation of "political *ecology*" with its emphasis on identifying the social and environmental causes of resource depletion. The *regional political ecology approach* employed in this study questions the complex processes of social structure, government policies, institutions, and their combined effects on groundwater depletion ecology. This approach makes two significant arguments responsible for groundwater depletion: first, the decreasing size of land holdings resulting from disintegration of the joint family system in the farming communities under the law of inheritance; and second, government policies, which encourage farmers to grow water-intensive crops, such as rice and sugarcane. Resultantly, to maximize their profit from small-sized land holdings, farmers developed an unhealthy completion to grow more cash and cereal crops from the same piece of land by utilizing more groundwater for irrigation. This practice has given rise to changes in cropping pattern. In short, this reduction in the size of land holdings is viewed as a primary cause of groundwater depletion.

The current study used both secondary and primary sources of data. A combination of quantitative and qualitative methods of data collection was employed. Quantitative data regarding access to and use of groundwater by farmers were obtained with the help of a questionnaire. Similarly, qualitative data concern the experiences and perceptions of farmers involving political, ecological, and economic processes that invaded the use of groundwater resource for irrigation. Such qualitative data are strongly recommended in the regional political approach, which suggests that researchers put farmers at center stage for the explanation of groundwater use (Blaikie and Brookfield 1987; Awanyo 2001).

16.5 Political Ecology: A Research Framework

Reduction in the stock of groundwater resources in terms of quantity and quality in utilization by farming communities has been a traditional theme addressed by geographers. For geographers who have long been interested in human-induced modifications of nature, this amplifies the importance of these themes (Parson 1971; Kates 1987; Turner 1989; Simmons 1981). The human modifications of nature caused curiosity among geographers to develop a new concept of political ecology. But it is only during the past decade that it has both effloresced and, to a degree, coalesced around a set of particular propositions (Page 2003). The idea captures fundamental aspects of human existence as both natural and political beings, which include the energetic and material exchange that occurs between human beings and their natural environment. A political ecological perspective highlights not only the impact of political-economic relationships on the pattern of resource use but also the significance of environmental variables and how their interaction with political-economic forces influences human–environment relationships (Grossman 1984; Watts 2000; Page 2003; Walker 2006; Zimmerer 1991, 1994, 2006). Walker (2006) argued that political ecology has become a firmly established and a dominant field of human–environment research in geography. The term political ecology represents an attempt to develop an integrated understanding of environmental and political forces that interact to mediate social and environmental changes (Bryant 1992), which means considering how relationships of dependence and control between people develop out of and into relationships of dependence and control between people and their natural environment. The common themes that recur in political ecology analysis include the need to set a problem or phenomenon into its broader social and economic context and the need to relate both the phenomenon and its socioeconomic context to a variety of scales ranging from local to global. However, equally important is the setting of political ecology at the local scale. Therefore, what is maintained is that there are connections and influences that rise above any particular scale. The local scale is usually the focus for political ecology, and hence the term *regional political ecology* is applied to appraise the local resource base and its state of utilization, providing insights into the context and process underlying resource depletion.

In a political ecology context, regional political ecology (Blaikie and Brookfield 1987) as conceptual framework provides better understanding for considering regional politico-economic forces alongside the local ecology of resource depletion. The approach has already been used in a variety of studies (Bassett 1988; Mayer 1996). At one point in time, this controversial approach had been applied mainly to the agricultural system and food systems. Its controversy was rooted in the resistance to integration into cultural ecology of the political economy and its associated concepts. Recently, the political ecological framework has been used to understand the patterns and processes of resource utilization.

Recent attempts to identify the social and environmental causes of resource depletion and environmental degradation are clear in their complexities.[5] Regional political ecology, as led by Blaikie and Brookfield (1987), represents a broad-based approach encompassing a variety of scales, methodologies, and conclusions regarding the causes of land degradation. A variant of this approach has been applied in this study with two distinct aspects: first, the introduction of a small size of farmland is incorporated into the capitalistic mode of production; and second, this in turn may force the farmers to overdraw unsustainable groundwater.

The empirical regional analysis in this study clarifies, elaborates, and extends a number of key propositions of the regional political ecology framework. First, the study demonstrates how the small farmers are in a trap of production and competition. The farmers are trying to produce more agricultural production from small-size holdings by purchasing water from the groundwater markets for irrigation. In examining this phenomenon, the study deepens the *regional political ecology approach* by paying attention to the social relationships of the communities within which the resources are utilized.

When the ecological perspective is applied in the context of irrigation, a number of interesting implications arise (Coward 1980). In agricultural communities, such variables as society, technology, and population characteristics determine the adaptability of irrigation practices. At the microlevel, the political ecological perspective suggests the apparently irrational, unarticulated, and random activities in a particular irrigation system may be unmatched. This phenomenon when examined in relationship to the environment and sociopolitical context of its occurrence can exemplify a deeper understanding in which the events occur.

---

[5]The significance of recent work by Blaikie (1985), Rees (1985), Blaikie and Brookfield (1987), Redclift (1984, 1987), Emel and Peet (2002) and others is in the development of linkage between resources, economies, institutions, individuals, and societies. These works do not contribute particularly to management-level issues in terms of offering specific approaches. Instead, the major emphasis is on the social, political, and economic origins of environmental problems and the consequences of resource depletion.

Further, the approach adopted in this chapter assumes a concept of social structure that allows decrease in the size of farmers' land holdings that influences groundwater exploitation. The study emphasizes that society-made features reinforce the base for local groundwater markets for small and marginal farmers.[6] Second, the study puts weight upon social relationships within and between the suppliers and users of groundwater. It examines the social behavior of relationships among farmers and inquires how these relationships affect social behavior and human interaction, resulting in groundwater irrigation. The study offers empirically rich insight on how social identities are critical for comprehending the farmers' ability to take advantage of inquiring about irrigation water from a neighboring farmer who has irrigation facilities. A third argument is to examine the rationalization of institutions such as property rights and administrative systems. Broadly, both formal and informal institutions are interesting areas of research in natural resource management.

16.6 Results

16.6.1 Change in Cropping Patterns

The study area includes Punjab, Haryana, and Western Uttar Pradesh located in Northwest India. This region emerged as the most responsive region in the country in terms of agricultural production after the inception of the Green Revolution in the early 1960s. Groundwater irrigation, because of its suitable water quality, has been the most significant factor explaining the success of the revolution. Moreover, farmers to some extent were organized through political parties even in the pre-Independence era.[7] The study area as a part of the Indus Basin shows clear signs of groundwater depletion in many districts. Pressure on groundwater resources has produced a few major developments in the cropping system of Punjab, Haryana, and Western Uttar Pradesh in recent years. The most widely noted development has been the extension of paddy cultivation. In semiarid areas of Punjab, paddy remained a marginal crop of the *kharif*[8] season until about the mid-1960s. Since then, the explosion of tube well irrigation has increased the area under this crop. In addition,

[6] In the northwestern part of India, a mosaic of different shapes and sizes of operational holdings exists; there are, nevertheless, regional characteristics that are of some importance in accounting the regionalization of groundwater irrigation.

[7] The most important factor for the emergence of the farmer's organizations in the *Green Revolution Belt*: a long history of peasant struggle, favorable conditions, and awareness among the farmers. Consequently, the highest level of agricultural development at an all-India level has been recorded in Punjab, Haryana, and western Uttar Pradesh.

[8] In India, generally, three crop seasons are distinguished, namely, *kharif*, *rabi*, and *zaid*. *Kharif* season crops are those crops that are grown during the southwest monsoon period. It is the principal wet season of the country. Its duration is from July to October (the agriculture year begins in July in India). Likewise, *rabi* season pertains to the winter period, beginning in late October and ending in March. Similarly, the *zaid* season is related to crops that are grown between these two seasons.

Table 16.1 Shift in cropping patterns in Punjab, 1960–1961 to 2003–2004 (percent of area)

Crop	1960–1961	1970–1971	1980–1981	1990–1991	1999–2000	2003–2004
Rice	4.79	6.88	17.49	26.86	33.29	33.07
Maize	6.91	9.78	5.65	2.51	2.08	1.95
Cotton	9.46	6.99	9.59	9.34	6.07	7.72
Sugarcane	2.82	2.26	1.05	1.35	1.37	1.57
Wheat	29.58	40.48	41.58	43.63	43.18	43.57
Pulses	19.08	7.29	5.05	1.91	0.79	0.55
Oilseeds	3.91	5.19	3.52	1.38	1.26	1.08
Potato	0.19	0.29	0.59	0.31	0.97	0.83
Other	23.26	20.84	15.48	12.71	10.99	11.66
Total	100	100	100	100	100	100
Total cropped area (in thousands of hectares)	4,732	5,678	6,763	7,502	7,847	7,905

Source: Bhullar and Sidhu 2006 (*Economic and Political Weekly*)

the area under sugarcane also shows an increasing trend over time (Table 16.1). The area under rice, a crop with high water consumption, which was 4.79 % of the total cropped area in 1960–1961 increased to 17.49 % in 1980–1981 and further to 33.07 % during 2003–2004. Similarly, the area under wheat increased from 29.58 % of the gross cropped area in 1960–1961 to 41.58 % in 1980–1981 and 43.57 % in 2003–2004. Because the water requirements of these crops are relatively higher, total water requirements therefore increased (Singh 1992). The changing cropping pattern has a negative effect on groundwater resources. In fact, as fertilizer consumption increased and the irrigated area expanded, the growth rate of yield has slowed down and depletion of groundwater has been pervasive. Similarly, in Haryana and Western Uttar Pradesh (Meerut, Moradabad, Bulandshahar, Aligarh, Agra, Madhura, and Roorkee) during the past two decades, overexploitation of groundwater for rice and wheat along with sugarcane has adversely affected groundwater storage. The area under cultivation for wheat, paddy, and sugarcane has increased many fold. The fact is that the area under sugarcane increased without any impact of the Green Revolution because of price incentives. Therefore, production and acreage are largely determined by price and profitability.

During 1960–1961, Punjab State had only three sugar mills operating, with a daily sugarcane crushing capacity ranging between 2,950 and 3,200 tons. By 2001, the number of sugar mills increased to 19 with the per day crushing capacity increasing eightfold. Consequently, the area under sugarcane also shows increasing trends over time. The changing cropping pattern has a tremendous negative effect on groundwater storage. The latest assessment of water utilization compared to its availability shows that of 118 administrative blocks in Punjab, 73 are ranked as excessively depleted. Similarly, in Haryana more than 40 administrative blocks have been declared highly depleted in terms of overexploitation.

The states of Punjab and Haryana have practiced multiple cropping from one field during one cropping year. By converting these crops into areas, one can arrive at gross or sown areas in 1 year. Hence, these areas are directly proportional to the water consumed by the crops grown. In 1992, the gross irrigated area stood at 72 million ha served equally by surface and groundwater resources. This figure rose to 81 million ha in 1997. Disaggregated data are not available; however, it is reasonable to suppose that new water came from wells (Kochhar 2000). Continuous exploitative agriculture by adopting cereal-based (rice–wheat) crop rotation and application of heavy irrigation has caused an adverse impact on groundwater resources. In Haryana and Punjab States, production of cotton has sharply declined in both States. It is one of the several crops grown during *kharif* that was replaced by rice. Because rice is a more profitable crop than cotton, farmers preferred its cultivation.

16.6.2 Government Policies

More than other parts of India, the three States of Punjab, Haryana, and Western Uttar Pradesh have been responsible for overexploitation of groundwater. The expansion in groundwater irrigation in this area is the result of several factors such as improvement in technology of drilling and raising water, rural electrification program funded by the State, liberal loan facilities, and subsidized supply of electricity. On all these grounds, the government's policies of supporting and encouraging private groundwater development has been until recently widely acclaimed. There is now growing apprehension that these policies are leading to overexploitation of the groundwater. Evidence of progressive decline in the groundwater table is accentuating many socioeconomic and environmental problems (Bilas 1980; Pant 1987; Janakarajan 1996; Jeet 2001).

India faced the problem of chronic food shortage up to the end of the 1960s, which forced the government to identify high productivity regions to increase the production of food grains. Punjab State was identified as the potential food grain basket of India as it was endowed with fertile land and sufficient irrigation sources. Government intervention in the market led to remunerative price and assured marketing of the produce through a minimum support price policy backed by public stocking of food grains. Sufficient funds were invested in agricultural research to increase the productivity of wheat and rice (Sidhu et al. 2005). The area under wheat and rice increased at a very sharp rate (Table 16.1). Assured procurement of rice and wheat acted as a strong catalyst to shift the area toward these crops. The Food Corporation of India purchased whatever quantity was offered for sale and ensured market clearance. The minimum support price (MSP) (although is declared for 24 crops) is effective only for these 2 crops. In this way, the farmers were induced to divert their land from other crops to wheat and rice (Sidhu and Johl 2002).

The rice and sugarcane crops are in need of more water from sowing to harvesting periods. Such a high intensity of irrigation cannot be met with canal water alone even if it is available. So, the demand is met through pumping groundwater with electricity-operated tube wells. Further, the agriculture sector was supplied electricity at a subsidized rate or even free of cost for several years. The government had followed a distorted price policy. In the 1990s, the average annual electricity subsidy to the agriculture sector was rupees 10,470 million. Rice and wheat together consumed 66 % of the total irrigation water used in the agricultural sector of Punjab State (Hira et al. 2004). In Haryana, rice has been cultivated even in those areas where the soils are not suitable. A recent study conducted in Haryana has shown that until a few years ago the Rohtak District did not have any land under rice cultivation because the soils were not suitable for rice cultivation. Because of the remunerative price policy of the government, however, farmers have started growing paddy in this district (Jeet 2009).

16.6.3 Decrease in the Size of Land Holdings

The term agricultural holding or land holding refers to the amount of land held by one farmer. The average size of a holding in India is about 2 ha (or about 6.2 acres). The vast majority of farmers in India have very small and uneconomic holdings; only a small majority of farmers have large farms. Decrease in the size of land holdings is another significant factor in groundwater depletion of the three states in Northwestern India. Following disintegration of the joint family system, land holdings are equally divided among all the male members. In this process of division, the size of land holdings decreases from one generation to another. Table 16.3 provides information regarding land ownership patterns in the three states. Nearly 57 % of families in villages of these states own tiny land holdings, only up to 1 ha. Another 18 % of families own farms between 1 and 2 ha. In other words, 75 % of farmers own small land holdings consisting of up to 2 ha or about 5 acres, and they own only 26 % of the total area of the land of India. These patterns are for marginal and small farmers only.

Much the same is true of farmers of Haryana, Punjab, and Western Uttar Pradesh (Table 16.2). The size of land holdings decreases very quickly, which is an indication of the division of the joint family system of the farmers (Table 16.3). The plains of Northwest India were pioneers in the adoption of the new technology of the

Table 16.2 Land holdings in Punjab, Haryana, and Uttar Pradesh

		1990–1991	2000–2001
Size	Class	Percent	Percent
Marginal	Less than 1 ha	57	69
Small	1–2 ha	18	23
Medium	2–10 ha	23	07
Large	More than 10 ha	02	01

Source: Directorate of Economics and Statistics (2004)

Table 16.3 Average size of land holdings in Punjab, Haryana, and Uttar Pradesh

State	Average size of land holdings (ha)		Percent decrease
	1990–1991	2000–2001	
Punjab	3.79	3.11	21.86
Haryana	2.13	2.01	5.59
Uttar Pradesh	0.86	0.84	2.3

Source: Statistical Abstract of Haryana, Department of Economic and Statistical Analysis, 2003–2004

Green Revolution in India. Consequently, the highest level of agricultural development has been recorded in Punjab, Haryana (except for the districts of Hisar, Sirsa, Mahendragarh, and Dadri and Loharu Tehsil of Bhiwani District), and many districts of Western Uttar Pradesh.

The success of the Green Revolution in this region was itself made possible by assured and constant irrigation, which was provided initially by water obtained from a canal system harnessed from the river network and, subsequently, by water obtained as a result of the installation of electric tube wells and pump sets. With the disintegration of the joint family system, every individual farmer wants to get more yield from the small operation holdings. In doing so, small and marginal farmers install their individual tube wells and pump sets, although the efficacy of these water-extracting machines is questionable. Presently, a large number of tube wells extract groundwater in an unsystematic, uncontrolled, and injudicious manner. Interestingly, during 1966–1967 in Haryana State, one tube well used to supply water to nearly 2 ha (1.76 ha) of agricultural land, but this supply decreased to 0.07 ha in 2001–2002. Consequently, the geographic density of tube wells/hectare has increased tremendously. As a result, the interfering cone of each individual tube well has created a cone of depression leading to excessive groundwater depletion.

Second, in areas where operation holdings are too small and farmers are not in a position to install individual tube wells, groundwater markets are in operation. In this case, small farmers purchase water from the nearby tube well owner. As a result, widespread groundwater markets have arisen in this area. As a result, there exists a positive correlation between farm size and ownership of irrigation assets. Moreover, groundwater purchase can be expected to be inversely related to the farm size.

The nexus of agriculture and groundwater irrigation in the Northwest region of India has brought benefits but it is also characterized by problems and weaknesses: (1) there has been overexploitation of groundwater, leading to depletion, (2) water markets tend to emerge in the context of groundwater extraction through tube wells, and (3) there is danger of unsustainable extraction and inequitable relationship between sellers and buyers. Free and highly subsidized pricing of electricity has become the major reason for groundwater overdepletion, and (4) the response to the aforementioned three problems resides in groundwater regulation, but this has so far not been found feasible because of political and legal constraints.

16.7 Conclusions

In Northwestern India, the cropping pattern has changed drastically since the inception of the Green Revolution. Farmers grow more water-consuming crops such as rice and sugarcane. Hence, they are very enthusiastic about installing shallow tube wells. The advantage of a tube well relative to the canal supply lies in the fact that the source of water is owned by the individual farmer. Moreover, small farmers are in competition with each other for obtaining greater yield from a small piece of land in terms of growing more crops from the same field. It requires more water to practice multi-cropping from the same field. Therefore, because of the wholesale exploitation of groundwater, a tendency of falling water level leading to competitive deepening and larger financial losses has taken place. In many areas, shallow tube wells have gone dry and farmers now drill multiple bores alongside or within existing structures, and even this system of shallow tube wells had shown failure in terms of sufficient yields of water. Consequently, big farmers in this region install submersible pumps to tap the deeper aquifer beyond 200 m depth. This phase of water extraction has created various political, social, economic, and environmental problems among the farming communities and governmental organizations. However, challenges arise in three interrelated contexts:

1. This part of the country has witnessed changes in social structure, emerging from the split in landholdings caused by the increase in population. Consequently, large-sized landholdings are being converted into small pieces of land. This change in social structure encourages small farmers to grow cash crops such as wheat, paddy, and sugarcane for maximum profit and economic standards. This change in society is perennial and unmanageable.
2. The nexus among groundwater suppliers (big farmers) and users (small farmers) has led to the development of exploitation in groundwater markets. This nexus is strong and irrevocable so long as the ownership of the resource remains in the hands of landholders.
3. Present laws and regulations are not efficient enough to control overexploitation. Even political will is lacking in the execution of these regulatory measures. This concept raises a fundamental question: What went wrong with the existing regulatory mechanism?

References

Awanyo L (2001) Labor, ecology and a failed agenda of market incentives: the political ecology of agrarian reforms in Ghana. Ann Assoc Am Geogr 9(11):92–121

Bassett TJ (1988) Political ecology of peasant–herder conflicts in northern Ivory Coast. Ann Assoc Am Geogr 78:453–472

Bell M, Roberts N (1986) Development theory and practice in human and physical geography. Area 18:3–8

Bell M, Roberts N (1991) Political ecology of dambo soil and water resources in Zimbabwe. Trans Inst Br Geogr 16:301–318

Bharadwaj K (1990) Irrigation in India: alternative perspectives, research in economics. Indian Council of Social Sciences, New Delhi
Bhullar AS, Sidhu RS (2006) Integrated land water use: a case study of Punjab. Polit Econ Weekly XLI:5353–5357
Bilas R (1980) Groundwater resource of Varanasi district, India: an assessment of condition, use and quality. Geogr Rev India 26(122):81–92
Blaikie P (1985) The political economy of soil erosion in developing countries. Longman, London
Blaikie P, Brookfield H (eds) (1987) Land degradation and society. Methuen, London
Bryant RL (1992) Political ecology: an emerging research agenda in Third World studies. Polit Geogr 11:12–36
Coward EW (1980) Irrigation and agricultural development in Asia. Cornell University Press, London
Dhawan BD (1995) Groundwater depletion, land degradation and irrigated agriculture in India. Commonwealth Publishers, New Delhi
Directorate of Economics and Statistics (2004) Indian agriculture brief. Ministry of Agriculture, Government of India, New Delhi
Emel J, Peet R (2002) Resource management and natural hazards. In: Peet R, Thrift N (eds) New models in geography: the political-economy perspective. Unwin Hyman, London, pp 49–76
Grossman LS (1984) Peasants, subsistence ecology, and development in the highland of Papua New Guinea. Princeton University Press, Princeton
Haque T (1996) Sustainability of small holder agriculture in India. Concept Publishing Company, New Delhi
Hira GS, Jalota SK, Arora VK (2004) Efficient management of water resources for sustainable cropping in Punjab. Research Bulletin, Department of Soils, Punjab Agricultural University, Ludhiana
Janakarajan S (1996) Consequences of aquifer overexploitation: a case of prosperity and deprivation. MIDS, Madras
Jeet I (2001) Water resource management. Manisha Publications, New Delhi
Jeet I (2009) Rainwater harvesting. Mittal Publications, New Delhi
Kates R (1987) Human environment: the road not taken, the road still beckoning. Ann Assoc Am Geogr 77:525–534
Kochhar N (2000) Negative effects of water intensive crops. Indian J Agric Econ 5:24–32
Kolavalli S, Chicoin DL (1989) Groundwater markets in Gujarat, India. Int J Water Resour Dev 5:38–44
Mayer JD (1996) The political ecology of disease as a new focus for medical geography. Prog Hum Geogr 20:441–446
Page B (2003) The political ecology of *Prunus africana* in Cameroon. Area 35:441–446
Pant N (1987) Groundwater depletion. Econ Polit Weekly 22(6):219–220
Parson JJ (1971) Ecological problems and approaches in Latin American geography. In: Lenteck B, Carmin RL, Martison TL (eds) Geographical research on Latin America. Indiana University Press, Muncie, pp 34–46
Punjab State Environment Report (2007) Punjab State Council for Science and Technology
Redcliffe M (1987) Sustainable development: exploring the contradictions. Methuen, London
Redclift M (1984) Development and environmental crisis: red or green alternatives. Methuen, London
Rees J (1985) Natural resources: allocation, economics and policy. Methuen, London
Saleth RM (1996) Water institutions in India: economics, law and policy. Institute of Economic Growth studies in economic development and planning, vol 63. Commonwealth Publishers, New Delhi. http://www.amazon.com/Water-institutions-India-Economics-development/dp/8171693776
Shah T (1985) Transforming groundwater markets into powerful instruments of small farmers development: lesson from Punjab Uttar Pradesh and Gujarat. Overseas Development Institute, London

Sidhu RS, Johl SS (2002) Three decades of intensive agriculture in Punjab: socio-economic and environmental consequences. In: Johl SS, Ray S (eds) Future of Punjab agriculture. Centre for Research in Rural and Industrial Development, Chandigarh

Sidhu RS, Bhullar AS, Joshi AS (2005) Income, employment and productivity growth in farming sector of Punjab: some issues. J Indian School Polit Econ 17(1-2):12–18

Simmons IG (1981) The ecology of natural resources. Edward Arnold, London

Singh B (1992) Groundwater resources and agricultural development strategy: Punjab experience. Indian J Agric Econ 47:105–113

The Tribune (2007) July 8, p 5

Turner BL (1989) The specialist-synthesis approach to the revival of geography: the case of cultural ecology. Ann Assoc Am Geogr 79:88–100

Vaidyanathan A (1994) Performance in Indian agriculture since Independence. In: Basu K (ed) Agrarian questions. Oxford University Press, New Delhi

Walker PA (2006) Political ecology: where is the ecology? Prog Hum Geogr 30:382–395

Watts M (2000) Political ecology. In: Johnston RJ, Gregory D, Pratt G, Watts M (eds) The dictionary of human geography. Blackwell, Oxford, pp 28–32

Zimmerer KS (1991) Wetland production and smallholder persistence: agricultural change in a highland Peruvian region. Ann Assoc Am Geogr 81:443–463

Zimmerer KS (1994) Human geography and new ecology: the prospect and promise of integration. Ann Assoc Am Geogr 84:108–125

Zimmerer KS (2006) Cultural ecology: at the interface with political ecology–the new geographies of environmental conservation and globalisation. Prog Hum Geogr 30:63–78

Chapter 17
Assessment of Human Vulnerability and Risk of Flood Hazards in Orissa, India

G.K. Panda

Abstract Odissa, which is one of the eastern coastal states of India, is affected by frequently recurring flood hazards, many of which have turned out to be disastrous. These floods have often brought large-scale human casualties as well as loss of property for the state. This paper aims to develop and test a set of models to assess the risk of human casualty of the flood hazards at the district level using multivariate linear regression analysis. This method has been used to estimate the human casualty at the district level using the available human casualty and other socioeconomic data from the Government of Odissa, Census of India, and the United Nations Development Programme (UNDP). For this purpose, a number of explanatory variables are used and human casualty has been a response variable. The observed data show that the districts from the coastal regions have high human casualty and population exposure in comparison to other parts of the state. Relative vulnerability is high for the non-coastal districts because the exposed population is less in comparison to human casualties. Model-predicted human casualty shows a nonlinear relationship with recorded human deaths. Results conclude that flood is an extreme event of nature and that its impacts can be predicted with greater accuracy using the models shown in this chapter if the data can be used at microlevel, preferably from the blocks and villages.

Keywords Flood hazard • Human risk • Physical exposure • Relative vulnerability

17.1 Floods: Need for Vulnerability Analysis and Risk Assessment

Orissa is endowed with a large network of rivers with many catchments extending beyond the state. Most of these rivers drain to the Bay of Bengal through an extensive coastal and deltaic terrain flooded by the rivers and their distributaries

G.K. Panda (✉)
Department of Geography, Utkal University, Vani Vihar, Bhubaneswar, Odisha 751004, India
e-mail: gopalkrishna_panda@yahoo.co.uk

A.K. Dutt et al. (eds.), *Spatial Diversity and Dynamics in Resources and Urban Development*, DOI 10.1007/978-94-017-9771-9_17

following cyclonic rains during the rainy season (Sinha 1985). There are frequent flash floods in the upper catchments of these rivers. The state has been recurrently affected by flood hazards, many of which have turned into disasters with widespread damage to property and loss of human life. Further, the impacts on livelihood and damage to housing and crops are also major concerns of the people and the state as well (Behera 2002).

This chapter describes some methodological aspects of the application of the Disaster Risk Index (DRI), a concept used in the World Disaster Report 2004 (UNDP 2004) to improve understanding of the relationship between development and disaster risk. The major assumption behind the DRI was that differences of risk levels faced by countries with similar exposures to natural hazards are explained by socioeconomic factors, that is, by population vulnerability. It can allow measurement and comparison of relative levels of risk, exposure to hazards, and vulnerability at different spatial levels. The DRI can also contribute more quantitative evidence for planning and decision making in the field of disaster risk reduction and management (Dao and Peduzzi 2004). The findings of this study can be a useful proposal for development of more effective hazard management policies, programs, and strategies in the state. This chapter aims at the evaluation of the flood hazard risk of Orissa at the district level using data on exposed population, various socioeconomic indicators, and past human casualties.

17.2 Conceptual Framework and Methodology

A statistical analysis was carried out to identify the socioeconomic indicators reflecting human vulnerability to flood hazards. The final output includes a set of indicators for measuring levels of risk, an evaluation of the population exposed, and the identification of socioeconomic parameters for estimating human vulnerability to flood hazards. The study has been taken up in four distinct stages of 'hazard analyses,' 'human impact analysis,' 'relative vulnerability and disaster risk index analyses,' and prediction of human risk' through multiple regression models. The multiple regression analysis has been undertaken identifying impact indicators and development indicators. The major components, data, methodology, and the possible outcome of the study are discussed next.

17.2.1 Risk of a Hazard

Following a definition by the United Nations, risk "refers to the expected losses from a particular hazard to a specified element at risk in a particular future time period. Losses may be estimated in terms of human casualty, or buildings destroyed or in financial terms" (UNDRO 1979; Burton et al. 1993, p. 34). Hazards are extreme events that may create risk and potentially turn into disasters if the exposed elements are vulnerable. The risk can also be measured in terms of loss of livelihood or in economic terms. However, such data are not available from the affected areas, even at the lowest spatial level of villages and blocks.

17.2.2 Study Area and Database

The present study is taken up for the state of Orissa, which comprises 30 districts. The data on flood hazards of the state are collected from the secondary sources of the Government of Orissa. The available data on flood hazard characteristics include the frequency and time of occurrence, affected population, loss of human life, loss of public property, houses collapsed and damaged, loss of crops, and area affected. The study is based on these data of the flood hazards for a period of 10 years from 1998 to 2007 (Government of Orissa 1992, 1999, 2008). These data are considered fairly reliable because of their consistency, collection through the public system, and the utility to support relief operations and their use in the disaster mitigation and management activity of the state. The socioeconomic data at the district level have been collected from the census organization, that is, the office of the Registrar General of India, Bhubaneswar.

17.2.3 The Choice of Risk Indicators

In this study, the choice of risk indicators is made from among the available disaster impacts. Loss of property, loss of crops, loss of domestic animals, area affected, or damage to houses do not reveal the magnitude of total loss of a geographic unit in a flood because of variation in their units of representation. Hence, the number of human casualties is chosen, which is less dependent on subjective evaluations. It is generally accepted that the loss of life is the most critical indicator of a disaster. When the total number of lives lost is considered, certain districts such as Ganjam, Baleswar, Kalahandi, and Cuttack always remain at the top of the list of the areas at risk. Rather, the number of human lives lost per exposed population generally gives a higher rank to the less-exposed and low-density districts. In the DRI, that is, the number of human beings who lost their lives per exposed population represents the relative risk faced by each district, whereas figures on total population killed highlight the districts facing severe impacts, emphasizing the need for disaster mitigation and management.

17.2.4 Modeling Risk of Flood Hazards

According to the UNDRO definition (UNDRO 1979), the risk of losses of a hazard results from three components: hazard occurrence, elements at risk, and vulnerability. In case of risk of death, the elements at risk are the exposed population. The hazard occurrence refers to the frequency of a hazard of a return period of a given magnitude, whereas the vulnerability is "the degree of loss to each element should a hazard of a given severity occur." If the hazard frequency or the population vulnerability increases, then risk will be augmented accordingly. Assuming that the risk follows a multiplicative function, the equation for estimation of risk is

$$R = H * Pop * Vul \tag{17.1}$$

where R = is the risk (measured in terms of people killed/year or property lost/year)
H = hazard [characterized by its magnitude and frequency (event/year)]
Pop = population living in a given exposed area (population affected/event)
Vul = vulnerability depending on socioeconomic factors (no units).

In computing the DRI (Disaster Risk Index), the combination of frequency of a hazard and its exposed population is called physical exposure: this is the average number of people exposed to a hazard per year. Hence, the formula 1 (Eq. 17.1) for risk can be simplified as follows:

$$R = PhExp * Vul \tag{17.2}$$

where R = risk of human life lost and Vul = population vulnerability
Ph. Exp = average number of people exposed to a flood hazard per year.

Using the foregoing equation, the relative vulnerability can be calculated as follows:

$$Vul = Risk / PhExp \tag{17.3}$$

17.2.5 Relative Vulnerability and Disaster Risk Index (DRI)

The Disaster Risk Index, which was developed by UNDP, is a mortality-calibrated index that measures the risk of death in a disaster. It is a function of physical exposure and vulnerability to a hazard. People are more or less vulnerable to a hazard depending upon a range of social, economic, cultural, political, and physical variables. The number of deaths is used as a proxy to manifest risk because of nonavailability of other aspects, which can represent the total disaster risk. Using this index, the DRI therefore is able to calculate the relative vulnerability of the districts by dividing the number of deaths by the number exposed. When more people are killed with respect to the number exposed, the relative vulnerability of a district is higher. Using Eq. 17.3, the relative vulnerability of the districts for flood hazards is calculated by using the past data on human casualties and affected population, which represent the physical exposure. The number of deaths of people caused by floods per millions of exposed population is called the DRI (Disaster Risk Index, Eq. 17.3©). The spatial pattern of disaster risk reveals that districts such as Mayurbhanj, Gajapati, Nawarangapur, Nawapara, Jharsuguda, and Sundargarh show higher values and come under the high and very high disaster risk index because of low exposed population. The coastal districts, which are traditionally flood prone, show low DRI because of higher number of exposed population in comparison to number of deaths.

17.3 Characteristics of Flood Hazards in Orissa

17.3.1 Frequency of Flooding

Being primarily caused by natural factors and often induced by human activities, Orissa (Odisha) is perpetually affected by flood hazards and disasters. The flood data for the period 1998–2007 indicate there has been no year without floods in Orissa except 2002. The number of flood incidents varied from two to four in a year (Fig. 17.1). During 1998–2007, there were more than two flood incidents per year during 1998, 1999, 2003, 2005, and 2007. The coastal and deltaic areas are most vulnerable to flooding because of their dense network of rivers and their distributaries. When the Orissa Coastal Zone is affected by floods, their impacts are felt not only on the economy of the region, but also on the entire state, because it is the "rice bowl of the state" (Panda 1989).

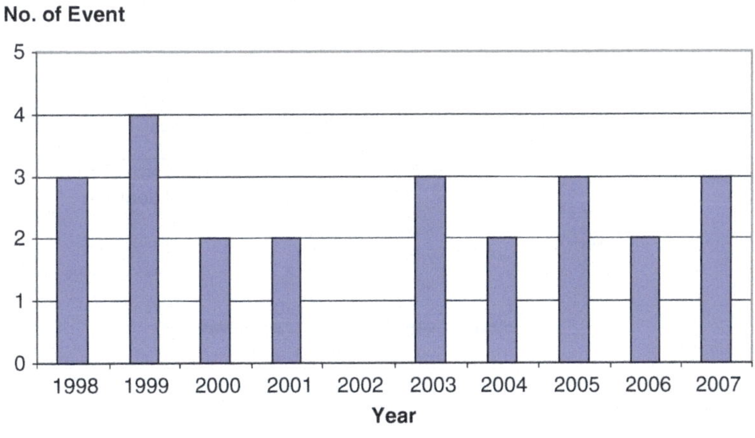

Fig. 17.1 Annual distribution of floods, 1998–2007, Orissa

17.3.2 Economic Impact of Flooding

17.3.2.1 Loss of Life and Property

The annual average property loss of the state caused by floods was Rs 7,053 million during 1998–2007. From 1981 to 2000, the state had spent nearly Rs 1, 000 million alone on relief against the tenth finance commission's assessment of the relief expenditure of Rs 140 million. In 1982, the state experienced the highest flood of the century, affecting an area of 34,000 km^2, and the loss of property at 1981–1982 prices was Rs 2,140 million. The state suffers an annual average crop loss of Rs 1,277 million. Annually, 1,720 km^2 of area in the state is affected by floods. During 1998–2007, the number of fully collapsed houses was 83,510. Although most of these houses were made of mud walls and straw thatching, some were built with brick walls and tile roofs. In spite of the prevalence of the hazard, the affinity of the people for the coastal and deltaic tracts and flood plains has continued unabated, partly because of the geographic inertia of settlements growing through peripheral accretion, reasons of socioeconomic compulsion, and as an environment of persistent appeal. The spatial pattern of human casualties in the state caused by floods can be seen from Fig. 17.2.

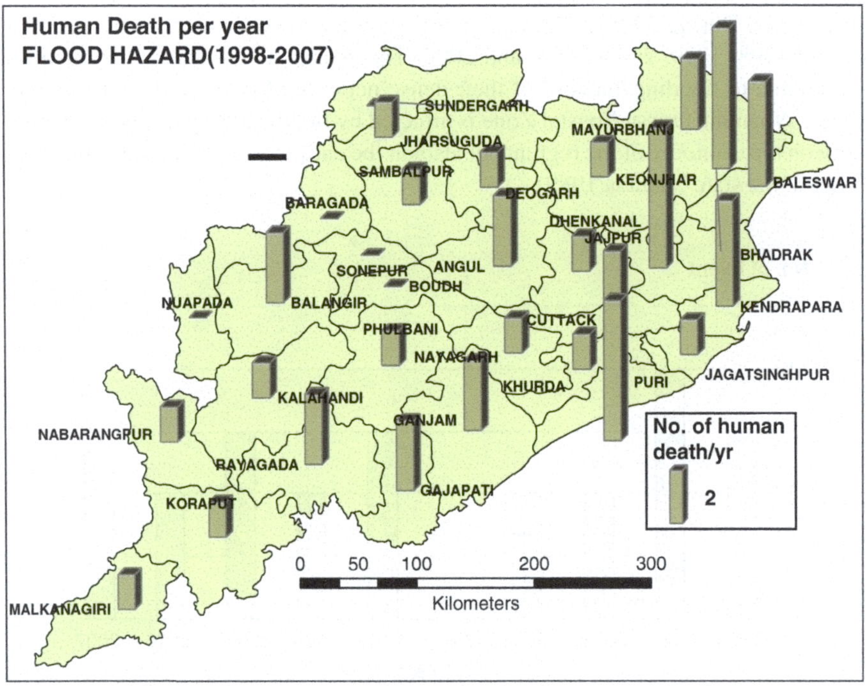

Fig. 17.2 Human deaths from floods, Orissa, 1998–2007

17.3.2.2 Population Affected and Vulnerability Pattern

About 3.5 million people in the state are exposed every year to floods, with a total of 483 deaths for the period 1998–2007. The number of people marooned during this period of 1998–2007 was 2.7 million. Despite massive expenditure on flood control and management, flood losses are continuing to rise (Suri 2000). Nearly 14 % of Orissa is prone to floods as per the Vulnerability Atlas of Orissa (BMTPC 2006). The coastal districts are eternally vulnerable to flooding. However, based on the current flood impact data, the vulnerability is spreading toward the western and southern districts of Rayagada, Bolangir, Sonepur, Sambalpur, Kalahandi, and Kendujhar, which are away from the coast (BMTPC 2006). The population affected by floods has varied from 0.6 to 7.6 million from 1998 to 2007. As more and more densely populated and flood-prone areas are coming under the grip of the floods, the number of people affected by floods has increased significantly since 2001 (Fig. 17.3).

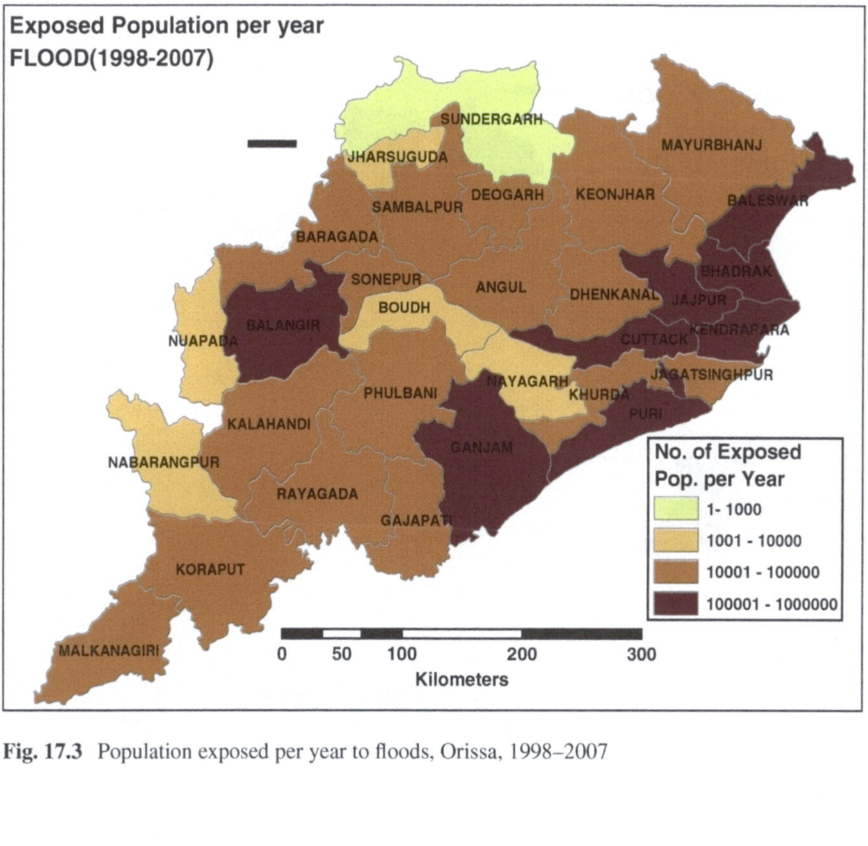

Fig. 17.3 Population exposed per year to floods, Orissa, 1998–2007

17.3.3 Causes of Flooding

The floods in Orissa result from heavy cyclonic and monsoon rainfall over the catchments of its rivers (Disaster Management Institute 1988). Because of interlinkage among the rivers in their lower reaches, the flood from one river passes on to the other rivers. When the floods occurring in all the major coastal rivers coincide, the devastation becomes catastrophic (Mahalonobis 1941). The Orissa Coastal Zone experienced such floods in 1982, 2001, and 2003. Flash floods are associated with the sub-montane tracts and the Eastern Ghats region of Orissa. In 1989 the Rushikulya River was flooded by a cyclonic rain of 40 cm in 1 day in its catchments, affecting the Ganjam District. Today these situations are experienced more in the coastal districts where poor drainage outlets create waterlogging and prevent the discharge of rainwater (Fig. 17.4).

Heavy rainfall in the interfluves of the Mahanadi delta and coastal backshore zone leads to the ponding of rainwater in depressions and low-lying areas. A high groundwater table also contributes toward this type of flooding. Extensive areas of the lower fluvial plains in the deltaic and coastal region, associated with the meandering channels, tidal creeks, Zora or Pata Lands (low-lying areas), oxbows,

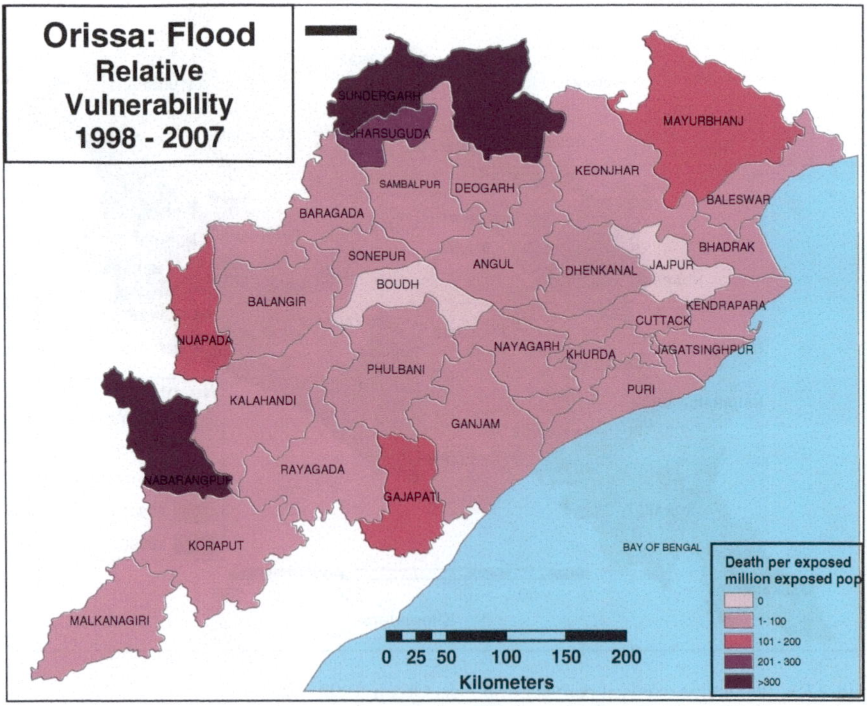

Fig. 17.4 Orissa: Disaster Risk Index for floods

and back swamps, are affected by this type of flooding, which is more prevalent in the districts of Bhadrak, Kendrapara, Jagatsinghpur, and Puri. The Orissa coast is frequently ravaged by the landfall of tropical disturbances. The storm surges push the seawater to a few kilometers inland along the creeks and river mouths, resulting in tidal flooding damaging the crops and contamination of groundwater, and often sweeping away the villages. The spatial spread of this zone varies from 5 to 15 km during depressions and storms, whereas high storm waves in the range of 2 to 3 m sweep inland to a distance from 20 to 25 km during severe storms and cyclones.

17.4 Modeling Human Risk of Flood Hazards

It is now a globally accepted paradigm that the nature of vulnerability and the magnitude of risk are intimately connected to poverty. The poor have been the most vulnerable and their level of risk to natural hazards is relatively high. It is presumed that poverty can contribute to enhancing the disaster risk with lesser capability to recover quickly and recoup the damages inflicted on the people (UNDP 2009). Based on the presumption that poverty contributes toward greater disaster risk, the chosen socioeconomic indicators were reflective of poverty and status of development.

17.4.1 Socioeconomic Factors Associated with the Risk of Floods

Orissa is an agrarian state where a large part of the total population lives below the poverty line (49 %). Of the people, 38 % belong to the Scheduled Caste and Scheduled Tribes (SC/ST). The majority of the population is engaged in agriculture, the predominant economy of the state. As per working classification, the population engaged in primary economic activities such as forestry, fishing, hunting, agriculture, and mining comprises nearly 70 % of its people. The level of urbanization is relatively low (16 %) in comparison to the national average, and literacy of the state is 68 % (Director of Census Operations 2005). The people living in kutcha houses (houses with mud walls and straw-thatched roofs) vary from 65 to 85 % across the districts. People living in such houses are often the most vulnerable because of the risk of death from the collapse of their homes. Thus, while choosing the independent variables that may configure disaster risk in Orissa, choice is based on the premise that poverty is the overriding factor. Flood hazards are noted for aggravating poverty in two ways: through destruction of food stocks and the meager assets of the poorer households, and through loss of livelihood, making employment opportunities scarce. Poverty is directly linked to the poor household infrastructure, which makes them more vulnerable to flood damages.

In Orissa the people suffer more from wage loss because both farm and non-farm employment opportunities are reduced after a major flood. They own fewer assets to cover the expenditures needed during the disaster and the recovery phases. The poor people depend on borrowing, principally from moneylenders, and face difficulty in buying food because of decreased income and increased prices. They suffer relatively more from diseases and malnutrition. The compounding effect of closely following disaster shocks or concurrent disaster and non-disaster shocks on the poor contributes to increased poverty in the hazard-prone areas. The flood hazards destroy the local food security mechanisms temporarily. The poorer households usually settle on less desirable high-risk marginal land and are unable to afford disaster-proof housing. Consequently, they are compelled to lose their basic investments in housing infrastructure after a disaster. They have lesser access to the social and economic support needed for recovery. The socially marginal groups, such as the Scheduled Tribes (STs) and the Scheduled Castes (SCs), are more adversely affected than the other castes because of their higher incidence of poverty.

Further, people with low income have less command over their resources to handle disasters. Literacy is extremely important for disaster preparation measures. Hence, lower literacy levels are reflected through higher vulnerability to disasters. In the Indian situation, when income (at the district, state, and national level) does not translate into poverty alleviation, achievements in human development are extremely important in improving access to various opportunities in life and thereby reducing poverty and vulnerability. Hence, the Human Development Index (HDI) can represent poverty and vulnerability more than income (P & C Department 2006). The higher the HDI values, the higher is the ability to face disasters successfully. Pucca houses (burnt brick and concrete) provide greater protection during floods and also present the possibility of saving lives during submergence. Kutcha houses, made of mud walls with wooden support and straw thatching, provide little or no protection. Pucca houses provide the possibility of overcoming disaster impacts much faster and can withstand flood hazards, but kutcha houses enhance vulnerability and risk. Keeping all these factors in mind, the following independent variables were chosen, at the district level, to explain disaster risk: the Human Development Index (HDI), percentage of BPL (below poverty line) population, population density, percentage of SC and ST population, literacy, percentage of population under primary economic activities, and percentage of households living in kutcha houses.

A multiple correlation between DRI and the aforementioned variables was carried out taking their log-normal values (Chow 1964). The conversion of data to the log-normal form was to reduce their variations. The multiple correlations reveal that DRI shows a positive correlation with BPL population, primary workers, SC and ST population, and percentage of kutcha houses. However, the '*r* value' is only significant for the SC and ST population at a 99 % confidence limit. There is negative correlation with affected population, literacy, population density, and HDI, of which the first three are significant at a 99 % confidence limit. The affected population, population density, and literacy show negative correlation because the DRI is high in the districts where deaths reported for floods have occurred with a very low exposed population.

17.4.2 The Multiple Regression Model

Multiple logarithmic regression models are used with district-level data to study the human risk of flood hazards and disasters against a grouping of socioeconomic attributes that define disaster risk. The socioeconomic variables as discussed earlier, that is, the Human Development Index (HDI), percentage of BPL population, population density, percentage of population belonging to SC and ST, percentage of literates, percentage of population under primary workers, and percentage of households living in kutcha houses, are taken as a set of independent variables and DRI as the dependent variable. To reduce diversity in the variance of different independent variables and change over time, their log values have been considered and the following general model is used:

$$R = A_1X_1 + A_2X_2 + A_3X_3 + \cdots A_nX_n + C \qquad \text{(Model 17.1)}$$

where R is the dependent variable
$X_1, X_2, X_3 \ldots Xn$ are the independent variables
$A_1, A_2, A_3 \ldots An$ are the multiple regression coefficients
'C' is the residual that follows normal distribution with mean 0 and variance 1.

The output of the models is assessed based on the value of R^2. To work out the different models, stepwise regression has been considered eliminating the least significant variables at each step of the analysis. The eight different models based on this step are given in Table 17.1 along with the values of R and R^2.

17.4.3 Analysis of Results

The first model uses eight independent variables to predict the DRI. The value of R^2 is 0.56; that is, the model with all eight variables explains 56 % of total variability for human risk. There seems to be high multicollinearity among the independent

Table 17.1 The different multiple logarithmic models and their corresponding values of R, R^2, adjusted R^2, and the standard error of the estimate

Method	Model	R	R^2	Adjusted R^2	Standard error
Enter/remove/backward	1	0.75	0.56	0.39	0.67
Backward	2	0.75	0.56	0.42	0.65
Backward	3	0.75	0.56	0.44	0.64
Backward	4	0.74	0.55	0.46	0.63
Backward	5	0.73	0.53	0.46	0.63
Backward	6	0.72	0.51	0.46	0.63
Backward	7	0.69	0.47	0.43	0.65
Backward	8	0.65	0.43	0.40	0.66

variables, which suggests either looking for other appropriate variables or removing some of the variables in subsequent models. In the second model, the backward regression method is applied with seven variables, removing literacy, which was least correlated (Table 17.2). Although the values of R and R^2 are the same, adjusted R^2 improves, which suggests that literacy does not have any significant role in predicting human risk. The third model is similar to the second except that six variables are used, removing HDI. It is observed that the value of R^2 remains unchanged, but the adjusted R^2 improves to 0.44. In the fourth model, five variables were used, taking out primary workers; the value of R^2 is 55 % compared with 56 % in model 3, but adjusted R^2 improves. In all the other four models, the values of R and R^2 decrease. The foregoing models show the importance of different independent variables in explaining the variability in human risk based on the value of R^2, which varies from 56 % to a low of 43 %.

A review of the regression coefficients in the first model reveals that human casualty increases with increase in affected population, which is significant at the 99 % level of confidence (Table 17.2). Although not significant, the higher the illiteracy, the greater is human death. A negative relationship exists between human death and the log of HDI, meaning that human deaths decrease with improvement in the HDI of the districts. As expected, human casualty is positively associated with the log of the density of population; the effect is significant at the 87 % level. When log of BPL population and log of kutcha houses are higher, human risk is decreased, showing significance at 33 % and 76 %, respectively, which reveals a contradictory situation, because a greater BPL population and more kutcha houses are likely to increase human risk in a flood.

Whenever primary workers and SC and ST population variables are higher, human risk is also high. These results are statistically significant at the 63 % and 92 % level of significance, respectively, possibly because they are a lower income group of people who normally dwell in vulnerable locations with greater risk of death and damage. Similar situations are found in the second and third models, taking out literacy and HDI. In order of their decreasing significance to explain human casualties, the variables are literacy, HDI, BPL population, primary workers, kutcha houses, population density, and SC and ST population. However, in the sixth model, which best explains the predictability of human casualties, these variables are affected population, population density, and SC and ST population. Using these models, human casualties as estimated for the districts are presented in Table 17.3.

Of all eight models tested, adjusted R^2 is maximum in the fourth, fifth, and sixth (Table 17.1) with 63 % of the standard error. The variables selected by the statistical analysis in this model are physical exposure, population density, and percentage of SC and ST population. The equation for the estimation of risk is as follows:

$$\ln(R) = 0.50\ \ln(Popln.Exp) + 0.46\ \ln(\text{Popln. Density}) + 0.90\ \ln(\text{SC \& ST Popln,}) - 11$$

Table 17.2 The different multiple regression models, beta coefficients, and their level of significance

Model	Variables	Unstandardized coefficients		Standardized coefficients		
		Beta	Standard error	Beta	*t*	Significance
1	(Constant)	−6.79	10.23	–	−0.66	0.51
	Affected population	0.47	0.16	0.71	2.92	0.01
	Literacy	0.03	1.03	0.01	0.03	0.98
	HDI	−0.42	1.15	−0.09	−0.37	0.72
	Population density	0.66	0.42	0.49	1.58	0.13
	BPL population	−0.51	1.16	−0.09	−0.44	0.67
	Primary workers	0.94	1.02	0.24	0.92	0.37
	SC and ST population	1.06	0.59	0.58	1.81	0.08
	Kutcha houses	−1.14	0.95	−0.26	−1.21	0.24
2	(Constant)	−6.72	9.69	–	−0.69	0.50
	Affected population	0.47	0.15	0.71	3.05	0.01
	HDI	−0.40	0.94	−0.08	−0.43	0.67
	Population density	0.66	0.40	0.49	1.63	0.12
	BPL population	−0.51	1.12	−0.09	−0.46	0.65
	Primary workers	0.93	0.94	0.24	0.98	0.34
	SC and ST population	1.05	0.50	0.58	2.10	0.05
	Kutcha houses	−1.14	0.89	−0.26	−1.27	0.22
3	(Constant)	−10.08	5.61	–	−1.80	0.09
	Affected population	0.49	0.15	0.74	3.31	0.00
	Population density	0.63	0.39	0.47	1.61	0.12
	BPL population	−0.50	1.10	−0.09	−0.45	0.66
	Primary workers	1.02	0.90	0.26	1.13	0.27
	SC and ST population	1.12	0.47	0.61	2.40	0.03
	Kutcha houses	−1.12	0.88	−0.25	−1.28	0.21
4	(Constant)	−11.59	4.43	–	−2.62	0.02
	Affected population	0.48	0.14	0.73	3.34	0.00
	Population density	0.68	0.37	0.51	1.82	0.08
	Primary workers	0.93	0.86	0.24	1.08	0.29
	SC and ST population	1.11	0.46	0.61	2.42	0.02
	Kutcha houses	−1.20	0.85	−0.27	−1.42	0.17
5	(Constant)	−9.01	3.74	–	−2.41	0.02
	Affected population	0.52	0.14	0.79	3.77	0.00
	Population density	0.46	0.31	0.34	1.46	0.16
	SC and ST population	1.08	0.46	0.59	2.35	0.03
	Kutcha houses	−0.69	0.70	−0.16	−0.98	0.34
6	(Constant)	−11.00	3.15	–	−3.50	0.00
	Affected population	0.50	0.14	0.75	3.65	0.00
	Population density	0.46	0.31	0.34	1.48	0.15
	SC and ST population	0.90	0.42	0.49	2.13	0.04
7	(Constant)	−8.06	2.49	–	−3.23	0.00
	Affected population	0.57	0.13	0.86	4.42	0.00
	SC and ST population	0.55	0.36	0.30	1.54	0.13
8	(Constant)	−4.54	1.04	–	−4.38	0.00
	Affected population	0.43	0.09	0.65	4.55	0.00

HDI Human Development Index, *BPL* below poverty level

Table 17.3 Estimation of risk of human casualty caused by floods using multiple regression models (M)

		Estimated human risk through different models								Observed
Sl. no.	District	M1	M2	M3	M4	M5	M6	M7	M8	human risk
1	Anugul	0.69	0.69	0.66	0.68	0.72	0.67	0.79	0.98	1.70
2	Baleswar	5.06	5.04	4.70	5.46	4.87	4.71	4.18	3.41	4.60
3	Baragarh	1.32	1.32	1.26	1.25	1.24	1.44	1.47	1.38	0.70
4	Bhadrak	2.64	2.64	2.54	2.80	2.52	2.77	2.52	2.55	5.10
5	Bolangir	1.78	1.79	1.70	1.68	1.87	2.11	2.46	2.06	1.60
6	Boudh	0.28	0.28	0.26	0.29	0.30	0.38	0.48	0.63	0.20
7	Cuttack	3.02	3.03	2.85	3.00	3.31	2.80	2.48	2.60	2.10
8	Deogarh	0.62	0.61	0.63	0.72	0.64	0.55	0.71	0.72	0.90
9	Dhenkanal	0.87	0.86	0.80	0.84	0.90	0.95	1.01	1.17	0.70
10	Gajapati	1.19	1.19	1.04	1.04	0.91	0.80	0.87	0.79	2.40
11	Ganjam	3.05	3.05	2.70	2.81	2.10	1.60	1.68	1.96	1.60
12	Jagatsinghpur	1.59	1.59	1.37	1.40	1.35	1.47	1.14	1.42	0.90
13	Jaipur	3.51	3.50	3.13	3.34	3.51	3.60	2.93	2.56	4.10
14	Jharsuguda	0.53	0.53	0.51	0.50	0.58	0.56	0.42	0.49	0.60
15	Kalahandi	1.15	1.16	1.15	1.16	1.00	0.98	1.02	0.97	1.10
16	Kandhamal	0.99	0.98	0.84	0.92	1.10	1.05	1.33	1.00	1.10
17	Kendrapada	2.26	2.26	2.14	2.21	2.18	2.62	2.63	2.79	3.40
18	Kendujhar	1.30	1.29	1.21	1.36	1.40	1.43	1.35	1.10	1.00
19	Khordha	1.01	1.01	0.91	1.07	1.46	1.26	1.01	1.39	0.90
20	Koraput	0.75	0.75	0.65	0.77	0.74	0.73	0.69	0.63	1.30
21	Malkangiri	0.89	0.89	0.76	0.84	0.87	0.91	1.05	0.79	0.50
22	Mayurbhanj	0.96	0.96	0.97	1.10	1.16	1.26	1.02	0.84	2.30
23	Nawarangpur	0.69	0.69	0.60	0.65	0.53	0.59	0.43	0.42	1.30
24	Nayagarh	0.70	0.70	0.63	0.70	0.72	0.71	1.01	1.37	1.10
25	Nuapada	0.88	0.89	0.87	1.02	0.99	1.02	1.15	1.04	0.20
26	Puri	1.84	1.84	1.74	2.00	1.92	1.95	2.08	2.40	3.80
27	Rayagada	1.36	1.37	1.23	1.31	1.36	1.31	1.41	1.04	1.70
28	Sambalpur	0.74	0.74	0.70	0.71	0.88	0.81	0.85	0.80	0.70
29	Sonepur	0.58	0.58	0.53	0.59	0.50	0.58	0.56	0.72	0.30
30	Sundargarh	0.84	0.84	0.84	0.91	0.89	0.78	0.65	0.62	0.40
Variance		1.15	1.14	0.97	1.21	1.04	1.01	0.78	0.64	1.77

The regression shows that exposed population (*Popln.Exp*), SC and ST population (SC&ST Popln), and high population density (Popln.Density) areas are more subject to suffering casualties from floods. The estimated risk of death from floods as shown in Table 17.3 indicates that the districts showing high risk are Baleswar, Bhadrak, Cuttack, Jaipur, Kendrapada, and Bolangir; the medium-risk districts are Bargarh, Ganjam, Jagatsinghpur, Kondhamal, Kendujhar, Khurda, Mayurbhanj, Nuapada, Puri, and Rayagada; and the districts showing low risk are Anugul, Deogarh, Dhenkanal, Gajapati, Jharsuguda, Kalahandi, Koraput, Malkangiri, Nayagarh, Nawarangpur, Sambalpur, Sonepur, Sundargarh, and Boudh.

17.5 Conclusion

This study has two main findings: the calculation of the average risk of death per district, and a set of indicators that point out the districts that are most at risk, vulnerable, and exposed to floods. Another important feature of the DRI is that it is based on the datasets with district-level resolution. The method used in this statistical analysis proved to be appropriate, and of the correlations observed among the data variables, physical exposure appeared to be the most significant factor leading to risk. In a sense this also validates the methodology developed for estimating the number of people exposed to flood hazards. The research has highlighted a relationship between higher level of development, higher literacy, and low casualties. This relationship can be understood in both ways: lower development may lead to higher casualties, or higher development may lead to lower casualties, but high hazard occurrence may also lead to lower economic development because it destroys infrastructures and crops.

Such models should not be used as predictive models when the precision of the data sources is not sufficient. A database with block-level resolution and risk defined in terms of monetary loss can be better substitutes for this exercise. The study indicates a need to improve the socioeconomic impacts of floods in terms of precision and completeness. There is also a need to improve data on hazard characteristics and exposure in addition to inclusion of indicators on disaster risk management and reduction. In the present study, DRI only measures the levels of risk and their associated factors, but not the actions taken to reduce risk. This study underlines the usefulness of continuing the improvement of data collection for a better identification of populations at risk.

References

Behera A (2002) Creating sustainable and disaster resistant communities. J Gopabandhu Acad 1(1):32–34, Bhubaneswar

Building Material and Technology Promotion Council (BMTPC) (2006) Vulnerability Atlas of India, first revision. Ministry of Housing and Urban Poverty Alleviation, Government of India, New Delhi

Burton I, Kates RW, White GF (1993) The environment as hazard. Guildford Press, London

Chow VT (1964) Hand book of applied hydrology: a compendium of water resources technology. Tata McGraw Hill Book Company, New York

Dao H, Peduzzi P (2004) Global evaluation of human risk and vulnerability to natural hazards. Enviroinfo, Sharing, Editions du Tricorne, Genève

Director of Census (2005) General population and housing table. Bhubaneswar

Disaster Management Institute (1988) Flood control and management proceedings volume. Paryavaran Parisar, Bhopal

Government of Orissa (1992) Cyclone advance preparation in Cuttack District. Unpublished report. Collectorate, Control Room, Cuttack

Government of Orissa (1999) Flood and cyclone advance preparations in Jagatsinghpur District. Control Room, Collectorate, Jagatsinghpur

Government of Orissa (2008) Annual reports of the Department of Revenue and Disaster Management. Bhubaneswar

Mahalonobis PC (1941) Rainstorms and river floods in Orissa. I. & P. Department of Orissa, Bhubaneswar

Panda GK (1989) Drainage and flood problems of coastal plain: a study in applied geomorphology. Ph.D. dissertation. Department of Geography, Utkal University, Bhubaneswar

Planning and Coordination Dept. (P & C Dept.) (2006) Orissa human development report. Government of Orissa, Bhubaneswar

Sinha BN (1985) Geography of Orissa. NBT Publications, New Delhi

Suri S (2000) Orissa disaster, agony of the living. Jawahar Park, Laxmi Nagar, Hyderabad

UNDP/BCPR (2004) Reducing disaster risk: a challenge for development. UNDP/Bureau for Crisis Prevention and Recovery, New York. http://www.undp.org/bcpr/disred/rdr.htm

UNDP (2009) Preliminary presentations on world disaster report. Bahrain

UNDRO (United Nations Disaster Relief Coordinator) (1979) Natural disasters and vulnerability analysis in report of expert group meeting (9–12 July 1979). UNDRO, Geneva

Chapter 18
Irrigation Water Demand for the Hindon Basin: Uttar Pradesh, India

Preeti Sachar

Abstract Water is an invaluable natural asset. The efficient and optimal use of global natural resource bases such as air, land, and water have emerged as an area of universal concern during recent decades. To create an economic system that is environmentally friendly and nonpolluting requires a significant mobilization of resources. Water is vital in a modern economy, not only in the urban-industrial context, but also in rural areas through its more widespread use in productive activities and its potential to improve living conditions. This chapter is a study of the Hindon Basin in Uttar Pradesh (India), which is a part of Upper Ganga-Yamuna Doab. The purpose of the chapter is to assess the requirement of irrigation water in the region for estimating the future projection of water demand. The study is based on primary as well as secondary data and used a stratified random sampling technique for selecting 9 villages from 4,000 villages from three sub-basins delineated from the Hindon Basin region. The Dastane et al. (1970) method is utilized for estimating water requirements of various crops in the region. The study provides useful understanding of water needs and efficient utilization of water for crop production in the region using scientific methodology.

Keywords Water demand and supply • Water management • Hindon-Basin • Evapotranspiration • Irrigation • Crop production • Dastane Methods for determining water requirements

P. Sachar (✉)
Department of Geography, Swami Shradhanand College, University of Delhi, Alipur, New Delhi, Delhi, India
e-mail: preeti_sachar@rediffmail.com

A.K. Dutt et al. (eds.), *Spatial Diversity and Dynamics in Resources and Urban Development*, DOI 10.1007/978-94-017-9771-9_18

18.1 Introduction

Water is an environmental resource, and the principles of sound natural resource utilization and its management necessitate the understanding of the relationship between supply of water and the demand that exists. The efficient and optimal use of our global natural resource base, including air, land, and water, has emerged as an area of universal concern during recent decades. In the water sector, resources are needed to restore degraded environments, to clean up water systems and polluted land, to replant forests, to regulate water flows, and to recreate damaged habitats. To create an economic system that is environmentally friendly and nonpolluting requires a significant mobilization of resources. In many current social systems that are based on the concept of continuous growth, a shift to a more sustainable development path will involve profound changes in social structure. The efficient utilization of water resources, their availability along with intergenerational equity questions, are critical because sustainable development in water and other sectors involves, in essence, a shift of resource allocation from current generations to those of the future. Although the major issues vary widely over space and time, here, for the purpose of analysis, the irrigated water demands as per the available water resource potential needs specification.

Water has a critical function in every aspect of human activity. In more recent times, water has emerged as a key natural resource to be efficiently managed for environmentally sustainable development. Water is vital in a modern economy, not only in the urban-industrial context, but also in rural areas through its more widespread use in productive activities and its potential to improve living conditions.

The micro-basin irrigation water demands for Hindon here refer to both theoretical as well as empirical demands. In this study, the micro-basin irrigation water demand refers to water requirements for different crops during different seasons. This chapter highlights the basic water requirements of important crops grown in the basin in different agriculture seasons, namely, kharif (during autumn), rabi (during spring), and zaid (during March to June). In other words, the aim is to assess the micro-basin water demand in terms of total water required to grow the crops in the basin.

Water is one of the most important inputs for assured crop production in an area, especially where production suffers because of its scarcity or irregular rainfall distribution. Water is required in large quantities and also needs to be supplied at regular intervals to meet crop demands. Thus, adequate and timely water supply is one of the basic inputs for obtaining desired crop yields. Very often, it is also a limiting factor in crop production for three reasons: first, water is required in huge quantities; second, it needs to be supplied intermittently throughout crop growth because of evapotranspiration and the limited water-holding capacity of the soil; and third, it affects yields not only directly, but also indirectly, by influencing sowing time, responses to fertilizers, and other management factors.

In a tropical country such as India, with erratic, inadequate, and unevenly distributed rainfall, assured irrigation is the only way for permanent and profitable farming.

It is here that most of the natural water harnessed by man is used in agriculture, but it is seldom realized that irrigation water is the most expensive input that farmers are using. However, this costly input can be used judiciously and economically for producing optimum crop yield per unit of irrigated land or per unit of water used in irrigation by working out a precise water management technique. The economic and efficient utilization of water, therefore, becomes essential in water use for which the knowledge of the water requirements of crops and its relationships with other input factors is imperative.

The Hindon Basin, in spite of being in a favorable position with respect to water potential, continues to import colossal quantities of food. The basin's miserable food output results from a wasteful system of irrigation, caused by lack of proper technical guidance and inefficient administrative control. Irrigation water supply along with efficient utilization of available irrigation water is an important factor for an assured crop production. It permits better utilization of other production factors and leads to increased yield per unit of land. The integrated development of water resources, judicious method of water application, proper soil and crop management practices, and scientific scheduling of irrigation according to the developmental rhythm of planting form important aspects of a comprehensive irrigation development program. This chapter attempts to assess the requirements for irrigation water in the Hindon Basin and also to make future projections for water demand.

18.2 Study Area

The delimitation of the study area, namely, the Hindon Basin, is on the basis of the existing Hindon River system located in the northwestern portion of Uttar Pradesh. The Hindon River flows in an interfluvial area between the two great canal systems, namely, the Eastern Yamuna Canal and the Upper Ganga Canal. These two canal systems are taken as the western and eastern boundaries, respectively, for the Hindon Basin. The Hindon River in its upper reaches and that of the lower reaches forms the meeting point of the northern and southern parts of the basin, respectively. The area thus selected for the study is the Hindon Basin, which forms part of the Upper Ganga-Yamuna Doab. It lies between 28 and 31°N and 77°10′E and 79°10′E. The Hindon Basin on the whole contains 16 tehsils and 48 blocks and four districts, that is, Saharanpur, Muzaffarnagar, Meerut, and Ghaziabad, each having four tehsils. A tehsil is an administrative area in India, also known as a taluk or mandal, an administrative division of some countries of South Asia. Gradually, under the British Raj these terms replaced the earlier terms pargana, pergunnah, and thannah (Sharma 2012, p. 53).

Nature has endowed the basin with plentiful supplies of both surface and subsurface water. The area is drained by the Hindon River, which is 258 km long and is an important perennial tributary to the Yamuna River that originates from the Shiwalik foothills. Its major tributaries are Krishni and Kali, which form two major sub-basins for the main Hindon Basin, along with small streams, namely, Dhamola and Nagdeo, on its upper reaches (Fig. 18.1).

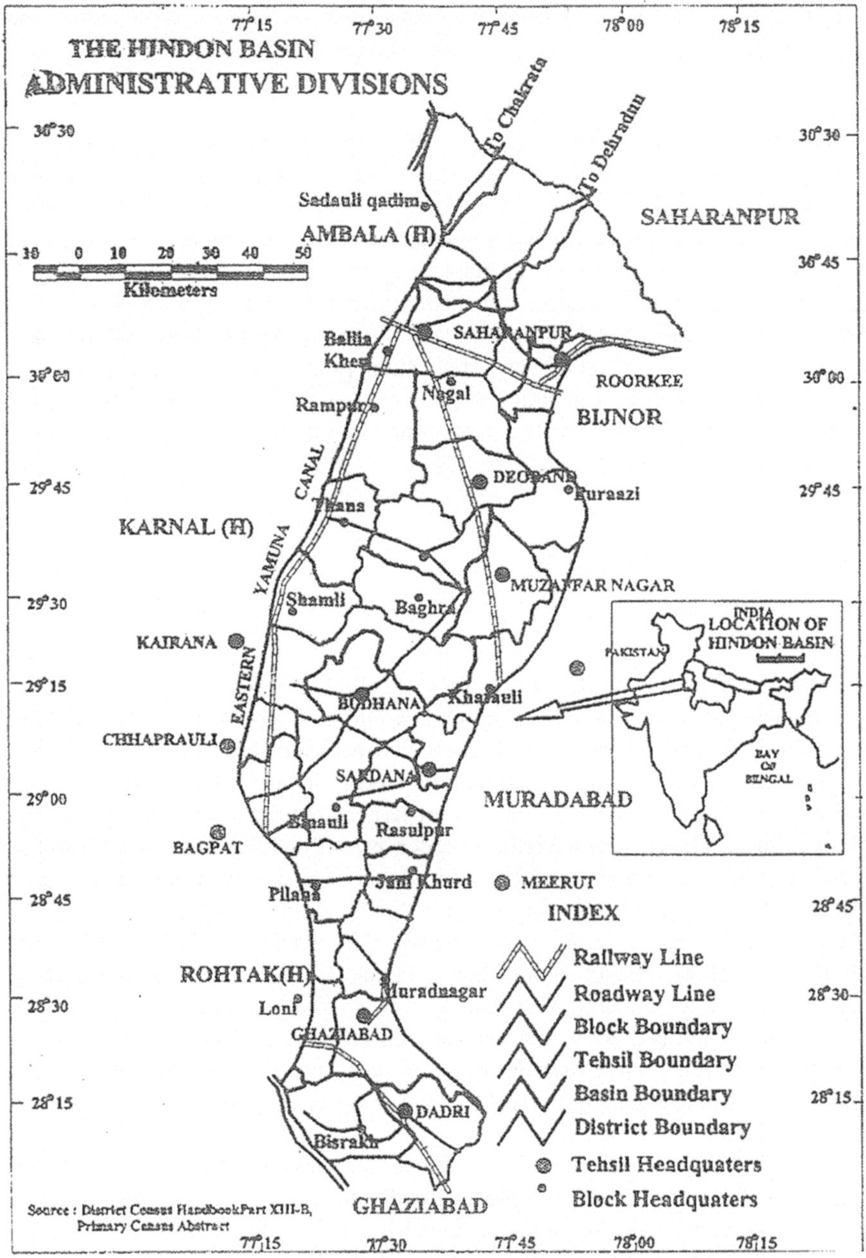

Fig. 18.1 The Hindon basin administrative divisions

The basin is spread over an area of 13,065 km^2 covering parts of Saharanpur, Muzaffarnagar, Meerut, and Ghaziabad Districts. The basin slopes from north to south or southeast with reference to its alignment of major rivers such as Ganga and Yamuna, which is also the flow direction of main drainage in the area. Most of the cultivated lands on either side of the canals cover more than 60 % of the net sown area of the watershed.

Geologically, the basin is a part of the Indo-Gangetic Plain. It is formed of both consolidated and unconsolidated alluvium composed of sand, silt, and other soils brought down by the rivers from the foothills of Sivalik, formed during the Pleistocene and recent geologic times. The basin is characterized partly by Sivaliks, Bhabar, Tarai, Khadar, and Bhangar. It is a fertile gentle sloping plain, except the Khadar area, which is ravinous terrain and sandy. The study area has an elevation of 200–300 meters (m) above sea level. It is an almost flat but gently undulating featureless plain.

The basin enjoys a subtropical monsoon climate, which indicates the seasonal rhythm of the weather. It is characterized by general dryness except during the brief span of the monsoon. Normal rainfall varies from 400 to 700 mm, and increases from west toward east. However, rainfall does not exceed 250 cm annually.

Being agriculturally developed, the basin is a major producer of wheat, paddy, and sugarcane. It depends on irrigation for development of the agriculture sector. The cultivated crops change seasonally from kharif to rabi, dominated by food crops, such as cereals covering 80–90 % of the irrigated area, and less than 10 % by fodder crops. The irrigation pattern varies from use of tube wells to canals among 4,000 villages. However, because of the small catchment area of the river, optimal water resource utilization and its management become prime requisites for the long-term development of the region.

18.3 Data and Methods

The study is based on primary as well as secondary data. For ascertaining the water demands of various crops, the entire agricultural year inclusive of kharif as well as rabi irrigation was considered. The primary survey was conducted through a structured questionnaire and village schedule through spot observations, whereby queries were made regarding water quantities used and the output that resulted per hectare depth of water in inches applied for different crops grown in the basin.

For the convenience of the study, the basin was divided into three sub-basins, namely, the Hindon sub-basin, Krishni sub-basin, and Kali sub-basin, to analyze actual water demand in the basin. Thus, for conducting the survey, a stratified random sampling technique was adopted with a village as the primary unit and beneficiaries in a village as the ultimate unit of sampling. Of 4,000 villages in the basin, 9 villages were selected, 3 from each of the three sub-basins. The selected villages are listed in Table 18.1.

Table 18.1 List of selected villages

Ghaziabad	Ghaziabad	1	Hindon	Bikhanpur
Meerut	Baghpat	1	Krishini	Mansoorpur
Meerut	Sardana	1	Kali	Chandana
Muzaffarnagar	Muzaffarnagar	1	Hindon	Dholra
Muzaffarnagar	Muzaffarnagar	1	Kali	Saidpur Khurd
Muzaffarnagar	Kairana	1	Krishini	Sainta
Saharanpur	Roorkee	1	Kali	Saidpura
Saharanpur	Saharanpur	1	Hindon	Sona
Saharanpur	Saharanpur	1	Krishini	Ismailpur

18.4 Review of Methods

The basic concepts of water requirements are multifarious. Proper techniques are not employed to determine the water needs of crops, which results in obtaining little information, causing enormous misuse of water and earning poor returns from the major irrigation projects. There are several irrigation research stations in India, such as Pusa, Roorkee, Shajahanapur, Banaras, Coimbatore, and New Delhi, which have carried out water use research on water requirements for rice, wheat, cotton, and sugarcane. Several individuals and agencies have also worked on this problem. The approaches employed in determining water requirements for different crops are broadly grouped into the following four categories (Dastane 1970, p. 2; Thornthwaite 1948): (1) transpiration ratio, (2) depth-interval yield, (3) soil moisture deficit, and (4) climatological.

Initially water requirement was taken as transpiration ratio, that is, the quantity of water transpired by a crop to produce a unit amount of dry matter. Plants were grown in a pot culture house; evaporation was minimized by sealing the soil surface. Water was added at intervals on the basis of periodic weighing of the pots. Records were maintained on quantity of water added and dry matter produced by plants. Later, the effects of different species, duration, manure, and applied fertilizers on transpiration ratio were also studied. The values, however, reported were far from those obtained under field conditions and thus had limited practical utility insofar as irrigation scheduling for field crops was concerned. The values of transpiration ratio varied from one place to another and for one crop to another. For example, the wheat transpiration ratio varied in water requirement from 200 to 1,000 mm and that of sugarcane from 250 to 1,500 mm. It was observed that addition of manure reduced the ratio.

Based on the first two approaches, several studies commenced as early as 1910 at Pusa and Cuttack and at the Institute of Agriculture Research at Varanasi (Leather 1911; Singh et al. 1985; Ganguli 1952; Choudhary and Mahapatra 1963). Penman (1948) established a theorem that the water requirement of all crops must be the same if they are grown on the same soil and in the same growing season. It was based on the law of physics that inside the plants, functions performed by water are innumerable, but that the quantity of water directly required by plants is limited. It is now recognized that for a given growth of dry matter, the amount of water transpired may vary greatly according to climatic conditions. The foregoing approach was found to be of no use as less attention was paid to different components of water requirements, namely, consumptive use, effective rainfall, irrigation efficiency, percolation loss, and contribution from the water table.

This effort was followed by field experiments known as the depth-interval yield approach. In this, the delta (depth of irrigation) and interval of irrigation were tried for scheduling irrigation and finding the optimum water requirements of crops. Different depths of irrigation such as 150, 225, or 300 mm were applied at intervals of 2 to 4 weeks. The irrigation treatment that gave the maximum yield with the minimum delta of water was taken as the optimum water requirement of that particular crop. The bulk of work conducted in this field has been done by Anonymous (1950), Ramiah and Vachhani (1951), and Hiranandani (1953).

A real breakthrough occurred when field experiments based on the soil water regime concept were conducted for various crops. This approach uses the depth-interval yield approach by fixing the treatment of depth of irrigation on the basis of soil moisture deficit. In this concept, water content at field capacity (upper limit of regime) was considered as 100 % available for crop growth and that at the permanent wilting point as 0 %. The safe limit of allowable soil water depletion (lower limit of regime) for a crop was determined by experimentation and was taken as a criterion for scheduling of irrigation. The formula used under this method is defined by Dastane et al. (1970):

$$d = \frac{FC - A}{100} \times ASG \times D$$

where

d = moisture deficit in soil layer
FC = field capacity in percent
A = actual moisture content in percent
ASG = apparent specific gravity of soil
D = depth of soil layer

The net deficit in the root zone thus worked out was corrected by the value of irrigation efficiency that estimated the depth of water to be applied at each irrigation. Experiments of this nature have been in progress since 1956 at the Indian Agriculture Research Institute (IARI), New Delhi and many other agricultural colleges. This technique is found satisfactory for all practical purposes for determining precise scheduling, depth of irrigation, and water requirements for various crops.

The consumptive use of water by a crop, which is the main part of the water requirement in most field crops, is governed primarily by meteorological parameters. Climatic parameters have a predominant role in governing the water needs of crops and the criteria of soil water availability for scheduling of irrigation. In this approach, consumptive use of water by a crop is first worked out by recording soil moisture losses throughout growth, correlating it with the values of meteorological parameters, and thus deriving quantitative relationships. It is concluded that the climatological approach is simple, rapid, and reliable and has a high extension value. Using this approach, work has been carried out in India mainly at IARI, New Delhi. The most recent and more widely adopted for scheduling of irrigation is plant water status, considered as an ideal criterion as the plant is a good integrator of soil, water and climate factors.

18.5 Estimation of Water Requirement of Crops

Estimation of water requirement of crops is one of the basic needs for crop planning on a farm. Water requirement may be defined as the quantity of water, regardless of its source, required by a crop or a diversified pattern of crops in a given period of time for normal growth under field conditions at a specific location.

Evapotranspiration refers to the total water loss from evaporation from soil and transpiration from a crop for a particular area during a specified time, whereas effective rainfall is total precipitation or a fraction thereof that forms part of the water requirement. The soil moisture deficit at the rooting depth of a crop before precipitation mostly decides the value of effective rainfall. According to the climatological concept of water requirements of crops, even light showers are determined as effective rainfall. Only that rainfall which is lost by runoff and deep percolation is ineffective. Thus, the role of evapotranspiration and effective rainfall for estimating water requirement of crops in the basin becomes important.

For the proper development of agriculture, the quantum of irrigation supply is important. The Hindon Basin, in spite of being in a favorable position in respect of water potential, continues to import colossal quantities of food. The basin's miserable food output results from a wasteful system of irrigation from lack of proper technical guidance and inefficient administrative control.

Irrigation water supply along with efficient utilization of available irrigation water is an important factor for assured crop production. It permits better utilization of other production factors and leads to increased yield per unit of land. The integrated development of water resources, judicious method of water application, proper soil and crop management practices, and scientific scheduling of irrigation according to the developmental rhythm of the plant form important aspects of the comprehensive irrigation development program. Efficient water management requires thorough study of not only soil properties but also of plant–water relationships, climate, agronomic practices, and economic assessment. An effort, thus, has been made in this chapter to highlight the importance of the soil water–plant relationship in crop production through the demand factor in terms of water requirements.

Irrigation water is thus needed mainly to meet the demands of evapotranspiration (ET) and metabolic activities of plant, together known as consumptive use (Cu or u). Because water used in the latter part is negligible, ET is considered practically equal to Cu. Water requirement, thus, includes losses from ET (or Cu) + losses during application of irrigation water (unavoidable), and the quantity of water required for special operations such as preparation, transplanting, and leaching. It is formulated as

$$\text{WR} = \text{ET or Cu} + \text{Application losses} + \text{special needs.}$$

Water requirement is, therefore, demand and supply and would consist of contributions from any source of water, the major source being irrigation water (IR), effective rainfall (ER), and the soil profile contribution(s) including that from the shallow water table.

To assess the relationship between irrigation supply and yield, various experiments were tested at Varanasi for different crops. Yields for different crops were obtained and plotted against the amount of water they consumed. The results showed a linear relationship or proportion between the yields and the amount of water consumed provided additional manure is used. Further, the length of the life cycle of a plant has great influence on its water requirements. Again, a linear correlation exists between them, provided plants are irrigated in periods that ensure maximum possibilities of their growth and yield. It is only then, when we have reliable information about the exact number of times when plants need maximum water, that any scientific irrigation of crops is possible. The optimum yield, namely, highest yield with most economic use of water, was obtained by a supply of a 3-in. depth of irrigation provided at intervals of 2 to 4 weeks. The Irrigation Research Institute at Roorkee (UP) has conducted such experiments for different crops grown in the Hindon Basin (Table 18.2).

18.6 Requirements of Water by Major Crop

While reviewing work on different techniques adopted for water requirements for different crops, it was found that each of the techniques is relevant. Although some of the essential details are required to utilize the aforementioned methods, water requirements for different crops as assessed are summarized next.

18.6.1 Cereals and Millets

18.6.1.1 Rice

Rice occupies 20–30 % of the irrigated area in the basin. Its water requirements are many times greater than those of other food crops. Being a semiaquatic plant, it is a major consumer of water and needs careful water management to increase water use efficiency. It is grown under varied soil and climatic conditions; one to three crops

Table 18.2 Water requirements of different crops grown in the Hindon Basin

Crop	Dry matter produced/ac in laboratory	Water transpired by crop in in./gallon	Depth in in./acre	Water evaporated from soil (gallons/acre)	Depth in inches	E_t water required by crops in depth in inches	Total requirements of crops in depth in inches	Duration of crops in days	Daily water requirements of crop in depth in inches
Cotton	4,025	169,590	7.5	473,772	20.9	28.4	44.0	230	0.19
Rice	4,932	256,346	11.3	365,904	16.1	27.4	45	105	0.43
Potato	8,014	435,969	19.2	27,583	1.2	20.4	27	88	0.30
Sugarcane	20,594	444,009	19.6	582,146	25.7	45.3	88	310	0.28
Wheat	4,079	166,645	7.4	27,582	1.2	8.6	16	168	0.10
Barley	2,664	150,161	6.6	27,582	1.2	7.8	14	88	0.16
Maize	1,150	86,162	3.8	93,638	4.2	8.0	20	105	0.19
Mustard	1,358	71,063	3.1	27,582	1.2	4.3	11	88	0.12
Linseed	1,260	118,798	5.2	27,582	1.2	6.4	13	88	0.14

Source: Irrigation Research Institute (1993), Roorkee, Uttar Pradesh

per annum are taken in the basin. Irrigation depth as great as 300 mm and an interval as long as 28 days were selected in Uttar Pradesh by the Public Works Department. Values of water requirements vary from 750 to 2,500 mm as obtained by different workers. Rice offers the maximum scope and challenge for economizing water resource use. The percolation loss within the field can be reduced by following various practices, such as selection of heavy soils and shallow depths, land leveling, puddling to permeability, and intermittent drying of the field at the proper stage (Table 18.2).

18.6.1.2 Wheat

Wheat occupies a considerable area in the basin. Nearly 50 % of this crop is irrigated. The basin constitutes one of the best wheat-producing regions where the cool and dry climate prevails during the growing season (rabi). Studies on water requirements of wheat are easier because of its short duration and medium rooting depth. The frequency of irrigation required for this crop is lower than for other crops. Using the empirical depth-interval yield, with application in fixing depth and frequency of irrigation, trials have been conducted by Singh et al. (1985), Singh and Nijhawan (1951), and Singh (1945). The work in general shows that for tall Indian wheat varieties, three to four irrigations are required for obtaining maximum yield in low rabi rainfall tracts (50–100 mm) on medium-textured soils. The consumptive use of water is about 250–300 mm during growth. The yield of wheat increased with the number of irrigations, but the response per irrigation decreased gradually. For deciding the policy of intensive versus extensive irrigation in the command area, a comparative assessment is reported in Table 18.3.

18.6.1.3 Maize

Meager work has been reported on the water requirements of maize. In the Hindon Basin, maize is grown mostly in kharif (June–October) as rain fed or irrigated, and on a smaller scale during the summer season (February/March/May/June) under

Table 18.3 Intensive versus extensive irrigation in wheat

Number of irrigations/ha	Irrigated crops		Unirrigated		Total irrigations (A×B)	Total yield (g/100 ha) (B×C) + (D×E)
	Area (ha)	Yield (a/ha)	Area left unirrigated	Yield (q/ha)		
5	20	47	80	9.3	100	1,700
4	25	42	75	9.3	100	1,768
3	33	35	66	9.3	100	1,799
2	50	34	50	9.3	140	2,175
1	100	30	–	–	100	3,040

Source: Dastane et al. (1970), p. 45

Table 18.4 Effect of soil moisture regime on grain yield of maize on sandy loam soil of the basin

Treatment	Yield q/ha	Irrigation requirement		Seasonal consumptive use (mm)	Water use efficiency kg/ha (mm of consumptive use)
		Number	Amount		
25 % ASM	305	4	300	531	5.74
50 % ASM	37.6	6	450	573	6.56
CD at 5 %					

Source: Indian Agricultural Research Institute (1977), p. 198
ASM available soil moisture as measured at 15–30 cm depth of soil layer

Table 18.5 Irrigation requirements of major crops in the Hindon Basin

Crop	Soil type	Season	Number of irrigations	Requirements amount (mm)
Maize	Sandy loam	Kharif	4	225–275
Sorghum	Sandy loam	Kharif	4	250
Barley	Sandy loam	Rabi	4	300
Ragi	Loam	Kharif	2–3	150
Gram	Sandy loam	Rabi	1–2	150–180
Mustard	Sandy loam	Kharif	1–3	75–80
Sugarcane	Sandy loam	Kharif/rabi	11	660

Source: Sharma and Upadhyay (1973), pp. 220–221

irrigated conditions. The improved varieties of maize require 100–120 days to mature, except in hilly tracts of the basin where they take 10–20 days more to mature (Table 18.4).

The irrigation requirements of maize vary with the type of soil and the season it is grown. Maize is normally sown in the basin after a pre-sowing irrigation as the monsoon arrives late, and the crop requires two or three irrigations during the pre-monsoon per iod (Table 18.5).

18.6.1.4 Jowar (Sorghum)

Sorghum is next in area and importance to rice and wheat in the basin. It is grown best in semiarid areas with well-distributed rainfall of about 300–1,000 mm. Favorable temperature is in the range of 16–40 °C. The crop is grown mostly in the kharif (June–November) season and is also cultivated in the rabi (October–March) season on moisture-retentive deep soils. According to Patil (1963), irrigation at 25 % soil water availability in the surface 30-cm layer on heavy clay soil was adequate for sorghum grown for grains in kharif season (Table 18.5).

18.6.1.5 Bajra

Bajra is well suited to warm areas of low rainfall (250–500 mm). It is mostly grown in the kharif season under rain-fed conditions and takes 85–90 days to mature. Bajra is preferred when the rainfall is inadequate for maize and jowar, as it is more tolerant to drought. Regarding soil moisture requirements, bajra tolerated 75 % depletion of available water from 0 to 30-cm depth of soil on heavy clay and on sandy loam soil, with best yields obtained with irrigation at a 50 % depletion level. The irrigation requirement of bajra as given in Table 18.5 shows that irrigation requirement during kharif varies from 150 to 200 mm in most areas.

18.6.1.6 Barley

Next to wheat, barley is an important rabi cereal (October–March) under irrigation. It is hardier than wheat and can tolerate saline soil or water. It has wide ecological adaptations, needing less water than wheat or oats for growth. It is, however, grown as a rain-fed crop in areas having rainfall of 400–500 mm in the growing season. In dry areas, this crop needs irrigation for good growth. It is generally grown on sandy loam soils of medium fertility level; saline and alkaline soils are also suitable. Yield of this crop is lower under drier compared to irrigated conditions. The crop is taken after rice and with a pre-sowing common irrigation. Barley is grown on sandy loam soil with irrigation at 50 % available soil moisture. Treatments of one irrigation at 30 days or at pre-flowering were equivalent, but both treatments were superior to no irrigation in yield responses (Table 18.5).

18.6.1.7 Ragi

Ragi is grown on small areas and is one of the drought-tolerant crops suited only to dry-farming areas. The crop is grown as rain fed in the rainy season (June–October) and is irrigated all year round. It can be grown on a variety of soils including saline and alkali soils. Patil (1963) observed soil moisture and irrigation requirements on heavy clay soils; at 50 % available moisture depletion in the top 30-cm soil layer, irrigation was necessary to obtain higher yields of ragi sown in the summer season. In comparison, the same crop sown in the kharif season did not require any irrigation. Irrigation at 50 % depletion of available soil moisture on sandy loam soil was found adequate for ragi grown in the rabi season.

18.6.1.8 Pulses

Work on pulses is scanty because they are grown during kharif as a mixed crop with cereals. They utilize soil moisture very efficiently by virtue of a deep root system and thus require less irrigation as compared to cereals. Legumes in kharif season are normally grown without any irrigation in most areas unless the rain is unfavorable. Most of the legumes grown in rabi and summer season depend on irrigation.

18.6.1.9 Gram (Chana)

Gram is grown extensively in rabi season on alluvial sandy loam to loam soil. It is best suited in low-rainfall areas, as it has been found to respond to irrigation in the absence of winter rain, especially on the lighter soils. One irrigation of 6 cm was found adequate after planting, but irrigation may be required if no winter rains are received until flowering. Frequent irrigation did not benefit the crop except for the vegetative growth.

18.6.1.10 Red Gram (Arhar, Tur)

This crop is mostly grown on alluvial sandy loam soil in the basin. It benefits from irrigation if dry spells prevail for a longer period. Two irrigations, one at flower initiation (75 days after sowing) and the other at pod filling (100 days after sowing), were found adequate for short-duration red gram to give a higher yield; irrigation at 40 % available soil moisture was required.

18.6.1.11 Mustard (Edible Oilseed or Yellow Sarson)

Mustard is grown in rabi season (September/October–February/March) on sandy loam to clay loam soils. These crops benefit from irrigation if there are no or low winter rains. For this crop at Roorkee on sandy loam soil, one 60-mm irrigation was found best as compared with rain-fed mustard (Table 18.5).

18.6.2 Fiber Crops

18.6.2.1 Cotton

Cotton is a subtropical crop grown in areas with 750–2,500 mm rainfall. It tolerates temperatures to 45–46 °C but does not do well below 28 °C. Being a deep-rooted crop, it requires less irrigation water for its growth after establishment as compared to many grain crops having shallow roots. It is grown on a variety of soils, but in the Hindon Basin it is grown on sandy loam to loam soils, where it requires irrigation water for good growth. The optimal time for sowing cotton is during the month of May. Work on the water requirements of cotton shows a wide range of results as there are several species with different lifespans and growing seasons. Also, as cotton is a kharif crop that can grow on light as well as heavy soils, frequency and depth of required irrigation vary considerably with the soil type. During the rainy season, no irrigation is normally required if the rainfall is well distributed. The water requirement varies from 250 to 260 mm. More irrigation is required on light soil and in low-rainfall areas than on medium soil and in moderate-rainfall areas such as the Hindon Basin.

18.6.2.2 Sugarcane

Nearly 70 % of the area under sugarcane is irrigated. Sugarcane occupies the land for about 10–18 months and thus necessitates adequate irrigation for realizing its yield potential. Its commercial importance has prompted considerable work on the water requirements of this crop. It grows best on clay loam to loam soil, requiring more irrigation on sandy loam. Heavy clay soil is also suitable provided it is well drained. Sugarcane requires a warm humid climate for good growth, with temperatures ranging from 20 to 45 °C. In the basin, planting is mostly done with the commencement of the spring season. Sugarcane crop tolerated 50 % of depletion of available soil moisture on clay loam soils. It was found that five irrigations during the pre-monsoon period and one to two irrigations during the post-monsoon period were conductive to good growth. However, pre-monsoon irrigations from February to June are important. The rainfall is about 700–800 mm during the growth period (Table 18.5).

18.6.2.3 Sugar Beet

Sugar beet is grown in rabi season under cool temperatures (up to 18–22 °C), which favor its rapid growth and better sugar accumulation. Sugar beet required irrigation at 0.20 bar tension measured at the 25-cm depth of soil. The moisture loss by sugar beet varies from 950 to 1,010 mm during the entire growth period.

18.6.3 Projected Water Requirements

The total water requirement up to the year 2,025 A.D. for three districts of Uttar Pradesh, namely, Meerut, Muzaffarnagar, and Saharanpur, comprising the Hindon Basin, is provided in Table 18.6. These studies were carried out by the Central Ground Water Board. Thus, it can be seen from Table 18.6 that the demand as estimated is showing continuous increase from 0.942 million ham/year in 1980–1981 to 1.330 million ham/year in the projected year 2011–2021 for the basin.

However, among various demands, irrigation constitutes the major component as the basin is agriculturally developed. A major part of the demand for water is for the variety of crops that are grown in the basin. Table 18.7 presents total irrigation water requirements for the entire basin. Along with the requirement, cropping intensity,

Table 18.6 Estimated water use and projected requirement for the basin (million ha/year)

Place	1980–1981	1990–1991	2000–2001	2011–2021
Hindon Basin	0.942	1.128	1.229	1.330

Source: Central Ground Water Board (1984), pp. 37–38

Table 18.7 Total irrigation water requirement for the Hindon Basin

Year	Net sown area	Cropping intensity	Total cropped area (million ha)	Average consumptive use (mm)	Total requirements (million ha/year)
1980–1981	1.387	1.55	2.143	400	0.857
1990–1991	1.387	1.85	2.566	400	1.026
2000–2001	1.387	2.00	2.774	400	1.110
2011–2021	1.387	2.15	2.982	400	1.194

Source: Central Ground Water Board (1984), pp. 37–38

total cropped area, and the average consumptive use are also shown. It is thus inferred that all the parameters show continuous increase, except net sown area, which remains constant throughout. The total water requirement for irrigation has grown 0.857 million ham/year in 1980–1981 to 1.194 for the projected year 2021, which indicates intensive land and water utilization practice as demand is continuously increasing (Fig. 18.2).

A close observation of the foregoing projections for the rational basis of water allowance on a canal system aids in estimating the irrigation demand on a channel during a crop season. The estimates are summarized by working out the daily requirements of the principal crop in Table 18.8. Table 18.8 also provides the mode of applying irrigation to the key crops during a season. According to this table, average total daily requirement of crop in summer is at a depth of 0.25 in. against 0.13 in. in winter. Thus, winter crops mature on 52 % of the supply required by crops in summer. As there are lower winter discharges in canals, the percentage of losses is higher. The actual winter supply at the head of the canal is only 66 % of its designed supply. Actually, the capacity factor, which is the ratio of the mean supply in a canal during a period to its designed full capacity, varies from 0.60 to 0.70 in. depth in winter and from 0.90 to 0.95 in. depth in summer.

The peak of the water requirement curve could be seen as maximum demand for 1 day on the canal with the supply being 67.7 in./acre, which is equivalent to 5.64 acre-feet (ac ft)/daily or 5.64/1.98 = 2.87 cubic foot/sec (cu s) in the fields, corresponding to a water allowance of 3.60 cu/thousand acres at the outlet head, allowing for 20 % loss in the watercourses. The ratio of average water requirement or average discharge utilized to the capacity of the canal is the capacity factor for the season. If the channel is run in rotation during winter so that every channel is used

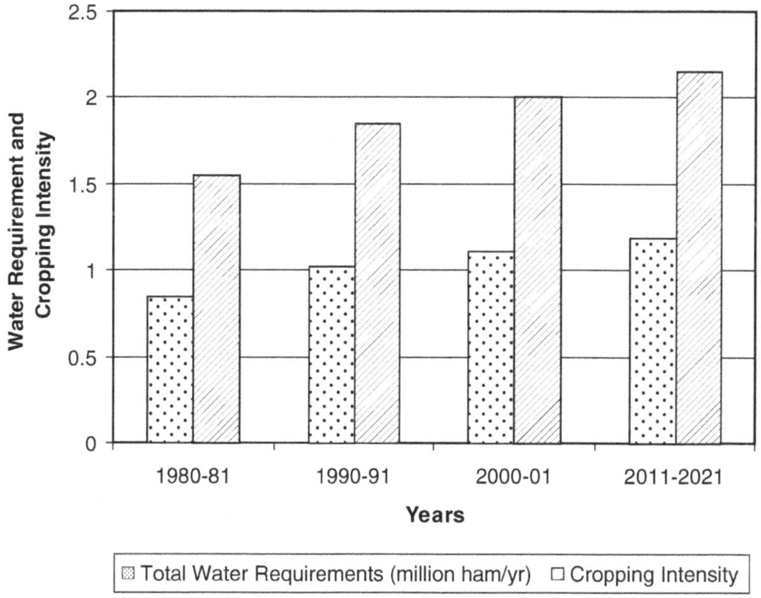

Fig. 18.2 Total irrigation water requirement for Hindon Basin

for a week after a run of 2 weeks, the capacity factor for rabi will be 0.66, which will allow an average discharge of $3.60 \times 0.66 = 2.40$ cu s at the outlet head for rabi crops. This quantity of water available per day in fields will vary, given in Table 18.8 as 0.25 for kharif with the average daily water requirement of rabi crops accepted as 0.13 in. depth/day.

The water requirement of crops during kharif of the Indian summer is plotted graphically to present a water requirement curve in the last column of Table 18.8. The demand for water starts from nil on first date and becomes full on the second date as indicated from the watering period mentioned in the table. All these requirements, plotted in Fig. 18.3, are added up to obtain a water requirement curve for the season in inch-acres in the fields. The peak of the water requirement curve could be seen as maximum demand on 1 day on the canal supply being 67.7 in./acre; this is equivalent to 5.64 ac ft daily or $5.64/1.98 = 2.87$ cu s in the fields, which corresponds to a water allowance of 3.60 cu per thousand acres at the outlet head, allowing 20 % loss in water courses. The ratio of average water requirement or average discharge utilized to the capacity of the canal is the capacity factor for the season. It will be given by the ratio of areas ABDE to ABCDEF. If the channel is run in rotation during winter so that every channel is used for week after a run of 2 weeks, the capacity factor for rabi will be 0.66, which will allow an average discharge of $3.60 \times 0.66 = 2.40$ cu s at the outlet head for rabi crops. This quantity of water available per day in the fields will vary (Table 18.8) as 0.25 for kharif with the average daily water requirement of rabi crops accepted as 0.13 in. depth/day.

Thus, water allowances from the available canal system in the Hindon Basin form a vital portion to meet the specific total requirements of principal crops, for

Table 18.8 Daily water requirements of principal crops for designing water allowance on a canal system

Crop	Life of crop in days	Frequency in days	Number	Inches of depth each time	Total depth in inches	Daily requirement in inches of depth	Total in inches per acre
Kharif							
Rice	105	7	15	3	45	0.42	25.8
Sugarcane	310	14	22	4	88	0.28	14.0
Cotton	230	21	11	4	44	0.19	1,701
Maize	105	21	5	4	20	0.19	9.5
Fodders	100	14	7	3	21	8.21	6.3
Miscellaneous	100	21	5	4	20	0.20	4.0
			Average: 0.25 in.				
Rabi							
Wheat	160	42	4	4	16	0.10	
Mixed grains	150	49	3	4	12	0.08	
Fodders	70	14	5	3	15	0.21	
			Average: 0.13 in.			0.39	

Source: Sally (1968)
Note: As major crop in *rabi* are wheat and mixed grains, the daily requirement of water by average crop will vary much, from 0.10 in. depth, over a lifespan of 150 days

which the existing cropping pattern and required water frequencies are needed. These data will further help to meet the inefficiencies existing in irrigation water scheduling and, subsequently, the reduction in crop yields. Thus, it can be concluded that with the very fact of water allowances made in the basin, it will be quite possible to double the production in the irrigated areas, especially with the present availability of the water resources.

18.7 Conclusion

The local economy of the study area is based greatly on the production and processing of agricultural products. The primary resources of land and water are not the limiting factors for the region's agricultural growth, but overexploitation of groundwater resources and inefficient utilization of surface water resources may bring fundamental changes in the value of both land and water in the basin. It is evident that the micro-basin water demand in terms of water requirements for different crops varies throughout the basin area. Even though considerable and commendable work has been done in this country on various dimensions of water requirements of crops, it has a limited value because some of the basic data on experimental conditions

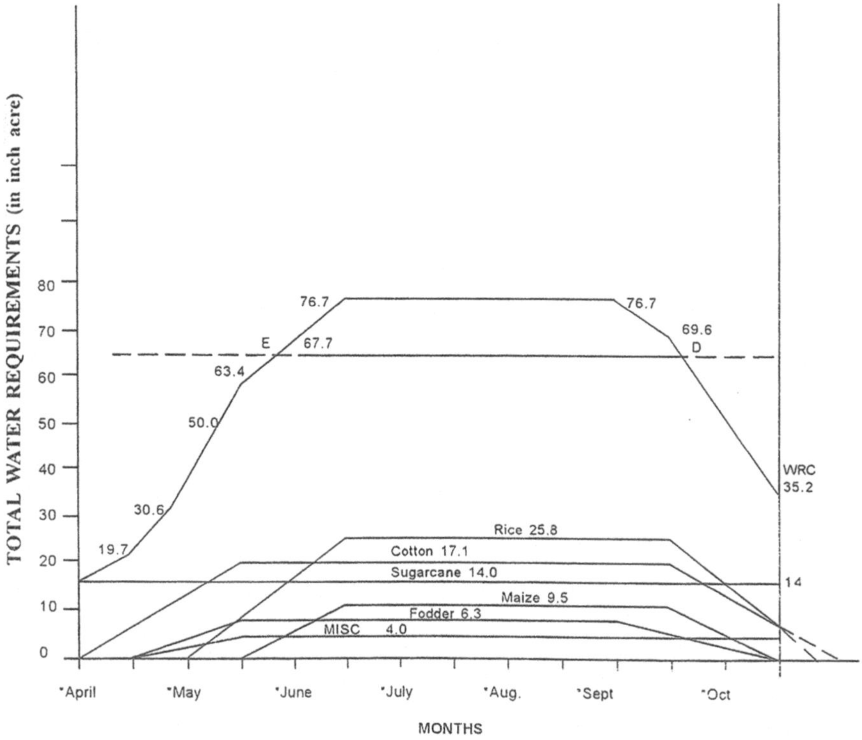

Fig. 18.3 Daily water requirement of crops during kharif crop season of Indian summer (in.) (From Sally 1968, p. 91)

were not either fully reported or recorded. The precise knowledge of water requirements of different crops, and its relationship with other input factors, thus becomes imperative. There is surely a need to appraise the correct approach/latest concept of the soil water–plant–climate relationship to avoid any layout of faulty experiments and to achieve both the economic and efficient utilization of the water resources. Last, it becomes evident that the available information and findings will help in further assessment of water use efficiency, as demand and supply patterns already exist.

References

Anonymous (1950) Annual report (technical). Central Board of Irrigation and Power, India 2(50), p 406

Central Ground Water Board (1984) Groundwater studies in upper Yamuna Basin. Project findings and recommendations. Annual report for Ghaziabad, Meerut, Muzaffarnagar, Saharanpur, pp 37–38

Choudhary MS, Mahapatra IC (1963) Effect of manuring on water requirement of rice. In: Proceedings, 36th annual meeting. Central Board of Irrigation and Power, New Delhi

Dastane NG (1970) New concepts in irrigation – necessary changes for new strategy. Polit Wkly India 4(13)4:27–30
Dastane NG, Singh M, Hukkeri SB, Vamadevan VK (1970) Review of work done on water requirements of crops in India. Navbharat Prakashan, Pune, pp 1–2
Ganguli PN (1952) Water requirement of rice plant. In: Proceedings, 9th meeting, Crop soils. Central Board of Agriculture, India
Hiranandani MG (1953) Duty of irrigation water proposed to be adopted for various crops in delta area of Mahanadi river. Central Board of Irrigation and Power, New Delhi
Indian Agricultural Research Institute (1977) Water requirement and irrigation management of crops in India. Caxton Press Private Limited, New Delhi, p 198
Irrigation Research Institute (1993) Annual report 1992–93. Roorkee, Uttarakhand
Leather JW (1911) Water requirements of crops in India. Department of Agriculture. 1:133–134, 205–281
Patil NP (1963) Optimum utilisation of water for irrigated crops in red soil areas of Bangalore district, Mysore State. In: Proceedings of symposium. Central Board of Irrigation and Power, New Delhi
Penman HL (1948) Natural evaporation from open water, bare soil and grass. Proceedings of the Royal Society of London. Series A, Mathematical and Physical Sciences 193(1032):120–1445
Ramiah K, Vachhani MN (1951) Farming under irrigation. Indian J Power River Valley Dev 1:5–8
Sally HL (1968) Irrigation planning for intensive cultivation. Asia Publishing House, New Delhi
Sharma AK (2012) Population and society. Concept Publishing Company, New Delhi
Sharma HC, Upadhyay SC (1973) A note on irrigation requirement of maize. Indian J Agron 18:220–221
Singh D (1945) Water requirement of crops. In: Proceedings, crops and soil wing. Board of Agriculture and Animal Husbandry, India, pp 166–168
Singh D, Nijhawan SD (1951) Water requirement of crops. Technical annual report. Central Board of Irrigation, India, pp 976–977
Singh BN, Singh BR, Singh K (1985) Investigations into the water requirements of crop plants. Proc Indian Acad Sci Sect B 1:471–495
Thornthwaite W (1948) An approach towards a rational classification of climate. Geogr Rev 38(1):55–94

Chapter 19
A Comparative Case Study of Small Hydro Development in the Indian Himalaya

A. John Sinclair, Alan P. Diduck, and Matthew McCandless

Abstract The hilly and mountainous regions of the Indian Himalaya offer vast potential for hydro development because of the number of fast-flowing rivers and streams. Several state governments, along with the central government, continue to promote the harnessing of this power potential to correct energy deficits, encourage industrial development, and improve social well-being. For their part, the states of Himachal Pradesh and Uttarakhand have encouraged developers to take advantage of various incentives to exploit their hydro potential. Our research focuses on one aspect of these initiatives in these two states, the burgeoning small hydro industry. Small hydro development has become very popular in these and other regions in India because of its potential to provide local benefits, such as electrification, reduction in fossil fuel use, and enhanced local economic development opportunities, especially in rural and remote areas. Using a qualitative approach involving a review of documents, field observations, and interviews with government representatives, industry officials, and community members, the research investigated the development activities for two small hydro projects. Our findings show that the most successful project was the one where the local community was fully engaged in planning, construction and operation. They also reveal the need for significant policy change in relation to the approval of hydro developments that are not community based. Additionally, the work revealed that community members were concerned about the vast extent of micro-hydro development in their regions.

Keywords Small hydro • Himachal Pradesh • Uttarakhand • Public consultation • Equity

---

A.J. Sinclair (✉)
Natural Resources Institute, University of Manitoba, Winnipeg, Manitoba, Canada
e-mail: john.sinclair@umanitoba.ca

A.P. Diduck
Department of Environmental Studies and Sciences, University of Winnipeg, Winnipeg, Manitoba, Canada
e-mail: a.diduck@uwinnipeg.ca

M. McCandless
International Institute for Sustainable Development, Winnipeg, Manitoba, Canada
e-mail: mmccandless@iisd.ca

A.K. Dutt et al. (eds.), *Spatial Diversity and Dynamics in Resources and Urban Development*, DOI 10.1007/978-94-017-9771-9_19

19.1 Introduction

The hilly and mountainous regions of the Indian Himalaya offer vast potential for hydropower development. More than 75 % of the country's estimated 148,700 MW of hydropower potential is found in the Himalayan region (Water for Welfare Secretariat 2008a; Agrawal et al. 2010). The Government of India has actively promoted the harnessing of this potential to correct energy deficits, encourage industrial development, and improve social well-being (Water for Welfare Secretariat 2008a). Of the 162 hydro projects in the Central Electricity Authority's 50,000 MW initiative, 133 are in the Indian Himalayan region (Agrawal et al. 2010). The 50,000 MW initiative, announced in 2003, is a program to prepare preliminary feasibility reports for hydro projects in 16 states for the purpose of increasing national hydropower capacity by at least 50,000 MW (Central Electricity Authority 2004). The government's goal is increasing hydropower's share of the country's total generating capacity from 25 to 40 % (Water for Welfare Secretariat 2008a).

An important strategy for the central government in increasing hydropower capacity has been to work with state governments and the private sector. For example, an objective of the central hydro policy outlines the additions to capacity expected from the central, state, and private sectors. The states have the largest expected contributions; for example, for the tenth national plan period, 6,537 MW are planned with 990 coming from the center, 4,498 from the states, and 1,050 from the private sector (Water for Welfare Secretariat 2008a). For their part, Himachal Pradesh and Uttarakhand have joined with the central government in encouraging development of hydro potential. The 50,000 MW initiative includes 33 projects in Uttarakhand and 15 in Himachal Pradesh (Central Electricity Authority 2010). Uttarakhand has an estimated 20,000 MW of potential power, of which only 1,400 MW have been harnessed (Asian Development Bank 2005; Joshi 2007). The state has adopted an explicit policy to develop the remaining capacity "in the shortest possible time," and it too is relying on a strategy of encouraging private sector investment (Uttarakhand Jal Vidyut Nigam Limited 2010, p. 1). Himachal Pradesh, with an estimated 21,000 MW of potential and more than 6,311 MW of harnessed power, has also adopted a policy of rapid growth and privatization in the hydro sector (Himachal Pradesh State Electricity Board 2010).

This chapter focuses on one component of hydropower policy in these two states: the burgeoning small hydro industry. Small hydro refers to projects with less than 25 MW of capacity, and includes micro (<100 KW) and mini (100 KW – 5 MW) projects (Water for Welfare Secretariat 2008b; Ministry of New and Renewable Energy 2010). Small hydro has become very popular in these states and other regions of India because of its potential to provide local benefits, such as electrification, reduced dependence on fossil fuels, Kyoto Protocol Clean Development Mechanism (CDM) investments, and enhanced economic development opportunities, especially in rural and remote areas (Saxeena 2007; Tanwar 2008; Bhattacharya and Jana 2009). Uttarakhand has adopted various policy measures (e.g., wheeling of electricity, provision of transmission lines, a single window clearance system) to catalyze growth in the small hydro sector, seeking to harness 600 MW from small hydro by the year 2020 (Water for Welfare Secretariat 2008b). Himachal Pradesh has its own detailed incentive package (including purchasing, wheeling and banking of power, a streamlined clearance process, and waiver of royalties for usage for a period of 12 years), and the state "attaches significant importance to the exploitation of small hydro potential" (Himachal Pradesh Energy Development Agency 2010).

Given the policy importance of the small hydro sector in Himachal Pradesh and Uttarakhand and given the importance of small hydro for local and regional economies, this research investigated planning processes for small hydro development in those states. Part two of the chapter summarizes our methodology, including a rationale for our choice of case studies. Part three describes the two hydro facilities we examined. Part four presents the local people's perceptions of the consultations for the facilities, along with their perceptions of the potential and real impacts of the two projects. The final part of the chapter discusses the results and considers their policy implications.

19.2 Research Design and Method

Our research design was a qualitative comparative case study. The primary data collection methods were qualitative interviews, document reviews, and field observations. The case study sites were the 50-KW facility at Malari village in the Chamoli district of Uttarakhand and the 3-MW facility at Solang village in the Kullu district of Himachal Pradesh (Fig. 19.1). These sites were chosen for several reasons:

- We opted for projects under 5 MW because this covers the majority of small projects being undertaken
- We had considerable background information about the projects
- The host communities had similar cultural and social demographics
- There was significant community interest in the research
- The timing of the projects presented favorable field opportunities (they were under development or very recently completed when we conducted the research).

From the numerous small hydro projects we have studied, we selected these two cases for analysis because of their characteristics and because the ways the projects were initiated and developed were quite different. In the case of Malari, reports indicated the project was a successful example of broad public participation and community-led development. In contrast, the Solang site was developed as a demonstration facility through the United Nations Development Project–Global Environmental Facility (UNDP-GEF) India Hilly Hydro program. Such Hilly Hydro projects are "intended to be a showcase of the latest developments in terms of technology, project execution and people's participation" (Himachal Pradesh Energy Development Agency 2000).

The interviews were scheduled and nonstructured and included residents from in and around Malari and Solang, workers who took part in project development and

Fig. 19.1 Map of Himachal Pradesh and Uttarakhand, India, showing the locations of Malari village, Chamoli district, and the Dauliganga River in Uttarakhand, Solang village, Kullu district, and the Beas River in Himachal Pradesh

subsequent operation of the facilities, and officials representing government energy and environmental approval agencies, non-government organizations (NGOs), and project proponents. Interviews were undertaken at locations convenient to the participants, and multiple interviews were carried out with some respondents as required. Interviews and field observations regarding the Malari project occurred over a 4-month period. Data concerning the Solang project were collected in numerous field seasons spanning more than a decade.

In the case of Solang, 20 individuals, representing 11 households, were interviewed during the first field season, supplemented with interviews from 13 to 11 individuals, representing 7 and 5 households, respectively, in subsequent field seasons. In addition, interviews were done with local (three) and state (two) government officials and the project proponent (two). In Malari, interviews were conducted with approximately ten residents in individual interviews. As well, groups of five or more were interviewed on six occasions. Additionally, interviews were conducted with the three people employed by the Malari hydro plant, the village headman, and two employees of the NGO Society for the Promotion of Wastelands Development (SPWD), which was directly involved in the Malari project.

Interview data were recorded in field notebooks, and observations were recorded with notes and photographs. Data sources for the document review were primarily public records such as policies, government reports, and project approval documentation. Preliminary analysis of the data began in the field and continued in the laboratory with the assistance of QSR NVivo qualitative data analysis software. Sorting and coding of the data into data segments allowed the development of families of codes or themes that were both grounded in the data and based on constructs derived from the literature.

19.3 The Solang and Malari Small Hydro Projects

The Malari and Solang facilities are both run-of-the-river projects that required the development of hydraulic structures to divert water to a small powerhouse. Each consists of a diversion and intake structure, a pipeline, a desilting chamber, a penstock, a power house composed of turbine, generator, etc., and a tail race channel (see Kumar 1996 or Badarinath 1996 for descriptions of micro-hydro engineering in India). During construction, both sites (but particularly Malari) suffered from issues of remoteness, rugged terrain, and, particularly, the lack of road infrastructure. This last issue is a general concern that makes it difficult to reach many potential small hydro sites in both states.

The Solang project is located on the Beas River, 60 km upstream from the district headquarters in Kullu. It is situated near the village of Solang on a link road from the Manali-Keylong-Leh highway. The Solang facility is a large project, with a generating capacity of up to 3 MW. A reinforced concrete weir was built on the left bank of the Beas River to divert water; the water conductor system consists of a steel pipe covering a distance of 2 km with a pipe diameter of 1.1 m. The head is approximately 80 m, and a maximum of 2.6 m^3/s of water can be removed from the river, which has a minimum flow rate of 3.5 m^3/s.

Solang itself is a small village located above 2,000 m. To reach the village one must cross over the Beas River via footbridge to the left bank and wind up a steep hill on a walking path for about 10 min. Road access now extends to the hydro site on the right bank of the river below the village. The village has approximately 20 homes that are more or less clustered together on the side of the mountain. Many families in Solang are shepherds, farmers, and employees in the tourism sector when work is available. Solang is on the local power grid, but this energy is only used in very limited ways, in part the result of cost and the availability of "free" wood. Villagers indicated power is used in their houses for running one or two lights, televisions, and, in a few cases, electric rods for heating water, but not for home heating or cooking. They also indicated that the power supply was intermittent, especially in winter "when it could be off for days."

The Malari project is much smaller than the Solang facility, having been designed to power only the homes in the isolated village, with the possibility that some energy might be made available for industrial development. According to the Detailed Project Report (DPR) (a document developed for the regulatory approval process), the Malari project features a 1,000-m rock and concrete diversion channel, approximately 400 m of steel penstock, a turbine and 50-KW generator, and a concrete tailrace channel exiting the plant and flowing through the village. In the village the tailrace channel is used as a water source for drinking and household use. Below the village the channel continues to the villagers' fields where it is used for bucket irrigation (Composite Engineers 2001). As noted earlier, it is a run-of-the-river facility, diverting water with concrete weirs approximately 50 cm in height from two mountain streams, the Tiburcha and the Tungcha Nallas. The diversion occurs immediately upstream of where the nallas come together to form the Kunti Gadhera, which flows for approximately 1 km into the Dauli Ganga. The catchment area of the diversion works is approximately 2.5 ha (Composite Engineers 2001).

Malari is located in the Niti Valley at an elevation of approximately 3,100 m. The village has intermittent road access (which is dependant on weather; the road is closed in winter), the trip is difficult, and there is only one bus per day. Villages in this area have been inhabited for around 500 years, and the locals believe that the original settlers came from Tibet; however, the literature suggests that they may be Mongol in origin from Afghanistan or Mongolia (Handa 2002). As are most high-elevation villages in the Chamoli district, Malari is inhabited only in the summer months. The villagers migrate to lower-altitude villages before winter snowfall in early November. Before 1962, many families in this region were traders, leading mule trains loaded with goods across the border into Tibet. Since the closure of India's borders with China in 1962, their two main means of sustaining livelihoods have been agriculture and animal husbandry. The main crops are beans, peas, lentils, potatoes, millet, and peppers. Goats are raised for meat and sheep are raised for shearing, and some families own cattle. There was no electricity in the village prior to the development of the micro-hydro project, and cooking and space heating is done almost exclusively with wood. The villagers had been relying on kerosene lanterns, solar-rechargeable lanterns, and candles for lighting.

19.4 Local Views of Project Implementation and Impacts

19.4.1 Project Planning Consultation

Given the widespread importance placed on public consultation, especially for UNDP-GEF-funded projects, respondents in the case of the Solang project were asked how they "first heard about the hydro project." This query garnered some interesting responses. Most people said that they found out when "the government people came and marked the land" – "2 years ago when they marked the land – that was how I found out." Others told us that they did not find out until the project was under construction: "I found out last year when they started work on the project." A third group remained unsure of the project purpose even after construction was well underway: "When I saw the construction start I was told they were building a water tank."

Given these comments it was not surprising that when asked, respondents universally indicated that they had not been consulted before the construction of the project, or after construction began, as captured in the following quotations:

> Nobody was consulted on this project. Only the power people will know what is going on.
> We knew that the project was being built, but no one came and asked us about it.
> Nobody talked to us about the project – it has just come up.
> No one came and asked us. We are not even sure where they will send the power.
> There was no consultation. The company who is building has permission from the state government. When the village president began questioning someone from the company, he was shown a permit from the government.

Even local village leaders indicated that that they had not been consulted. It seems that the only group that was eventually contacted were the landowners whose land was going to be affected by the project so that compensation could be arranged. One of these landowners complained bitterly that he had "not been consulted" and was only told of the project after his land had been marked. He also said that the government and proponent had not yet compensated him as promised despite the project nearing completion.

A much different approach to consultation was taken in the Malari case. The planning process for the hydro facility began when the SPWD, a British-funded NGO specializing in rural development, came to Malari as part of a participatory rural appraisal (PRA) development project. The people of Malari expressed their desire to have the village electrified. A design for a small hydro project was then formulated through collaboration between the SPWD and the villagers. Design of the plant was done with significant input from the villagers, in particular the late former headman. Further, all steps in the development of the plant had to be approved by the villagers of Malari at regular meetings.

Respondents described extensive collaboration between the NGO and village leaders with respect to project planning and implementation. The planning program was administered jointly between the village and the SPWD. Ideas were exchanged, and deals were negotiated. The SPWD operates using PRA tools, for which sharing of information is a cornerstone. We were told that information flowed between the village (with many of the ideas coming from the late village headman) and the SPWD project officers. The village was responsible mainly for local arrangements, while the SPWD handled the business end of the project, negotiated deals with suppliers, and obtained necessary regulatory approvals. Information was shared between village leadership and the SPWD regularly, and village meetings were held where the village population was updated on progress. Some of the comments from respondents included these:

> There was a meeting with [the NGO] in August 1999. Everyone expressed interest in going ahead with the project. The project was agreed to in the year 2000. It was approved by SPWD headquarters in Delhi. The money was then released. The total amount was 26 lakh, 10 lakh from villagers in contribution of work.

The community were willing participants.
Firstly, there was a public meeting. There were 60 persons at this meeting.
There were seven or eight meetings with SPWD prior to the selection of the micro-hydro project.

It was also noted that in the beginning there was opposition to the idea of building a hydro plant from some villagers; however, as major decisions in these villages are made by consensus, the dissenters reluctantly agreed to follow the popular opinion. None of the research participants in Malari admitted to being against the project in the early going, nor did they indicate who was.

In the beginning there were only five or six people who were pushing hydro. Later, up to 50 % were actively pushing the idea. Around 25 % of the village was not convinced that it would work, but they were convinced after the lights came on. Before that they thought they were wasting money.

The decision by dissenters to go along with the project may have been cultural:

One good thing about Malari is that meetings are thorough and argumentative, but always end in agreement – this is traditional in Bhotia management systems – community cohesiveness is very strong.

In the final analysis, villagers were involved in planning, development, and construction, and continue to be involved in the operation of the plant. While the vision and impetus for much of the project came from the late headman, there were consultations in the village and the project was endorsed in these consultations.

19.4.2 Anticipated Project Benefits

Given the wide range of benefits identified by government agencies such as the Himachal Pradesh Energy Development Agency, organizations such as the UNDP-GEF, project proponents, and the scholarly literature (Bastakoti 2006; Reddy et al. 2006), villagers were asked what they thought the benefits of the hydro projects would be. As the comments below show, the sole and almost universal benefit identified in the case of Solang was more reliable power in the winter months. The only other benefit noted was the potential for new "small industries."

The benefit would be in winter. Now the wires get blown down and the power people do not come and fix them in the snow.
The benefits about this would be in the winter – 2-3 months – for lighting.
The only benefit would be a more sure supply of power in the winter months.
If someone wants to build some industry there may be a better supply of power.
The only benefit we could get is a more sure supply of power in the winter. Now the lines go down often. If the power is closer the lines will be shorter and may not go down as often. Local people might be able to fix the lines.

Because the proponent indicated that new small-scale cottage industries were a stated benefit of the project, Solang respondents were asked specifically about this issue if they did not identify it. The majority believed there would be no change in the industrial make-up of the community. They thought this because of the location of the community and the fact that the village already had power and such industries have not yet chosen their communities for business: "People can't build mills here because the winter is too hard."

However, a few respondents did see the potential for new business in their community: "If the new power is there, more business might open, like a thrashing mill or flour mill" – "The power that we have now is fine for our needs but if we want to build something like a sawmill we will need power." Two apple growers were asked specifically about the potential to develop a cold storage facility for local apples. Neither thought that such a facility was likely:

> Nobody in the area stores apples currently. They may be stored in Delhi depending on the price available to the buyer when the apples arrive in Delhi. We don't need cold storage. Why would we need it? We just pluck apples and send them directly. Who says that we will need this here? By fall if we just keep them in a dark room they are fine. Where would the apple farmers get money to pay for a cold storage? You can see the temperature now (September), so why do we need cold storage?
>
> There is a cold storage in Patlikul 20 km south. I am not sure that the cold storage is good for the apples. They can't dry properly. I would not use cold storage.

In addition to the availability of power in the winter and the potential development of small-scale industries, Solang villagers also indicated that there were rumors of other benefits for the community, such as streetlights throughout the village and a new bridge to replace the wooden footbridge across the river. Follow-up visits to Solang village as recently as 2013, 16 years after project commission, revealed that little had changed in the village, as some participants had predicted. Villagers indicated that power has become a bit more reliable in the winter but that "it is provided by the government not the project." None of the power generated by the project comes to the community. The only physical change is the addition of eight streetlights in the community – "delivery on one of the initial promises," as one villager indicated. However, even this has been a problematic benefit, as the following quotes indicate:

> Nothing has changed in the village other than the installment of occasionally functioning streetlights.
>
> The streetlights seem to work (as in they are not broken) since they are occasionally turned on for a wedding or a festival but they are off most of the time.
>
> Of these promises, only the streetlights were fulfilled and these are only switched on if there is a wedding or a festival.

Visual inspection of the streetlights in 2013 found them in disrepair and unlikely to work if needed. A number of respondents also commented on the bridge that had been promised, indicating that the proponent had promised "Rs. 5 lakh to build a bridge across the river that never materialized." Villagers also indicated that no new small-scale industries had been developed. Villagers did indicate, however, that there was "an open relationship between Mr. Sharma (the plant developer) and the local community." It was noted that "he walks and talks with people and gives money to the local Devta."

Because villagers had no access to electricity in Malari before the development of the micro-hydro project, the main project benefit anticipated was the provision of power to each household. However, some villagers did not believe this would occur, as already noted. In addition to power, two other main benefits were anticipated from the project: irrigation of crops and the development of a berry processing plant for managing locally collected berries. At the time of the research, 80 % of village homes had an electrical connection and consumers were paying Rs. 20 per household per month for power: "Power in Malari costs Rs. 20 per month per bulb. The money collected goes to maintenance. The money is enough to cover operations. 80 houses have electricity."

The first year of operations revealed, however, that the plant was not economically viable and that its operations had to be subsidized with money that the village received from leasing land to the army. The plant paid its two operators Rs. 2,600 per month each (for a total cost of Rs. 5,200), and took in around Rs. 2,400 per month in consumer fees. It was running a deficit of Rs. 2,800 per month, validating the suggestion in some of the literature that many of small hydro plants are not economically sustainable on their own (Alternate Hydro Energy Centre 2003; Kumar, 17 November 2004, Personal communication with Dr. Arun Kumar, Director AHEC-IIT, Roorkee).

Villagers were using power only for a light bulb or two in their homes. Malari residents also spoke to us about the psychological benefit of having electricity in their homes. They noted in particular that they no longer had to spend their evenings in the dim light of candles and kerosene lanterns because they now had bright light around which to do things at home in the evening: "The light has given us much confidence, we do not feel confined by darkness at night, and we are able to work at night doing various things, such as making rugs, spinning wool, making clothing, etc."

The people of Malari also noted that they were very proud of their plant, and it was the envy of the other villages in the Niti Valley, especially among those villages that do not have electricity. This instilled a sense of pride and empowerment in the community: "The micro hydro has given us a sense of confidence, we feel that if we work collectively we can achieve big things."

The plant also had the added benefit of providing irrigation in the community, realizing another anticipated benefit. As noted earlier, the tailrace from the plant was directed into a channel flowing to the fields, where it was used for bucket irrigation. The vision of developing a berry processing plant, had not, however, been realized. This gap is important because the berry plant was meant to help offset the costs of running the hydro facility. It was also supposed to address the issue of using low off-peak loads, as the plant was planned to run during the day when the power generated was not being used for lighting.

19.4.3 Environmental Effects

Local comments about the potential environmental effects of the projects generally agree with the statements of project proponents and government agencies to the effect that "not much damage is being done." In the case of the Solang facility, most respondents felt that the project was having little effect if any on the local environment. The most common concern expressed was about the removal of trees, although some respondents also noted that blasting and walking paths affected field crops:– "I think that they have cut down lots of small trees" – "Till now there has not been much impact on the environment – just lost trees" – "To this point there have been no real bad things, just the cutting of trees" – "Some trees have been damaged, but we don't know the owner of the project." It is interesting to note that one respondent thought that people were too busy to notice: "Lots of people are in tourism and are too busy to notice any damage from the construction activities."

As indicated at the outset, the Solang hydro project was promoted on the positive environmental benefits it was going to generate, primarily through getting people to stop using wood for heating and cooking. Villagers were asked about both their fuel wood consumption needs and the likelihood of electric power replacing their current fuel wood uses. All respondents indicated that they used wood to heat their homes and cook their food, as reflected in the following comments:

> Wood is a necessary thing for heating in the winter.
> Mostly we use wood to cook with – some gas; 1 cylinder last year – but not power.
> Due to the extreme cold in this area we must use wood for heating.
> For six months we need wood. Wood is the necessary thing for heating.
> Without wood we would not be able to stay in this area.

Respondents in Solang were universal in their rejection of using hydropower instead of wood for heating and cooking. Most indicated that they already had hydro and did not use it for these purposes and they would not switch even if the hydropower were more reliable. Generally hydropower is not currently used for heating and cooking because of the "high cost" of the electricity. Most respondents were "fearful" of these costs going higher: "the power is much more expensive this year." Some respondents did note that they could see hydro replacing gas for those who had already made the switch from wood to gas for some of their cooking needs, if the hydro were cheaper than the gas. There was 100 % agreement, however, that "hydro would not replace wood for heating." "This is not a plains area where we can use a heater. My family could not sit and stay warm by an electric heater."

When asked under what conditions people might use hydro for heating, all indicated that if the hydro were free or of lower cost than other options, they might consider using it: "Maybe, if it were free, some people would use hydro for cooking and heating." But even if free, some thought that hydro would not replace wood for heating: "Even if the hydro were free it would be very hard to heat our homes." Solang village has a forest or "jungle" area that is part of the village commons and is used for firewood collection. All viewed wood as a free resource and did not place a value on the time that it takes to collect this wood. Even the female respondents did not see value in the potential time gain that would be created if they did not have to collect wood.

> There are lots of rotten and fallen trees around here in the forest and we would still use this wood for heating and cooking. The income of people here is low and most people can not afford to pay electricity bills for cooking and heating, so they would still use wood because it is free.

> We need a stove to stay warm and wood is easy to get. Just look at the forest all around us – wood is easy.
> Very few people buy wood, only people in towns buy wood – village people never buy wood.
> Right now people do not have much work so they are bringing wood and they get it for free. Maybe in the future they will use electricity for something else or if there are no trees.

Given the small size of the Malari plant, villagers had almost no concern about the environmental impact of the project. During the planning of the project a lot of attention was paid to its location because of landslide hazards in the area, so a site was selected that the villagers believed was safe from landslides and avalanches. They have found though that some minor repairs to the concrete diversion channel are required periodically. They noted to us that the plant is run-of-the-river, and as such the water is drawn from high mountain streams and only a portion of the flow is removed. As well, they told us that there are no aquatic communities being adversely impacted, as these streams are intermittently flowing and the flows are diverted approximately 100 m above where they would enter a larger watercourse.

Stream flows are an issue for plant operation, and these may become more problematic with climate variability. Since the plant was first put into operation in 2003, its operation has been sidelined by low flows in the two mountain streams from which water is drawn (Prasad and Prasad 2007). These low flows are experienced in the late summer and fall. There are plans to construct a diversion channel from a third stream to provide sufficient flow; this would involve constructing approximately 100 m of additional concrete channeling to the third stream, and a concrete weir similar to the two existing ones. The weir would be stone and concrete, approximately 50 cm high and 5 m wide.

It was anticipated that there would be environmental benefits from the project in the form of reduced greenhouse gas emissions from reduced combustion of wood and fossil fuels. These benefits had been partially realized. Residents with hydro no longer used kerosene lanterns, but continued to cook over wood fires and with kerosene stoves, often indoors. Malari villagers estimated that the 80 or so families with electricity had reduced their monthly kerosene use from 4 or 5 l per month to 1 or 2 l per month, for a typical reduction of 3 l per month per family.

19.4.4 Local Jobs

As with most proposed development projects, government agencies and project proponents predicted that "many local jobs" would be generated through the construction and operation of the Solang and Malari facilities. With Solang, more than 150 skilled and unskilled workers were required to build the project. Most respondents indicated that "All the jobs went to outsiders." A few respondents were aware that at least one local person had received a job as a "chaukador" (watchman): "Only one job went to a local, the rest are all outsiders." The proponents of the Solang project indicated that "The local people do not want to work as laborers." For the most part, the villagers confirmed this and attributed it to the tourism jobs available in the district:

> Everybody is doing tourism.
> There were laborer jobs offered but no one wants to work as a laborer because it only pays Rs. 70 a day, and they have their own work to do. Many are involved in tourism and make more money – like Rs. 500-600 a day.
> We have a festival after one week, and then there is not too much work. Otherwise people are busy with tourists and make more money.
> If they had asked we might have worked in the off-season, but it is very hard work.

The plant in Malari, on the other hand, was constructed entirely using local labor. Outsiders were only brought in for highly specialized tasks, such as for the final installation of the generator and installation of the control panel. Villagers did the work as a communal effort. They were given nominal compensation by the village for their contributions during construction and their efforts offset what would have otherwise been project costs that would have been difficult for the SPWD to cover: in other words, it was the village's contribution to the project. A reason why community engagement in construction was different between the Malari and Solang plants is that in Malari the villagers were better able to contribute their time because they generally have little in the way of employment opportunities.

> I helped in building the channel. Everybody was a part, everybody was involved in the rotation; someday some people, some days others.
> The community was a willing participant in the construction – everyone was a part of it, including myself.
> The work included building and hauling materials over landslides on the road.

In addition, an elected committee of villagers administered the plant in Malari, and the plant operators and manager were community members. The general manager was a local shopkeeper, and the two plant operators were local village men who were selected and hired by the management committee: there was 100 % civic involvement in the plant's operation. Training for the operators was organized by the SPWD. As part of the training, the operators were sent to see a similar community micro-hydro project in Nepal, where they observed operations and learned basic repairs and maintenance. Since the Malari plant was inaugurated, there has been no outsider involvement, and the SPWD has had no role in the operation of the plant. The village handles all plant operations: "Two people work full time on the plant: Mr. [name withheld] and Mr. [name withheld]. Mr. [name withheld] is the president of the Malari micro-hydro."

19.5 Discussion and Implications

The Malari example confirms how effective and equitable micro-hydro projects can be when they are community led, when there is sufficient community capacity to support such an initiative, and when the socioeconomic circumstances are suitable. In this case, although the SPWD facilitated the project's development, it did not become involved in Malari with the intent of seeing a micro-hydro project built. The very idea for the project resulted from a generic PRA process. That is, the hydro project idea was, at least in part, a normative expression of a community-based aspiration for the village. In contrast, the Solang project followed an externally driven development model within which the local people had little or no involvement in the vision for the project or its normative planning phase. Rather, an outside entrepreneur recognized the opportunity for development. In fact, the results show that there seemed to be few requirements put on the project proponent to consult with local people, beyond the people whose land was needed for the project. There was also little local involvement even in the operational phase of development despite this facility being a demonstration project meant to highlight good public participation.

As well, in the case of Malari, although the SPWD facilitated the project's development, the leadership, decision making, and human resource capacity in the village were essential ingredients in the project's success. First, the village headman was highly supportive of the project and had considerable authority and legitimacy in the community. Second, the consensus-oriented decision making the model used in the village reduced overt conflict and helped sway dissenters in the community. Third, the local labor pool was sufficiently skilled and otherwise up to the task of fulfilling a major role during construction and operation. On this last point it should not be overlooked that a significant reason why there was such extensive local involvement in construction and operation (and even planning) of the plant and why the village enjoyed such significant economic benefits was because the socioeconomic circumstances were suitable. Plainly put, a high proportion of villagers needed and wanted local employment. In contrast to Solang, there were significantly fewer opportunities to make a living in tourism or other economic sectors.

Another notable point of contrast with Solang relates to the size of the two projects. The Solang plant was considerably larger, and in some respects this difference makes it difficult to compare the two facilities. At the same time, the difference in scale and its attendant financial, technical, and political complications highlight the challenges in taking a community-led approach to developing a project such as that built in Solang. Taking a ground-up approach in Solang would have been exponentially more challenging than in Malari. The externally driven development model witnessed there, along with its flawed community engagement process and its resulting imbalance in the distribution of costs and benefits, is much more typical of projects of Solang's size. Experiences similar to Solang's can be seen in numerous small hydro projects and other developments in the rural mountain regions of India (Dutta 2009; Saxeena 2007; McCandless 2007; Sinclair 2003). These experiences point to another challenge for continued energy development in Himachal Pradesh and Uttarakhand, namely, how to proceed with small hydro projects in a manner that includes meaningful community engagement and an equitable distribution of costs and benefits, while still achieving national and state goals for power generation and economic development.

The reasons for constructing the two generating facilities also differ. The Solang plant was constructed to meet the growing energy needs in Himachal Pradesh by increasing the load available on the electrical grid, despite what the locals might have understood for the project goals, whereas the Malari plant was constructed to provide electrical energy for the exclusive use of the previously unelectrified village of Malari. Both projects were successful in their ultimate objectives, but they differ significantly when the distribution of positive and negative impacts is considered. In Solang, the benefits were mainly accrued by the project proponent who had more reliable power available for his steel manufacturing plant, through supplementing the grid, while Solang residents bore the brunt of the negative impacts. Malari had few negative impacts on local residents. As well, the benefits were accrued at the local level, in terms of better water supply for domestic and agriculture use and, most significantly, energy.

One of the main challenges for continued energy development in Himachal Pradesh and Uttarakhand is to ensure that negative and positive impacts are adequately balanced. The impacts of hydro development are primarily at the local level, while the benefits can be realized at any scale: at the local level in the case of Malari and at the regional level in the case of grid-connected projects such as Solang. To properly understand and address potential negative impacts and amplify positive impacts, it is critical that a proper environmental assessment (EA), including a publically available DPR, be completed. Both the projects we considered appeared to suffer from data gaps and process flaws in the pre-approval assessment stage. In the case of Solang, even limited pre-consultation and site visits would have revealed that the likelihood of the demonstration project fulfilling the goals of UNDP-GEF were quite slim. Local residents made it quite clear that the only benefit they saw from the project was more reliable electric power in the winter. Although this is obviously very important to the people affected (almost everyone mentioned it), it was not one of the stated goals of any of the agencies involved. Further, local people made it quite clear that under the current circumstances they are not going to reduce their use of wood, which is a basic necessity for the people in the region. In a similar way, although the DPR and pre-approval processes for Malari were more complete, a finer assessment and better consultations would have revealed the likelihood of an insufficient water supply for the plant and how unlikely was the development of a berry plant.

Given the pace of hydro development in both these regions, it also critical for decision makers to start to consider the cumulative effects of these projects as part of the pre-approval assessment. Local respondents expressed concern about the number of projects "coming up" in the Kullu valley, where the Solang project is located, and the potential synergistic effects these might have on local streams and rivers. In fact, one respondent noted that the government should be "keeping track of the effects of these projects and informing the people." It was also noted that often the biggest benefit of small hydro is the avoidance of large-scale hydro development. However, both states are aggressively pursuing hydro development of all sizes, as already noted. This means small hydro development is occurring as well as large – the small is not displacing the large, so the impacts of hydro development are intensified. Soon this part of the Himalaya may be like parts of Europe where virtually every mountain river is affected by hydro development (Tuffer et al. 2001).

It is also apparent that government agencies need to review their policies on public participation for all types of hydro projects. The results presented here, as well as research on other projects we have considered (McCandless 2007; Sinclair 2003), reveal serious deficiencies in public participation processes for projects that are not community based as is Malari. Also, even with community-based projects, significant efforts need to be made to be inclusive, as the Malari findings show. These findings are supported by others (e.g., Dutta 2009; Diduck et al. 2007; Paliwal 2006; Sinclair and Diduck 2000), which show that despite legal mechanisms for participation, such as those found in EA legislation, there is often, on the ground, a lack of actual involvement because of poor information dissemination and documentation, weak hearing processes, and lack of integration of consultation results with formal planning and approvals. There is a dire need for opportunities for earlier (at the site approval stage), more decentralized (e.g., a greater role for village panchayats), and more active (e.g., engagement in monitoring) public participation in hydro development decisions.

In conclusion, despite the pace of hydro development, and the desire on the part of the Himachal and Uttarakhand governments to encourage more such development, there is still opportunity to change the direction of policy in the way we have outlined here. In the years ahead, many new projects will be commissioned and undertaken, and these should have the benefit of more thorough pre-approval assessment and more meaningful public participation. There are many agencies, both governmental and nongovernmental, in the Himalayan regions of these states that have studied environmental approval and public consultation and understand these issues and the need for policy change, but they need a clear vision at the political level and adequate resources to make this change in direction a reality; these include organizations such as The Himachal Pradesh Science and Technology Department, the EIA cells of the Wildlife Institute of India, the G.B. Pant Institute (active in both states), and many other NGOs. Such changes in policy will, however, require local pressure to encourage political will.

Acknowledgments We want to gratefully acknowledge the research participants for making the time to meet with us, the Shastri-Indo Canadian Institute and the Social Sciences and Humanities Research Council of Canada for their financial support, and Weldon Hiebert, Cartographer in the of the University of Winnipeg Department of Geography, for preparing Fig. 19.1.

References

Agrawal DK, Lodhi MS, Panwar S (2010) Are EIA studies sufficient for projected hydropower development in the Indian Himalayan region? Curr Sci 98(2):154–161

Alternate Hydro Energy Centre (AHEC) (2003) Small hydro power: initiatives and private sector participation, 3rd edn. Indian Institute of Technology (IIT), Roorkee

Asian Development Bank (2005) Uttaranchal power sector project in India, summary environmental impact assessment. Asian Development Bank, Manila

Badarinath HS (1996) Design of small hydropower projects: hydraulic structures and water conductor system. In: Varma CVJ, Rao ARG (eds) International course on small hydro power. International Association for Small Hydro, New Delhi, pp 109–145

Bastakoti BP (2006) The electricity–livelihood nexus: some highlights from the Andhikhola Hydroelectric and Rural Electrification Centre (AHREC). Energy Sustain Dev 10(3):26–35

Bhattacharya SC, Jana C (2009) Renewable energy in India: historical developments and prospects. Energy 34(8):981–991

Central Electricity Authority (2004) Hydro power development: executive summary of preliminary feasibility reports of hydroelectric schemes. Ministry of Power, Government of India, New Delhi

Central Electricity Authority (2010) Status of 50,000 MW hydro electric initiative 2010 [cited March 13 2010]. Available from http://www.cea.nic.in/hydro/Status%20of%2050,000%20MW%20Hydro%20Electric%20Initiative/Status%20of%2050000 %20 MW%20 HydroElectric%20Initiative.pdf

Composite Engineers (2001) Detailed project report of Malari Micro Hydel Project, capacity 50 kW. Composite Engineers, Dehra Dun

Diduck AP, Sinclair AJ, Pratap D, Hostetler G (2007) Achieving meaningful public participation in the environmental assessment of hydro development: case studies from Chamoli District, Uttarakhand, India. Impact Assess Proj Apprais 25(3):219–231

Dutta R (2009) Environmental democracy in the Himalayas: an assessment of access to justice, public participation and access to information in Himachal Pradesh and Uttarakhand. The Access Initiative Himalayan Coalition, New Delhi

Handa OC (2002) History of Uttaranchal. Indus Publishing Company, New Delhi

Himachal Pradesh Energy Development Agency (HIMURJA) (2000) Status note on small hydro power development programme up to 3 MW capacity in Himachal Pradesh. H.P. Energy Development Agency, Shimla

Himachal Pradesh Energy Development Agency (2010) Small hydro development programme in Himachal Pradesh. Government of Himachal Pradesh Energy Development Agency, undated [cited March 15 2010]. Available from http://himurja.nic.in/smallhydro.html

Himachal Pradesh State Electricity Board (2010) Highlights of investor friendly new hydro policy on privatization. Government of Himachal Pradesh State Electricity Board undated [cited March 15 2010]. Available from http://www.hpseb.com/hydro_policy.htm

Joshi MC (2007) Hydro power potential in Uttarakhand. In: International conference on small hydropower. International Centre for Hydropower, Norway, Kandy, Sri Lanka

Kumar A (1996) Small hydro power–civil works. International course on small hydro development. In: Varma CVJ, Rao ARG (eds) International association for small hydro. International Association for Small Hydro, New Delhi, pp 158–175

McCandless M (2007) Community involvement in the development of small hydro in Uttaranchal, India. Unpublished masters thesis. Natural Resources Institute, University of Manitoba, Canada

Ministry of New and Renewable Energy (2010) Small hydro power programme. Government of India Ministry of New and Renewable Energy undated [cited March 15 2010]. Available from http://mnes.nic.in/prog-smallhydro.htm

Paliwal R (2006) EIA practice in India and its evaluation using SWOT analysis. Environ Impact Assess Rev 26(5):492–510

Prasad J, Prasad Shiva HJ (2007) Performance evaluation of existing mini hydro power projects of Uttarakand: a case study. In: International conference on small hydropower – Hydro Sri Lanka, 22–24 October 2007. Available: http://ahec.org.in/links/International%20conference%20on%20SHP%20Kandy%20Srilanka%20All%20Details%5CPapers%5CTechnical%20Aspects-A%5CA31.pdf

Reddy VR, Uitto JI, Frans DR, Matin N (2006) Achieving global environmental benefits through local development of clean energy? The case of small hilly hydel in India. Energy Policy 34(18):4069–4080

Saxeena P (2007) Small hydro development in India. In: International conference on small hydropower. International Centre for Hydropower, Norway, Kandy

Sinclair AJ (2003) Assessing the impacts of micro-hydro development in the Kullu District, Himachal Pradesh, India. Mt Res Dev 23(1):11–13

Sinclair AJ, Diduck AP (2000) Public involvement in environmental impact assessment: a case study of hydro development in Kullu District, Himachal Pradesh, India. Impact Assess Proj Apprais 18(1):63–75

Tanwar N (2008) Clean development mechanism and off-grid small-scale hydropower projects: evaluation of additionality. Energy Policy 36(1):2000–2015

Tuffer B, Markard J, Bratrich C, Wehrli B (2001) Green electricity for alpine hydropower plants. Mt Res Dev 21(2):19–24

Uttarakhand Jal Vidyut Nigam Limited (2010) Policy on private sector investment in hydropower project above 100 MW capacity. Uttarakhand Jal Vidyut Nigam Limited undated [cited March 15 2010]. Available from http://www.uttarakhandjalvidyut.com/sitemap.htm

Water for Welfare Secretariat (2008a) Government of India policy on hydro power development. In: Water for Welfare Secretariat (ed) Hydro power policies and guidelines. Indian Institute of Technology, Roorkee, Roorkee

Water for Welfare Secretariat (2008b) Government of Uttarakhand policy for harnessing renewable energy sources in Uttarakhand with private sector/community participation. In: Water for Welfare Secretariat (ed) Hydro power policies and guidelines. Indian Institute of Technology, Roorkee, Roorkee

Chapter 20
Floodplain Management in Rapti River Basin

Narendra Kumar Rana

Abstract Floodplains are resources of immense value. They are the lowland adjoining the channels of rivers, streams, and other watercourses or the shorelines of oceans, lakes, or other bodies of standing water. The common framework for floodplain management comprises four general types of activities: modifying flooding, modifying susceptibility to flooding, modifying the impact of flooding, and preserving the natural and beneficial functions of floodplains. Given this overview, this chapter is a study of the floodplain management issues of Rapti River, flowing through eastern Uttar Pradesh. Regular floods in general and unprecedented floods in known pockets are common features of the Rapti River Basin. The study suggests the utilization of performance indicators to evaluate the progress and success of floodplain management programs. A decentralized resource allocation at the local level with standardization of terminology, collection of appropriate data concerning the socioeconomic profile of the flood-prone community, and flood behavior is necessary for efficient flood management.

Keywords Flood plains • Flood plain management • Rapti river basin • Planning • Uttar Pradesh (India)

20.1 Introduction

Floodplains are resources of immense value. Since the beginning of civilization, floodplains have been favored locations for living. The rivers provide flat and fertile land for crops, abundant underground as well as surface water, a natural site for defense, and easy routes for trade and communication. Thus, intensive use of floodplains is a common development strategy often chosen by society. Many important

N.K. Rana (✉)
Department of Geography, D.D.U. Gorakhpur University, Gorakhpur, Uttar Pradesh, India
e-mail: nkrana.in@gmail.com

A.K. Dutt et al. (eds.), *Spatial Diversity and Dynamics in Resources and Urban Development*, DOI 10.1007/978-94-017-9771-9_20

cities in Asia, Europe, and America have been established on the floodplains. Delhi, Varanasi, Allahabad, Kolkata, Chennai, and Cuttack are a few examples in India. The intensive use of floodplains for commercial and residential purposes has undoubtedly contributed to the economic success and sustainability of larger parts of India. However, increasing population pressure along with encroachment of marginal floodplains is now putting those residential and commercial activities at severe risk of flood damage. Major floodplains along the Rivers Ganga, Yamuna, Brahmaputra, and their tributaries in the north and notable peninsular rivers in the south are known for their productive and carrying capacity. They have also their distinct history of flooding and accompanying human misery in terms of loss of lives and property. Despite all these drawbacks, modern society places extreme demand on those resources. As a result there is a growing debate among policy makers, planners, and researchers that floods and floodplains now need to be managed because the floodplains are areas of primary environmental significance and their well-being is essential to the survival of many ecosystems.

20.2 Flooding and Floodplains

Floodplains are the lowlands adjoining the channels of rivers, streams, and other watercourses or the shorelines of oceans, lakes, or other bodies of standing water. These are lands that have been or may be inundated by flood water. Floodplains develop on the lower course of the rivers. From time to time excess rainfall or snow-melt cause rivers to overflow their banks and reclaim the floodplains. It is an integral part of the river system, even though the river uses it occasionally to pass down flood flows. Floods are critical factors in the health of the floodplain itself. Whenever the floodplain is free from floodwater, it is beneficially used as a part of the land system for diverse economic activities.

Depending upon the objective of the study, floodplains may be defined and mapped in different ways. Geomorphologists define floodplains as an area inundated by flood events of a particular magnitude and frequencies; to planners and lawyers, it may be an area defined by statute. From the aspect of risk assessment and hazard problems, floodplains often are categorized and termed as floodways, flood fringe, probable maximum flood-prone area (PMF), and defined flood area (DFA) by disaster managers. A particularly useful distinction on many floodplains from the point of view of hazard problems is that between the floodway and the floodway fringe. The floodway is the area of the floodplain, usually marginal to the main channel, in which land filling or flow concentration as a result of construction would significantly increase flood level. The floodway fringe area lies adjacent to the floodway and is where water is stored during flood. The distinction is formally recognized by the U.S. Army Corps of Engineers and other agencies in the United States (U.S.) for flood zoning purposes (Fig. 20.1). Similarly, the best practice principles and guidelines of floodplain management in Australia define floodplains in terms of the probable maximum flood (PMF). The area defined by a PMF event

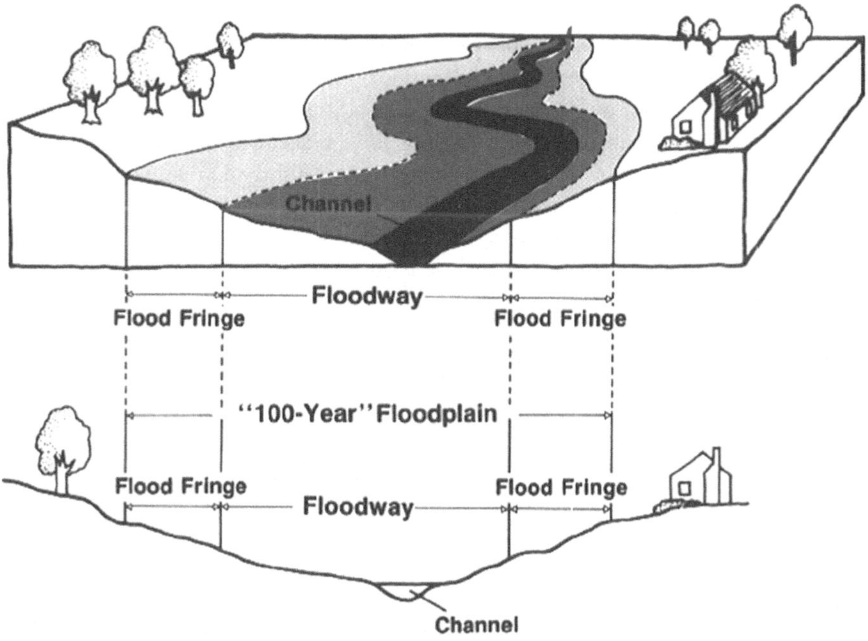

Fig. 20.1 The floodplain with floodway (From FEMA 1992, p. 8)

is flood prone. Land outside the PMF is truly flood free. However, for practical feasibility it is allocated for defined flood events on which planning and development controls are based (Fig. 20.1).

20.3 Floodplain Management

Parker (2000) and Smith (2013) have given a brief description of the hazard paradigm. For much of the twentieth century, floods and other hazards were primarily viewed as natural phenomena. This approach usually focuses upon the physical agents in the natural and human-modified environment that pose a threat to society. It is an agent-specific approach through which we view "the environment as hazard" (Burton et al. 1978). Here storm, precipitation, and runoff processes that generate a flood become the principal focus of attention. This approach has been increasingly criticized during the latter part of the twentieth century and another approach has evolved in opposition to the dominant hazard paradigm. The main competing approach focuses on flood hazard as a social phenomena. This approach is based upon the view that social, economic, and political conditions, and the variable quality of material life, are overwhelmingly important in the incidence and distribution of damage in disasters, including flood disasters (Parker 2000, p. 10). Thus, flood disasters are viewed not as natural phenomena but as being produced by society. Central to this social agent approach is the concept of vulnerability. This approach was reinforced by the works of Hewitt (1983) and Watts (1983). Different studies show that social agents may generate vulnerability to flood disasters in various forms: through physical exposure to floods as a result of occupying floodplains; through living in a dwelling having little resistance to flood; or through weaknesses that come with gender, age, health, etc. A third approach, known as the environmental approach to flood hazards and disasters, builds upon both of these approaches. The environmental approach is based on the view that both social and physical environments influence the creation of flood hazards and disasters.

Based upon these approaches, the common framework within which floodplain management operates comprises four general types of activities: modifying flooding, modifying susceptibility to flooding, modifying the impact of flooding, and preserving the natural and beneficial functions of floodplains (Federal Interagency Floodplain Management Task Force 1994).

Modifying flooding refers to structural flood control measures that aim to keep the flood water away from settlements. The purpose of this measure is to control flood water movement to keep it from doing damage to the built environment: this is popularly known as the "levees only" approach, which is still regarded as the dominant form of structural measures. Other forms of "modifying flooding" include dams and reservoirs, floodwalls, channel alteration, and on-site detention ponds.

The second approach, that of modifying susceptibility to flooding, connotes a set of activities designed to keep people and the built environment away from the flood hazard. Commonly known as nonstructural mitigation measures, these include floodplain zoning, detection, and warning, floodproofing, building codes, growth management policies, and community awareness programs through education. Floodplain zoning is a regulatory tool used to control type and development of land use. Zoning ordinances can prohibit new development that is not suitable for a particular type of land use. It also prohibits the development of critical facilities such as hospitals, schools, and hazardous industries in hazard-prone areas. Flood forecasting and warning systems can also reduce a community's susceptibility to flooding. Given proper warning with adequate time, there are several options that the communities can take to minimize their loss in a flood. Building codes can be designed and apply to ensure that new construction built in hazardous areas is resistant to damage.

The third type of activities that constitute floodplain management are those that modify the impact of flooding on people by sharing the financial losses caused by flood damage. Flood insurance, tax adjustment, and disaster relief are some of the ways in which monetary loss from floods is spread to a much larger population. The fourth set of floodplain management activities are those that seek to preserve the natural and beneficial functions of floodplains from an ecological perspective. The main aims of this approach are to reduce flood losses and to preserve and restore the natural resource and functions of these lands (Federal Interagency Floodplain Management Task Force 1994). It is recognized that floodplains do more than simply provide places to hold excess floodwaters: they can also help maintain groundwater supply and quality, provide fish and wildlife habitat, offer productive agricultural lands, and provide places for recreational uses and open space (Myers and Passerini 2000, p. 245).

20.4 Rapti River Basin

The River Rapti flows in the subhumid to humid monsoon region of the middle Ganga Plain. It is the largest tributary of the River Ghaghara, which in turn is a major constituent of the Ganga. The Rapti River Basin extends from 26°18′00″N to 28°33′06″N and 81°33′00″E to 83°45′06″E and covers an area of 25,793 km^2, of which 44 % (11,380 km^2) lies in Nepal and 58 % (14,413 km^2) in Eastern Uttar Pradesh (Fig. 20.2). The Rapti River flows through the districts of Rukum, Salyan, Rolpa, Gurmi, Arghakhanchi, Dang, and Bank of Nepal territory and the Bahraich, Shrawasti, Balrampur, Siddharthnagar, Santkabirnagar, Gorakhpur, and Deoria districts of Eastern Uttar Pradesh. It rises at an elevation of 3,048 m in the Dregaunra

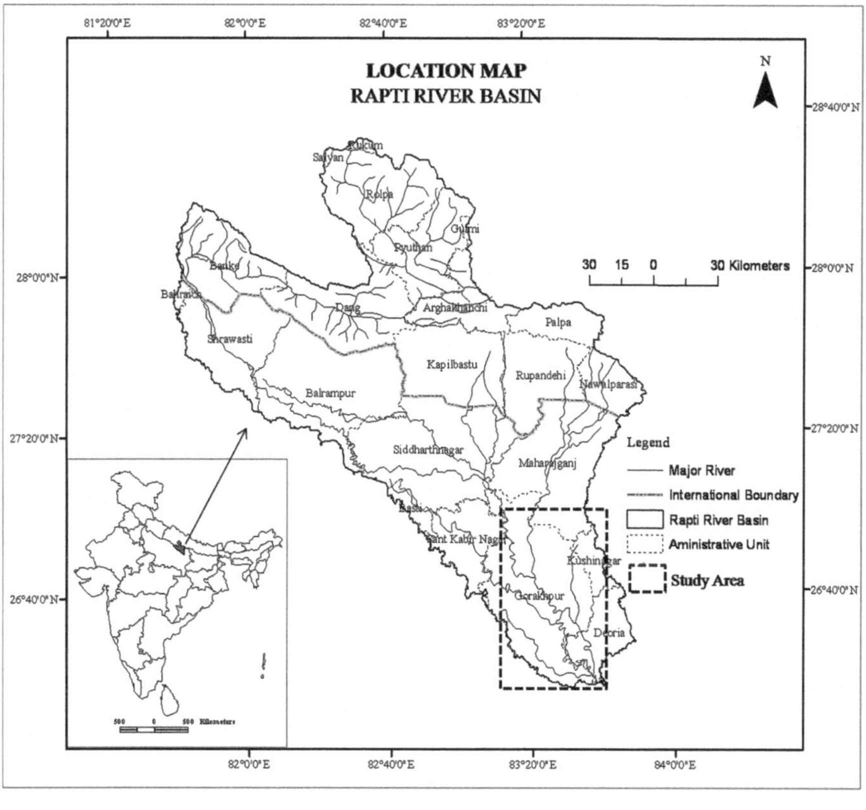

Fig. 20.2 The Rapti River Basin

range of Nepal Siwalik and covers a total distance of 782 km, of which 331 km lies in Nepal, before joining the Ghaghara at Barhaj in Deoria District of Eastern Uttar Pradesh. River Rapti is fed by numerous tributaries and affluent. Those of the northern or left bank originated from Siwalik and the *Bhabar* region; those on the south represent merely old beds of the river. Important left bank tributaries on the Rapti are Burhi Rapti, Ghonghi, Kain, and Rohin. The Bhakla, Ami, and Taraina are the notable right bank tributaries (Fig. 20.2). The basin consists geologically of two distinct portions: structurally it is a segment of the great Indo-Gangetic trough and it has also some marginal portion of the Himalaya's foothills region of the Siwalik. The Indo-Gangetic trough portion consists entirely of the alluvium, a composition of sand, silt, and clay in varying proportions. In respect of their geologic age, these deposits correspond with two main divisions of the quaternary era, the Pleistocene and the Recent. The alluvium is found in two broad groups: the older alluvium known as "*Bhangar*," the age of which is estimated as Middle Pleistocene, and the newer alluvium, *Khadar*, which is more recent and undergoing formation by the aggregational work of the rivers. Recent findings of the chrono-association of the Gandak megafans areas revealed that the alluvia of the Old Rapti Plain (Burhi Rapti) is that of 5,000 years before present (b.p.) and that of Rapti is more than 500 b.p. (Mohindra et al. 1992, p. 651).

The Rapti River Basin can be divided into two sub-catchments, namely, Lower Rapti and Upper Rapti (Central Ground Water Board 1984). Subsequently, both the sub-catchments can be divided into a number of watersheds (Fig. 20.2). The catchment areas of these watersheds vary from as small as 42,000 ha (LB-Ghaghara) to 145,000 ha (Ami) (Table 20.1).

Table 20.1 Watershed divisions of Rapti River Basin

Sub-catchment	Watershed number	Name of stream	Area (thousands of hectares, ha)	District covered
Lower Rapti (C) (682)	C1	LB-Ghaghara	42	Deoria, Saharsa
	C2	Taraina	55	Gorakhpur
	C3	Gaura	144	Gorakhpur, Deoria
	C4	Lower Rapti	123	Gorakhpur, Basti
	C5	Rohin	112	Gorakhpur
	C6	Chillua	91	Gorakhpur
	C7	Ami	145	Gorakhpur
Upper Rapti (D) (732)	D1	Main	63	Basti, Gorakhpur
	D2	Rapti	84	Gonad, Basti, Bahraich
	D3	Bhakla	93	Bahraich
	D4	Ghonghi	103	Basti, Gorakhpur
	D5	Burhi Rapti	88	Basti
	D6	Bhainbar	92	Gonda, Basti
	D7	Pera, Bahuwa	120	Gonad
	D8	Kain, Gholia	90	Bahraich, Gonda

Source: Department of Agriculture and Cooperation (1988): *Watershed Atlas of India*

20.5 Flood Hazards and Risk in the Floodplains of the Rapti River Basin

Regular floods in general and unprecedented floods in known pockets are common features of the Rapti River Basin. Every year, devastating floods leave several lakh hectares of cropland inundated and that many villages were either marooned or washed away. The floods were largely caused by the River Rapti and the Ghaghara besides other minor river systems such as Rohin, Ami, Burhi Rapti, Ghonghi, Kain, Bhakla, and Taraina. Heavy rainfall accompanied with breaching of embankments, man-made breaches to allow the release of excess water at some locations, coupled with main channel migration in some places are the main causes of devastating floods. The recorded history of floods in the basin dates back to the nineteenth century A.D. The available literature on past floods indicates that the basin has witnessed massive floods once or twice in every 50 years (Table 20.2). The basin witnessed the most devastating floods of the twentieth century in the year 1998, which surpassed all the earlier records. Almost all the notable rivers, namely, Rohin, Ami, Gorra, Kuwano, and Rapti, along with Ghaghara, exceeded their maximum flood level at once. The basin witnessed two successive floods in 1998. The first phase started on July 22–July 27/28, 1998. In the second phase, which started on August 20, 1998, the water level continued to increase and reached a peak at 77.54 m, that is, relatively more than 2.50 m from the danger level and 0.70 m more than the previously recorded highest flood level.

In addition to this, the Rivers Rohin, Ami, and Gorra also exceeded their recorded highest flood level. Also, notable small tributaries (called 'Nala' in local language) such as Faren Nala, Kalan Nala, Gobrahia Nala, Bar Nala, Taraina Nala, and Pon Nala, and major wetlands (called Tal in local langrage) such as Ramgarh Tal, Bakhira Tal, and Chillua Tal, inundated adjacent villages. Gauge level data revealed that within a period of 60 days the River Rapti showed the maximum number of peak flows and a quick or short response time. Within that time it continued to flow between the danger level and the highest recorded flood level of 1974. Surprisingly, the severity of the flood situation was confined to the lower hydrological regime of the Rapti basin. From the reported loss of life and properties damaged, the worst affected area of the flood was identified as the lower Rapti basin, comprising the districts of Gorakhpur, Deoria, Maharajganj, and Kushinagar; this designation was also verified from the satellite imagery taken during the time of floods. The magnitude of the catastrophe was exceptionally high in Gorakhpur division. The probable causes of the flood were identified as heavy rainfall and embankment failures followed by inadequate relief and rescue operations.

The southwest monsoon is the usual time of high floods in the basin. The general time for the onset of monsoons is June 17 and their retreat is as late as September 25, but sometimes this deviates by 30 days. The rainfall is not uniformly distributed in this period; there may be more than one intensive rainfall period. This period is found sometimes too early and sometimes too late. Therefore, floods in the basin occur as early as the first week of July and as late as the third week of September. The normal flood period is August 15 to September 15, which includes 70–80 % of

Table 20.2 Characteristics of flood hazard in Rapti River Basin

Sl. number	Time period	No. of recorded floods/year	Characteristics
1	1800–1850	2	Severe floods in the years 1823 and 1839
			Maximum damage to Ghaghara-Rapti Doab was caused by floods of 1823
			Floods in the year 1839 were caused by sudden rise in water levels of the Rivers Ghaghara, Rapti, and Ami; Gorakhpur City remained as an island
2	1850–1900	4	Floods of 1873 caused maximum damage to agricultural lands and settlements
		1871, 1873, 1889, 1892	Floods of 1892 recorded maximum inundation and inflicted heavy losses to Gorakhpur urban area
3	1900–1950	11	Flash floods of 1903 caused maximum damage to Rapti-Rohin Doab
		1903, 1906, 1910, 1922, 1924,1925, 1927, 1928, 1929, 1930, 1932	In 1906, the basin experienced the highest flood of the past 50 years
			Maximum inundation and damage were caused by failure of embankments
			Floods caused by continuous and heavy rainfall
			Flood water stagnated for a longer period of time, causing immense difficulty for cattle grazing
			Maximum damage to crops; some villages located on the riverbank totally wiped out
			The basin witnessed continuous floods from 1927 to 1932 in the Ami and Rapti Rivers that inundated kachhar areas of Sadar and Bansgaon Tehsil
4	1950–2000	19	Severe floods in the years 1974 and 1998
		1953, 1954, 1955, 1956, 1957, 1958, 1959, 160, 1961, 1967, 1968, 1970, 1971, 1973, 1974, 1993, 1998, 1999, 2000	Floods of 1974 surpassed early records, where simultaneous rise in water level of major rivers such as Rapti, Rohin, Ami, Mohab, Ghonghi, Bafela, Basmania, Pyas, Dudhi, Chandan, and Burhi Gandak caused flood havoc on the entire part of Eastern Uttar Pradesh
			About 27,913 ha of land was inundated, affecting 1,390 villages
			Total flood-affected population was estimated to be 631,045
			About 17 embankments were breached
			The floods of 1998 were caused by heavy rainfall in July (910.045 mm) and August (949.100 mm)
			Two peaks witnessed: first phase of flood inundation recorded July 22, 1998 and second phase August 20, 1998
			Embankments failed to withstand water pressure and breached at 20 sites, thereby affecting 1,414,790 of the populations of 1,594 villages in Gorakhpur district
			Total flood-inundated areas of 171,316 ha of which 92,804 ha was net sown area

the total flood frequency. The early flood frequency is 15 % and the late flood frequency is only 5 % of the total annual frequency (Yadav 1999, p. 47).

The Rapti brings down with it an immense quantity of silt that has gradually raised its bed at places above the level of the surrounding country, resulting in heavy floods and the adoption of a new channel. Thus, it shows extensive meandering and braided structure throughout its courses. Heavy downpour accompanied with distinct basin characteristics and human modifications to the natural drainage are the most important and intense causes for serious inundation and floods in the Rapti basin.

20.6 Floodplain Management in Rapti River Basin

State governments are responsible for flood and floodplain management in India; central funding only supports state programs. In the post-independence period, different committees were set up and partial studies were undertaken to minimize the flood risk. Even if those committees suggested an elaborate plan of structural and nonstructural flood mitigation measures from time to time, a generalized framework of disaster management process was visualized only after the enactment of the Disaster Management Act of 2005. With this act a major paradigm shift in public policy (from relief and rehabilitation to prevention, preparedness, response, and recovery) on disaster management took place. Accordingly, the states develop their own plans at state and local level for different types of disasters, including floods. In the Rapti River Basin, state agencies adopted different types of measures to manage flood risk. They can be grouped into two principal categories for discussion: structural flood mitigation measures, and nonstructural or flood emergency measures.

Further, experienced floodplain dwellers have developed their indigenous mode of flood risk management.

Structural flood mitigation measures refer to the works such as levees, detention basins, or channel improvements that are aimed at modifying flood behavior (i.e., keeping water away from people). Construction of levees or embankments, locally known as *bandh*, is the dominant form of structural measure adopted in the study area. Extensive embankments (especially on the lower Rapti basin) were constructed on both sides of the principal river, the Rapti, to protect urban as well as village settlements and agricultural lands. The general height of the embankments varies from 5 to 6 m. In some places freeboards are incorporated as a safety factor to embankment designs. Also, *ring bundh* or circular embankments are constructed to provide safety to those villages situated between the river and embankments. Most of those embankments were constructed during the late 1960s and mid-1970s. The embankments created a false sense of security against flood, as a result of which new settlements are coming up adjacent to those embankments. Similarly, the floodplain community is now putting land having a higher level of flood risk to intensive use. The embankments are designed to withstand low to moderate flood risk having a return period of 5–10 years but cannot withstand floods with a return period of

25 years or more as evident in the flood of 1998. The association between the water level of River Rapti at birdghat (gauge-level data) and the day of embankment failures in 1998 revealed that the embankments did not withstand the water pressure at 77 m of water level or the accompanying discharge of approximately 6,512.50 m^3/s. However, the embankments are capable of managing existing flood risk, enhancing future as well as residual flood risk. Table 20.3 shows some important features of those embankments.

Those embankments are periodically facing the problems of erosion, seepage of water, embankment failure from a heavy flood, and holes created by rats and other animals. Annual maintenance of the embankments, which extend up to 555 km, is becoming a difficult task for the local authority.

20.7 Nonstructural Measures

The nonstructural measures adopted in the study area are basically flood emergency measures such as flood forecasting and warning, evacuation, and recovery plans. These plans are aimed at reducing flood hazards by modifying the response of the population at risk so that they will be better able to handle actual flood events.

20.8 Flood Forecasting and Warning

Flood forecasting is a scientific evaluation of an event in real time leading to the issue of a general alert about hazardous conditions (Smith and Ward 1998, p. 265). The practical aim of flood forecasting is to reduce the loss of life and the economic damage caused by floods. Thus, the accurate forecasting of flood conditions is an essential prerequisite for the provision of reliable flood warning schemes. In recent years, flood forecasting accuracy has increased greatly with improvements in telecommunications and computerized data handling and processing. The need for reliability in flood forecasting has been stressed frequently. Errors in the forecast of flood stage or of the time of arrival of flood conditions may lead to underpreparation and loss from avoidable damage or to overpreparation, unnecessary expense and anxiety, and a subsequent loss of credibility. Further, social acceptability of the flood forecasting and warning is becoming a major issue.

The flood forecasting network of the Rapti River Basin comes under the supervision of Middle Ganga Division No. 1, Lucknow. The entire network of the basin is administered through two subdivisional headquarters. The upper region of the Rapti comes under the upper Rapti-Ghaghra subdivision with headquarters at Gonda. The subdivision has four gauging sites on the River Rapti, namely, Kakardhari, Bhinga, Balrampur, and Bansi, and one gauging site at Kakarauli on the River Burhi Rapti. Similarly, the lower Rapti region is administered by the lower Rapti-Ghaghra subdivision with headquarters at Gorakhpur, with two sites (Regauli and Birdghat) on the River Rapti and only one site on the River Rohin (Trimohanighat). All these sites

Table 20.3 Embankment failures in Rapti River Basin

			Embankments			Embankment failure					
						1974			1998		
Sl. no.	District	River	Number	Total length (km)	Protected area (in hectares, ha	Number	Number of affected villages	Total affected population	Number	Number of affected villages	Total affected population
1	Gorakhpur	Rapti, Rohin, Kuwana, Ghaghara	65	439.95	102,610.0	19	1,390	6.3 lakh	20	1,594	14.14 lakh
2	Deoria	Rapti, Ghaghara, Gurra	14	115.22	40,449.00	N.A.	N.A.	N.A.	10	585	7.50 lakh
				555.17							

Source: Office of the District Magistrate (2001a, b)

Table 20.4 Rapti River Basin: details of all sites for flood forecasting and warning

Name of river	Gauging site	Type of site	Coordinate		Danger level (m)	Highest flood level recorded (m)	Year
			Latitude	Longitude			
Rapti	Kakardhari	G	27°51′	81°49′	131.00	131.35	1979
	Bhinga	GDQ	27°40′	81°51′	119.50	120.10	1997
	Balrampur	GDSQ	27°27′	82°12′	104.62	105.25	2000
	Bansi	G	27°11′	82°58′	84.90	85.82	1998
	Regauli	GDSQ	26°46′	83°18′	80.30	82.12	2000
	Gorakhpur (Birdghat)	GDSQ	26°46′	83°21′	74.98	77.54	1998
Burhi Rapti	Kakrahi	G	27°14′	82°59′	85.65	88.97	1998
Kunhra	Uska bazar	G	27°12′	83°09′	83.52	85.62	1998
Rohin	Trimohani Ghat	G	27°06′	83°25′	82.44	85.43	2001

Source: Central Water Commission (2008); G=Gauge; D=Discharge; S=Silt; Q=Water quality

are operated and maintained by the Central Water Commission. Even if the basin has a good network of gauging stations and flood forecasting and warning performance, their practical utility remains a cause of concern during emergency management because of the low level of social acceptability, lack of faith, accompanying rumor, unsatisfactory evacuation, rescue, and relief camps, and a sense of insecurity for property and belongings among the floodplain dwellers (Table 20.4).

20.9 Flood Emergency Plans

The Rapti basin has a well-established flood emergency plan to cater to any exigency; it was strengthened with the enactment of the Disaster Management Act of 2005. Different nodal units operate at the grassroots level to minimize the flood risk. Some important units with their specific functions are given next.

20.9.1 Barh Chawki

Barh Chawki is the local level flood emergency unit located in remote villages. The primary task of this unit is to collect information on an actual flood situation at the village level and to report to the district administration on a regular basis. It also provides information on flood forecasting and warnings issued by the concerned department to the villagers. Each Barh Chawki serves six to seven villages in flood-prone areas. It caters to the needs of those villages in flood relief, rescue, and information dissemination. This unit is headed by an assistant development officer and supported by a lekhpal, multipurpose workers, and workers from health and animal husbandry departments. All the Barh Chawki operate under the direction of a

Subdivisional Magistrate (SDM) and automatically come into operation when the water level of the nearby river crosses the danger mark. The assigned works of the Barh Chawki are (1) to communicate flood-related information to the control room on a daily basis, (2) to keep information on the condition of the embankments and to inform the control room in special cases, (3) to appraise flood-related local issues and to mitigate them in time with prior consultation with the *Barh Surakhya Samiti*, (4) to send health workers throughout the flood-affected villages to ensure first aid, immunizations, and treatments to the villagers, (5) to store equipment and tools needed for emergency management such as a boat, emergency lights, polythene sheets, etc., and (6) to function as a primary unit of maintaining records on floods and flood-related damages/loses.

20.9.2 Relief Center

During the time of flood disaster, relief materials such as food grains, fodder, other essential items, and money are distributed from this centre. Usually the Barh Chawki and the Relief Centre are located in the same place, but in case of waterlogging and complete inundation of the *barh-chawkis*, relief centers are located in nearby safer places, keeping in mind easy accessibility and transport network facilities. Each relief center caters to the needs of 10 to 15 villages. The center distributes the relief materials to the flooded villages through a locally made boat or motorboat. The nodal officer of this center is an additional development officer (ADO), or Naib Tehsildar.

20.9.3 Relief Camp

A network of relief camps is identified in every district. Public institutions such as colleges, schools, public buildings, embankments, roads, and other places situated on higher ground are considered, designated, and notified as relief camps. Before the flood events basic facilities such as drinking water, food, lights, fodder for animals, and medicines are ensured by the sub-divisional magistrate (SDM), who is in charge of all relief camps.

20.9.4 Barh Sector

To facilitate flood relief and rescue operation and survey, the districts are divided into different sectors. Generally, each developmental block (which is affected by flood) is designated as a flood sector or Barh sector. The block development officer (BDO) is the nodal officer; each is responsible for the smooth operations of the barh chawki, relief center, and relief camp that come under their jurisdiction. The BDOs work with the control and direction of an additional district magistrate (ADM).

20.9.5 *Flood Cell*

The flood cell is the highest decision-making body at the district level. The additional district magistrate for finance and revenue (ADM) heads the cell with representative members from the Department of Irrigation, Health, Police, Jal Nigam, Nagar Nigam, Veterinary, and Electricity. The ADM, finance and revenue, is designated as a full-time district disaster relief officer. He is supposed to communicate between different departments and to facilitate relief and rescue work at the district level. The primary work of the cell is planning for flood control and micro-planning for flood relief. The cell also coordinates with NGOs, civic society, government, and private institutions during the time of floods and ensures the distribution of work among them. The flood control room at the district level operates within their supervision.

20.9.6 *Flood Control Room*

The primary task of the flood control room is to collect and communicate flood-related information at the district level. Communication plays a vital role in rescue and relief operations. Real-time dissemination of information related to flood behavior facilitates rescue and evacuation work. The control room registers complaints regarding flood-related work even before the flood condition. The information officer informs about flood conditions through a press release to counter unnecessary rumors.

20.10 The Floodplain Community and Their Management Measures

The floodplain residents of the Rapti River Basin have developed an elaborate system of indigenous adjustments to the floods, consisting of measures that are traditional in type to minimize the negative impacts of recurrent flooding. During the time of field observations and questionnaire surveys, several indigenous adjustments to minimize flood risk were found in the basin that can be analyzed under the following headings: (1) indigenous flood-proofing of shelters, and (2) indigenous agricultural adjustments to floods.

Two distinct modes of flood-proofing of shelters were identified in the floodplain dwellers of the upper Rapti catchment who are living in temporary dwellings made of locally available branches of trees, straw, and grasses. The huts are usually surrounded by trees. This mode of adjustment has nothing to do with the socioeconomic conditions of the floodplain dwellers; rather, it is a function of the multiple risks perceived by them: (a) dynamic shifting of the stream channel accompanied by massive bank erosion and (b) recurrent flooding. They often abandon the home and built a fresh one at a location that is not liable to see a channel shift in the near future. The main flood adoption measure observed in floodplains of the lower catchment is marked by raising the house plinth above normal flood levels as well as by building different types of platforms for different uses as human and cattle shelters. Responses of the floodplain dwellers to a predetermined checklist and open-ended questions indicate that some adjustments represent successful adaptations to normal flood regimes, whereas others are intended to provide protection from abnormal floods. The courtyard and the plinth level of individual homes were maintained and raised above flood level on a routine basis in all sampled villages by digging earth from local depressions and ponds surrounding the homesteads.

Among the other measures, many respondents of the upper catchment floodplain indicated that during the catastrophic of floods of 1961, 1981, and 2000, when the ground floor of a large number of houses was inundated, the residents were forced to take shelter on different types of platforms. Often temporary shelters were constructed inside the home for the purposes of cooking, resting, and sleeping. Second, elevated grounds of historical importance, embankments, schools, and colleges built on safer locations and railway lines were also frequently utilized as emergency shelters. Another major contingency measure was the provision of platforms for the storage of grains and animal feed. Residents often used the roof tops and branches of tall trees to store flood grains and cattle feed.

Besides protecting infrastructures on the floodplain from high-magnitude floods, one of the main goals of the floodplain residents of the Rapti River Basin is to adjust their agricultural practices with respect to flood cycles. Different varieties of rice crops (main crops grown in the rainy seasons) to varied flood depths at different levels of the floodplains have been the most common indigenous agricultural adjustment observed in the sampled villages. The respondents practiced different types of adjustments, some of which were measures to minimize crop losses from an abnormal flood regime while others were adjusted to recurrent normal regimes. Combinations of structural and nonstructural measures were adopted by the floodplain dwellers to protect their agricultural land. The structural measures involved construction of an earthen ridge along the agricultural field as a means to prevent shallow to moderate flood water; these are maintained and the height of the ridge is increased on a yearly basis. Other important measures adopted by the floodplain dwellers are crop selection based on speculation about floods and the management of land.

Selection of crops based on speculation about floods is best reflected in the lower regime of the basin. Because flood waters arrive suddenly and leave rapidly, floodplain dwellers of the upper regime are hardly concerned about the selection of flood-resistant crops. They were very much influenced by other location factors (for example, a sugar factory in Balrampur) and mostly grow sugarcane and rice in kharif season and wheat, mustard, and *arhar* in rabi season. The floodplain dwellers have great consideration for higher agricultural return. They grow kharif crops with minimum investment because of the impending flood danger. The most common adjustment of rice cropping to uncertainties of natural flood regimes was evident from the practice of mixed cropping of arhar, kodon, maize, and jowar. Here, paddy and jowar may resist low to medium flood levels whereas other crops will perish quickly. But in years without flood, *arhar* and kodon give better results. Thus, this measure assured that at least flood-tolerant paddy would be secured during an abnormal flood regime, whereas during a normal to no flood regime all these crops would succeed, often resulting in a bumper crop. Rabi crops account for the major share of their total food grains. The rabi crops thrive well because previous floods recharge soil fertility and moisture. Wheat is the dominant rabi crop grown by the floodplain dwellers of both the upper and lower Rapti basin.

Zaid crops are common in the lowland and waterlogged area of the floodplain. Extensive uses of the lowlands along the river's course or river islands were found. These areas are put under zaid crops by the poor and marginal section of the floodplain dwellers having no land ownership, who mostly grow water cucumber, watermelon, nenua, and other summer vegetables and a special type of rice known as *Bora*.

Although the basin has an established floodplain management program consisting of structural measures and flood emergency plans, there are several drawbacks. Notable among these are the following. Other structural and nonstructural measures such as land use planning controls and development and building controls were not included in the ongoing floodplain management measures. As the wetlands encroach, the water-retaining capacities of those natural detention basins are decreasing. Proper implementation of the land use policy is needed, and there is also need for standardization of embankment construction and maintenance so as to minimize corruption. Public institutions such as schools, primary health centers, Gram Panchayat buildings, fertilizer distribution centers, etc. in the floodplains are located in low-lying areas, thus facing the twin problems of flooding and waterlogging even in flood-free years. The grassroots-level emergency centers lack minimum facilities, resources, and trained manpower, thereby making their presence meaningless. Lack of integration of roles and responsibilities was the main hindrance for relief and rescue operations during the times of emergency. A contingency plan for emergency operations if the embankments fail is greatly needed.

20.11 Integrated Floodplain Management

An integrated approach to floodplain management is required to bring together the diverse issues and stakeholders that affect or are affected by floodplain management. This approach takes flooding behavior, flood risk, and flood hazard into account, along with all relevant planning factors (ARMCANZ 2000, p. 6). The main objective of this approach is to develop a floodplain management plan that facilitates the use of the floodplain for appropriate purposes, that limits flood hazard and damage to socially acceptable levels, enhances the waterway and floodplain environment, and fosters flood warning, response, evacuation, cleanup, and recovery during the onset and in the aftermath of a flood.

Thus, incorporation of an integrated approach into the ongoing floodplain management process in the study area requires attention to the following points.

20.11.1 A Proactive Management Plan

Floodplain management in the Rapti Basin needs to be proactive. Past flood measures in the basin were taken after a series of floods had occurred. Generally this type of approach leads to an ad hoc attitude and is limited in scope and effectiveness: it does little to control the increasing level of flood hazard across the Rapti Basin. Thus, we have to develop a culture of proactive response and planning to mitigate the problems.

20.12 Integration of Policy, Legislation, and Floodplain Management Measures

A policy framework to support the management of floodplains should be developed to integrate all concerned agencies. The State and Central Government should work together to develop and implement integrated strategies against flood risk incorporating legislative, financial, and technical support. The policy should be supported by appropriate legislation, regulations, standards, guidelines, and planning policies that clearly and unambiguously define the responsibilities and liabilities of all agencies. The different departments such as Civil Administration, Irrigation, Road and Railways, Public Works, etc. have their own area of operation and plan of action to cater to flood risk. Their plans should not be antagonistic.

20.12.1 Flood Emergency Plan

The development and implementation of effective flood emergency plans are the only means of reducing the damage and hazard associated with residual risk. The floodplain management plans and the flood emergency plan should be complementary. The plan should assess a full range of floods, such as recurrence periods of 25, 50, and 100 years and, if possible, by defining a probable maximum flood, while developing the flood emergency plan. An emergency plan for handling potential embankment failures should be included.

20.12.2 Flood Study and Flood Risk Management

A detailed flood study is necessary to collect information on the extent, level, and velocity of flood waters and on the distribution of flood flows to enable us to define the extent of flood hazards across the floodplains. The study should incorporate both hydrological as well as hydraulic aspects of the flood waters. After this flood study, zoning on the basis of flood hazards should be undertaken to identify the floodplain dwellers exposed to or affected by the risk of flooding. Further, identification of public and private property, social systems, and environmental elements at risk of flooding should be incorporated while assessing flood risk and vulnerability analysis.

20.12.3 Flood Maps

Flood maps have a vital role in floodplain management. Flood maps that show the extent, depth, velocity, and hazard of flooding for designated flood events are an important tool in floodplain management as well as in emergency operations. During the time of emergency operations, an evacuation plan for each village indicating safe and accessible routes is needed by the armed forces. Similarly, for airdropping of relief materials, the geographic coordinates of submerged/marooned villages are needed. Thus, there is an urgent need to map the flood-inundated villages with precision and greater detail.

20.12.4 Community Expectations

Ideally, the community can expect that floodplains will be developed and used in an ecologically, economically, and socially sustainable fashion and in accordance with the broader principle of sustainable natural resource and environment management and of integrated or total catchment management. Thus, in every stage of planning, the community's aspirations and local needs should be incorporated through their direct involvement and participation in an inclusive manner.

Floodplain management needs to ensure that the following expectations of the community are met. People wish to be able to live and work on floodplains at no untoward risk to life and health or unacceptable risk of damage to goods, possession, and infrastructure because of flooding, which needs site-specific integrated management measures for existing, future, and residual flood problems. People can be secure in knowing that in the event of inevitable future floods, effective arrangements will be made to alleviate the economic and social costs of flooding, on both an individual and community basis, and recovery of the flooded area and its residents and occupants will be fostered. The community is to be actively involved in the floodplain management process, both in developing management plans and in meeting their obligations under those plans.

20.13 Policy Integration and Implementation

Effective policy and legislation are essential for providing a reliable social and legal foundation for floodplain management. An integrated policy framework is required within all agencies (national, state, local) to support the management of floodplains.

20.13.1 Risk Awareness

The local community or the floodplain dwellers need to be flood aware. For successful management of the floodplains, they need to understand the concept of flood risk and exposure to flood hazards. Their active participation in preventing unnecessary rumors and cooperation during relief and rescue operations is of outmost significance. For their understanding and awareness, appropriate flood risk terminology, and flood maps along with action plans in the form of informative brochures, should be given. Flood risk awareness should be promoted and communicated to the villagers through different plays (locally known as Nukaad Natak) in local languages, and in schools and colleges a flood risk awareness program should be included in course curricula as a part of extracurricular activity.

20.13.2 Appropriate Land Uses

Land use needs to be appropriate to the level of hazards. It must be matched carefully to both maximize the benefits of floodplain and minimize the risks and consequences of flooding. Table 20.5 shows, in principle, desirable locations of various land uses.

Appropriate land use can be ensured through land use planning controls such as floodplain zoning, which are aimed at ensuring that land use is compatible with flood risk; and development and building controls, such as minimum floor levels and flood proofing, aimed at reducing the risk of inundation and amount of damage that occurs when such a flood eventuates.

Table 20.5 Land use compatible with degree of hazard

Extreme	High	Medium	Low
Agriculture*	Agriculture	Agriculture	Agriculture
Recreation	Recreation	Recreation	Recreation
Open space	Open space	Open space	Open space
Environment (with special control)	Environment	Environment	Environment
	Commercial	Residential*	Residential
	Industrial	Commercial*	Commercial
		Industrial*	Industrial
		Schools*	Schools
		Public institutions	Public institutions
		Police station	Police stations
			Telephone exchange
			Hospitals
			Museums
			Army/Air Force station
			Airport

*With appropriate measures

Source: Agriculture and Resource Management Council of Australia and New Zealand (2000)

20.14 Conclusion

Floodplain management across India can only occur on an objective and equitable basis if appropriate performance indicators are defined and used to evaluate the progress and success of floodplain management programs. Monetary and other incentives should be given to the states and district authority. For this, we need standardization of terminology, and collection of appropriate data concerning the socioeconomic profile of the flood-prone community and flood behavior on an objective basis for the design and assessment of floodplain management programs.

Acknowledgments This research paper is part of a major research project supported by University Grants Commission (U.G.C.), New Delhi, No. F.32-345/2006 (SR). Their financial support is thankfully acknowledged.

References

Agriculture and Resource Management Council of Australia and New Zealand (ARMCANZ) (2000) Floodplain management in Australia: best practice principles and guidelines. CSIRO Publishing, Collingwood

Burton I, Kates RW, White GF (1978) The environment as hazard, 1st edn. Oxford University Press, New York

Central Ground Water Board (1984) Watershed Atlas of India, Ministry of Water Resources, Government of India, New Delhi

Central Water Commission (2008) Flood appraisal report monsoon season, 2008, Middle Ganga Division-1. Government of India, Lucknow

Department of Agriculture and Cooperation (1988) Watershed Atlas of India, All India Land Use Survey. Government of India, New Delhi

Federal Emergency Management Agency (FEMA) (1992) Floodplain management in the United States: an assessment report, vol 1, Summary report. Federal Interagency Floodplain Management Task Force, Washington, DC

Federal Interagency Floodplain Management Task Force (1994) A unified national programme for floodplain management, FEMA248

Hewitt K (ed) (1983) Interpretations of calamity from the viewpoint of human ecology. Allen & Unwin, Boston

Mohindra R, Prakash B, Prasad J (1992) Historic geomorphology and pedology of the Gandak Megafan, Middle Gangetic Plains, India. Earth Surface Process Landf 17(11):643–666

Myers M-F, Passerini E (2000) Floodplain management: historic trends and options for the future. In: Parker DJ (ed) Floods, vol I. Routledge, New York, pp 244–253

Office of the District Magistrate (2001a) District disaster management plan, Gorakhpur. Government of Uttar Pradesh, Gorakhpur

Office of the District Magistrate (2001b) District disaster management plan, Deoria. Government of Uttar Pradesh, Deoria

Parker DJ (2000) Floods, vol I. Routledge, New York

Smith K (2013) Environmental hazards, 6th edn. Routledge, London

Smith K, Ward R (1998) Floods: physical processes and human impacts. Wiley, Chichester

Watts M (1983) On poverty of theory: natural hazards research in context. In: Hewitt K (ed) Interpretations of calamity from the viewpoint of human ecology. Allen of Unwin, Boston, pp 229–260

Yadav RP (1999) Flood in eastern U.P. Radha Publication, New Delhi

Chapter 21
Artificial Recharge to Augment Groundwater Resources in Lucknow City

Archana Gupta

Abstract The importance of water is obvious to everyone: we cannot imagine existence of life in the form of flora and fauna without water. At present, space scientists are vigorously engaged in searching for water on other planets. The existence of life on other planets is not conceivable for the human mind unless there is evidence of water. Groundwater is the major source of water in rural as well as in urban areas. About 85 % of the drinking water supply and 50 % of the irrigation supply in rural areas is met through extraction of groundwater. Similarly, 33 % of the domestic supply in cities is also met through tapping of groundwater by dug wells or tube wells (Athavale, Water harvesting and sustainable supply in India. Center for Environment Education, Ahmedabad, Rawat Publication, New Delhi, p 8, 23, 2003). There are certain areas in the country where the table depth has increased by several meters during the past three decades because of overextraction of groundwater resources. The signs of overexploitation are a continuous fall in the water table, drying of wells, or deterioration of groundwater quality. Just as other major urban centers in India, Lucknow is also facing an ironical situation of water crisis today. On the one hand there is acute water scarcity and on the other the streets are often flooded during the monsoon period. This situation has led to serious problems with the quality and quantity of the groundwater. Most of the existing traditional water-harvesting system in Lucknow City also has been neglected and fallen into disuse, worsening the urban water scenario. One of the solutions to improve the urban water crisis is rainwater to enhance groundwater resources through artificial recharge techniques. Keeping in mind all these facts, a plan has been prepared to generate data regarding projected population for 20 years (2009–2029), per capita per day requirement of water, requirement of artificial recharge, and availability of rainwater for recharge.

Keywords Water depletion • Rain water harvesting • Artificial recharge • Water development • Lucknow

---

A. Gupta (✉)
Department of Geography, University of Lucknow, Lucknow, Uttar Pradesh, India
e-mail: guptaarchana30@yahoo.in

A.K. Dutt et al. (eds.), *Spatial Diversity and Dynamics in Resources and Urban Development*, DOI 10.1007/978-94-017-9771-9_21

21.1 Introduction

The rapid growth of population and consequent urbanization has led to stress on available water resources as the consequence of overuse of surface water supplies and overexploitation of groundwater. Dependence on groundwater, to meet various requirements, is so heavy that water levels are declining rapidly and the dug wells and bore wells are even drying up. Because of the rapid rate of urbanization, cities are converted into concrete jungles. Thus, the natural recharge of the groundwater has diminished considerably because of the loss of open land. Therefore, in the present scenario there is an utmost need to augment groundwater storage.

Groundwater is the major source of water in cities. Almost 33 % of the domestic supply is met through tapping groundwater by dug wells or tube wells. There are certain areas in the country where the table depth has increased by several meters during the past three decades by overextraction of groundwater resources. Overextraction or overexploitation of groundwater resources means withdrawing more groundwater than is replenished annually through the process of natural recharge, during which a percentage of the water percolates through the soil and is added to the aquifers. The signs of overexploitation are a continuous drop in the water table, drying of wells, and deterioration of groundwater quality in terms of increased brackishness or the concentration of harmful substances such as fluoride (Athavale 2003).

Urban centers in India are facing an ironical water crisis situation today. On one hand there is acute water scarcity and on the other the streets are often flooded during the monsoon period. Thus, there are serious problems with groundwater quality and quantity although the city receives good rainfall. However, this rainfall occurs for short spells and with high intensity, so that most of the heavy rain falling on surfaces tends to flow away rapidly, with very little time to recharge the groundwater. Most of the existing traditional water-harvesting systems in cities have been neglected and fallen into disuse, which worsens the urban water scenario.

Therefore, one solution to improve the urban water crisis is rainwater harvesting. This technique involves capturing runoff to enhance groundwater resources through the process of artificial recharge. This solution is being practiced on a large scale in cities such as Chennai, Bangalore, and Delhi, where rainwater harvesting is a part of the local policy. Elsewhere, countries such as Germany, Japan, the United States, and Singapore are also adopting rainwater harvesting and artificial recharge techniques.

21.2 Basic Requirement for Artificial Recharge

The basic requirement for recharging groundwater is source water availability. The availability of source water is basically assessed in terms of noncommitted surplus monsoon runoff. This component can be assessed by analyzing the following:

1. Monsoon rainfall pattern
2. Frequency pattern

3. Number of rainy days
4. Variation in space and time

Broadly, to calculate a monsoon surplus, 50 % of the monsoon rainfall (i.e., during July, August, and September) can be considered as monsoon runoff.

21.2.1 Rainwater Harvesting

The rainwater in an urban area can be conserved through open land, parks, paved areas, roads, and pavements and rooftop rainwater-harvesting techniques (RTRWHT) for artificial recharge to groundwater. Among these, RTRWHT requires connecting an outlet/drop pipe from the roof of the building to divert the rainwater to either existing wells/tube wells/bore wells or specially designed structures.

Advantages of Rainwater Harvesting There are several benefits of this technique: rainwater is bacteriologically pure, free from organic matter, and soft in nature; it will help in reducing flood hazard; it will improve the quality of existing groundwater through dilution; rainwater may be harnessed at a place of need and may be utilized at time of need; and the structures required for harvesting the rainwater are simple, economical, and eco-friendly.

Practical Advantages of Rainwater Harvesting The several advantages of rainwater harvesting include availability not subject to outside utility control and not subject to pipeline interruption (seismic); quality is controlled by the consumer; available even when power is interrupted; reduces runoff and erosion; available even when storms and disaster strike; available immediately for fire suppression; reduces mosquito breeding grounds (dengue fever); thermal mass can naturally cool buildings; and ideal for people on low-sodium diets or with health concerns (weakened immunity systems).

21.3 Techniques of Artificial Recharge

Artificial recharge is the process by which the groundwater is augmented at a rate much higher than that occurring under natural conditions of replenishment through human intervention. Augmentation of groundwater resources through artificial recharge can be considered as an activity that supplements the natural process of recharge, which takes place through the percolation of a fraction of the rainfall through the soil to the water table. It is a form of water harvesting in which the surface water stored or flowing out of a watershed or basin during the rainy season is transferred to the aquifers and can be utilized in other months of the hydrological year. This water would be otherwise lost through evaporation or outflow from the watershed. The techniques of artificial recharge can be broadly categorized as follows (Fig. 21.1):

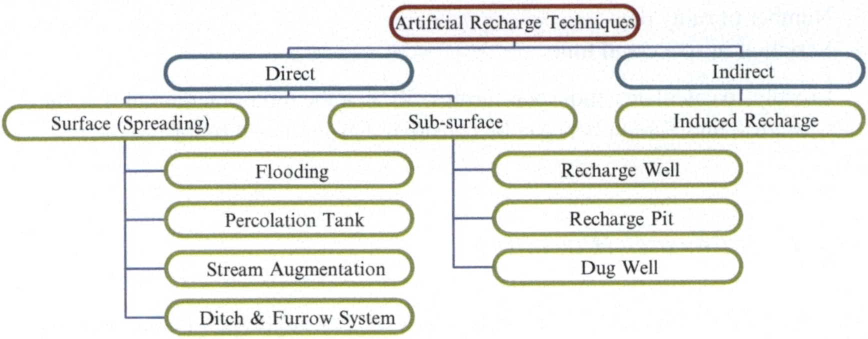

Fig. 21.1 Artificial recharge techniques

(a) Surface Spreading Methods: These methods are suitable where a large area of basin is available and aquifers are unconfined without an impervious layer above them. The rate of infiltration depends on the nature of the topsoil. If the soil is sandy, the infiltration will be higher than in silt-laden soil. The presence of solid suspensions in water used for recharge clogs the soil pores, leading to reduction in the infiltration rate, that is, the recharge rate. Water quality also affects the rate of infiltration.
(b) Subsurface Methods: In this method the structure lies below the surface and recharges the groundwater directly.
(c) Induced Recharge: This is an indirect method of artificial recharge involving pumping from an aquifer hydraulically connected with surface water such as perennial streams, unlined canals, or lakes. The heavy pumping lowers the groundwater level and a cone of depression is created. Lowering the water levels induces the surface water to replenish the groundwater. This method is effective where a stream bed is connected to the aquifer by a sandy formation.

21.4 Area of Study

Lucknow, the state capital of Uttar Pradesh is located in the Middle Ganga Plain and extends between 26°51′ north latitude and 80°36′ east longitude. It lies on the banks of the River Gomti and occupies an area of 340 km^2 in parts of three blocks: Sarojini Nagar, Chinhat, and Bakshi Ka Talab. It stretches more than 10 km along the Gomti River on both banks. The total length of River Gomti is 23.98 km, which covers an area of about 216.33 km. The Kukrail *nala* (drain) is 17.07 km long. There are 37 water bodies in Lucknow City, which covers an area of about 135.21 ha. Depletion of forest cover has left the city with only 2,086.61 ha under forest. According to Census 2001, the total population of Lucknow City was 2,245,509 (Statistical Abstract Uttar Pradesh 2008). The population density of 6,600 persons/km^2 with a high decennial growth of 39.98 % has put tremendous pressure on groundwater resources for drinking and other purposes (Kumar Arun et al. 2011) (Fig. 21.2).

Lucknow City depends heavily on groundwater resources and approximately 70 % of its water requirements are being met through this resource. Heavy withdrawal from

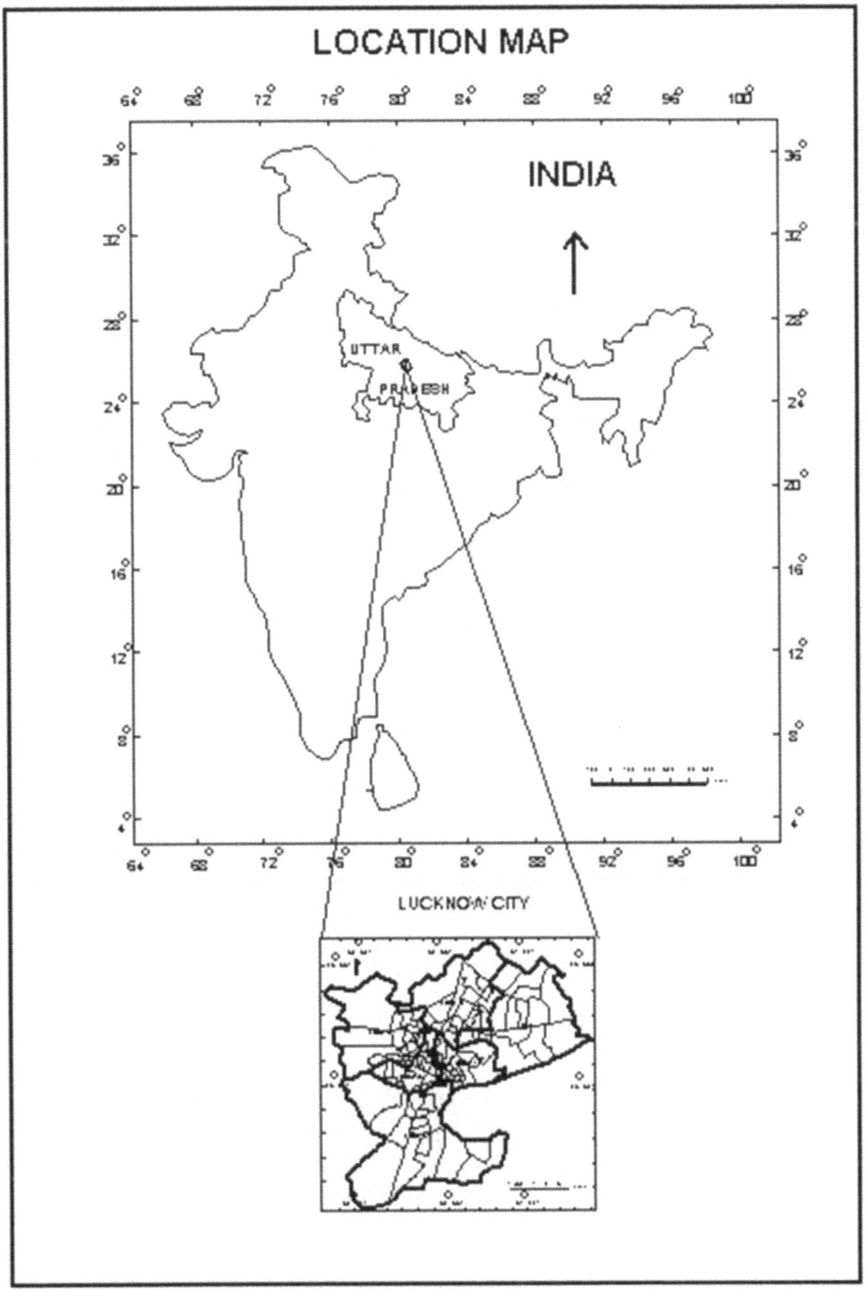

Fig. 21.2 Area of study

the first aquifer has led to a fall of the water table to 30 m or more in the heart of the city, and also the water table is declining at the weighted average rate of 0.73 m/year, which means it loses 6 lakh liters of groundwater daily from mining of groundwater.

Therefore, there is urgent need to evolve a mechanism for judicious use of groundwater and surface water resources to adopt a technique to surmount the present-day situation by augmenting groundwater resources through artificial recharge and conservation.

21.4.1 Climate and Rainfall

The climate of the city is subtropical, with three distinct seasons: monsoon, summer, and winter. The normal maximum mean temperature is 40.5 °C during May and the minimum is 6.9 °C during January. The long-term normal annual rainfall (1901–1970) recorded at Lucknow and Amausi observatories was 1,019 mm and 1,004 mm, respectively. The monsoon normal and annual rainfall recorded by the same observatories is 902 mm and 803 mm, and non-monsoon normal annual rainfall is 117 mm and 116 mm, respectively. The area has a subhumid climate; the hottest month is June and the coldest is January. The average annual rainfall recorded in Lucknow is shown in Table 21.1 and represented graphically in Fig. 21.3.

21.4.2 Physiography and Drainage

Lucknow City occupies the interfluvial region of the Gomti and Sai Rivers of the Middle Ganga Plain in the Ganga basin. It is an almost flat region with a conspicuous natural depression in the northeastern part around Jankipuram. The general

Table 21.1 25 years average (1981–2005) in Lucknow

Months	Avg. annual rainfall (mm)	No. of rainy days
Jan	20.5	1.6
Feb	15.5	1.3
Mar	8.4	0.9
Apr	4.1	0.4
May	20.6	1.9
Jun	106.5	5.2
Jul	236.9	11.2
Aug	213.6	10.5
Sep	218.4	9.0
Oct	43.5	1.6
Nov	6.0	0.5
Dec	14.3	0.9
Total	*908.4*	*45.0*

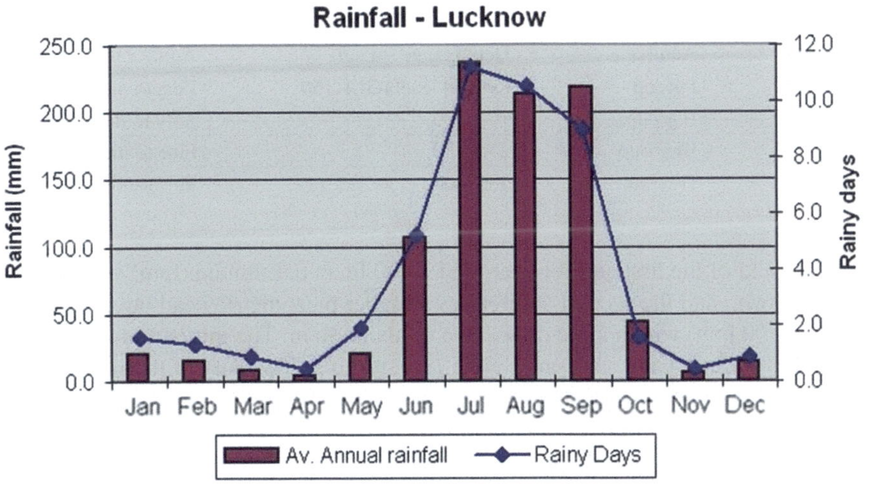

Fig. 21.3 Average annual rainfall and no. of rainy days in Lucknow, 1981–2005

slope of the city area is toward south and southeast. The elevation of the city area varies from 110 to 124 m above mean sea level in general.

The drainage system of the area is controlled mainly by River Gomti, which flows to southeast and Kukrail *Nala*, the only prominent tributary, which joins on the northern bank of the Gomti River. There are 23 *nalas* that drain into Gomti between Gaughat and Gomti Barrage. The drainage exhibits a dendritic to subdendritic pattern that is highly sinuous in nature. River Gomti originates from the spring line in the Terai belt of Pilibhit District. It flows southeasterly over a distance of 490 km before joining the River Ganga in Ghazipur District. The catchment area of the river is 30,500 km^2. Geomorphologically, the city area can be divided into three broad units: Younger Flood Plain, Older Flood Plain, and Interfluve Plain or Upland.

21.4.3 Hydrogeology

Lucknow urban and adjoining areas are underlain by quaternary alluvial sediments containing clay, silt, various grades of sand, and kankar (pebbles). Alluvium forms the most potential repositories of groundwater. There are four aquifer systems within the explored depth of 600 m. The first aquifer is spread within the depth of 150 m under unconfined to semi-confined conditions. Water transmission and flow from the aquifer varies from 1,000 to 1,500 m^2/day.

The Central Ground Water Board has carried out 27 exploratory drillings (down to a maximum depth of 600 m) in the district under its normal and accelerated exploration programs. The results of these exploratory drillings reveal the existence of four tier aquifer systems in the city area within the depth of 600 m, as detailed in Table 21.2.

Table 21.2 Lucknow aquifer system

Sl. no.	Aquifer	Depth range (m bgl)	Aquifer material
1.	I group	Ground level (GL)–150	Fine to medium sand
2.	II group	160–200	Silty to fine sand
3.	III group	250–350	Fine to medium sand
4.	IV group	380–600	Fine sand

The yield of the first aquifer is around 1,200 liters per minute (lpm) with 5–8 m of drawdown, and the second aquifer has a higher piezometric head but poor yield (around 500 lpm) with a large drawdown of about 30 m. The third and fourth aquifers have, by and large, the same yield (i.e., up to 1,200 lpm) as that of the first aquifer but the drawdown is large (about 30 m).

21.4.4 Groundwater Level Depth

The behavior of the groundwater level in the Lucknow urban area is being monitored by the Central Ground Water Board through 22 piezometers. The piezometer data revealed that the depth of the water level varied between 15 and 30 m below ground level (bgl) during May 2008. The deeper water levels have been observed in the central part of the city and around the Cantonment area and southeastern parts of trans-Gomti areas. The water level gradually becomes shallow toward the western and northwestern parts in the outskirts of the city area (Singh et al. 2011) (Fig. 21.4).

21.4.5 Water Level Trend (Long Term)

Long-term analysis of data from a hydrograph station located at Aminabad in Cis-Gomti area indicates that the water table has receded from 17.00 to 30.00 m during the decade of 1998–2008 and that the rate of decline of water level varies from 0.20 m/year in the peripheral area to 1.20 m/year in the central part of the city. Urban areas have an estimated 0.73 m/year decline on average (weighted average) in the water table (Singh et al. 2011) (Fig. 21.5).

21.5 Development and Quality of Groundwater

Although the blocks comprising parts of Lucknow City are categorized as safe, the city area is excessively exploited. The groundwater draft in the city area exceeds the net utilizable resources. The groundwater is generally potable in all aquifers. The water is the $CaMgHCO_3$ type and is suitable for both drinking and irrigation purposes.

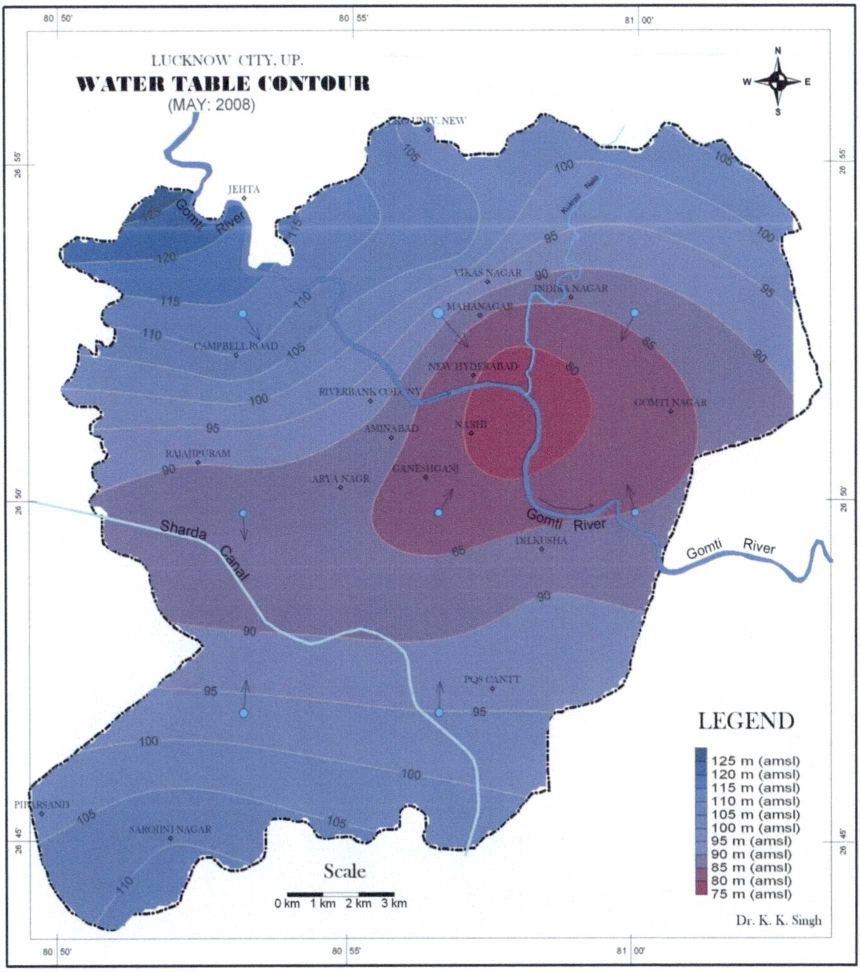

Fig. 21.4 Water table contour of Lucknow city, premonsoon: May, 2008

21.6 Effect of Urbanization on Groundwater Resources

An urban area evolves in the course of time, and population growth depends upon the rate of development. During the past ten decades (1901–2001), the area and population of Lucknow City have increased approximately eight times, that is, from 44.03 to 340 km^2 and from 256,239 to 2,245,509, respectively, causing severe stress on the groundwater (Dutta et al. 2011; Rai et al. 2004). The groundwater resource is an economical and readily available resource with ample supply in the aquifers. Urbanization puts extensive pressure on land and water resources. With increasing

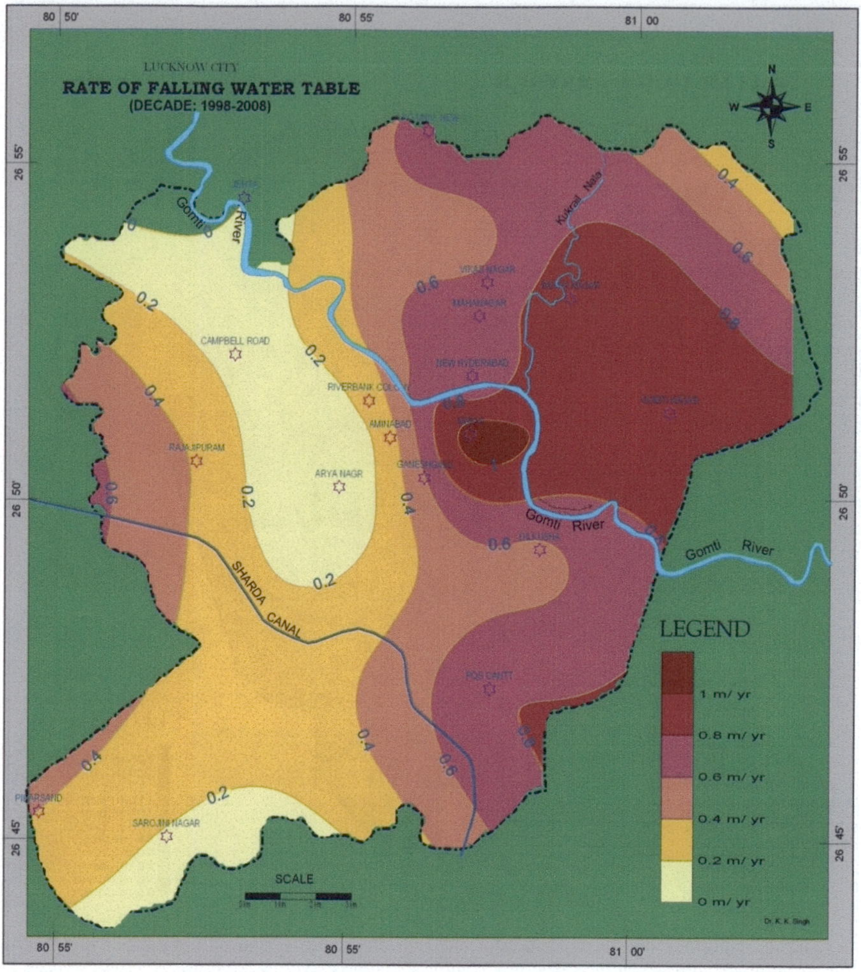

Fig. 21.5 Rate of falling water table in Lucknow city, 1998–2008

pace of urbanization, population density has increased considerably and is now 6,600 persons/km^2 with a decennial growth of 39.88 % in the Lucknow City area. Evolution of the urban area in Lucknow City has led to a decline of 20–30 m in water level in the areas of Gomti Nagar and the old city in particular and to general decline in other areas. The rate of the water table decline is alarming and is exceeding 1 meter per year (m/year), thus causing reduced efficiency or failure of tube wells. At present the water table is declining at a rate of 0.73 m/year (weighted average) in a city area of about 240 km^2 because of overexploitation of about 19.50 million cubic meters (MCM) of groundwater. To stabilize the present water level there is urgent need to artificially recharge the same volume of water through various recharge structures constructed in parks and open fields by utilizing techniques such as rooftop rainwater harvesting (RTRWH) and recharge from *Nala* through check dams and filtration wells (Rai et al. 2011).

21.7 Justification for Artificial Recharge

The all-round development of groundwater has put tremendous pressure on shallow as well as deep aquifers within the city. In general, the depth of the water level during a pre-monsoon period ranges between 10.00 and 32.00 m bgl (below ground level) and in the post-monsoon period between 6.00 and 32.00 m bgl (below ground level).

The water levels have receded during the past decade, and as a result all the dug wells having a maximum depth of 10–12 m have become dry. The trend of water level decline is almost as great as 1.20 m per year in some areas from overexploitation of groundwater, and the deepest water level observed is more than 32.00 m bgl (below ground level) in parts of the city area. The estimated available surplus surface water is of the order of 81.31 MCM (million cubic meters), which needs to be recharged artificially to groundwater through recharge structures.

21.7.1 Proposed Plan

As per the norm of the National Drinking Water Mission, the present per capita need of water in urban area varies between 130 and 150 LPCD (liters per capita per day). With the increasing population it is very difficult to maintain a balance between per capita need and availability of water.

Thus, artificial recharge to augment groundwater is a suggested measure in Lucknow City. Keeping in mind all these facts, a plan has been prepared to generate data regarding projected population for 20 years (2009–2029), per capita per day requirement of water, requirement of artificial recharge, and availability of rainwater for recharge. Based on Census 2001 population data, the projected population for 2009–2029 was calculated. Assuming 150 Lt/H/D (liters per head per day) or LPCD (liters per capita per day) of water need, the requirement of water is calculated. In the same way requirement of artificial recharge is calculated by subtracting 'supply of water from various sources' and 'proposed plan for surface water supply' from water requirements. Further, the availability of rainwater is calculated using the following formula:

$$\text{Availability of rain water} = \text{Area} * \text{Average Monsoon Rainfall} * .6$$

For calculating availability of surface water for artificial recharge, 50 % of noncommitted runoff is taken into consideration. In Lucknow City Jal Sansthan is now proposing a plan to increase surface water supply to minimize the supply of water from groundwater (Table 21.3).

Table 21.3 Annual requirement of groundwater after deduction of surface water supply and groundwater recharge from various sources

Year	Projected population (in thousands)	Requirement of water (in MCM)	Natural recharge from various sources (in MCM)	Supply of water from surface water (in MCM)	Supply of water from groundwater (in ham)	Proposed plan for surface water supply (in MCM)	Requirement of artificial recharge (in MCM)	Area of Lucknow city (in km^2)	Average monsoon rainfall (in mm)	Availability of rainwater (60 % of 9*10/1,000 in MCM)	Availability of rainwater for recharge (50 % of 11 in MCM)
		Assuming 150 Lt/H/D	(CGWD)	(Jal Sansthan)	(Jal Sansthan)	(Jal Sansthan)	(3-4-5-7)				
1	2	3	4	5	6	7	8	9	10	11	12
2009	2,962	162.16	21.53	91.25	87.6		49.38	337.50	802.00	162.41	81.3
2010	3,080	168.63	21.53	91.25	87.6		55.85	337.50	802.00	162.41	81.3
2011	3,203	175.10	21.53	91.25	87.6		62.31	337.50	802.00	162.41	81.3
2012	3,331	182.35	21.53	91.25	87.6	73	−3.43	337.50	802.00	162.41	81.3
2013	3,463	189.62	21.53	91.25	87.6	73	3.84	337.50	802.00	162.41	81.3
2014	3,602	197.18	21.53	91.25	87.6	73	11.40	337.50	802.00	162.41	81.3
2015	3,745	205.05	21.53	91.25	87.6	73	19.27	337.50	802.00	162.41	81.3
2016	3,895	213.25	21.53	91.25	87.6	73	27.47	337.50	802.00	162.41	81.3
2017	4,050	221.73	21.53	91.25	87.6	73	35.95	337.50	802.00	162.41	81.3
2018	4,211	230.57	21.53	91.25	87.6	73	44.79	337.50	802.00	162.41	81.3
2019	4,379	239.77	21.53	91.25	87.6	73	53.99	337.50	802.00	162.41	81.3
2020	4,554	249.33	21.53	91.25	87.6	73	63.55	337.50	802.00	162.41	81.3
2021	4,736	259.27	21.53	91.25	87.6	73	73.49	337.50	802.00	162.41	81.3
2022	4,924	269.61	21.53	91.25	87.6	73	83.83	337.50	802.00	162.41	81.3
2023	5,121	280.36	21.53	91.25	87.6	73	94.58	337.50	802.00	162.41	81.3
2024	5,325	291.55	21.53	91.25	87.6	73	105.77	337.50	802.00	162.41	81.3

2025	5,537	303.17	21.53	91.25	87.6	73	117.39	337.50	802.00	162.41	81.3
2026	5,758	315.26	21.53	91.25	87.6	73	129.48	337.50	802.00	162.41	81.3
2027	5,988	327.83	21.53	91.25	87.6	73	142.05	337.50	802.00	162.41	81.3
2028	6,227	340.91	21.53	91.25	87.6	73	155.13	337.50	802.00	162.41	81.3
2029	6,475	354.50	21.53	91.25	87.6	73	168.72	337.50	802.00	162.41	81.3
Total		5,177.21	452.23	1,916.25	1,839.6	1,314	1,494.72			3,410.51	1,707.3
Average		246.53	21.53	91.25	87.6	73	71.18			162.41	81.30

MCM (million cubic meters), *Lt/H/D* (litre per head per day), *CGWD* (central ground water department), *ham* (hectare meters), *mm* (millimeters)

21.8 Conclusion

Water resources are extremely vulnerable and once depleted take 100 years to replenish. The shortage of groundwater is more pronounced from urbanization and limited open areas. The groundwater level has reached a high level and has brought problems such as declining water table, failure of bore wells/tube wells, and deterioration in groundwater quality and quantity. Water has more than often been seen as a cause for social conflicts, protests, demonstration, and road blockage. In the current circumstances, rainwater harvesting could prove to be a viable solution for overcoming water scarcity.

The need for a policy framework for rainwater-harvesting systems arises mainly from a lack of a policy statement on the issue. There is a clear need to evolve a decentralized legal regime with regard to water that empowers people and makes them the real managers of the water resource. For promoting urban water harvesting, a policy should include a mix of carrot and stick approaches. Measures that need to be undertaken are these: rainwater harvesting to recharge the groundwater system should be essential, and town planning requirements and prerequisites for permission for development of new colonies should be required; provision of rainwater-harvesting structures in all planned buildings should be mandatory for issuing building permission; appropriate rebates on property/fiscal incentives should be granted for effective implementation of rainwater-harvesting systems; development of grazing fields, afforestation, and similar other work should be encouraged; priorities should be given to the scarcity areas in the planning of water resources development and a special water management system should be developed for economic use of water in these areas; and artificial recharge structures are to be constructed mostly with the objective of augmenting groundwater resources or to improve its quality. Assessment of impacts of the artificial recharge schemes is essential to assess the efficacy of structures constructed for artificial recharge and to evaluate the cost-effectiveness recharge mechanisms for optimal recharge into the groundwater system. This plan will also help to make necessary modifications in site selection, design, and construction of structures in future.

References

Athavale RN (2003) Water harvesting and sustainable supply in India. Center for Environment Education, Ahmedabad/Rawat Publication, New Delhi, pp 8, 23

Dutta V, Sinha RS, Nupoor P, Yunus M (2011) Land use change and management of stressed urban aquifers in U.P. with special reference to Lucknow. In: Proceeding of regional workshop on management techniques in ground water stressed urban areas. C.G.W.B., N.R., Lucknow, pp 46–53

Kumar Arun, Bhartariya SG, Kudesia S, Srivastava SK (2011) Lucknow City, Uttar Pradesh, Ground water scenario in major cities of India. CGWB, Faridabad, pp 124–134

Rai JN, Kumar Arun, Bhargava AK, Mehrotra S (2004) Management for sustainable development of ground water in Lucknow urban agglomeration. Our Earth, Lucknow 1(1):18–22

Rai JN, Kumar Arun, Bhartariya SG (2011) Sustainable development and management of ground water in stress aquifers of Lucknow City. In: Proceeding of regional workshop on management techniques in ground water stressed urban areas. C.G.W.B., N.R., Lucknow, pp 67–77

Singh KK, Singh Suresh, Biswas KB (2011) Artificial recharge to augment ground water resources in stressed part of Lucknow City, Uttar Pradesh. In: Proceeding of regional workshop on management techniques in ground water stressed urban areas. C.G.W.B., N.R., Lucknow, pp 26–32

Statistical Abstract Uttar Pradesh (2008) Economics and Statistics Division, State Planning Institute, Uttar Pradesh, Lucknow, pp 2–5

Chapter 22
Inter-State Water Sharing in India: From Conflict to Cooperation

Nina Singh

Abstract Perception of the threat of scarcity and unjust distribution of water, a finite, preeminent natural resource, has caused its connection with conflict. It is an area of continued interest and debate in both the policy literature and popular press at the international level. In India too, the states are in conflict with each other on a variety of issues over sharing of inter-state river waters. This paper attempts to explore different facets of this issue based on an exhaustive literature survey. The more valuable lesson of international experience is that water is a resource whose characteristics tend to induce cooperation and incite violence only exceptionally. In India, however, the situation is different: conflict–resolution mechanisms have to be addressed from different angles. To begin with, these disputes have to be disentangled from the more general State–State conflicts and from everyday political issues. The possibility of a National Water Commission independent of daily political pressures should be explored. River water resources require sustainable integrated planning for the larger ecological systems of which rivers are a part by way of State–State cooperation and extending beyond centralized technology-driven planning.

Keywords Inter-state river water • Conflict • Cooperation • Regulation • India

22.1 Introduction

Inter-state cooperation is required to efficiently and effectively design transboundary infrastructure projects such as generation and distribution of power, transport networks, traffic flows, urban development, implementation of flood and drought mitigation strategies, control of the spread of diseases, promotion of education, planning irrigation systems, rationalization of cropping patterns, tax structure, environment, watershed, and biodiversity management, protection against crime and

N. Singh (✉)
Department of Geography, Maharshi Dayanand University, Rohtak, Haryana, India
e-mail: ninasingh99@gmail.com

A.K. Dutt et al. (eds.), *Spatial Diversity and Dynamics in Resources and Urban Development*, DOI 10.1007/978-94-017-9771-9_22

terrorism, protection of water sources (rivers, lakes, mountains, forests, aquifers) from pollution, and degradation or denudation, to name only a few.

Natural units and urban zones do not conform to political boundaries and require inter-state collaboration. In particular, although river basins seem the natural unit for dealing with issues of water sharing, investment and management, they have been the focus of conflict rather than cooperation in India. For instance, the early years after independence saw water projects such as the Damodar Valley Corporation (DVC), modeled on the lines of Tennessee Valley Authority of the USA. It was set up in 1948 to carry out a comprehensive development of the entire valley region covering parts of Bihar and West Bengal. However, it could never develop into an autonomous regional river valley development corporation because of conflicts between the corporation and participating governments (Ramana 1992). The story of the Narmada Valley project is also known for its inter-state contentions rather than cooperation. Krishan (2006, p. 873) has highlighted several geographic parameters of this inter-state context, which call for a collaborative effort. Two instances may be quoted here. A collaborative strategy of watershed development would be most suitable for the Shivalik Hills, which are common to the entire northwestern region. In the same vein, management of the Ghaggar stream is best done if Himachal Pradesh, Haryana, and Punjab work together to their mutual advantage. Such cooperation can be at the level of governments, NGOs, academic institutions, or 'people-to-people' contact.

Water nurtures life, without which human beings would perish. It has acquired a new meaning beyond its traditional uses for irrigation, navigation, municipalities, industrial developments, and hydropower. It has come to be associated with a certain kind of lifestyle—entertainment, sports, and leisure. Water also holds a communal value: the meaning of the Ganges River to the Hindu culture or Niagara Falls and the canals of Venice to tourists and lovers. In every case, the waters are not just naturally constructed, but a socially constructed entity also (Reuss 2002, p. 6: accessed on 11 October 2008 from webworld.unesco org/water/wwwap/cd/pdf).

Encompassing a wide canvas, the scarcity of water, its distribution, and its quality are most likely to lead to intense political pressures. It is the only scarce resource for which there is no substitute and the need for which is overwhelming, constant, and immediate (Wolf 1998, p. 251). As a consequence of 'water stress' in many parts of the world, conflicts over water within and between nations are endemic where the demand for water is high because of sharp population growth and development needs. Although water quantity has been a major issue in the last century, water quality has been neglected to the point of catastrophe: more than 1 billion people lack access to safe water supplies; almost 3 billion do not have access to adequate sanitation; 5 million people die each year from water-related diseases or inadequate sanitation; and 20 % of the world's irrigated lands are salt laden to the point of affecting production. Water demands are increasing, groundwater levels are dropping, surface water supplies are increasingly contaminated, and delivery and treatment infrastructures are aging (Wolf et al. 2003). Thus, water being preeminent among natural resources, requires judicious management and collaboration among the stakeholders.

22.2 Conflict Versus Cooperation: International Scene

Is water a cause for conflict or cooperation? This question has been examined by Molen and Hildering (2005) who observed two different contrasting scenarios at the international level. The conflict scenario foresees serious water scarcities and an increasing potential of conflicts between numerous countries. It sees access to water as a matter of national security. The other is the cooperation scenario: "while freely admitting the possibility of conflict, and it denies its inevitability. The co-operation scenario further points to the possibility of co-operative arrangements for sharing river resources between the upstream and downstream countries, including treaties and river administration" (Toset et al. 2000).

The water and conflict point of view is also echoed in the writings of Westing (1986), Gleick (1993), Homer-Dixon (1994), Butts (1997), Samson and Charrier (1997), and Priscoli (1998). However, it must be realized that a war over water seems neither rational, hydrographically effective, nor economically viable (Wolf 1997). In contrast, there is a smaller body of work that argues more strongly for the possibilities and historic evidence of cooperation between co-riparian states. Libiszewski (1995), Wolf (1998), and Salman and Boisson de Chazournes (1998) have worked on historic assessments of cooperation, and Priscoli (1998) describes water as 'humanity's great learning ground for building community.'

Kliot et al. (2001) examined the nature, characteristics, and shortcomings of cooperative arrangements for the management of 12 transboundary river basins: the Mekong, Indus, Ganges, Nile, Jordan, Danube, Elbe, Rio Grande, Colorado, Rio de la Plata, Senegal, and Niger. They indicate that cooperative water resource management faces several obstacles, such as the critical nature of water for human existence; the multiple uses of water; and the sheer scale and the gap between policies and implementation of these policies, but conclude that in many river basins countries are able to overcome their differences and cooperate for the benefit of all. Similar findings have been presented by Wolf (1999b), Wolf et al. (2003), and Yoffe et al. (2004).

These investigators conducted research on factors that contribute to conflict or cooperation, including biophysical, socioeconomic, and geopolitical variables at multiple spatial and temporal scales using a geographic information system (GIS) of international river basins and associated countries. These variables were tested by using a database of historical incidents of water-related cooperation and conflict across all international basins during the period 1948–1999. They found that international relationships over freshwater resources are overwhelmingly cooperative and cover a wide range of issue areas, including water quantity, quality, joint management, and hydropower. It was also observed that most of the parameters commonly identified as indicators of water conflict are actually only weakly linked to dispute. These parameters include climate type (tropical rainy, dry, humid mesothermal, and humid micro-thermal), water stress (freshwater availability per capita by country or by basin), population (population density by basin and population growth rate by country), dependence on hydropower, dams, or development per se, level of development or "creeping changes," such as gradual degradation of water quality or climate change-induced hydrological variability. Their study suggests that institutional capacity within a basin whether defined as water management bodies or treaties, or generally positive international relationships, are as important, if not more so, than the physical aspects of a system. Other issues, such as general friendship/hospitality over non-water issues, and stability and types of government as defined by their level of democracy/autocracy within a basin, are all components contributing to functional institutional capacity.

A similar conclusion was reached by (Toset et al. 2000). Importantly, there is a history of water-related violence at the subnational level, generally between tribes, water use sectors, or states or provinces. Internal water conflicts, in fact, are quite prevalent. They range from interstate violence and death along the Cauvery River in India, to California farmers blowing up a pipeline meant for Los Angeles, to much of the violent history in the Americas between indigenous peoples and European settlers. Water issues have contributed to ethnic, political, and cultural strife, whether among Dutch polders (land reclaimed from the sea), Californian mining camps, Italian city-states, or the states of modern India or the United States (Beach et al. 2000).

The stability of a region can be unsettling as water quality or quantity degrades over time within a local setting and acts as an irritant among ethnic groups, water sectors, or states/provinces. The internal instabilities can be both caused and exacerbated by international water disputes, for example, between India and Bangladesh over building up of the barrage at Farakka, which diverts a portion of the Ganges flow away from its course into Bangladesh, and toward Kolkata 100 miles to the south, to flush silt away from the city's seaport. Adverse effects in Bangladesh resulting from reduced upstream flow have included degradation of surface water and groundwater, impeded navigation, increased salinity, degraded fisheries, and danger to water supplies and public health. Migration from the affected areas has further compounded the problem. Ironically, many of those displaced in Bangladesh have found refuge in India (Wolf 1999a).

22.3 Conflict Tendency of Indian States

Why are the states, particularly the neighboring ones, in conflict with each other? The story of the creation of the states may give an insight to this question (Krishan 2000). The pre-independence provinces of British India were not shaped by any rational or scientific planning, but "by the military, political or administrative exigencies or conveniences of the moment" (Government of India 1918, para. 3). Very often heterogeneity was created to discourage the sentiment of provincial solidarity. For instance, the Madras Province was one such unwieldy administrative entity formed without regard to its internal linguistic, cultural, economic, and physical unity. The province was a conglomerate of Tamil-, Telugu-, Kannada-, and Malayalam-speaking areas.

The Independence in 1947 saw stupendous changes in the administrative map of India. The first change came in 1948 when the erstwhile 10 provinces and 562 princely states, inherited from the British, were consolidated into 28 units classified into A, B, C, and D states in order of the relative degree of autonomy rendered to them.

The year 1956 saw the second major change when the Indian States were primarily reorganized on linguistic lines (Government of India 1955). There was a radical recasting of the administrative map of India into 14 states and 6 union territories. The process did not culminate here. The necessary changes continued to be made from time to time. This process of territorial reorganization extended to the northeast and culminated in the creation of many states. Some of the former union territories were elevated to the state level, and a few of their kind were also reconstituted. States were formed on the basis of language (Gujarat, Haryana), history (Goa), and culture (Nagaland). Although essentially a by-product of the reorganization of old Punjab on a linguistic basis; the Haryana territory was an appendage to it in the nature of a characteristic periphery (Singh 1998).

The linguistic reorganization of the states in India has given rise to a number of inter-state water disputes as new states succeeding the parties to the previous agreements. Some of the noted examples are the dispute between Punjab and Haryana for allocation of Ravi-Beas waters. The distribution of the Ravi-Beas waters between Jammu and Kashmir, Rajasthan, and composite Punjab was effected through an inter-state agreement in 1955. In 1966, Punjab was reorganized on linguistic grounds and Haryana was formed. Under the Indus Waters Treaty with Pakistan, surplus waters of the Ravi and Beas Rivers became available in 1970 for utilization in India. Haryana claims a share of these surplus waters, which Punjab is disputing so much so that the issue has been 'ethnically biased' by politicians in Punjab. The inaction of the Central Government has exasperated Haryana, which has reacted by occasionally stopping the water supply from the Yamuna River to the national capital (Swain 1998, p. 171). The other noteworthy disputes are between Karnataka and Andhra Pradesh over the utilization of Tungbhadra under the 1944 agreement between Madras and Hyderabad; and the dispute between Karnataka, Kerala, and Tamil Nadu relating to the 1924 agreement over the utilization of waters of the Cauvery owing to the territorial changes as a result of the reorganization of the two states (Jain et al. 1971, p. 27).

The three most recent states created in the Indian Union—Jharkhand, Uttarakhand, and Chhattisgarh—came into existence in 2000 without the recommendation of any States Reorganisation Commission, more on considerations of economic backwardness and discriminatory treatment by political elites of the respective parent states than on linguistic, religious, or tribal considerations, and disregarding administrative rationality and financial viability. The creation of these states is more the result of the new coalitional governing framework in New Delhi since 1989 (Singh 2008, p. 72). Obviously, the subregional identity assumes importance when inter-regional disparities and discrimination surface: there are now 28 states and seven union territories in India.

The states vie with each other for grants of development and other funds from the limited national coffer. Politicians often play with the sentiments of their people vis-à-vis neighboring states on issues having bearing on both states, particularly if they have different party affiliations. The disputes may relate to land or sharing of resources (Krishan 2006). The territorial dispute over Chandigarh and the conflict of sharing of river waters between Punjab and Haryana is one such case. Similarly, Himachal laments that it is not getting its due share in hydropower generation at the Bhakra and Pong dams. Still another instance will bring home this point fully. The waters of Narmada have been flowing wastefully into the sea for many years without any utilization because of an earlier dispute between Madhya Pradesh and Gujarat, both having large drought-prone areas.

22.4 Inter-State River Water Conflict Dynamics

In terms of the concept of 'water stress,' India may not be among the most water-stressed countries of the world, but the national aggregates and averages could be misleading as freshwater is a very unevenly distributed natural resource, both temporally and spatially. Characterized by a monsoonal climate, India experiences rainfall that is highly erratic, and not only does it occur in only a few months during the year, but even during that period the intensity is concentrated within a few weeks. Rainfall in some regions contrasts sharply with prolonged drought in others. In a gradient from west to east, precipitation varies from 100 to 11,000 mm. Broadly speaking, the Himalayan Rivers are snow fed and perennial, whereas the peninsular rivers are dependent on the monsoons and are, therefore, seasonal.

All the major rivers of India transcend multiple states. Nearly 77.67 % of the country's geographic area falls within these basins (Chitale 1992, p. 32). When several states are jointly dependent on the same river system, disputes among the states regarding allocation and utilization of inter-state river waters are natural phenomena in a federal relationship, particularly when large areas of the country are relatively arid. Moreover, the country has adopted intensive irrigation-based agriculture to raise its food production to feed its teeming billions. About four-fifths of India's water is used to irrigate agricultural land; the industrial and domestic sectors consume the rest. There is increasing water scarcity in the face of growing demand in both towns and villages. Nearly 60,000 villages of India still remain without potable water, forcing women and children to walk several miles to fetch water. All this has led to much acrimony between states over the issue of sharing the river waters. These conflicts erupt when upstream states, through their water projects, affect the quantity and quality of the water flow in the basin and restrict the scope of downstream water use. Also, inter-basin river water transfers also disturb riparian rights and lead to inter-state disputes (Shiva 1991, p. 184). The word "riparian" is related to the bank of the river or stream. So, to be a riparian owner one has to own some land abutting on the natural stream concerned. In India riparian land must be confined to land that is on the bank of a stream and which extends from that bank to a reasonable depth inland (Amin and Sastri 1986, pp. 161–162). Often skirmishes, *dharnas*, and slogan shouting have erupted over transboundary water issues between riparian rivals. The conflict over the Cauvery River waters has seen violence erupt in Karnataka and spread to Tamil Nadu. The politicians of both states are using the Cauvery River issue to further their political interests.

The water-sharing issue is becoming increasingly politicized in India. In the 1996 general elections this issue figured prominently in at least nine states (Thakkar 1996, p. 15). Conflict may also occur between the people and the state. The Inter State Water Disputes Act assumes that 'inter state river-waters disputes' means inter-governmental disputes. It does not recognize that there could be conflicts between the concerns and interest of the people and the aims and purposes of the government, as has happened in the case of the Sardar Sarovar Narmada project and the Tehri Dam project. The process of water-sharing negotiations between the states has been further complicated with the emergence of grass roots organizations to defend the interests of the people.

In Rajasthan, when community initiatives resulted in water reappearing in rivers and streams that had been dry for years, the State claimed the right of control over those waters for the purposes of allocation, licensing, and fisheries. This eminent domain makes community initiatives problematic. Although some kind of a *modus vivendi* seems to have been worked out, the legal issue remains and could come up again in a future case. Another noteworthy instance of water management through community initiative can be observed in the case of Arvari, a small river in Rajasthan. The people have established a kind of 'parliament' for dealing with the waters of the Arvari River and resolving conflicts relating to those waters, a purely informal body without any statutory backing (Iyer 2000, accessed on 25 September, 2008 from http://Planning-commission.nic.in/reports/genrep/bkpap2020/7-bg2020.pdf).

22.5 India's Federal Water Mechanisms

India is a federal democracy, and because rivers cross state boundaries, conflicts over the issue of sharing of river waters are a persistent phenomenon. Therefore, constructing efficient and equitable mechanisms for allocating river flows has long been an important legal and constitutional issue. There are many actors involved, including State governments composed of professionals, politicians, political parties, and interest groups, the national parliament, central ministries, the courts, and ad hoc water tribunals. To have a better understanding of the whole scene, let us briefly trace the historical background of the laws and policies governing utilization of inter-state river waters from India's pre-Independence period.

Even before Independence there were water disputes among and between provinces and the princely states and, as in other federal nations such as the United States, these arguments gave rise to case law that governed water relationships (Verghese 1997). Some important and far-reaching policy changes with regard to the execution and financing of irrigation projects were made in 1866 during British

Rule. It was decided that construction should be taken up by the Government through its agencies, financed from public loans raised specifically for the purpose, and that river waters should be utilized optimally irrespective of the political boundaries of British India and the then Princely States. The Sirhind Canal, completed in 1882, was the direct culmination of this policy. This was the beginning of an era of inter-state cooperation in harnessing water resources on an agreed basis.

Irrigation works were virtually under the control of the Central Government until the introduction of the Montagu-Chelmsford Reforms in 1921, when irrigation became a provincial but reserved subject. Under this arrangement also, the Central Government was in ultimate control of major irrigation works and projects on inter-state rivers (Shah 1994). However, the position changed materially with the passing of the Government of India Act in 1935 when irrigation became a provincial subject wholly within the legislative competence of the provinces. The Government of India Act of 1935 drew attention explicitly to river disputes between one province and another or between a province in British India and a (federated) Indian State.

22.6 Present Constitutional Provisions

The partitioning of the country and the integration of princely states resulted in redrawing the map, giving rise to new disputes over water rights. Entry 17 in List II of the present Constitution is identical to Entry 19 of the 1935 Act. Under the Indian Constitution, all matters relating to rivers concerning foreign countries are entrusted to the Union.

Basically, 'Water' is a State subject, with the Union's role limited to the inter-state rivers. The existing legislative framework specifically concerned with water is based on Entry 56 in the Union List, Entry 17, of the State List and Article 262 of the Constitution, as given below.

Entry 56 of List I (Union List):

Regulation and development of inter-state rivers and river valleys to the extent to which such regulation and development under the control of the Union are declared by Parliament to be expedient in the public interest.

Entry 17 in the List II (State List):

Water, that is to say, water supplies, irrigation and canals, drainage and embankments, water storage and water power are subject to the provision of Entry 56 of List I. This is qualified by

Article 262: Disputes relating to Water. Adjudication of disputes relating to waters of Inter-State Rivers or river valleys.

- Parliament may by Law provide for the adjudication of any dispute or complaint with respect to the use, distribution or control of the water of, or in, any inter-state river or river valley.

- Notwithstanding anything in this constitution, Parliament may by Law provide that neither the Supreme Court nor any Court shall exercise jurisdiction in respect of any such dispute or complaint as is referred in Clause (I).

Thus, the powers of the States are not unqualified, but are subject to the powers of the Central Government as provided in Entry 56 of the Constitution of India. However, Parliament has not made much use of the enabling provisions of Entry 56 (Iyer 1994). Various River Authorities have been proposed, but not legislated or established as bodies vested with powers of management. Instead, river boards with only advisory powers have been created. Hence, the State Governments enjoy full powers even over the waters of the inter-state rivers.

Within the powers available under Entry 56 of the Union List and Article 262, Parliament has enacted two laws, namely these:

22.6.1 River Boards Act of 1956

River Boards were advisory in nature, set up by the Central Government at the request of the interested parties. Their function was to help bring about proper and optimal utilization of river waters and promote development of irrigation, water supply, drainage, hydroelectric power, and flood control.

22.6.2 The Inter-State Water Disputes Act of 1956 (Last Amended in 2002)

This Act was legislated to handle conflicts, and included provisions for the establishment of tribunals to adjudicate where direct negotiations had failed. However, arbitration by tribunals is not binding. Sometimes even the courts have not succeeded. There also have been instances of direct intervention by the Centre but that too has not been able to resolve the most intractable cases, such as the sharing of the Ravi-Beas waters among Haryana, Jammu, and Kashmir, Rajasthan, and Punjab. One of the essential features of Indian federalism is the constant involvement of coalition construction to create a government at the center. This feature also means a substantial ground for explicit or implicit negotiations among the Centre and the states that are important partners of a central coalition. Water could be an important element of bargaining. In some cases the Central Government has taken very long in appointing tribunals and has dragged on the negotiations. Many a time the inter-state water disputes have been entangled with other Centre–State conflicts and further entangled politically (Richards and Singh 2001, accessed on 11 October 2008 from http://people.ucse.edu/~boxjenx/indiawater.pdf).

The non-implementation of the water-sharing agreement, accord, or tribunal award of the waters of the Ravi, Beas, and Sutlej Rivers among Punjab, Haryana, Rajasthan, Jammu, and Kashmir is partly the result of different party affiliations between states per se and between the Centre and the States at various points in times. In summary, an unambiguous institutional mechanism for settling inter-state water disputes does not exist.

Inter-state river water disputes in India constitute an important federal issue. The Sarkaria Commission on Centre-State relations (Government of India 1988) devoted an entire chapter to the problem, and made a series of recommendations that included timely constitution of the Tribunal and grant of award, notification of the award in the Gazette and its implementation, creation of a data bank and information system, and entrusting the same force and sanction behind the Tribunal's award as an award or decree of the Supreme Court to make its award binding. Further on, Iyer (1999) has made some suggestions on the recommendations of the Commission.

22.7 Nature of Inter-State Disputes in India

Disputes began to emerge soon after the framing of the Constitution. Constitutionally and legislatively, Indian inter-state river dispute settlement procedures involve either of two processes: negotiations and legal adjudication. Some small and specific disputes were easily settled through mutual discussions and negotiations. Herein the major issues were related to sharing of the costs and benefits of specific projects or project proposals.

Many disputes involve large river basins. The major contentious issues relate to validity, interpretation, and implementation of agreements made before or after reorganization of states on linguistic bases, arguments over prior appropriation and equitable apportionment (Krishna and Godavari basins), riparian rights of a lower state vis-à-vis subsequent developments and utilisation in an upper co-basin state (Cauvery basin), or vice versa, inter-state utilization of untapped surplus water (Krishna-Godavari basin), intrusion of new competing use (Rihand sub-basin of Sone basin), inequitable operation of common facilities at control points (Yamuna and Chambal basins), seismic risk and safety (Tehri Dam), computation of costs and benefits among the states involved in a project (Bajajsagar Dam between Rajasthan and Gujarat, the Tungabhadra Dam between Andhra and Pradesh), submergence of lands and displacement and suitable rehabilitation of people affected by the implementation of a project (Narmada basin), excess withdrawals by a state, cultural conflict (challenging of hydro project in Sikkim), and quality of river water (Yamuna River) (Swain 1998; Iyer 1994; Shah 1994; Jain et al. 1971). A large number of states in India are engulfed in these issues: Maharashtra, Andhra Pradesh, Karnataka, Madhya Pradesh, Orissa, Kerala, Karnataka, and Tamil Nadu. Disputes that have been referred to Tribunals relate to Krishna, Godavari, Narmada, Cauvery, Indus (the Ravi-Beas waters) basins. There is a growing consensus that the existing mechanisms for adjudicating inter-state water disputes are inefficient and ineffective.

The threat point of no agreement has been the outcome in several major disputes (e.g., Cauvery, Ravi-Beas). A natural outcome is slow economic growth. Hence, these disputes have to be disentangled from the more general Centre–State conflicts, and with everyday political issues.

22.8 Conclusion

This chapter argues that competition is inherent in the ideology of market and hence inevitable. Also, it derives direction and stability from the vast systems of economic and legal cooperation (Nisbet 1968, p. 390). Cooperation is definitely a means of enhancing effectiveness in the struggle for existence. The deepening of global processes has necessitated intensification in the levels of interaction, interconnectedness, or interdependence among the states and societies that constitute the world community (McGrew 1992). The world is experiencing the integration of supranational regions to derive maximum benefit from mutual cooperation. In contrast, in India the states are working in conflict with each other on some issues of national importance. The sharing of inter-state river waters is a case in point. What are the possibilities or compulsions of cooperation?

How can inter-state water disputes be amicably and efficiently resolved? What types of policy recommendations can one make? How can inter-state cooperation be ensured in this context? These questions are not easy to answer, but we must continue to explore possibilities of reducing the conflicts if not completely eliminating them. It has been observed that any agreement reached by a state government allied to the central government and contending states become a source of continued protest by the political opposition and lobbies outside the formal political process.

If India is to be kept integrated, several objectives require addressing while planning for river water resources: economic development, joint management, water quality, quantity, and hydropower. For planning itself to be sustainable, it should be carried out for a larger ecological system of which rivers are a part. The question of how to reconcile the large, centralized, 'top-down,' technology-driven projects and local decentralized, community-based, people-centered alternatives needs to be adequately addressed (Iyer 2000, accessed on 25 September, 2008 from http://Planning-commission.nic.in/reports/genrep/bkpap2020/7-bg2020.pdf).

Therefore, the possibility of a National Water Commission independent of daily political pressures, a federated structure incorporating river basin authorities and water user associations, and fixed time periods for negotiations and adjudication can be explored. Also, the adjudication system provided under the constitutional and legislative framework for resolution of inter-state river water disputes must be made to work better.

The specific recommendations made by the Sarkaria Commission should be implemented; this has also been reiterated by the National Commission on Integrated Water Resources Development Plan in its report submitted in September 1999. There is a need for openness, transparency, and strong regulatory oversight, particularly in an administrative-political context where the institutional framework is weak and regulations are easily circumvented (Molen and Hildering 2005).

Universities and research agencies can contribute substantially to the mitigation of water-related conflicts in at least three major ways. First, they are to collate and interpret relevant primary data in a policy framework, second, they should attempt prognosis of future water disputes and the regions likely to be affected, and finally they should prepare a blueprint of action for preempting such situations.

References

Amin BK, Sastri CJ (1986) The law of easements (revised by G.C. Mathur), 5th edn. Eastern Book Company, Lucknow

Beach L, Hamner J, Hewitt J, Kaufman E, Kurki A, Oppenheimer J, Wolf A (2000) Freshwater dispute transboundary resolution: theory, practice and annotated references. United Nations University Press, New York

Butts K (1997) The strategic importance of water. Parameters 27:65–83

Chitale MA (1992) Development of India's river basins. Int J Water Resour Dev 8(1):30–44

Gleick PH (1993) Water and conflict: fresh water resources and international security. Int Secur 18(1):79–112

Government of India (1918) Report of Indian constitutional reforms. Government Printing Press, Delhi

Government of India (1955) Report of the States Reorganisation Commission. Government Publications Division, Delhi

Government of India (1988) Report of the commission on centre state relations. Sarkaria Commission. Government of India Press, Nasik

Homer-Dixon TF (1994) Environmental scarcities and violent conflict. Int Secur 19:5–40

Iyer RR (1994) Indian federalism and water resources. Water Resour Dev 10(2):191–202

Iyer RR (1999) Inter-state river water disputes: some suggestions. Mainstream, June 5, pp 7–10

Iyer RR (2000) Water: charting a course for the future (a discussion paper). Downloaded from http://Planning-commission.nic.in/reports/genrep/bkpap2020/7-bg2020.pdf on 25 Sept 2008

Jain SN, Jacob A, Jain SC (1971) Interstate water disputes in India. N.M. Tripathi (Under the auspices of The Indian Law Institute, New Delhi), Bombay

Kliot N, Shmueli DF, Shamir U (2001) Institutions for management of trans boundary water resources: their nature, characteristics and shortcomings. Water Policy 3:229–255

Krishan G (2000) How India stayed integrated? Ann Natl Assoc Geogr India 20(2):3–9

Krishan G (2006) A call for paradigm shift: from competition to cooperation. Econ Polit Weekly 11:872–875

Libiszewski S (1995) Water disputes in the Jordan basin region and their role in the resolution of the Arab-Israeli conflict. Occasional paper #13, August. Centre for Security Studies and Conflict Research, Zurich

McGrew AG (1992) Conceptualising global politics. In: McGrew AG, Lewis PG (eds) Global politics: globalization and the nation-state. Polity Press, Cambridge, pp 1–28

Nisbet RA (1968) Cooperation. In: Sills David L (ed) International encyclopaedia the social sciences, vol 3. The Macmillan Company/Free Press, New York

Priscoli DJ (1998) Water and civilization: using history to reframe water policy debates and to build a new ecological realism. Water Policy 1(6):623–636

Ramana MVV (1992) Inter-state river water disputes in India. Orient Longman, Hyderabad

Reuss M (2002) Historical explanation and water issues. Downloaded from webworld.unesco org/water/wwwap/cd/pdf on 20 Sept 2008

Richards A, Singh N (2001) Interstate water dispute in India: institutions and policies, downloaded from http://people.ucse.edu/~boxjenx/indiawater.pdf on 20 Sept 2008
Salman SMA, Boisson de Chazournes L (eds) (1998) International watercourses: enhancing cooperation and managing conflict. Technical paper 414. The World Bank, Washington, DC
Samson P, Charrier B (1997) International freshwater conflict: issues and prevention strategies. Green cross international, Geneva
Shah RB (1994) Inter-state river water disputes: a historical review. Int J Water Resour Dev 10(2):175–189
Shiva V (1991) Ecology and the politics of survival: conflict over natural resources in India. Sage, New Delhi
Singh N (1998) Administration and development of Indian states. Anmol Publications, New Delhi
Singh MP (2008) Reorganisation of States in India. Econ Polit Weekly 15:70–75
Swain A (1998) Fight for the last drop: inter-state river disputes in India. Contemp South Asia 7(2):167–180
Thakkar H (1996) Water of contention. Down to Earth 5(3):5
Toset HPW, Gleditsch NP, Hegre H (2000) Shared rivers and interstate conflict. Polit Geogr 19:971–996
van der Molen I, Hildering A (2005) Water: cause for conflict or co-operation? ISYP J Sci World Aff 1(2):133–143
Verghese BG (1997) Water conflicts in South Asia. Stud Confl Terror 20(2):185–194
Westing AH (ed) (1986) Global resources and international conflict: environmental factors in strategic policy and action. Oxford University Press, New York
Wolf AT (1997) International water conflict resolution: lessons from comparative analysis. Int J Water Resour Dev 13(3):333–365
Wolf AT (1998) Conflict and cooperation along international waterways. Water Policy 1(2):251–265
Wolf AT (1999a) The trans boundary fresh water dispute database project. Water Int 24(2):160–163
Wolf AT (1999b) Water and human security. AVISO: an information bulletin on global environmental change and human security. Bulletin #3, June
Wolf AT, Yoffe SB, Giordano M (2003) International waters: identifying basins at risk. Water Policy 5:29–60
Yoffe S, Fiske G, Giordano M, Giordano M, Larson K, Stahl K, Wolf AT (2004) Geography of international water conflict and cooperation: data sets and applications. Water Resour Res 40:W05S04. doi:10.1029/2003WR002530

Part VII
Energy and Forest Resources

Chapter 23
Nonconventional Energy Resources in India

Jitender Saroha

Abstract Energy is an important input for economic development. Consumption of energy has increased globally manifold as the population has increased, energy consumption in developed countries has magnified, and its consumption in India and China has risen at an enlarged rate among emerging economies. The increasing consumption of energy has led to harmful impacts upon the environment such as fossil fuel consumption and the consequent problem of global warming. The need of the hour is to shift toward nonconventional sources of energy such as sun, wind, and biomass. These sources of energy are abundant, renewable, and environmentally friendly. Given this overview, this chapter emphasizes the following three objectives: (1) to highlight the potential of nonconventional energy sources in India, (2) to analyze the achievements and constraints in their development, and (3) to suggest measures of reducing the gap between potential and achievement. Analysis suggests that government should provide financial assistance, such as fiscal incentives by reducing import duty and excise duty. The Government should provide tax holidays such that industry and private sectors can utilize technology transfers and strive for becoming energy independent while enhancing the use of nonconventional sources of energy.

Keywords Non-conventional sources of energy resources • Solar • Wind • Hydro power • Biomass • Biogas • Ocean thermal energy • Wave • Tidal • Energy recycling and India

---

J. Saroha (✉)
B. R. Ambedkar College, University of Delhi,
Main Wazirabad Road, Yamuna Vihar, Delhi 110094, India
e-mail: Jeetu43@yahoo.com

A.K. Dutt et al. (eds.), *Spatial Diversity and Dynamics in Resources and Urban Development*, DOI 10.1007/978-94-017-9771-9_23

23.1 Introduction

Energy is an essential input for economic development and for improving the quality of life. The energy sector has become a matter of major concern over the years because of the fast-rising price of petrol and diesel. Heavy reliance upon conventional fossil fuels has given birth to indefinite headaches for economies all over the globe. Another major consequence of increasing use of fossil fuels has been the harmful impacts on the environment at both local and global levels (Pachauri 1998). The need of the hour is to rely more on nonconventional sources such as sun, wind, and biomass. Mitigation of global warming is possible only through transition to a low-carbon economy. In other words, large-scale development of renewable energy is the greatest imperative of our times for environmental protection (Pillai 2009). The objectives of this chapter are (1) to highlight the potentials of nonconventional energy resources in India, (2) to analyze the achievements and constraints in their development and, finally, (3) to suggest measures to reduce the gap between potential and achievements.

India is blessed with plenty of renewable energy resources, such as solar, wind, hydro, and biomass. This energy is abundant, renewable, pollution free, and eco-friendly. It can be more conveniently supplied to urban, rural, and remote areas. Thus, the nation is capable of solving the twin problems of supplying energy in a decentralized manner and helping in sustaining a cleaner environment. The country offers cleaner technologies, environmentally sound national growth, and energy diversification. Policy makers have decentralized applications, which are ideally suited for a large country like India to give an impetus to rural development, employment generation, and the empowerment of women. Therefore, they can take the country toward developing the energy sector and acquiring high levels of energy security (Thakur 1998).

The importance of renewable energy was recognized in the country in the early 1970s. The renewable energy program was started with the establishment of the Department of Non-conventional Energy Sources (DNES) in 1982. In 1992, DNES was converted into the Ministry of Non-conventional Energy Sources and was renamed as the Ministry of New and Renewable Energy (MNRE) in 2006. The broad aim of the Ministry is to develop and deploy new and renewable energy to supplement the energy requirements of the country. The Indian Renewable Energy Development Agency (IREDA) was established in 1987 as the only agency of its kind in the world dedicated to the promotion and financing of renewable energy projects. The main objectives of IREDA are (1) to operate a revolving fund for the development and deployment of New and Renewable Sources of Energy (NRSE), (2) to give financial support to specific projects and schemes for generating electricity and/or energy through new and renewable sources and conserving energy through energy efficiency, (3) to bring down the cost of renewable power, (4) to assist in upgrading technologies in the country through new and renewable sources of energy (NRSE), (5) to develop criteria/systems/concepts for financing projects based on new and renewable sources of energy and energy efficiency/conservation, and (6) to strive for improvement in customer satisfaction.

The two main forces impacting energy demand are population growth and economic development (Reddy and Balachandra 2003). To satisfy the needs and wants of the growing population and to sustain the demands of a diversifying and expanding economy, development of renewable energy resources is essential. The need to boost efforts for further development and promotion of renewable energy sources has been felt the world over in light of the high prices of crude oil. As a result of the shortages and increases in the price of oil, the worst affected areas are the rural areas (Chopra 2004). Decentralized renewable energy resources have potential to provide balanced regional development in India.

The renewable energy programs encompass the entire gamut of technologies, including biogas plants, biomass gasifiers, solar thermal and solar photovoltaic systems, wind mills, co-generation, small hydro plants, energy recovery from urban/municipal and industrial wastes, geothermal energy, hydrogen energy, electric vehicles, and biofuels. More than 12,800 MW of grid-connected renewable power capacity (up to 2008) has been installed, mainly through wind, small hydro, and bio-energy, which account for about 8 % of the total installed capacity in the country.

The National Action Plan on Climate Change (NAPCC) released in 2008 proposes to start eight missions, among which one is the National Solar Mission. The NAPCC also mentions that other sources of renewable energy would be promoted. Specific action points include promoting deployment, innovation, and basic research in renewable energy technologies; resolving the barriers to development and commercial deployment of biomass, hydro power, solar, and wind technologies; promoting straight (direct) biomass combustion and biomass gasification technologies; promoting the development and manufacturing of small wind electric generators; and enhancing the regulatory/tariff regime to mainstream renewable energy sources in the national power system. During the past two decades, several renewable energy technologies have been deployed in rural and urban areas. Various policy and support measures initiated from time to time in the country have given a big boost to the spread of a number of renewable energy technologies. But still, the gap between potential and utilization is very large (Table 23.1).

Table 23.1 Renewable energy grid-interactive power potential and achievements by 31.12.2010

Source/system	Estimated potential (MW)	Achievements (MW)
1. Solar photovoltaic power	20 MW/km^2	17.82
2. Wind power	45,000	13,065.78
3. Small hydro power (up to 25 MW)	15,000	2,939.33
4. Biomass/power co-generation	66,000	2,559.13
5. Energy recovery from waste	7,000	72.46

Source: Ministry of New and Renewable Energy, Report 2010–2011

23.2 Solar Energy

India lies in the sunny belt of the world. India receives solar energy equivalent to more than 5,000 trillion kWh per year (20 MW/km^2 per annum). The average solar energy incidence is about 5.5 kWh per square meter per day. There are about 300 clear sunny days in most parts of the country. The highest annual radiation is received in Ladakh, western Rajasthan, and parts of Gujarat; the northeastern region of the country receives relatively lower annual radiation. Research and development efforts to harness solar energy were initiated at the National Physical Laboratory in 1957.

Presently, solar energy is being utilized through two different routes: solar thermal and solar photovoltaic. Solar energy can be converted into thermal energy with the help of solar collectors and receivers. Solar thermal devices are being utilized for water heating, space heating, cooking, drying, water desalination, industrial process heat, steam generation for industrial and power generation applications, and operation of refrigeration systems. Low-grade solar thermal devices such as solar water heaters, air heaters, solar cookers, and solar dryers have been developed and deployed in the country. Efforts are also on making use of solar passive architecture principles to reduce energy consumption and improve comfort conditions in buildings. Solar air heating systems have also been deployed for drying of industrial and agricultural products, which has helped in saving a substantial amount of conventional fuels. The world's largest solar steam cooking system for cooking food for 15,000 people per day has been installed at Trimulla, in Andhra Pradesh. The focus of the solar thermal power program is on research, design, development, standardization, and demonstration on all aspects of the current and emerging technologies. A large solar dish has been set up under a research project at Latur in Maharashtra to provide process heat for milk pasteurization.

Solar photovoltaic (SPV) technology enables conversion of solar radiation into electricity through photovoltaic systems based on single crystal cells of silicon. Solar photovoltaic systems are becoming an increasingly attractive source of decentralized electricity generation, especially in remote and energy-deficient areas, for meeting the essential requirements such as lighting, water pumping, powering primary health centers, community centers, schools, and similar requirements. Solar photovoltaic systems are the most reliable power source for unmanned applications, such as TV transmitters and battery charging. Under the solar photovoltaic demonstration and utilization program, about 8.17 lakh solar lanterns, 6.56 lakh solar home systems, 122,697 solar street lighting systems, and 7,148 solar water pumping systems have been installed in the country up to March 2010. In addition, 8,033 remote villages and hamlets have been electrified using different solar photovoltaic systems (Table 23.2). More than 700,000 solar PV systems aggregating about 57 MW are operational in the country, which makes this the largest such deployment in the world. Solar lighting systems are now being used in 380,000 homes, contributing to substantial savings in kerosene. About 190,000 rural radio telephones are also being powered by solar energy.

Table 23.2 Decentralized energy systems

Family type biogas plants	4.31 million
Solar street lighting systems	122,697
Home lighting systems	656,707
Solar lanterns	817,369
Solar water heating systems	3.97 million m^2
Solar cookers	6.0 lakhs
Solar photovoltaic (PV) pumps	7,015
Wind pumps	1,111
Battery-operated vehicles	212

Source: Ministry of New and Renewable Energy, Report 2010–2011

A 50-kW solar thermal power plant as a research and design facility was installed in the Solar Energy Centre near Gurgaon in Haryana during 1978. Two 100-kW grid interactive power plants have been installed at Kalyanpur in Aligarh district and Saraisadi in Mau district of Uttar Pradesh. A 100-kW power system from SPV solar cells has been installed at Walwahan Dam in Lonawala district of Maharashtra by the private company Tata Electric. Apart from this, two projects of 25-kW capacity are being installed in the S.N. Palayam and S.G. Palayam villages of Coimbatore, Tamil Nadu. The Sagar Island in West Bengal has become a solar island capable of meeting all energy needs from solar energy. The world's largest solar pond is in India at Madhapar, near Bhuj, Kachchh in Gujarat in an area of 60,000 m^2, providing 80,000 l of hot water at 70 °C. The Kachchh dairy uses hot water for sterilizing its milk cans. An important application of a solar pond is to obtain potable water.

23.2.1 Jawaharlal Nehru National Solar Mission

In the 11th Five-Year Plan, the government of India in January 2010 has initiated the Jawaharlal Nehru National Solar Mission. The objective is to establish India as a global leader in solar energy by creating the policy conditions for its diffusion across the country as quickly as possible. It is a major initiative of the Government of India and State Governments to promote ecologically sustainable growth while addressing India's energy security challenge. It will also constitute a major contribution by India to the global effort to meet the challenges of climate change. It has a twin objective: to contribute to India's long-term energy security and to its ecological security. The Solar Mission recommends the implementation in three stages leading up to an installed capacity of 20,000 MW by the end of the 13th Five-Year Plan in 2022. It is envisaged that as a result of rapid scale-up as well as technological developments, the price of solar power will attain parity with grid power at the end of the Mission, enabling accelerated and large-scale expansion thereafter.

The Mission will establish a single-window investor-friendly mechanism, which reduces risk and at the same time provides an attractive, predictable, and sufficiently extended tariff for the purchase of solar power for the grid.

The Mission also includes a major initiative for promoting rooftop solar PV applications. A solar tariff announced by the regulators will be applicable for such installations. The power distribution companies will be involved in purchase of this power. There are several off-grid solar applications that are already commercially viable, or nearly so, where rapid scale-up is possible; this requires regulatory and incentive measures as well as an awareness campaign. Solar thermal heating applications, such as water heaters, fall in this category. Solar lighting systems for rural and remote areas are already being distributed commercially in several parts of the country.

Currently, the bulk of India's solar PV industry is dependent on imports of critical raw materials and components, including silicon wafers. Transforming India into a solar energy hub would include a leadership role in low-cost, high-quality solar manufacturing, including balance of system components. Proactive implementation of the special incentive package (SIPs) policy, to promote PV manufacturing plants, including domestic manufacture of silicon material, would be necessary. Indigenous manufacturing of low-temperature solar collectors is already available; however, manufacturing capacities for advanced solar collectors for low-temperature and concentrating solar collectors and their components for medium- and high-temperature applications need to be built. An incentive package, similar to SIPs, could be considered for setting up manufacturing plants for solar thermal systems/devices and components. The SME (small and medium enterprises) sector forms the backbone for manufacture of various components and systems for solar systems. It would be supported through soft loans for expansion of facilities, technology upgrading, and working capital. IREDA would provide this support through refinance operations. It should be ensured that transfer of technology is built into Government and private procurement from foreign sources.

The major research and development initiative of the Mission is to focus first on improvement of efficiencies in existing materials, devices, and applications and on reducing costs of balance of systems, establishing new applications by addressing issues related to integration and optimization; and second, on developing cost-effective storage technologies that would address both variability and storage constraints, and on targeting space intensity through the use of better concentrators, application of nano-technology, and use of better and improved materials. The Mission will be technology neutral, allowing technological innovation and market conditions to determine technology winners. A Solar Research Council will be set up to oversee the strategy, taking into account ongoing projects, availability of research capabilities and resource, and the possibilities of international collaboration. An ambitious human resource development program, across the skill chain, will be established to support an expanding and large-scale solar energy program, both for applied and research and for development sectors.

23.3 Wind Energy

Wind energy is the kinetic energy associated with movement of large masses of air resulting from differential heating of the atmosphere by the sun. Wind energy is renewable and poses no major environmental threats (Dayal 1989). Wind is commercially and operationally the most viable renewable energy resource and, accordingly, is emerging as one of the largest sources in terms of the renewable energy sector (Athawale 1994). A wind electric generator converts the kinetic energy available in wind to electrical energy by using a rotor, gearbox, and generator. The amount of energy produced by a wind machine depends upon the wind speed and the size of the blades in the machine. In general, when the wind speed doubles, the power produced increases eight times. Larger blades capture more wind. As the diameter of the circle formed by the blades doubles, the power increases four times.

The Wind Resource Assessment Program is being coordinated by the Centre for Wind Energy Technology (C-WET). This center was established in Tamil Nadu in 1998 as an autonomous institute under the administrative control of the Ministry of New and Renewable Energy. It has so far covered 31 States and Union Territories, involving the establishment of 1,244 wind monitoring and wind mapping stations. C-WET has identified 233 potential sites for developing wind farms in the country; as of January 2011, the installed capacity of wind power in India was 12,129 MW (Table 23.3). Wind power accounts for 6 % of India's total installed power capacity, and it generates 1.6 % of the country's power. A cumulative total of more than 54 billion units of electricity has been fed to the State Electricity Grids. The Ministry of New and Renewable Energy (MNRE) has fixed a target of 10,500 MW during 2007–2012, but an additional generation capacity of only about 6,000 MW might be available for commercial use by 2012. The promotional policies include a package of fiscal and financial incentives that includes concessions such as 80 % accelerated depreciation, concessional custom duty on specified items, excise duty exemption,

Table 23.3 Installed and potential wind power in India (2010)

State	Potential power (MW)	Installed power (MW)
Karnataka	11,513	1,517.0
Gujarat	10,645	1,934.1
Andhra Pradesh	8,968	138.5
Tamil Nadu	5,530	5,072.8
Rajasthan	4,858	1,095.9
Maharashtra	4,584	2,107.8
Kerala	1,171	27.8
Madhya Pradesh	1,019	231.2
Orissa	255	–
Total	48,561	12,129.20

Source: Ministry of New and Renewable Energy, Report 2010–2011

sales tax exemption, and income tax exemption for 10 years. In addition, State Electricity Regulatory Commissions (SERCs) are determining preferential tariffs and IREDA provides loans for setting up wind power projects.

In India, efforts to use the latent power of wind began in 1985 when in the joint sector a windmill of 1.6 megawatt (MW) capacity was established at Mandvi, Gujarat. Another wind farm of 28 MW is located in Gujarat at Lamba. The Natural Energy Processing Company established in 1986 has developed wind energy in Tamil Nadu in collaboration with MICON of Denmark. Asia's largest wind farm cluster is located at Muppandal in Tamil Nadu. Other major wind farms are located at Kayathar in Tamil Nadu, at Akal in Jaisalmer District, Rajasthan, Samana in Rajkot District, at Motisindholi in Kachchh District of Gujarat, at Kanjikode in Palakkad District, Kerala, and at Jagmin in Satara District, Maharashtra.

The Indian Wind Turbine Manufacturers Association (IWTMA) has a leading role in promoting wind energy in India. Although a relative newcomer to the wind industry compared with Denmark or the U.S., a combination of domestic policy support for wind power and the rise of Suzlon (a leading global wind turbine manufacturer) have led India to become the country with the fifth largest installed wind power capacity in the world. The worldwide installed capacity of wind power reached 120,798 MW by the end of 2008; the U.S. (25,170 MW), Germany (23,903 MW), Spain (16,754 MW), and China (12,210 MW) are ahead of India.

The short gestation period for installing wind turbines, and the increasing reliability and performance of wind energy machines, has made wind power a favored choice for capacity addition in India. In fact, during the past few years the wind energy industry has developed rapidly, thanks to the active participation of the private sector, the package incentives and promotional policies of the MNES, and financing through the Indian Renewable Energy Development Agency (IREDA). The legal clarity and certainty of regulatory principles together with a supportive policy framework has ensured continued developer interests in the wind sector, which has ultimately resulted into significant growth in harnessing wind energy across various States (Joshi 2009). The capacity addition by wind farms has far exceeded the tenth Plan target. India's potential for wind power has been assessed at 45,000 MW assuming 3 % land availability. Wind power has advantages over fossil fuels on social, economic, and ecological grounds (Table 23.4).

Against these advantages, wind power also has the following limitations. (1) Wind machines must be located where strong, dependable winds are available most of the time. The ideal wind speed would be between 8 and 23 meters per second (m/s); rotors stop moving beyond or below this range. India has a national wind average of 9.4 m/s. (2) Because winds do not blow strongly enough to produce power all the time, energy from wind machines is considered "intermittent," that is, it comes and goes. Therefore, electricity from wind machines must have a backup supply from another source. (3) As wind power is intermittent, utility companies can use it for only part of their total energy needs. (4) Wind towers and turbine blades are subject to damage from high winds and lightning. The rotating parts, which are located high off the ground, can be difficult and expensive to repair. (5) Electricity produced by wind power sometimes fluctuates in voltage and power factor, which can

Table 23.4 Advantages of wind power over fossil fuels

Concerns	Wind power	Fossil fuels
Availability	Usable as it exists	Have to be procured and made usable through laborious and environmentally damaging processes
Limitation on availability	Inexhaustible resource, highly decentralized	Limited in reserves, expected to become completely exhausted in the coming 60 years, centralized distribution
Transportation	Used where it is available	Have to be transported from the site for further processing, exposing environment to danger
Economic implications	Cheapest source of electrical energy (on a leveled cost over 20 years), requires least equity participation, low-cost debt availability, fastest payback period, lowest gestation period, low operation and maintenance, no marketing risks	Costly and cost increases over the period of time long gestation period, high operation and maintenance costs, more investment in manpower
Geo-political implications	Reduces our reliance on oil, safeguarding national security	Overreliance on oil as a resource has undermined our energy security, e.g., OPEC crises of 1973, Gulf War of 1991, and Iraq War of 2003; the Jasmine revolution 2010–2011
Effects on environment	There is no adverse effect on the global environment; the whole system is pollution free and environment friendly	Pollution, global warming, and acid rain

cause difficulties in linking its power to a utility system. (6) The noise made by rotating wind machine blades can be annoying to nearby neighbors. (7) People have complained about the aesthetics of wind machines and the avian mortality caused by these machines.

23.4 Small Hydro Power (SHP)

Small hydro power (SHP) projects have the potential to provide energy in remote and hilly areas where extension of the grid system is either not possible or is uneconomical. These projects are economically viable, environmentally benign, and have relatively short gestation periods. It is perhaps the oldest renewable energy technique known to mankind for mechanical energy conversion as well as electricity generation. In India, hydro projects up to 25 MW station capacity have been categorized as small hydro power (SHP) projects. Small hydro sources could yield

Table 23.5 Selected state-wise installed capacity of small hydro power (SHP) projects (as of 31.1.2011)

State	Potential (MW)	Installed capacity (MW)
Himachal Pradesh	2,267.81	375.38
Uttarakhand	1,577.44	134.12
Jammu and Kashmir	1,417.80	129.33
Arunachal Pradesh	1,328.68	78.83
Chhatishgarh	993.11	19.05
Madhya Pradesh	803.64	86.16
Karnataka	747.59	725.05
Maharashtra	732.63	263.82
Kerala	704.1	136.87
Tamil Nadu	659.51	94.05
Andhra Pradesh	560.18	189.93
West Bengal	396.11	98.9
Punjab	393.23	153.2
Total	15,384.15	2,953.58

Source: Ministry of New and Renewable Energy, Report 2010–2011

15,000 MW of power. The total installed capacity of small hydro power projects as of 31 March 2009 was 2,953 MW from 801 projects, and 271 more projects with an aggregate capacity of 914 MW are under construction (Table 23.5). Most capacity addition in this sector has been achieved through the private sector.

In the early 1990s, most of the SHP projects were set up in the public sector; from the last 10 years or so, most of the capacity addition has come through private sector projects. The beginning of the twenty-first century saw near commercialization in the small hydro sector. Private sector entrepreneurs found attractive business opportunities in small hydro, and state governments also thought that private sector participation may be necessary in tapping the full potential of rivers and canals for power generation. The private sector has been attracted by these projects because of their small adoptable capacity matching with their captive requirements or even as affordable investment opportunities. Some of the major small hydel projects are Kothi (100 kW) in Manali District, Titang (800 kW) and Juthed (100 kW) in Chamba District, Himachal Pradesh, and Kalmoni (200 kW) near Guwahati in Assam, and 8 MW at Tungabhadra Dam, Andhra Pradesh.

23.5 Biomass Power/Co-generation

The biomass power/co-generation program aims at the optimal utilization of a variety of biomass materials for power generation through the adoption of efficient and state-of-the-art conversion technologies. So far, a total capacity of 222 MW biomass power-generating systems has been installed in the country. Projects of capacity of 332 MW are under installation. The biomass power/co-generation program is implemented with the main objective of promoting technologies for optimal use of country's biomass resources for grid and off-grid power generation. Biomass materials successfully used for power generation include bagasse, rice husks, straw, cotton stalks, coconut shells, soya husk, de-oiled cakes, coffee waste, jute waste, groundnut shells, and sawdust. The technologies being promoted include combustion/co-generation and gasification either for power in captive or grid-connected modes or for heat applications.

The current availability of biomass in India is estimated at about 500 million metric tons per year. Studies sponsored by the Ministry have estimated surplus biomass availability at about 120–150 million metric tonnes per annum, covering agricultural and forestry residues corresponding to a potential of about 16,000 MW. This apart, about 5,000 MW additional power could be generated through bagasse-based co-generation in the country's 550 sugar mills, if these sugar mills were to adopt technically and economically optimal levels of co-generation for extracting power from the bagasse these produce.

The sugar industry has traditionally practiced co-generation by using bagasse as a fuel. With the advancement in the technology for generation and utilization of steam at high temperature and pressure, the sugar industry can produce electricity and steam for its own requirements. It can also produce significant surplus electricity for sale to the grid using the same quantity of bagasse. For example, if steam generation temperature/pressure is raised from 400 °C/33 bar to 485 °C/66 bar, more than 80 KWh of additional electricity can be produced for each ton of cane crushed. The sale of surplus power generated through optimum co-generation would help a sugar mill to improve its viability, apart from adding to the power generation capacity of the country.

The Ministry has been implementing the biomass power/co-generation program sincethe mid-1990s. A total of 203 biomass power and co-generation projects aggregating to 1,677 MW capacity have been installed in the country for feeding power to the grid. In addition, around 171 biomass power and cogeneration projects aggregating to 1,850 MW of electricity are under various stages of implementation. Co-generation projects in sugar mills include 82 projects with installed capacity aggregating to 690 MW. Another 107 projects are under implementation, aggregating to 1,280 MW. The leading States for biomass power projects are Uttar Pradesh, Andhra Pradesh, Karnataka, Chhattisgarh, Maharashtra, and Tamil Nadu (Table 23.6).

Besides central financial assistance, fiscal incentives such as 80 % accelerated depreciation, concessional import duty, excise duty, and tax holiday for 10 years are available for biomass power projects. The benefit of concessional custom duty and excise duty exemption on equipment is provided. In addition, the State Electricity Regulatory Commissions have determined preferential tariffs and Renewable Purchase Standards (RPS), and the Indian Renewable Energy Development Agency (IREDA) provides loan for setting up bagasse co-generation projects.

Table 23.6 Selected commissioned biomass power/co-generation projects by state (as of 30.11.2008)

State	Total (MW)
Uttar Pradesh	372.50
Andhra Pradesh	334.25
Tamil Nadu	308.70
Karnataka	274.28
Chhattisgarh	156.10
Maharashtra	155.50
Rajasthan	31.30
Punjab	28.00
Haryana	6.00
Madhya Pradesh	1.00
Gujarat	0.50
Total	1,677.13

Source: Ministry of Non-conventional Energy Sources, Report 2009

23.6 Biogas Energy

In India the National Biogas Program was initiated in 1981–1982 for the promotion of family-size biogas plants with the aim of providing a clean and cheap source of energy along with other benefits, such as enriched organic manure, improved sanitation and hygiene, and reduction in drudgery for women. Biogas can be offered as an excellent fuel for many energy applications such as cooking, lighting, and motive power. The biogas program is implemented through the State Governments and administrators, corporate/registered bodies, Khadi and Village Industries Commission (KVIC), and nongovernmental organizations (NGOs). More than 4 million family-size biogas plants have been installed so far. India is second only to China in biogas plants (India 2011). In addition, enriched organic manure is produced from biogas plants to supplement and complement environmentally degrading chemical fertilizers. The biogas plants and improved wood stoves presently in use are resulting in saving more than 13 million tons of fuel wood every year, besides producing 45 million tons of enriched organic manure.

To propagate large-scale use of biogas technologies, financial subsidy is provided for installation of biogas plants on a turnkey basis with free maintenance for the first 3 years. Against a potential of about 12 million biogas plants based on cattle dung, about 4 million plants have been set up, thus covering 33 % of the potential. The National Program on Improved Chulha (NPIC), launched in 1986–1987, plans to replace traditional stoves with thermally efficient ones. The program aims at conservation of fuel, reduction of smoke in kitchens, and checking deforestation. A total of 32 million thermally efficient *chulhas* have been installed, covering about 27 % of the estimated potential of 120 million households. These stoves are expected to save 10 million tons of fuel wood per annum.

23.7 Energy Recovery from Urban Wastes

The industrialization, urbanization, and changes in the pattern of life that accompany the process of economic growth generate increasing quantities of wastes, leading to increased threats to the environment. In recent years, technologies have been developed that not only help in generating substantial quantities of decentralized energy but also in reducing the quantity of waste for its safe disposal.

The Ministry is promoting all the technology options available for setting up projects for recovery of energy from urban wastes. Although incineration and biomethanation are the most common technologies, pyrolysis and gasification are also emerging as preferred options. A common feature in most developed countries is that the entire waste management system is being handled as a profitable venture by private industry or nongovernment organizations, with a tipping fee for treatment of waste being one of the major revenue streams. The major advantages for adopting technologies for recovery of energy from urban wastes is to reduce the quantity of waste and net reduction in environmental pollution, in addition to generation of substantial quantity of energy.

The main objectives of the proposed National Programme on Energy Recovery from Urban Wastes are as follows: (1) to accelerate the promotion of setting up of projects for recovery of energy from urban wastes, (2) to create conducive conditions and environment, with a fiscal and financial regime, to develop, demonstrate, and disseminate utilization of wastes for recovery of energy, and (3) to harness the available potential of MSW (Municipal Solid Waste) to energy by the year 2017.

According to a recent estimate, about 42 million tons of solid waste (1.15 lakh tons per day) and 6,000 million cubic meters of liquid waste are generated every year by our urban population. Three projects for energy recovery from Municipal Solid Wastes with an aggregate capacity of 17.6 MW have been installed at Hyderabad, Vijayawada, and Lucknow. Other urban waste projects include a 1-MW project based on cattle manure at Haebowal, Ludhiana; a 0.5-MW project for generation of power from biogas at a sewage treatment plant at Surat; a 150-KW plant for vegetable market and slaughterhouse wastes at Vijayawada; and a 400-KW plant for production of biogas from slaughterhouse wastes at Medak, Andhra Pradesh. Another 300-KW project based on vegetable market waste is under commissioning at Chennai.

The major advantages of setting up of waste-to-energy projects are (1) the quantity of waste is reduced by nearly 60–90 %, depending upon waste composition and the technology adopted; (2) demand for land, which is already scarce in cities, for land filling is reduced; (3) the cost of transportation of waste to far-away landfill sites is reduced; and (4) there is net reduction in environmental pollution and, apart from generating power from the waste, the slurry produced from biomethanation technology acts as a good fertilizer.

The growth of this sector has been affected by the following limitations/constraints. (1) Waste-to-Energy is still a new concept in the country; (2) most of the proven and commercial technologies in respect of urban wastes must be imported;

(3) the costs of the projects, especially those based on biomethanation technology; are high as critical equipment for a project must be imported; (4) in view of the low level of compliance of MSW Rules 2000 by the Municipal Corporations/Urban Local Bodies, segregated municipal solid waste is generally not available at the plant site, which may lead to nonavailability of waste-to-energy plants; (5) lack of financial resources with Municipal Corporations/Urban Local Bodies; and (6) lack of conducive Policy Guidelines from State governments in respect of allotment of land, supply of garbage, and power purchase/evacuation facilities.

The recent developments in Municipal Solid Waste (MSW) management are the following. (1) Notification of Municipal Solid Wastes (Management and Handling) Rules of 2000 necessitates all Class I cities to provide proper treatment and disposal facilities for MSW; (2) the Twelfth Finance Commission has recommended that at least 50 % of the grants provided to States for Urban Local Bodies (ULBs) should be utilized to support the cost of collection, segregation, and transportation (segregated wastes require relatively simpler and less expensive equipment and devices for conversion into energy); (3) initiatives being taken under the National Urban Renewal Mission are expected to give a major boost to the efforts for improving waste management in 60 large cities; and (4) the Parliamentary Standing Committee on Energy, while appreciating the installation of a 6-MW power plant based on MSW at Hyderabad, desired that similar projects should be taken up for all the major cities where huge quantities of solid waste are generated.

23.8 Ocean Thermal Energy Conversion

Basically, there are three ways of generating power from the high seas: (1) taming the waves, (2) harnessing tidal power, and (3) using the difference in temperatures between the layers of the ocean. The last technique is called Ocean Thermal Energy Conversion (OTEC). The Tamil Nadu government signed a memorandum of understanding in 1994 with the Sea Solar Power Company of USA to set up six units of 100 MW each at some 30 km off the coast of southern Tamil Nadu. The idea is simple: the OTEC operation exploits the temperature difference between the surface of the sea, and at a depth of 1,000 m or more, to extract energy. In tropical countries such as India, the strategy works even better as the temperature gradient in the sea is as great as 25 °C.

23.9 Wave Energy

Wave energy potential in India is estimated at about 40,000 MW. One wave energy power plant of 150 KW has been installed at Vizhinjam near Thiruvananthapuram, Kerala. The plant has been declared as a national facility for wave energy and wave application studies. The center has a facility for testing the design of turbines and metals against sea corrosion. The Ocean Development Department has taken up two new projects for breakwater integrated wave energy systems at Thangassery in Kerala and Mus Point in Car Nicobar.

23.10 Tidal Energy

Energy can be extracted from tides by creating a reservoir or basin behind a barrage and then passing tidal waters through turbines in the barrage to generate electricity. Tidal energy is extremely site specific; it requires mean tidal differences greater than 4 m and also favorable topographic conditions, such as estuaries or certain types of bays, to bring down costs. As India has a coastline of 7,516 km, its potential to harness tidal energy has been recognized by the Government of India. Potential sites for tidal power development have already been located. The most attractive locations are the Gulf of Cambay and the Gulf of Kachchh on the west coast where the maximum tidal range is 11 m and 8 m with an average tidal range of 6.77 m and 5.23 m, respectively. The Ganges Delta in the Sunderbans in West Bengal also has good locations for small-scale tidal power development. The maximum tidal range in Sunder bans is approximately 5 m with an average tidal range of 2.97 m.

The identified economic tidal power potential in India is of the order of 8,000–9,000 MW, with about 7,000 MW in the Gulf of Cambay, about 1,200 MW in the Gulf of Kchchh, and less than 100 MW in Sundarbans. The Kachchh Tidal Power Project with an installed capacity of about 900 MW is estimated to cost about Rs 1,460/-crore generating electricity at about 90 paisa (Indian monetary unit) per unit. A 3.75-MW capacity demonstrative tidal power project at Durgaduani Creek in the Sunderbans area was started by the West Bengal Energy Development Agency (WBREDA) in February 2008. The main objective of the project is to supply power to 11 remote villages in Gosaba and Bali Bijaynagar Islands located in the South 24 Parganas district of West Bengal.

There are vast possibilities of developing and exploiting geothermal energy in India. About 340 hot spring localities have been identified; many of these have a temperature nearing boiling point. Extensive surveys are being conducted to develop geothermal energy for direct heat and power generation. A 5-KW geothermal pilot power plant has been commissioned by the National Aeronautical Laboratory at Manikaran in Kullu district of Himachal Pradesh. A potential of 5 kilowatts (KW) geothermal energy has been estimated in Puga valley of Ladakh in Jammu and Kashmir. A project of mushroom cultivation and poultry farming using geothermal power is under implementation at the Regional Research Laboratory, Jammu. Studies carried out by the National Geophysical Research Institute (NGRI), Hyderabad have shown the existence of potential geothermal sites in Surajkhand in Jharkhand and in Tapovan in Uttrakhand.

In spite of the aforementioned achievements, the record for generating nonconventional and renewable energy has rather been poor to date. With the exception of biogas plants and improved *chullahs*, most of the other projects are still at the take-off stage. The major reasons for this state of affairs are the following. (1) Compared to conventional energy sources, the nonconventional sources are not economically viable because of various technological and other constraints. Economic viability is critical to the success of a renewable energy program. In India, this is particularly relevant, because currently several renewable energy technologies require large initial investments and, hence, with high interest rates, their cost in terms of present value turns out to be unfavorable in relationship to conventional alternatives, which may have lower initial costs but high annual costs (Pachauri 1998). (2) There is a lack of linkage of technology with industry, and there have always been low budgetary allocations for their promotion; for example, in the Eighth Five-Year Plan allocation for the renewable energy sector represented only 0.8 % of the total funds allocated to energy. (3) The major external barriers are price distortions, differential taxation regimes, inadequate decentralized delivery mechanisms and infrastructure, and grid limitations in the electricity sector (Padmanabhan 2004). (4) Lack of product maturity in terms of user-friendly features, reliability, and marketing and maintenance support are other drawbacks (Padmanabhan 2004). And, (5) it is envisaged that India's electricity sector is likely to face serious challenges in meeting the demand for a skilled and semiskilled workforce to match the massive capacity addition during the 11th and 12th Five-Year Plans. The need would range from skilled workers who can manufacture, build, and operate the facilities to graduates who understand the technical, commercial, and social context of the industry (Joshi 2009).

23.11 Conclusion

The nonconventional energy resources available in India are abundant, renewable, and environment friendly. They have the potential to sustain a high gross domestic product (GDP) growth rate and provide a base for balanced regional development. The importance of renewable resources was recognized in the country in the early 1970s, but the gap between potential and utilization is huge. In the case of wind power, only one fourth of the potential, and in small hydro power only one fifth of the potential, have been achieved. In family-type biogas plants, one third of the potential has been achieved. In wave energy, OTEC, tidal energy, and geothermal energy, the majority of projects are still at the experimental level. To make a real beginning in the solar power sector, in 2010 the Jawaharlal Nehru National Solar Power Mission was started. To reduce the gap between potential and utilization, central financial assistance, fiscal incentives, concessional import duty, excise duty, and tax holiday schemes are required. Linkage of the technology with industry needs to be established. Technology transfer, upgrading technology, and involvement of the private sector should be encouraged. Government should play the most significant role by providing budgetary support for research, development, and demonstration of technologies, facilitating institutional finance through various financial institutions, and promoting private investment through fiscal incentives, tax holidays, depreciation allowance, and remunerative returns for power fed into the grid. Energy security, energy availability, and energy access and affordability are key concerns to be addressed while developing new and renewable energy resources in India.

References

Athawale SB (1994) Wind energy. National Book Trust, New Delhi

Chopra SK (2004) Sustainable energy security, vol 48. Yojana, Ministry of Information and Broadcasting Publication, New Delhi, pp 5–9

Dayal M (1989) Energy: today and tomorrow. Publication Division, Ministry of Information and Broadcasting, Government of India

India (2011) A reference annual. Publication Division, Ministry of Information and Broadcasting, Government of India, pp. 310–318

Joshi B (2009) Wind power development: regulatory and policy perspectives, vol 6, IREDA News. IREDA Publications, New Delhi, pp 27–32, No. 1, January–March, 2009

Pachauri RK (1998) Renewable sources of energy: prospects and policies, vol 42. Ministry of Information and Broadcasting Publication, New Delhi, pp 23–27, No. 1

Padmanabhan BS (2004) Renewables geared to play pivotal role, vol 48. Yojana, Ministry of Information and Broadcasting Publication, New Delhi, pp 41–44

Pillai GM (2009) Renewable energy law, regulation and tariff, vol 6, IREDA News. IREDA Publications, New Delhi, pp 21–26, No. 1, January–March, 2009

Reddy BS, Balachandra P (2003) A sustainable energy strategy for India revisited. Econ Polit Wkly 37:5264–5276

Thakur B (1998) The energy resources of India: problems and planning. Regional Symbiosis 6:76–91

Chapter 24
Joint Forest Management in India

Punyatoya Patra

Abstract Joint Forest Management (JFM) is a partnership in forest management among state forest departments and local communities in India. The policies and guidelines of the JFM were enunciated in the Indian National Forest Policy of 1988 and JFM guidelines of 1990 proposed by the Government of India. These guidelines were further revised in 2000. The basic rationale of the JFM approach is the cooperation of local communities and the state government in the protection of forest resources from fire, illegal grazing, and timber cutting, in exchange for which they receive non-timber forest products. Indian JFM programs are characterized by a joint venture approach; they are different from other community management schemes, and are a cross between the top-down and bottom-up approaches. This chapter discusses two case studies, Morni-Pinjore and Yamunanagar Forest Division (Haryana) and Janaram Forest Division in Andhra Pradesh, to shed light on the problems and prospects of JFM in India. It was concluded that there is spatial disparity in the success and failure of JFM programs between states and within states in India.

Keywords Joint Forest Management (JFM) • National Forest Policy • History of Forest management in India • Morni-Pinjore and Yamunanagar Forest Division • Janaram Forest Division

24.1 Introduction

In the Joint Forest Management (JFM) system, the Forest Department and local communities jointly protect and manage forest resources and share the costs as well as the benefits thereof. India is one of the pioneering countries in the world in which forest management regimes stress partnerships between the State Forest Department and local communities. The main driving force behind this is the '1988 Forest Policy,' which envisages people's involvement in the development and protection of

P. Patra (✉)
Department of Geography, Aditi Mahavidyalaya, University of Delhi,
New Delhi, Delhi, India
e-mail: punyatoyapatra@gmail.com

A.K. Dutt et al. (eds.), *Spatial Diversity and Dynamics in Resources and Urban Development*, DOI 10.1007/978-94-017-9771-9_24

forests. Acknowledging this, the Ministry of Environment and Forest, Government of India issued policy guidelines, on 1 June 1990, that involve both Forest Department and village communities in the management of forests, popularly known as Joint Forest Management. In February 2000, these guidelines were revised to strengthen the program. To date, 27 states have issued enabling orders for the implementation of JFM.

As JFM is a major change/shift in managing forests in India, from state control to community control, a number of issues of importance have emerged, mainly because JFM is viewed differently by different people. Proper understanding of the concept of JFM is necessary to make the program successful. Therefore, this chapter critically reviews the JFM concept, its characteristics, the perceptions of different groups of people, etc., along with two case studies. These two case studies from two different states of India have also been reviewed to understand the problems and prospects of JFM in India.

24.2 How JFM Works

At the initial stage, the JFM programme has been implemented in degraded forest lands, especially in the vicinity of habitation/villages. The main objective behind it was to afforest and develop degraded forest areas. However, gradually the program was extended to good forest areas for better resource planning and management. To implement the JFM program, the funding is arranged in two ways: first, Government of India finances in the shape of a grant-in-aid from the National Afforestation and Eco-Development Board, and second, the village community also obtain funds from other government agencies, industrial houses, and different foreign agencies for undertaking these activities. Non-government organizations (NGOs) are also involved in the JFM program. They motivate local people to participate in a newly framed community management program. They also arrange workshops to train both Forest Department and local communities, as the Forest Department officials are not trained work with a community and the local people also have always feared Forest Department officials. The NGOs have also a great role in settling conflicts among different user groups.

Access to forest land and usufruct benefits are given to the beneficiaries, who are organized into the village institution known as the Forest Protection Committee (FPC). Each FPC consists of the residents of one village or of a group of hamlets who are interested in protecting and regenerating the forest. Each FPC is given a particular patch of forest land for which they care. The members of the FPC form a core committee, popularly known as the Executive Committee, for a term of 2 years. The Executive Committee prepares a micro plan for execution. The committee has 10 to 15 members, including the village head, ward member, forest officials, and elected members from the FPC, nominees of concerned NGOs, etc. The membership rule varies from state to state. The beneficiaries (members of FPC) are given usufructs such as grasses, lopped-off branches, branch tops, and other minor forest produce, free of cost. Further, if they successfully protect the forests, they are given a portion of the share from the sale of trees when they mature; the other portion of the share goes to the Forest Department. The share division between Forest Department and village communities varies from state to state.

24.3 Origin of the Concept in India

The concept was originated during the late nineteenth century, when Deitrich Brandis, founder of the Indian Forest Service, had anticipated that a system of forestry founded exclusively on state control would lead to great dissatisfaction in the countryside (Johnson 1995). Brandis had also suggested a system of forest management based on a collaborative relationship between the state and local communities during that time. However, Brandis's suggestion was not accepted by Indian forestry, which preferred the principle of a state monopoly over forest land. However, his idea has a strong contemporary resonance in the form of JFM. After a gap of almost 100 years, in the early 1970s, Ajit K. Banerjee, a Divisional Forest Officer, and silviculturists in the Arabari Forest range of West Bengal motivated people from ten villages in Arabari, involved them in protecting 1,250 ha of totally degraded natural sal forest, and were successful. The officer had realized that it would be difficult to manage the forest without the cooperation of the local people who depend on it for their livelihood. The entire idea gained further ground in the late 1980s when the same officer, working in the World Bank, was able to sell the idea. Later, Dr. Amitabh Mukherjee, the then Minister in Charge (MIC) of Forests, was able to convince the government; as a result, the Government of India passed the JFM resolution in 1990.

Almost during the same time period of Ajit K. Banerjee's Arabari Forest Protection, P.R. Mishra, a retired soil scientist, became famous for his pioneering work in protecting and developing the catchment of Sukhna Lake in the Sukhomajri area of Haryana. The model he developed for the same is popularly known as Chakriya Vikas Pranali (CVP). The basic idea of CVP is the pooling of private land/individual land or even village common land and putting them together as a Common Property Resources (CPR) land. After pooling the land, its management is predominantly based upon an agro-forestry system. The management is done by the workers, who are enrolled among the residents of the village. Last, the net profit is divided by the landowners and the workers. The workers and owners in CVP can be compared with communities and Forest Departments in the JFM Programme, respectively.

24.4 Preexisting Environment in Which JFM Is Implemented

JFM is not being implemented in a vacuum; it is being implemented in a specific context, in a preexisting environment. Therefore, knowledge of forest management in India is necessary to understand the present-day forest management system. Here, the history of forest management in India has been divided into three broad periods: Pre-Colonial, Colonial, and Post-Colonial. During the *Pre-Colonial Period*, forests of the region were almost treated as an 'open-access resource,' that is, open to everybody. The local people used to cut trees for their basic needs such as food, fodder, and fuel with traditional management practices. The *Colonial Period* witnessed rapid exploitation of the forests. During that period forests were recognized as commercially important resources and monopolized by colonial rulers. Large-scale supply of commercial timber from the country was undertaken. After *Independence* (Post-Colonial), the forests were considered as a National Resource. The Government also realized that there was local need of forests besides the national need. In reserved forest area, the general public was not allowed to enter. Only mature trees marked by the Forest Department were cut and sold by the Department itself. In protected forest areas, the nearby villagers were allowed to enter to take forest products such as bamboos, firewood, and small timbers for making agricultural implements, fencing fields, constructing houses, etc. In return, they paid a nominal forest tax. They were allowed to take a limited quantity according to their needs, not for sale or any other purposes: this was the setup.

However, gradually, because of the increase in population, large-scale industrialization, and urbanization, the increasing demand for timber could not be met by the legal timber trade. Therefore, organized looting of forests by an exclusive group of people such as contractors, smugglers, politicians, and bureaucrats was a regular feature even in reserved areas. Sometimes forest officials also became involved in organized looting from fear of the contractors, smugglers, and politicians who were the so-called gangsters of the area. Sometimes they even took bribes from them. For all these reasons, deforestation took place to such an extent that the local people could not even get dried leaves and wood for fuel. So in some areas the local people started protecting the forests by forming informal protection committees such as the Youth Club, Mahila Mandal. In the meantime, the Government also revealed two things: first, it is very difficult to manage a vast forest area with the existing number of forest staff, and second, the local people also do not feel their belonging in the forest. Then, the Government realized if these local people who reside in forest fringe areas became involved in forest management, then only the forest could be protected and no outsider could enter into the forest. As a result, in 1990 the Government of India passed a resolution to encourage cooperation between the Forest Department and local communities in forest management, with the partnership basis that was popularly known as Joint Forest Management (JFM).

24.5 Main Characteristics of Joint Forest Management

24.5.1 JFM Is a Joint Venture Between the Government and the Community

According to Ottiger (2003), JFM is the merger of two traditions, the local community on one hand and the Forest Department on the other. In the following paragraphs, aspects of the forest management system of local communities, of the Forest Department, and that of JFM have been analyzed. The needs of local communities are broadly divided into two: economic needs and sociocultural needs. Economic needs include their daily livelihood such as food, fodder, fuel, manure, and medicines. The sociocultural needs include hunting, cremation, and religious practices in the forest. Similarly, the need of the Forest Department is predominantly a special material need, which includes marketing timber species and a few minor forest products such as the leaves, bark, and roots of different species (None Timber Forest Products), which are used for different purposes. The Department has also ecological needs such as the protection and conservation of the forest. JFM fulfils the economic needs of both the country and the local people. It provides material to the Forest Department in the form of raw materials for forest-based industries in the country. It also provides food, fuel, and fodder to local people. Under this scheme, protection and conservation of the forest are also accomplished.

The local community has indigenous knowledge about the forest. Local communities manage a smaller forest area, so they have intimate first-hand knowledge about the flora, fauna, and local habitat that may go back several generations. On the other hand, the Forest Department manages a large forest area so it is difficult for them to have detailed information about the condition of the forest resources. Again, because of frequent transfers the forest officers cannot gather knowledge of a particular area, so they have a very generalized kind of knowledge. Of course, they have expert knowledge based on modern science. In the JFM program both indigenous and highly specialized knowledge work together.

The system that the forest-dwelling communities follow in managing the forest is a common property system, based on locally accepted norms and values. Access to the forest is via membership, such as kinship and caste. The system is decentralized and flexible. The Forest Department follows the State Property System, wherein the main characteristic is custodial management based on Government laws. Joint Forest Management is a new institutional framework based on local structure and backed by official legislation. Here, the framework is decentralized within a centralized system.

24.5.2 JFM Is Different from Other Community Management Systems

There is much confusion among the concepts: community forest management, participatory forest management, social forestry, and JFM. In *community forest management* the community takes the lead, manages the resources, and the Government is a passive supporter or an observer, whereas in *participatory forest management* the Government takes the initiative, manages the resources, and the community participates in various forms, most commonly as hired labor (1995).

According to the 1993 Forest Act, the Nepal Government gives exclusive rights to villagers to protect and manage forest lands. The community also operates as an autonomous body in using and selling forest products. The Forest Department facilitates the functioning of the community and gives technical assistance if needed: this is an example of community forest management. Earlier in India (before implementation of JFM), to work with the working plan of the Forest Department most of the time the local people were hired to work as laborers in the forest because of shortage of staff: this is an example of participatory forest management.

Social forestry was implemented in India during 1980s. The main objective of the program was to protect forest resources by keeping the villagers away from the forest areas, which could be possible by planting degraded lands, open lands, and roadside lands outside the forest areas. However, the JFM program is totally different from the other forest management programs. Here, both the government and the communities manage the forest resources and share the costs and benefits.

24.5.3 JFM Is Between a Top-Down Approach and Down-Top Approach

Natural resources can be managed in three different ways: the top-down approach, the down-top approach, and the co-management approach. In the top-down approach, the state holds all power exclusively and in the down-top approach the community has full autonomy in natural resources management. In co-management, the sharing of responsibilities between the state and users becomes effective. JFM is nothing but a co-management approach in which responsibilities, costs, and benefits are shared between the Forest Department and local communities.

24.5.4 The Dual Nature of the JFM Program

On the one hand, JFM is seen as a major shift that has transferred government-controlled policies and attitudes from centralized management to decentralized management, from revenue orientation to resource orientation, from a production motive to a sustainability motive, from target orientation to process orientation, and from restricting people to working with people (1992). On the other hand, it is seen as an effort by the Forest Department to garner increased financial outlay and expand its territories and spheres of jurisdiction to more and more areas, which are normally under the jurisdiction of the Revenue Department (protected Forest Areas) and other government development agencies. A number of paper mills, for newsprint, lose leaseholds in the forest areas.

24.6 JFM as Perceived by Different Groups of People

As practiced in India for more than 15 years, JFM means different things to different groups of people. In the following paragraphs, the views of four groups of people who are involved directly in the JFM Programme are discussed. These groups are Forest Department officials, NGOs, funding agencies, and local communities. Again, varying views within each group can be observed.

According to the Forest Department, JFM is a mode of generating/rehabilitation of degraded forests and conservation of good forests. R.S. Pathan (1998), formerly in the IFS, Gujarat, reaffirms his faith in JFM as a tool to enrich vast degraded lands in the country. Here, the forest officials are more concerned for the conservations of forest resources than for community interest. According to S. Palit (2002), who was earlier IFS, West Bengal, JFM is widely viewed as a cost-effective means of securing reforestation. Rao et al. (2006) think as does Palit. They consider JFM as a cost-effective approach for the regeneration and conservation of forests. Actually, without the JFM Programme, protection of the vast forest lands in India from smugglers, industrialists, and contractors needs strict policing by a large number of Forest Department staff, which would require a very large amount of money for their salary and is not feasible economically. Through the JFM Programme it is possible to renew and conserve the forest by involving people in forest protection. Also, the view of the Principal Chief Conservator of Forests, Madhya Pradesh (M.P.), toward JFM is interesting. According to him, without the concerted effort of specialists in participatory approach, gender sensitivity, and community organizing, JFM in M.P. is nothing more than career advancement for Forest Department officials (Andrea 2006). It means that in M.P. most of the forest officials are involved in JFM for their own career, not for community development.

According to some NGOs, JFM means empowerment of people in the management of their own natural resources. For other NGOs, the JFM program not only improves the forests but also has immense potential for uniting people and for working together. Many international-level NGOs state that it is the right of the community to participate in the local resource management system, which could be possible through JFM in India. According to the pioneers of Hill Resources Management Societies, the main aim of JFM is to initiate mobilization for collective action, empowerment, and institution building.

The different funding agencies also perceive JFM in different ways. For industrialists who are engaged in mining/industrial activities within the forest area and funding for the program, JFM is the common ground where they can be friendly with local people to avoid any future conflict regarding setting up an industry in their native land. For other donor agencies, such as corporate sector people, JFM is an opportunity where they can invest money to get an exemption from income tax.

The perception of JFM also varies from community to community. For the community that manages good forest land and has a regular income from the first day of management, JFM is a mode of solution to their problems of food, fuel wood, and fodder or a means to ensure their daily requirement of forest products and/or a way to increase income. The community that manages totally degraded forest land and are paid wages by the government for planting trees and protecting them in the beginning views the JFM Programme as an employment-oriented program. The community that manages poor-quality land, where there is no regular short-term income such as food, fuel, and fodder, and who wait for a long-term return (such as timber), view JFM as an approach in which the government take services from them as a favor, keeping returns in future consideration (they may or may not give these returns to them in future).

These different perceptions regarding JFM of different groups of people may create two types of contradictory situation. First, it may lead toward a conflicting situation among them. Foresters, NGOS, donor agencies, and communities may have obvious difficulties in adjusting to each other. Second, they may be encouraged to have different, complementary roles in the JFM Programme. For example, the role of the Forest Department includes formulating policies, implementing various schemes, demarcating village forest boundaries, and imparting training. The community role includes protecting forest areas, planting trees, collecting minor products, logging, etc. NGOs motivate people, arrange workshops to train both Forest Department staff and local communities, settle conflicts among different user groups; a funding agency invests funds for a cause. The second situation is the better situation. The communities, NGOs, and foresters should realize that JFM is probably the last opportunity for them to exhibit learning and cohesive working regarding their forests. Otherwise, the economic pressures on the Government may compel opting for privatization of such lands, where the gainers would be few.

24.7 Problems and Prospects of JFM

From the foregoing paragraphs one can apprehend that although the JFM program has great prospects in India, it would have a number of problems at the field level where a number of groups and organizations with different bents are involved. So, the problems and prospects behind the JFM Programme have been explained through two different case studies from different parts of India, one from the northern part of India and the other from the southern part of India: the first case study was taken from Morni-Pinjore and Yamunanagar Forest Division of Haryana, and the second case study from the Janaram Forest Division of Andhra Pradesh.

24.7.1 Case Study 1: Morni-Pinjore and Yamunanagar Forest Division, Haryana

A tract of land covering 3,000 km^2 in the Morni-Pinjore and Yamunanagar Forest Division, located toward north and northeast Haryana, has been taken as a case study. Physiographically, this region comes under the Siwalik Himalayas and has a hilly and rugged topography. The rivers and seasonal torrents that originate from the Himalayas become wider when they enter the plain area. The vegetation is of the tropical moist deciduous type with high-quality bhabbar grasses. The local population consists of Gujjar, Ramdaisya, and Banjar; Bhanjda castes primarily depend on forestry and livestock. The Bhanjdas community makes baskets, the Banjaras make ropes from bhabbar grass, and the Gujjars depend on livestock. Some people in the foothills also practice agriculture. In the 1970s, reckless felling of trees, frequent forest fires, and increasing biotic pressure destroyed the vegetation in this area, leading to soil erosion and flash floods. Ultimately, the livelihood of people who depend on forestry, livestock, and agriculture was affected. The economic condition became very noticeable. Realizing this problem, the Energy and Resources Institute (TERI) took the initiative in introducing the JFM Programme in this region in 1990 with the collaboration of the Haryana Forest Department. The Ford Foundation financially supported this program. With the initiative of TERI, village-level committees were formed in about 55 villages to protect the forests. These Forest Protection Committees are popularly known as Hill Resources Management Societies (HRMS). Different patches of forest lands were leased to different HRMSs for growing and protection of bamboos and other forest products. Water-harvesting structures and check dams were constructed in places suitable for soil and water conservation. Grass leases and bamboo permits were given to the Hill Resources Management Societies. As a result, within a year or two, both agricultural and forest products increased and grazing problems were solved. The economic condition of the local people also improved. The benefits accruing to the HRMSs from the sale of the surplus of bhabhar grass are substantial. This money was invested in village development such as forestry work, repairing water-harvesting structures, the development of infrastructure, etc. The overall condition, that is, the environmental, social, and economic condition of the region, was improved. Further, to bridge the gap between the local villagers and Forest Department officials, three separate working groups at the range, division, and state levels were set up. The range- and division-level groups facilitate and monitor the program; the state-level group is the apex body that primarily recommends and proposes policy guidelines for the JFM Programme.

Here, the main reason behind the success of forest resources management were its ecology, hardworking community people, the committed NGO, and helpful forest officials. The general ecology of this Tarai region very delicate. Little carelessnesses such as felling of trees lead toward soil erosion, flash floods, and development of rills and gullies, etc. Little careful acts such as construction of water-harvesting structures and check dams along with plantation leads toward the development of vegetation within a short span of time. Culturally, the people of Haryana are very hard workers and entrepreneurs. So, when the local people were motivated by the TERI officials, they worked hard for their economic betterment and ultimately for the success of the JFM program. The committed officials of TERI regularly monitored the program from time to time through different techno-feasibility studies. Also, the strategy to strengthen the organizational structure was very appreciable. It worked at three levels, namely the micro or community level, the meso or regional level (here, range and division level), and the macro or state level.

24.7.2 Case Study 2: Janaram Forest Division, Andhra Pradesh

A tract of land covering 2,236 km^2 of the Janaram Forest Division, located in the upper reaches of the Godavari Watershed in the Adilabad District of Andhra Pradesh, has been taken as one of the case studies. Physiographically, this region is in the southern part of the Deccan Plateau and has an undulating and rugged topography. The vegetation is of the dry deciduous type with very high quality teak and a number of None Timber Forest Products (NTFP). The Forest Division is surrounded by tribal villages where Gond tribes constitute the majority of the population. Gonds are landless people and mainly depend on forests for their basic needs such as food, fodder, and fuel. However, the Forest Division had reached its peak of degradation during the late 1980s and early 1990s as a result of illegal logging of commercially valuable teak by the smugglers. The tribal people did not get even their basic requirements.

With this background, a village, Behroogunda, initiated the first Forest Protection Committee (FPC) in 1990 in Andhra Pradesh. As a result, one of the Gond leaders of the village sacrificed his life to the timber smugglers. Still, the villagers did not back away. Influenced by their boldness, togetherness, and community feelings, the Divisional Forest Officer of the Janaram Forest Division, along with other high-level government officials, helped them in initiating JFM. Behroonguda received official recognition of its FPC from the district administration in 1993. The FPC in Andhra Pradesh is popularly known as Vana Samrakshyana Samiti (VSS). The JFM Programme was funded by World Bank. As their effort succeeded, more VSSs were formed in the neighboring villages and protect 2,236 km^2 of forest area around the Kawal National Park.

All these villages that formed VSSs were located around Kawal National Park of Janaram Forest Division. So, when the villagers protect the forest patches allotted to them surrounding Kawal National Forest, automatically the whole area of Kawal National Forest along with the forest patches allotted to the villagers was protected. Joint night patrolling by forest guards and villagers all along the Reserved Forest border, both on foot and in jeeps, helped in reducing the incidence of timber smuggling. To complement this effort, employment opportunities were created for the members of VSS groups for 100 days during lean periods. Women members established nurseries for tree species producing NTFP. All this paid work as well as silvicultural operations and NTFP production provided substantial income to the local people. By 1998, many of the villagers could increase their VSS funds by selling timber from their allotted patches.

Unfortunately, by early 2000, most of the VSSs stopped operation and timber smuggling rapidly returned. Thus, the villagers had protected the forests for the past 7 years to receive the final harvest but ultimately that final harvest went to the smugglers. By the middle of 2004, valuable teak forests in this division were being extensively logged by illicit timber felling. There are a number of reasons behind the collapse of this promising community-based forest protection: termination of funding and project resources (as the World Bank Project was for only 5 years), transfer of a popular Divisional Forest Officer, lack of involvement of the NGO, etc. Because VSS operations were entirely dependent on wages from forest work funded by the JFM project (here, World Bank), when the project ended so did the activities of VSS. Although the wages generated 100 days of employment for the average VSS member, it did not build group savings, create a credit base, or enhance their livelihood assets.

Only the VSS who possessed teak and bamboo in their JFM areas benefited from silvicultural operations. The involvement of the NGO could have helped the illiterate tribal people in building community institutions and in empowering their leaders to act independently. Because no attempt was made to legally empower the VSS as autonomous bodies, with clear rights to the forest assets, the groups' primary incentive to participate was to secure short-term employment. Once the project ended, the employment opportunities ceased. The VSS had little capacity, authority, resources, or incentives to continue operation. Further, the transfer of the popular DFO who supported community efforts also affected the JFM program negatively. It is assumed that the transfer of a committed Divisional Forest Officer resulted from pressure from powerful people such as politicians, bureaucrats, and contractors who have a vested interest in timber smuggling.

Again, the JFM project was initiated toward the end of 2003 in the Janaram Forest Division with second-round funding by the World Bank. This time the involvement of the local NGO, Velegu, improved. The NGO's main focus is on a livelihood enhancement action plan. It does not focus on community building and empowering their leaders, which is required more than is a livelihood enhancement action plan. This time the forest officials were not sincere, were not friendly with the local people, although they acted in a very diplomatic way so that they could not be caught according to law. For example, they utilized the World Bank Fund for wages to local people for their labor. They would give them a lesser amount than was shown on the papers the people had to sign for their pay. Second, they spent money unnecessarily in fencing forest areas, where social fencing (i.e., protection by patrolling by forest fringe villagers) is required, but physical fencing is not. Third, Forest Department officials also used to show fewer trees on paper than their actual number. This time also the JFM program did not succeed well.

24.8 Conclusion

It has been found that the Joint Forest Management Programme is very successful in some states and not at all successful in others. It is also more successful in some regions and less successful in other regions of the same state. At this juncture, it is necessary to make it successful to avoid privatization of forest lands, which ultimately worsens the situation. Therefore, it is necessary to understand the concept of joint forest management, its characteristics, and the perceptions of different groups of people about JFM, to make it more successful. Also, building the capability of FPC institutions and Forest Department staff, innovative micro-planning, and involvement of NGOs should be encouraged.

References

Andrea CD (2006) Raising and realising the real issues: the World Bank Forestry Project, Forests. Society for Promotion of Wasteland Development Board, New Delhi, p 281

Johnson S (1995) Clarified JFM concepts. In: Roy SB (ed) Enabling environment for joint forest management. Inter-India Publications, New Delhi. Society for Promotion of Wasteland Development (1992)

Ottiger N (2003) Joint forest management: merging of two traditions? In: Seeland K, Schmithusen F (eds) Forest management and forest policy in South Asia: proceedings of an international seminar held in Kathmandu, Nepal in 1998. D.K. Printworld (P) Ltd, New Delhi, pp 121–143

Palit S (2002) Exploring for empowerment: plantation and non timber forest products in Joint Forest Management. *Wasteland News*, May–July, pp 40–43

Pathan RS (1998) The community action in development leads to sustenance.....let us prove it. Wasteland News, May–July

Rao RJ, Murali KS, Ravindranath NH (2006) Joint Forest Planning and Management in Karnataka, enabling, empowering status and potential, forest, vol I. Society for Promotion of Wasteland Development Board, New Delhi, p 116

Chapter 25
Community Forestry and Management of Forest Resources in Bhutan

Ramashray Prasad

Abstract Community forestry is that part of forest management in which local rural people are involved in strengthening, spreading, managing, utilizing, and increasing the productivity of the forest under their custody, as well as protecting and preserving the reserved government forests by reducing human pressure. The concept of community forestry came because of the drastic reduction in forest cover near or in the vicinity of human reach and the deteriorating productivity of the forest. It is not only the loss of forest but also of other organisms found in the region; many of the species of forest flora and fauna are on the verge of extinction. The rainfall and climate are changing rapidly, and there is a large extent of deteriorating forest cover. The forests also play a greater role in reducing greenhouse gases such as carbon dioxide by consumption through photosynthesis, apart from releasing oxygen. The vegetative cover is pumping out more and more moisture into the atmosphere to regulate the overall climate at a local level to the global scale. The forest is indispensable as it has multifaceted importance and cannot be further exploited without giving any second thought. When forests are to be utilized, such use must be done very judiciously and cautiously so that every proper care is taken. In another words, the forests should be managed sustainably. Almost all of Bhutan is mountainous by physical character with the very weak, fragile, susceptible, and shattered newly formed high Himalaya. The people are very much attached to the forests for their livelihood, and here it becomes even more important to manage the forest. Community forestry offers great hope and is growing very rapidly. Therefore, an attempt has been made to study community forestry in Bhutan in general, with particular emphasis on Eastern Bhutan.

Keywords Carbon sink • Community forestry • Forest products • Rural poverty • Sustainable forest management

---

R. Prasad (✉)
Bhim Rao Ambedkar College, University of Delhi, Yamuna Vihar, Main Wazirabad Road, Delhi 110094, India
e-mail: drrprasad2007@gmail.com

A.K. Dutt et al. (eds.), *Spatial Diversity and Dynamics in Resources and Urban Development*, DOI 10.1007/978-94-017-9771-9_25

25.1 Background

From the very advent of human evolution, human beings were closely related with nature. In fact, nature guided every activity of human beings. They lived with what was available in nature. The environment was in quite its original state. It was untouched, unharmed, and undamaged, and somehow, people were surviving. Those days, a simple roof or cave, the shade of a tree or hut over their heads, food from forest products or animal meat to fill their bellies, and some other requirements such as the skins of animals or large-sized leaves of trees to protect them from the harsh physical conditions of nature were used. For everything, humanity was dependent on nature. The gift of nature available on the Earth was lying there in almost pristine condition. With the introduction of agricultural activities, human beings started interacting with nature more forcefully. With the advancing scope of the human mind, they developed various types of tools and implements. These tools and implements helped make their life easier by developing and renovating their basic needs. Those basic needs kept on changing, particularly in terms of quality and comfort.

Now, the needs of modern-day people, especially in urban areas, demand more from nature. Rural areas are also transformed but to lesser degree. The ever-growing human population is the biggest culprit in pressuring nature as it demands more and more. Thus, natural resources are in great danger, and managing these is not easy. Forest resources, these days, are really very hard pressed in comparison to other resources.

The past century has witnessed a tremendous growth in world population. The growing population needs various resources from nature. Nature/Earth is limited and so are available resources. Many resources and species are threatened with extinction because of overutilization, because the level of consumption is greater than the replacement level. To feed the increasing population, people have used almost all the available land for cultivation on the plains as well as on hilly slopes. The mounting demand for food items from agriculture has led to cultivation of critical areas also. Such results are again detrimental for the survival of other resources at the place of criticality as well as downstream. It is time to think positively about proper planning and utilization of resources before expanding our uses. Increasing demand by people has resulted in overexploitation of forest resources. To overcome this problem, community forestry is one approach to manage the forest and forest resources.

25.2 Understanding Community Forestry

The term community is used for a group of people living in a particular area and sharing certain basic interests. For successful functioning of the community or society, the people must develop diverse institutions and organizations (Barua 2009). Generally, community forestry is seen as a way of sustainable forest management. Through community forestry, local people are involved in supervising the part of the forest allotted to them. They are concerned with the forest in terms of planting trees, looking after their growth, allowing them to mature, and using them when needs arise for their own purpose. If the forest provides in excess of their requirements, they sell its products in the open market to fetch money to meet their other needs. Community forestry is local-level forest activity in which collective action is taken to plan, launch, monitor, evaluate, administer, obtain, and harvest various types of trees, bushes, herbs, fruits, plants, and vegetables, on communal lands. Various products from these forests are consumed by the community.

Rural poor people rely very greatly on the forest for various needs such as many types of wild vegetables, edible plants and roots, and a variety of fruits. The Secretary of the Ministry of Agriculture and Forest, Bhutan, Sherub Gyaltshen (Department of Forest and Park Services 2010) concludes that community forestry is a program that contributes to achieving many development goals: empowerment of communities, sustainable forest management, good governance, and poverty alleviation. Through community forestry, there is a reversal of the forest management paradigm from nationalization and centralization to devolution and decentralization in forest resource management.

Forestry is often defined as the manipulation of forests to achieve a desired objective, and it is the objective that distinguishes different types of forestry. In this context, managing forests with the express intent of benefiting neighboring communities constitutes community forestry (Brendler and Carey 1998). Forested or degraded forestland owned by the government, but formally handed over to a group of villagers for its protection, management, and utilization, is known as community forestry. All members of a community that regularly use a forest to meet their household needs organize themselves as a group, termed a Community Forest Management Group (CFMG), to protect, manage, and utilize the forest. Many rural communities depend on nearby forests as a source of economic well-being. Three attributes are commonly found with community forestry efforts.

25.2.1 Access of Forest Land and Its Resources by the Community

Community forestry is closely associated with the benefits from forest resources, both timber and non-timber products. Poor local people receive a wage for collecting various items for their livelihood. These forest products are sold for money or may go to a factory as raw material. They also collect some edibles for their own consumption.

25.3 Decision-Making Authority with Community

A Community Forestry Management Group (CFMG) has the authority to make decisions for the betterment of the forest and its resource management, which includes the harvesting of forest products and selling in the open market whatever is in excess to the needs of the community, as well as regeneration of the forest.

25.3.1 Protection and Restoration of Community Forest

The CFMG takes the responsibility of protecting and restoring the community forest as and when required. A balance between the use of land and the effects of such use is also controlled by a CFMG.

By promoting community forestry, indigenous knowledge of protecting and promoting forestry becomes very pertinent. Before this, forest resource management was in the hands of an educated generation who came through the bureaucratic system and used to tell the local people how to manage their forests. With community forestry, even the local knowledge that is passed from generation to generation verbally is now acknowledged. Rural people are not only getting the information from forest officials but also imparting their familiarity with nature to the officials. Now, it is not only a one-way traffic but works both ways.

25.4 Needs and Goals of Community Forestry in Bhutan

The community forestry in Bhutan is in great need with many goals starting from simply meeting the need of local people to the sustainable management of forests at the national level through this endeavor. The development goal for community forestry in Bhutan has been formulated as rural communities empowered to manage their own community forests sustainably to meet the majority of their timber demands and other forest goods and services, to derive economic benefits from the sale of forest products and services, and contribute to a reduction in rural poverty. Many key principles have been identified to guide the formulation of strategies for future focus and development of community forestry in Bhutan (Department of Forest and Park Services 2010): (1) ensure 60 % of the country's geographic area is under forest for all times to come; (2) balance conservation with sustainable utilization and management of biodiversity and forest and water resources; (3) support decentralization and devolution by empowering communities to manage their local forests; (4) improve governance of community forests, leading to improved forest conditions and the equitable distribution of benefits; (5) generate income for local communities through commercial harvesting of timber and non-wood forest products; (6) contribute to poverty reduction; and (7) provide, insofar as possible, timber for rural construction and maintenance.

The main goals are the establishment of a decentralized qualified forestry extension service, protection of aquifers and biodiversity, and income generation through the sale of excess forestry products from the community forests (Temphel and Beukeboom 2007). All these goals are based on the basic sustainability principle that the communities have the management rights and the responsibilities of the surrounding forest. In this way, the community feels the ownership and meets the challenge to manage it properly. Different communities in the nearby villages feel very competitive to manage the resources better than other villages. This competitiveness makes them successful. Thus, they develop local pride, and this ultimately leads to better produce from the forest, both timber and non-wood forest products. Finally, the result reflects on the members of the community and enhances their livelihood and the betterment of their life, which helps in achieving the goal set for the 10th Five-Year Plan in reducing poverty.

25.4.1 A Brief Geoecological View of Bhutan

Bhutan is a landlocked country, small in size and further smaller in terms of population. It lies in the Eastern Himalaya between Tibet to the north and India elsewhere. It encompasses an area of 38,394 km^2 and a population of only 672,425 (Census of Bhutan 2005), with an average density of 17.5 persons per square kilometer (km^2). It has a rugged and dissected mountainous topography of which the larger part is uninhabitable because of steep slopes and harsh climatic conditions. Nevertheless, the Himalayan Kingdom of Bhutan is endowed with rich natural resources such as forest, water, wildlife, and various species of flora and fauna. The vegetation varies from broad-leaved, wet subtropical occurrences in the south to alpine forest, bushes, and shrubs in the north.

Of the total area of the country, 72.5 % is covered with forests (PPD 2008). From Table 25.1, it is obvious that only 15 % of the area is below 1,200 m elevation. Even this area is hilly and not very much suited to agriculture. The remaining 85 % area is mountainous and susceptible to erosion and landslides. Any human activity in this zone is very disadvantageous in terms of environment and sustainable management. Table 25.1 shows only 89 % of Bhutan's area but the remaining area is above the altitude of 4,600 m and is devoid of vegetation cover, lying barren, because of low temperatures and high-altitude glaciers.

A large part of the forested areas in Bhutan is protected by National Parks (41 % of the forest cover) and biological corridors (8.6 % of the forest cover). About 14 % of the total forest is devoted to timber production and other commercial uses, which is the amount economically accessible with present-day available technology. Roughly 35.4 % of the forested area is still not usable for harvesting at present. Currently only about 1 % of the forest cover is under community forestry, whereas the target is to bring 4 % of the forest area under community forestry by the end of the country's 10th Five-Year Plan (2008–2013) of the country.

About 69 % of the population of the country lives in rural Bhutan. As per the Bhutan Living Standard Survey (National Statistics Bureau 2003), about 31 % of the population exists below the national poverty line. Of these, 94 % are in rural areas, and they are heavily dependent on forest resources. The major priority of the 10th Five-Year Plan of Bhutan is to reduce and overcome poverty, particularly in rural areas. To make the forest sustainable and reduce rural poverty, the community forestry program was adopted in the country.

Table 25.1 Agro-ecological zones of Bhutan

Agro-ecological zone	Altitude (meters, m)[£]	Inter-altitude area in percent[¥]	Annual rainfall (in cm)[£]	Annual mean temperature (°C)[£]
Alpine	3,600–4,600	20.8	Less than 65	5.5
Cool temperate	2,600–3,600	27.4	65–85	9.9
Warm temperate	1,800–2,600	13.4	65–85	12.5
Dry subtropical	1,200–1,800	12.6	85–120	17.2
Humid subtropical	600–1,200	9.8	120–250	19.5
Wet subtropical	150–600	5.3	250–550	23.6

Source: [£]PPD (2008), p. 2; [¥]TFDP (2000), p. 47

25.5 Birth and Growth of Community Forestry in Bhutan

The birth of the concept of community forestry started with the Royal Decree of 1979 termed as social forestry in which it was said: "people's participation is key to conservation and utilization of forest resources" (Chhetri et al. 2009). In 1969, all forests were nationalized and termed government reserve forests. According to Section 10 of the Bhutan Forest Act 1969, "the Government reserves the rights to the absolute ownership of trees, timber and other forest produce on private land" (Dhital 2002). Before this, no restrictions existed on use of forest resources except poaching of wild animals. This provision was repealed and invalidated by Royal Decree of His Majesty and the Land Act of 1979. In the beginning after the Royal ruling in 1979, there was almost no progress in the field of community forestry, because there was no legal provision for transferring the government reserve forest land to a private community.

Agroforestry is one of the favored land management systems in the mountains (Sharma et al. 2007). The real progress in the development and growth of community forestry started with the beginning of the twenty-first century. Since then growth has picked up momentum. There has been a shift from a primary focus on protection and conservation toward a focus on balancing conservation with sustainable utilization (Peljor 2009). Some of the laws and policies followed by the Government of Bhutan concerning the forests are worth mentioning to understand the need for community forestry (Table 25.2) in Bhutan.

With the adoption of a forest and nature conservation act in 1995, the legal hurdles for the enterprise of community forestry were almost overcome. The first of its kind is Dozam community forest, which was established in 1997 in the Drametse block of Mongar District in Eastern Bhutan. It was the beginning of accepting the traditional rights of the people to access and use forest resources for the fulfillment of their needs.

The rural economy of Bhutan, where most of the people below the poverty line are in rural areas, is not very good. They rely heavily on the forest and forest products for their livelihood. The 10th Five-Year Plan of the government is also emphasizing poverty alleviation in which community forestry has a larger role. The policy is aimed at using a participatory approach to contribute to poverty reduction and

Table 25.2 Chronology of forest regulations and development of community forestry in Bhutan

Laws and policies	Decisions	Remarks
Department of Forestry (DoF) 1952	Department of Forestry (DoF) created but within Ministry of Agriculture	With Ministry of Trade and Industry
Thrimzhung Chenmo 1957	No restriction on use of forest resources except poaching of wild animals	Open access
Forest Act 1969	All forest resources nationalized as state property	Forest protected with restrictive use only
	All forests declared as Government Reserved Forest (GRF), including trees on private land	
	Traditional rights and customary laws withdrawn	
National Forest Policy 1974	Prescribed long-term national goal for forests and their utilization	Forest protected
	Scientific management of forest	Scientific use permitted
Royal Decree 1979	Social forestry around villages	Way paved for community forestry
	Plantation of trees on private as well as communal land	
	Managing local resources by local people	
Land Act 1979	Forest resources on private land were allowed to be used by people for their domestic needs but not to be sold to get money	Leniency apparent in accessing resources
Forest Policy 1991	Forest resources to be used on sustainable principles	Emphasis on proper resource management
	To ensure conservation of the environment	
	Obtain benefits from the forests with rational resources management	
	Renewable Natural Resources (RNR) was established	
Decentralization of Community Forestry 1993	Community Forestry at district (Dzongkhag) level activities	Decentralization came into action
	Dzongkhag Forestry Extension Section (DFES) created	
	Dzongkhag made responsible for implementing activities	
Forest and Nature Conservation Act 1995	Separate chapter on forest management appeared	Scientific participatory forest management by local community
	Traditional rights of community restored	
	GRF lease accepted for sustainable management by community	
First CF established in 1997	First certified community forest in Mongar district created mainly on degraded land with an area of 300 ha and 109 households	Start of new era in community forestry
Land Act 2007	Lease of GRF further clarified for economic activities	Forest use emphasized
Constitution 2008	Every citizen considered the trustee of natural resources of Bhutan	Citizen's role amplified

socioeconomic development (Chhetri et al. 2009). The revised rules concerning community forestry are recommended to be liberalized and simplified so that the community forestry program can benefit even more local people.

During the early 1990s, the decentralization policy was adopted in which the people's participation in protection and management of forests was entrusted. To make this system functional, various steps were adopted such as the creation of forestry extension units at the district (Dzongkhag) with trained block forestry extension officers to support the implementation of decentralized forestry activities. The extension officers at district and block levels started facilitating the implementation of community forestry and community forest management groups at the village level. Gilmour (2009) observed that the shift from centralized forest management toward more decentralized local management has resulted in the development, testing, and institutionalizing of a wide range of community-based management approaches throughout Asia and, in more recent years, in Africa and Latin America.

In the initial years there was a problem of conceptualizing and translating it into reality at the ground working stage. Apart from that, even the common people were also not very well informed about this approach, as well as doubtful whether the government would hand over the reserved forest land to them for management and utilization. Both the forest and government staff also were not very sure about the skills of the community in managing the forest resources properly and sustainably (Temphel and Beukeboom 2006). Community forestry is now considered to be the major component of the overall social forestry program in Bhutan (Chhetri 2009). For these reasons the pace of community forestry development in its initial years was slow, especially until 2006. After this, the growth was very fast (Table 25.3).

The district-wise spatial distribution of community forests and the number of households associated with them can be seen from Fig. 25.1. From 2007, the increase in the number of community forests sanctioned has been tremendous.

Table 25.3 Community forest (CF) growth (to December 2009) in Bhutan

	CFs handed over to Community Forest Management Group (CFMG)		Area of CFs handed over		Households involved in CFMGs	
Year	Numbers	Cumulative	Hectares	Cumulative	Numbers	Cumulative
Until 2001	3	3	1,546	1,546	530	530
2002	5	8	228	1,774	116	646
2003	7	15	1,052	2,826	413	1,059
2004	9	24	1,020	3,846	475	1,534
2005	7	31	1,411	5,257	709	2,243
2006	7	38	509	5,766	277	2,520
2007	19	57	2,089	7,855	845	3,365
2008	61	118	8,334	16,189	2,965	6,330
2009	82	200	8,808	24,997	3,433	9,763

Source: Department of Forest and Park Services (2010)

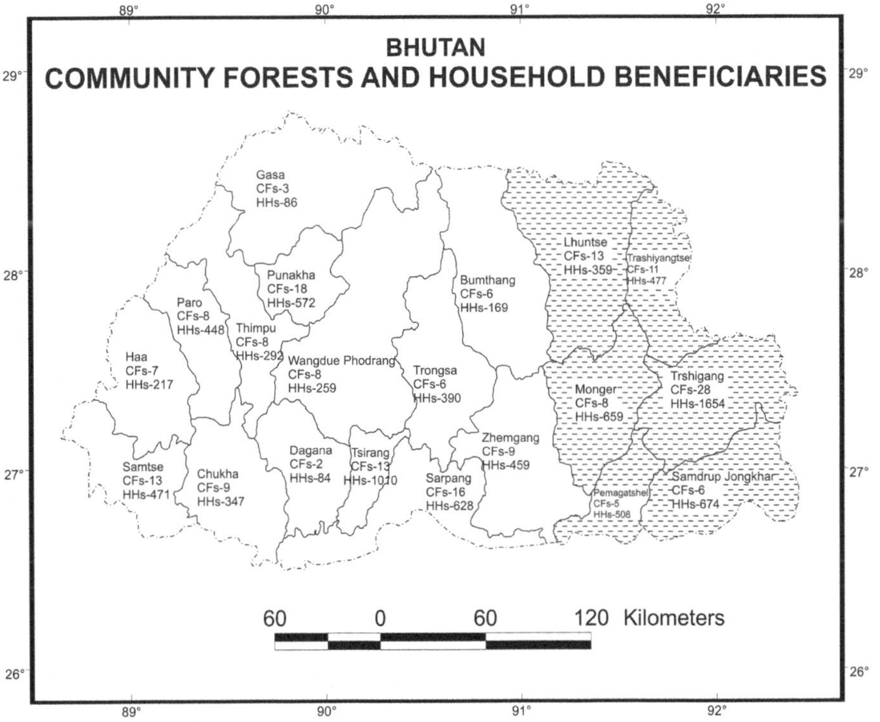

Fig. 25.1 Community forest and household beneficiary in Bhutan

Community forest has now become an important part of the national forest policy. Before 2007, there were only 38 community forests, but within 3 years (2007–2009) the number rose to 200 (Table 25.3). A total of 9,763 households have been brought under this scheme, which is 7.74 % of the total households in all Bhutan. At present, these community forest management groups have possession of 249.97 km^2 of community forest land, comprising 0.9 % of the total forest area in the country. The target is to bring 4 % of the forest area under community forest management by mid-2013.

25.6 Selection of Community Forests

In the beginning of the concept of the community forestry development scheme, the priority was on planting trees on barren and degraded land, popularly known as community plantation. The Forest and Nature Conservation Rule, 2000, Chapter IV, Section 36, Clause 1(e) established the requirement of securing "both degraded and good forest with equal ratio" (RGoB 2000) for allotment of community forest. This condition became an obstacle to smooth growth of community forests because getting both degraded as well as good forest land in equal proportion was not easy for every village. This obstructing clause was withdrawn by Forest and Nature Conservation Rule, 2006 Chapter IV, Section 27, Clause 1 and stressed that the community forests should be "in and around villages and human settlements" (RGoB 2006). This provision has been found to be very practical (Tshering 2009), and it is also reflected through the growth of community forestry in Bhutan (Table 25.3).

Now, community forests are allocated near to villages and village people are the beneficiaries. The households who manage the nearby forest given to them are known as community forest management groups (CFMG). To form a CFMG, a minimum of ten households has to be there as per the rule. Many villages in Bhutan have fewer than ten households. Thus, although they also are interested, they are not eligible to form a CFMG.

On average, the area of a community forest is limited to a maximum of 2.5 ha per household. This area is very small to provide a substantial benefit for a rural poor community. The country's 10th Five-Year Plan focuses on a poverty alleviation scheme through community forestry. Therefore, this smaller allocation of forest land is not going to pave the way for significant reduction of poverty. A smaller size of household villages and a smaller allotment of forested land sometimes do not go hand in hand. In most cases, admissible area and physical limit/boundary do not match. It is always better to define a community forest boundary using natural features such as ridges, valleys, and rivers/streams. When natural boundaries are used, it is generally difficult to fit within low area limits. Hence, there is a need to relax these provisions to improve the effectiveness of community forests (Tshering 2009).

25.7 A Brief Profile of Eastern Bhutan in the Country

Eastern Bhutan constitutes a little more than one fourth of the area, population, and households of the country and a slightly greater share of the area under forest (Table 25.4). However, the density of population is slightly lower than the national average, because two northern districts, Lhuentse (5.33 persons/km^2) and Trashiyangtse (12.34 persons/km^2), are low in population density. The main reasons for lower density in these two districts are steep slopes, high altitude, reduced accessibility, rough terrain, and harsh climatic conditions. Pema Gatshel in the south, with relatively better facilities, moderate altitude, and better agricultural development supporting a larger population, has a very high density (26.76 persons/km^2) in comparison with the national average (16.54 persons/km^2).

Eastern Bhutan is reasonably high in forest cover in comparison to the rest of Bhutan. The overall 8,860.9 km^2 (77.85 %) area of Eastern Bhutan is covered with forest, which includes 7,941.3 km^2 of true forest and 919.6 km^2 of scrub forest. Pema Gatshel is the only district that has a lower forest cover (53.57 %) than the national average. The highest forest cover is in Monger district (88.47 %). Hence, Eastern Bhutan is blessed with greater forest and forest resources in comparison to the rest of Bhutan.

True forest is dense forest in which the growth of vegetation is very high. Trees are very tall with a good canopy coverage. These forests are superior in commercial

Table 25.4 Some facts about Eastern Bhutan and the country

District/region	Area (in km²)[£]	Population[¥]	Density/km²	Number of households[¥]	Forested area (in km²)[£]	Percent of forest cover
Lhuentse	2,888	15,395	5.33	3,001	2,173.5	75.26
Monger	1,947	37,069	19.04	7,348	1,722.6	88.47
Pema Gatshel	518	13,864	26.76	2,937	277.5	53.57
Samdrup Jongkhar	2,308	39,961	17.31	8,363	1,783.6	77.28
Trashigang	2,283	51,134	22.40	10,813	1,802.7	78.96
Trashiyangtse	1,438	17,740	12.34	3,764	1,101.0	76.56
Eastern Bhutan (EB)	**11,382**	**175,163**	**15.39**	**36,226**	**8,860.9**	**77.85**
EB percent	**29.65**	**27.59**	–	**28.72**	**30.51**	–
All Bhutan	**38,394**	**634,982**	**16.54**	**126,115**	**27,912**	**72.70**

Source: [£]TFDP (2000); [¥]Census of Bhutan (2005)

production of timber for generating revenue through sale. These forests are also popularly known as woodland forest. In these forests, the growth of vegetation is multilayered, such as grasses, bushes, and smaller trees as well as big trees. These forests are concentrated at lower altitudes where the geographic condition of the vegetative growth is more favorable, such as soil development, rainfall, and temperature. True forest constitutes 89.62 % of the total forest area, or 69.77 % of the entire Eastern Bhutan.

Scrub forest is lower-quality forest cover, found particularly at the higher reaches in the northern part of the region. In this type of forest, the growth of vegetative cover is not very good for two reasons: first, the forest cover is destroyed by human activities and the growth is retarded by this as well as shifting cultivation; second, the climatic conditions are not conducive for vegetative growth. The first reason is more applicable in the lower altitudes where human activities are more prominent, such as Samdrup Jongkhar, Pema Gatshel, and Mongar. The second reason is appropriate for northern districts such as Lhuentse, Trashiyangtse, and Trashigang, where the climate is very harsh and the temperature is very low. Even in these districts, shifting cultivation is practiced, although it is banned today. The vegetative growth is retarded and few large-size trees are found. Most of the land cover is in the form of grasses and bushes with sparse smaller trees. Scrub forest represents 10.38 % of the total forest area or 8.08 % of all Eastern Bhutan.

25.7.1 Characteristics of Eastern Bhutan

Eastern Bhutan is characterized by a wet subtropical climate in the southern parts of Pema Gatshel and Samdrup Jongkhar to the alpine zone of northern Lhuentse and Trashiyangtse. The wet subtropical conditions prevail in the areas lying below

600 m altitude; above this, up to an elevation of 1,200 m the humid subtropical zone is found. At higher elevations, a dry subtropical and a warm temperate zone are present at heights between 1,200 and 1,800 m and from 1,800 to 2,600 m, respectively. Between 2,600 and 3,600 m is a cool temperate zone, and up to 4,600 m, alpine vegetation occurs. Above 3,600 m, all vegetation is reduced because of the severe cold. The characteristics of vegetation continue to change with altitude. In the lower altitudes, the vegetation is very dense with high rates of growth because of the high temperature and high rainfall. Eastern Bhutan receives a high annual rainfall, more than 500 cm per year. Geologically, this area is composed of unconsolidated sedimentary rocks formed with excessive compression during the Himalayan orogeny. In this process, the sedimentary rocks are crushed to such an extent that they cannot withstand the load lying above then. Therefore, landslides are very common, especially at higher elevations during the monsoon season when torrential rainfall lubricates the rocks.

High rainfall is also associated with higher erosion. Rivers have carved out very deep valleys and gorges. The slopes are extremely steep. All the rivers flowing from north run southward to enter Indian Territory. Important among these are the Kuri Chhu and Kulong Chhu Rivers. Kuri Chhu passes through Lhuentse and Mongar Districts; Kulong Chhu traverses through Trashiyangtse, Trashigang, and the eastern edge of Mongar. Both these rivers merge to form the Dangme Chhu, which finally joins the Manas River, and the Manas confluences with Brahmaputra in Assam. The Nyera Ama Chhu passes through Trashigang and Samdrup Jongkhar to meet directly with the Brahmaputra.

25.7.2 Status of Community Forestry in Eastern Bhutan

Community forestry is well developed in Eastern Bhutan in comparison to the other parts of the country. Eastern Bhutan constitutes roughly 29.5 % of the area of the country, but it has 37 % (74 of 200) of the community forests, which consititutes 47 % of the total area (11,864 km^2, of 24,997 km^2) under community forests. The total number of households benefited in entire Bhutan is 9,763, of which 4,331 households (44 %) are in Eastern Bhutan (Table 25.5).

The benefits from forestry are seen in two ways: first, timber, and second, non-wood forest products. The timber-based community forests in Eastern Bhutan represent 40 % of those of the entire country, and still more of non-wood forest products, more than two thirds (68 %) of the whole country. Taken together, roughly half (47 %) of the entire community forest area lies in Eastern Bhutan. Eastern Bhutan is the most privileged in the development and growth of community forests in Bhutan. The distribution by districts of the area under timber forest and forest for non-wood forest products is shown in Fig. 25.2.

The total area under forest in Eastern Bhutan is 8860.9 km^2, equal to 30.51 % of the total for all Bhutan. The community forest area in Eastern Bhutan is 1.34 % of

Table 25.5 Community forest status (to December 2009) in Eastern Bhutan

Districts/Region	Number of CFs	Number of households benefited	Community forest types and area in hectares		
			Forest for timber	Forest for NWFP	Total
Lhuentse	13	359	654	18	672
Monger	8	659	582	1,909	2,491
Pema Gatshel	8	508	396	299	695
Samdrup Jongkhar	6	674	447	1,455	1,902
Trashigang	28	1,654	4,760	317	5,077
Trashiyangtse	11	477	486	541	1,027
Eastern Bhutan	**74**	**4,331**	**7,325**	**4,539**	**11,864**
EB percent	**37**	**44**	**40**	**68**	**47**
Entire Bhutan	**200**	**9,763**	**18,297**	**6,700**	**24,997**

Source: Department of Forest and Park Services (2010)

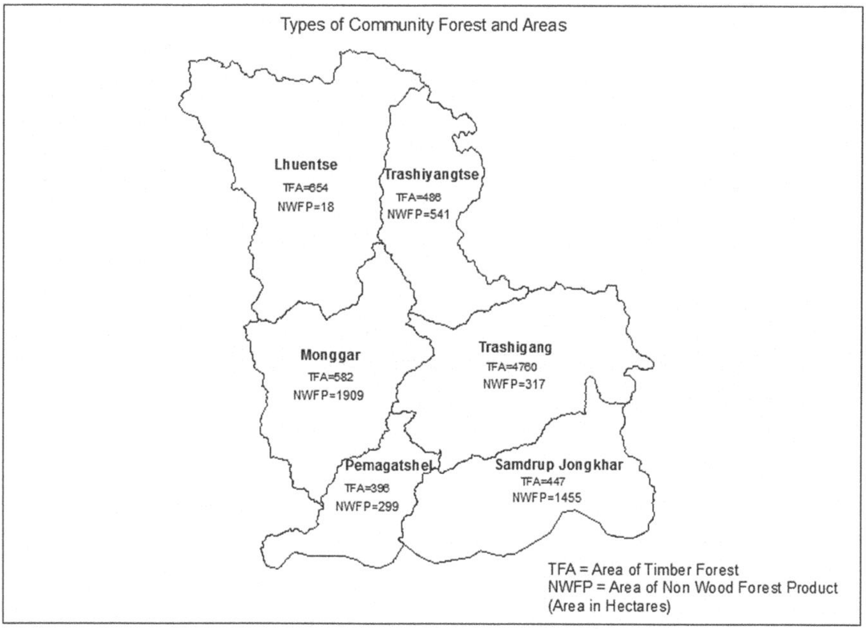

Fig. 25.2 Types of community forest and areas in Eastern Bhutan

the total forest area in the region, and for all Bhutan this figure is just 0.9 %. The country's 10th Five-Year Plan (2008–2013) emphasizes that 4 % of the area of the forest should be managed by community or private forestry (Department of Forest and Park Services 2010).

25.7.3 Dependence of Rural People on Forests in Eastern Bhutan

Rural people are heavily dependent on the forest in Eastern Bhutan. The economy is mainly based on forest and agriculture. Except for a narrow strip of the southern part, all of Eastern Bhutan is mountainous. Whatever development of agriculture is there is confined to the slopes in patches with small fields. Although the main occupation of people is agriculture, it is also not well developed, and cannot be compared with the agricultural practices of the plain areas. Therefore, the rural people use the forest to fulfill their requirements by collecting various items. Most rural people rely on the forest for firewood, used for cooking food as well as keeping their houses warm, particularly in winter. The overall dependence of people on firewood in all Bhutan is 90.7 % and in Pema Gatshel District (Eastern Bhutan) it is 98.2 % (Table 25.6). It is the same case in terms of collection of wild mushrooms, lemon grass, resin, and much more in different districts. People in Eastern Bhutan are more dependent on the forest in comparison to the rest of Bhutan.

More dependence on forest means more and more exploitation of nearby forests. Once community forest is allotted and the members are advised that they have to meet their needs from the defined units only, they will be more economical in use and better managers. Otherwise, they will consume more than their minimum needs such as burning wood to keep the house warm. In one way, this saves fuel/forest products and on the other hand it reduces carbon emission in the atmosphere. The reserved forest will be more protected while the productivity of the community forest will increase. Hence, it is management in two ways with double benefits.

25.8 Benefit Accrued to the Community Members and the Nation

Various benefits exist from community forestry, starting from the local grass roots level to the national and projecting to the global level. Community forestry has great potential in forest resources management. The beginning stage of the development

Table 25.6 Dependence of Eastern Bhutan people on selected forest products

	Percentage of household involvement		
Forest products	National average	Highest percentage in district	Name of district in eastern Bhutan
Firewood	90.7	98.2	Pema Gatshel
Wild mushroom	20.7	49.3	Samdrup Jongkhar
Lemon grass oil	0.8	5.0	Lhuentse
Dyes	0.7	1.8	Pema Gatshel
Resin	0.3	1.6	Monger

Source: Turkelboom et al. (2001); derived from Renewable Natural Resources–Census 2001

of community forestry in Bhutan has momentum. The benefit accrued can be seen in different perspectives such as economic gain, employment opportunities, environmental protection, and overall the carbon sink.

25.8.1 Economic Gain

The members of the management group are authorized to use the excess produce of the community forest for their own benefit. The members are expected to meet their needs from the allotted forest area only. They get the timber as well as non-wood forest products, which helps in meeting the requirements of the family and their livelihood. Apart from this, the excess products are sold in the open market, and in this way they acquire money to reinvest for appropriate management of the forest. The gain from community forestry is distributed equally among the members, thus providing the income of the community. Some of the community forest management groups are thinking of developing their forest as a recreational site by proper management so that local people or outsiders will be attracted to visit, but all will have to pay an entry fee. It is a sort of tourism and recreation but at a smaller scale.

25.8.2 Employment Opportunity

Not all the people of the community may need employment, but some poor people who do not have any work may be employed on a daily wage. The forest has many items that can be collected such as wild mushroom, cane shoots, lemon grass, fern tops, fodder, resin, and dyes. These items are known as non-wood forest products. Most of these products are not consumed, but are sold in the market for consumption by others, and some are sold as the raw material for some industries. Therefore, some local people collect the items from the forest and the person who is employed for this collection gets a wage. Somebody also will be employed to maintain the recreational site if it is developed as such. Therefore, the generation of employment is also "on the card" along with proper management of the community forestry. Many of these happenings will depend on the enthusiasm, caliber, and entrepreneurship of the community forest management group.

25.8.3 Environmental Protection

It is already a recognized fact that forests have very important roles in protecting and conserving primary resources such as water, soil, air quality, and biodiversity. Community forests are allocated by officials at the district and division levels on application by a community forest management group. The allotted area includes both good-quality forest and nearby degraded land. The community forest management group treats the degraded land by planting trees, shrubs, and grasses to protect the soils from further degradation. Trees, shrubs, and grasses are selected on the basis of suitability to the site, location, aspect, and altitude. These activities by community forest management groups enhance the overall environmental condition of an area. The degraded land is brought under forest cover. The needs of the local people are met from the community forest, which they are also maintaining: they are not going into the reserved forest for their requirements. Hence, the reserved forests are not further exploited, and they remain in pristine condition except for use by the forest department in a controlled manner. Therefore, the development of community forestry in Bhutan maintains a very good environmental balance. In this way the forests of the country are protected, a very positive indication. Environmental benefits in community forests are (1) the community rehabilitates degraded areas; (2) water resources are protected in the concerned areas; (3) careful harvest according to management procedures is determined; (4) forest fire reduction is expected; (5) the community keeps a watch on illegal activities by mischievous persons; and (6) the community protects the wildlife in the forested areas.

25.9 Contribution Toward Carbon Sink

The concept of a carbon sink became popular after the Kyoto Protocol. Carbon sink is the reduction of atmospheric carbon dioxide gas, done in both natural and artificial ways. An artificial carbon sink (landfills and carbon capture and storage) is very much limited in comparison to the natural sink through utilization by natural vegetation, a natural storage of carbon dioxide for a considerable period of time through the utilization of carbon dioxide gas by photosynthesis. But, removal of forest cover and its burning pump carbon dioxide into the atmosphere. The initiation of community forestry is very much related to the expansion of forest cover, even on degraded land. The increase in forest and its wise management definitely will consume more and more carbon dioxide, leading to a carbon sink. The carbon sink in one territorial limit is not only good for the country but for the whole of the Earth.

The benefits mentioned here are not only in the interests of the people or the local community, but also the government of Bhutan is relieved from carrying out some contingency work. People get employment, they become self-reliant to some extent, their better livelihood is ensured, the financial health of the community is improved, and overall the environment is protected. Ultimately, their surroundings are improved, leading toward achieving the goal of poverty reduction of the 10th Five-Year Plan. This work is also in accord with the Kyoto Protocol for reducing the amount of carbon in the atmosphere. Although the amount of carbon sink through community forestry is very low, it is still of some value, which must be acknowledged.

25.10 Benefits from the Forests

Scientific studies reveal that forests offer a number of benefits that ultimately lead to a higher level of economic gain to the communities who manage it as well as to all the living organisms. Forests provide a variety of ecological functions and interactions. Forests enable and facilitate rainwater to percolate down through the soils and finally recharge subsurface water. Thus, they help in providing a regular supply of valuable water for the survival of life. The forests are important in various ways. The primary and most valuable product of the forest is timber, which is used in many ways. Apart from timber, human beings have depended since time unknown on the forests for other products. Some of these products are firewood, soft wood for papermaking, bamboo, cane, rubber plant, herbs, wild nuts and fruits, medicinal plants, aromatic plants, wild edible plants, and fodder for domestic animals.

Numerous small-scale to large-scale industries have been developed today. They were also the source of livelihood in terms of hunting and gathering in olden days, and still are today in some remote areas. There are many more invisible benefits, for example, climate control, pollution check, shelter for wildlife, and various plant species. Still more benefits are water-retaining capacity, flood control, check on soil erosion, recreational site development, aesthetic revival for the people, unending supply of oxygen, carbon sink, and supply of soil nutrients. The loss of forest and forest cover is detrimental to the aforementioned benefits accrued and finally will mean the loss of the forests. It is estimated that a large healthy tree produces enough oxygen each day to support 18 people (Preston et al. 2006). A study of the Puget Sound area (in Washington State, USA) found that the tree canopy cover removed 38,990 tons of air pollutants in 1 year (American Forest 1998).

25.10.1 Forest Management

Forest management is the process by which forests are preserved in such a way that the benefits are accrued continuously without compromising quality, which involves protection, improvement, preservation, and maintenance of forest in a judicious manner. Proper knowledge and understanding of the forest as well as planning, decisions, and implementation are ensured, and this will lead to sustainable forest management.

Forest management has undergone a sea change since the mid-twentieth century. Previously forests were mercilessly exploited and many areas became completely devoid of forest cover. This loss has initiated global warming, exacerbated by the heavy use of fossil fuel in the recent past. Greenhouse gases are on the rise and the loss of forest is also continuing. Therefore, management of the forest and its proper utilization needs, urgently, to be addressed.

The scientific study of forest species and their interaction with the environment is commonly known as forest ecology whereas the management of forests is gener-

Table 25.7 Steps taken by community and the government to manage forest resources

Managing rules	Management by community	Government initiative
Stipulated date for harvesting	Decided by committee or community forest regulator	Decided by agency and accepted by local leaders
Monitoring	Community forest regulator	Forest guard
Applicable fine, if not obeyed	Fine to be paid to the community	Fine to be paid to government agency
Accountability	Accountable to the community	Accountable to agencies
Equal rights	Community forest used in equality	Equal rules to all
Leadership	Community officials or in charge	Manager of director of agency

ally referred to simply as forestry. These days the emphasis is on sustainable forest management. The forest department focuses on the integration of ecological, social, and economic values, mostly in consultation with local communities and other stakeholders. The forest should be utilized in such a manner that its quality and quantity are not affected in decades and centuries to come; indeed, it should rather improve. Damage has already occurred, but now there should not be further deterioration.

Human factors are the biggest culprits in the loss of forest cover. They represent the greater demand for forest wood for expansion of agricultural fields, road construction through pristine forest, slash-and-burn (shifting) cultivation, forest fire (human triggered and natural), urban sprawl, and overall human encroachment toward the natural environment and forested areas. Controlled human activities, particularly along the slope and in mountainous regions, is the demand of the time. Indirectly, population control is the foremost initiative to be taken up by society because the sharply growing number of persons means increasing needs for resources. Many resources on the earth are limited, even if they are renewable. Some of the management steps taken by the community and the government are listed in Table 25.7.

25.11 Challenge for the Future

Community forestry in Bhutan is very recent. In fact, the increase in allotment by district and division officials has gained momentum in recent years (since 2007). Many of the allotted forests are in the process of growth and development. The benefits from those forests will accrue after some time when they mature and the produce can be used by the people. These products may be timber or non-timber from the forests. Therefore, the success of community forestry will depend on many factors.

25.11.1 Support for Community Forestry

Community forestry emerged because of government enactment and support to enable the people to manage the forest allotted to them and have their needs fulfilled, and also receive the benefits from selling forest products with some regulations. Government support must continue so that community forestry develops properly; otherwise, a change in policy may cause harm before reaching maturity. The long-term vision for sustainable management of community forests is to build a strong institutional, political, and social path for a great future (Chhetri 2009). This development will have a significant contribution to rural livelihood, poverty reduction, and improved forest conditions, which ultimately lead to control in climate change.

25.11.2 Creating Awareness About Community Forestry

Many rural people are not very much aware of the concept of community forestry and its benefits, which is why awareness of community forestry is needed among the rural masses, so that the people may come forward in large numbers to engage in this work. People must be made aware of the functioning of community forestry, as well as the subsequent benefits. In this way, participation of rural people in this endeavor will be maximized. Thus, awareness programs are needed to be carried by the governmental agencies on a regular basis.

25.11.3 Relaxation of Some Rules to Form Community Forestry Management Groups

As per the rules laid down for the formation of community forestry management groups (CFMG), there must be at least ten households to form a CFMG as well as the average maximum area of 2.5 ha per household. These conditions create some problems, as many villages have fewer than ten households, and according to existing rules, they cannot form a CFMG. The average maximum cap of 2.5 ha per household is a very small area. When demarcating the area for the forest, many times the natural boundary does not match, creating a difficulty to be faced. Therefore, there is great need to relax the limitations while considering the CFMG. Relaxation of rules favors the people so they may form a smaller CFMG, as well as to obtain a larger area that will provide a better return from the community forests.

25.11.4 Ensure Sustainable Forest Management

The development and conservation objectives of community forestry can be achieved only by following fundamental principles of sustainable forest management. Hence, the forest should be managed according to scientific knowledge about the ecology and silviculture of the main forest type of the country (Chhetri 2009). In the absence of scientific insight, forest management will not be sustainable and long lasting. The varieties of tree and plant species should be selected on the basis of their adaptability and suitability to the place, location, altitude, and latitude. The utilization of forest resources should be based on their productivity and renewability. It must be ensured that forests are not further degraded, and that areas already degraded are put under forest cover with suitable treatment and management.

25.12 Conclusions

Community forestry in Bhutan is very new, but recently its growth has acquired momentum. It has been growing very steadily and has taken the shape of a movement, particularly in Eastern Bhutan. The 10th Five-Year Plan of the nation places more emphasis on poverty reduction, which is mostly associated with a rural population. Rural people are more dependent on forests and in various ways such as the collection of wild food items, fodder, fruits, nuts, herbs, fuel wood, and wood for the construction of houses, grazing, and many more. Community forestry has multiple roles in the life of rural people, starting from meeting the needs of rural people for household demands and extending to income generation. The government has great hope that community forestry will reduce poverty through proper management of allotted forests to CFMGs. It will be achieved through proper implementation and initiation of the community forests in various areas, with slight relaxation of rules to form the CFMGs on local merits and the demarcation of forest areas so that a maximum number of people are brought under the umbrella of the CFMGs. It also important to continue reviewing the benefits accrued to the community as well as to help them in managing the forest resources in scientific manner.

It is equally essential for CFMGs to be in contact with the Department of Forest to acquire the proper methods and training, in addition to their indigenous knowledge, to go ahead in maintaining, preserving, and increasing the productivity of the forest. Forest management is essential not only for the proper livelihood of the people but also for the overall global benefit by maintaining the balance of greenhouse gases. Carbon dioxide is consumed by trees while preparing their food through photosynthesis under sunlight and in this process oxygen is released. Community forestry confines the local people to the restricted zone only, which leads to the protection of other reserved forests managed, looked after, and utilized by the Department of Forest. Hence, trees and forests are climate-modifying agents, and their proper cover and balance will make this planet a livable place for all types of organisms.

References

American Forests (1998) Regional ecosystem analysis Puget Sound Metropolitan Area, calculating the value of nature. http://www.americanforests.org/downloads/rea/AF_PugetSound.pdf. Last accessed 14 Dec 2010

Barua I (2009) Conservation and management of community and natural resources: a case study from north east India. Stud Tribes Tribals 7(1):39–46

Brendler T, Carey H (1998) Community forestry, defined. J For 96(3):21–23

Census of Bhutan (2005) Population and housing census of Bhutan. Office of the Census Commissioner, Royal Government of Bhutan, Thimphu

Chhetri BB (2009) Overview of community forestry in Bhutan. In: Proceedings of the national community forestry workshop 2009, Thimphu, 16–17 Apr, pp 24–32

Chhetri BB, Schmidt K, Gilmour D (2009) Community forestry in Bhutan: exploring opportunities and facing challenges. Community forestry international workshop, Pokhara, Nepal, 15–18 Sept, pp 1–9

Department of Forest and Park Services (2010) National strategy for community forestry: the way ahead. Ministry of Agriculture and Forest, Royal Government of Bhutan, Thimphu

Dhital DB (2002) Overview of forest policy reviews in Bhutan. In: Proceedings of the forest policy workshop, Kuala Lumpur, 22–24 Jan, pp 22–23

Gilmour D (2009) Lessons learned from community forestry in Asia: a strategic perspective. In: Proceedings of the national community forestry workshop 2009, Thimphu, 16–17 Apr, pp 11–15

National Statistics Bureau (2003) Poverty analysis report. Royal Government of Bhutan, Thimphu

Peljor N (2009) Community forestry in Bhutan: a policy perspective. In: Proceedings of a national workshop, community forestry in Bhutan: directions for the future, Thimphu, 16–17 Apr, pp 16–23

Policy and Planning Division (PPD) (2008) Compendium of renewable natural resources statistics. Policy and Planning Division, Ministry of Agriculture, Thimphu

Preston M, Frazier L, Hudson L, Morse S, Rice T (2006) Community forest management plan. City of Cambridge Island, pp 1–10

Royal Government of Bhutan (RGoB) (2000) Forest and nature conservation rules of Bhutan. Department of Forest, Ministry of Agriculture, Thimphu

Royal Government of Bhutan (RGoB) (2006) Forest and nature conservation rules of Bhutan. Department of Forest, Ministry of Agriculture, Thimphu

Sharma R, Xu J, Sharma G (2007) Traditional agroforestry in the eastern Himalayan region: land management system supporting ecosystem services. Trop Ecol 48(2):189–200

Temphel KJ, Beukeboom JJ (2006) Community forestry contributes to the national and millennium development goals without compromising the forestry policy: case studies on community-based forest and natural resources in Bhutan. Department of Forest, Thimphu. http://www.moa.gov.bt/moa/downloads/downloadFiles/MoADownload4ng10456es.pdf. Last accessed 15 Dec 2010

Temphel KJ, Beukeboom JJ (2007) Community forestry in Bhutan contributes to poverty reduction while maintaining the sustainability of the resources. International conference on poverty reduction and forest tenure, market and policy reforms, 3–7 Sept

TFDP (2000) Social forestry in eastern Bhutan, Ministry of Agriculture, 3rd forestry development project. Project Facilitation Office, Khangma

Tshering D (2009) Policy, institution and technical bottleneck that limit the expansion of community forestry: a Dzongkhag perspective. In: Proceedings of the national community forestry workshop 2009, Thimphu, 16–17 Apr, pp 33–49

Turkelboom F, Gurung TR, Dukpa D (2001) Role and use of common property resources in Bhutan Himalayas: between tradition and globalization. Brisbane Pacific Regional Meeting for Common Property Resources, 2–4 Sept. http://dlc.dlib.indiana.edu/dlc/bitstream/handle/10535/1294/turkelboom.pdf?sequence=1. Last accessed 15 Dec 2010

Part VIII
Land Cover and Rural Planning

Chapter 26
Land Cover Pattern and Road Types in Lop Buri Province, Thailand, 1989–2006

Risa Patarasuk

Abstract
Road development leads to several benefits to a region, such as accessibility for people and resources to market areas, farmland, forests, and residential areas. Such a development has an important role in the economic and social development of a nation. Road network development fosters economic growth by providing a means of transportation for people and goods. Although the development of roads brings an overall benefit to society by improving overall connectivity, its impact on land cover is less well understood. This study aims to determine the relationships between road types and land cover in Lop Buri Province, Thailand. The research questions are (1) what is the pattern of land cover associated with different road types? and (2) what is the pattern of change in terms of land cover distribution within a road type? The analysis relies on descriptive statistics and the use of GIS and remote sensing technology. The study concludes that with the improvement of the road network development that aimed to improve social and economic development over the past two decades, the distribution of land cover along these types of roads has changed over time. The loss of forest and increase in upland crops has led to changes in local climate and local hydrology as well as the loss of biodiversity.

Keywords Road network • Land cover • Lop Buri Province • GIS • Economic development

26.1 Introduction

Road development is an agent of land cover change that results from disturbances within a landscape caused by human activities (Almeida 2005; Glover and Simon 1975; Hawbaker et al. 2005; Lugo and Gucinski 2000). This change occurs because

---

R. Patarasuk (✉)
School of Life Sciences, Arizona State University, Tempe, AZ 85287, USA
e-mail: risa.patarasuk@asu.edu

A.K. Dutt et al. (eds.), *Spatial Diversity and Dynamics in Resources and Urban Development*, DOI 10.1007/978-94-017-9771-9_26

road development increases access for humans to the land (Saunders et al. 2002). However, the impact of roads on land cover is unclear. Road network development has contributed to land cover change in Thailand, particularly in the northern, northeastern, and central regions (Cropper et al. 1999; Hafner 1970; Hughes 1971).

The development of a road network that evolves from foot trails to complex highway systems brings benefits to human civilizations (Lugo and Gucinski 2000). As roads provide access to resources such as market areas, farmland, forests, and residential areas, they play an important role in economic and social development. First, road network development fosters economic growth by providing a means of transportation for goods (e.g., agricultural products) to local, regional, or international markets in less time and with lower costs. Second, roads supplement social development by providing access to services such as education, employment centers, and healthcare centers (Queiroz and Gautam 1992; Windle and Cramb 1997). Although roads bring benefits to human civilizations, they fragment habitats, facilitate the spread of invasive species, alter hydrology, alter the climate, and influence the patterns of land cover (Dale 1997; Geist and Lambin 2002; Hawbaker et al. 2005; Lambin et al. 2001; Trombulak and Frissell 2000). Thus, impacts of roads on the ecology of a region are of concern in tropical areas where the road infrastructure needed to handle economic and social development in a developing country such as Thailand is expanding rapidly into areas of high biodiversity such as the forests.

Previous studies have used many measures such as distance to roads (Chomitz and Gray 1996; Coffin 2009; Patarasuk and Binford 2012), connectivity indices (Coffin 2007; Rojnkureesatien 2006), and type of road cover (Layton 1996; Verburg et al. 2004) to assess the relationships between road network and land cover pattern. The most commonly used as a measure of land cover change associated with roads is distance to roads. Distance to a road is a measure of accessibility to the land. The closer the distance to roads, the better the accessibility to land. Thus, more land conversion occurs near roads. This phenomenon follows the concept of distance decay (Li and Yeh 2004; Walsh et al. 2001). For example, in the Brazilian Amazon region, road network development results in a fishbone structure. Road development contributes to vast deforestation as a result of agricultural activities. For example, Alves et al. (1999) estimated that 79 % of the forests in the Central Rondônia in the Western Brazilian Amazon was cleared within 12.5 km of roads between 1977 and 1995. Next, the link between distance to roads, road connectivity, and land cover change has been studied by Rojnkureesatien (2006). As road networks develop, the connectivity usually increases. Rojnkureesatien (2006) showed that larger patches of forests are cleared within 500 m of roads compared to those within 1,500 m.

The type of roads, which includes hierarchy and surface type, can be used as a proxy for land accessibility and patterns of land cover (Forman et al. 2003; Layton 1996; Verburg et al. 2004). Hierarchy of roads refers to national highways (or interstates), provincial highways, and local roads. Layton (1996) states that accessibility to land is less on a highway compared to a local road. A highway or an interstate, which is usually more than two lanes wide, provides the highest travel speed, serves longer distances, and usually connects major cities or large urban areas. This type of road connects high population and major trade centers (e.g., urban/suburban development) where the land cover usually impervious with less vegetation cover found. In comparison, local roads, which are usually two lanes wide, constitute the lowest hierarchy. Local roads connect residential houses or farmlands; these roads are usually surrounded by agricultural landscapes (Forman et al. 2003; Layton 1996). Furthermore, road type by surface type, either loose (unpaved) or hard (paved), portrays different degrees of accessibility to the land. For example, loose-surface roads typically are not easily accessible during the rainy season because the ground is too soft to accommodate the large vehicles that transport agricultural products in agricultural areas. In contrast, hard-surface roads are usable in all types of weather and can accommodate larger vehicles (Laurance et al. 2002; Verburg et al. 2004). Evidence from previous studies shows that when road surfaces are upgraded from unpaved to paved, land cover conversion is intensified. Examples can be seen from case studies in the region of Rondônia State, Brazil. The total length of paved roads increased from 1,434 km in 1979 to 25,324 km in 1988 (Dale et al. 1994). Paved roads encouraged new settlers to migrate to Rondônia and ultimately led to land cover conversion. In particular, especially along BR-364, much of the land conversion was from forests to agriculture and built-up areas. Between 1978 and 1987, 660 % of the forests along this BR-364 had been converted because of this road surface upgrade (Imbernon 2007; Roberts et al. 2002).

The road network in Thailand has been extensively developed since World War II. The purpose of road network development after the war was to support military movement for national security as well as to enhance its neighboring country contacts, especially the international relationships with Laos (Cropper et al. 1999; Kakizaki 2002). Since 1961, as part of the National Plan, the goal of road network development is to foster social and economic development as well as aiming to alleviate poverty (Puri 2006). Particularly, road building was planned to facilitate moving commercial crops from farmlands to market areas and to major cities for distribution to overseas markets (Rojnkureesatien 2006; Kermel-Torres et al. 2004). In Lop Buri, road network development was started in the 1930s by General Por Piboonsongkram, who established Lop Buri as a military town. As a result, Lop Buri has become the center for economic development within the upper Central Region of Thailand (Office of Information and Technology: Thepsatri Rajabhat University 2008). For the past several decades, ongoing road construction and improvement have occurred within the province. For example, National Highway No. 1, which is one of the most important of the four national highways in Thailand, passes through the province (Department of Highways District 1 Lop Buri 2008; Lop Buri Provincial Office 2006) (Fig. 26.1). This highway has a significant role in land transportation connecting Bangkok (the capital city of Thailand) with Lop Buri and the northern regions of Thailand; it is the main arterial for transporting agricultural products to the capital city.

This study aims to determine the relationships between road types and land cover in Lop Buri province. The main research questions are (1) what is the pattern of land cover associated with different road types? and (2) what is the pattern of change in terms of land cover distribution within a road type? The hypotheses are that

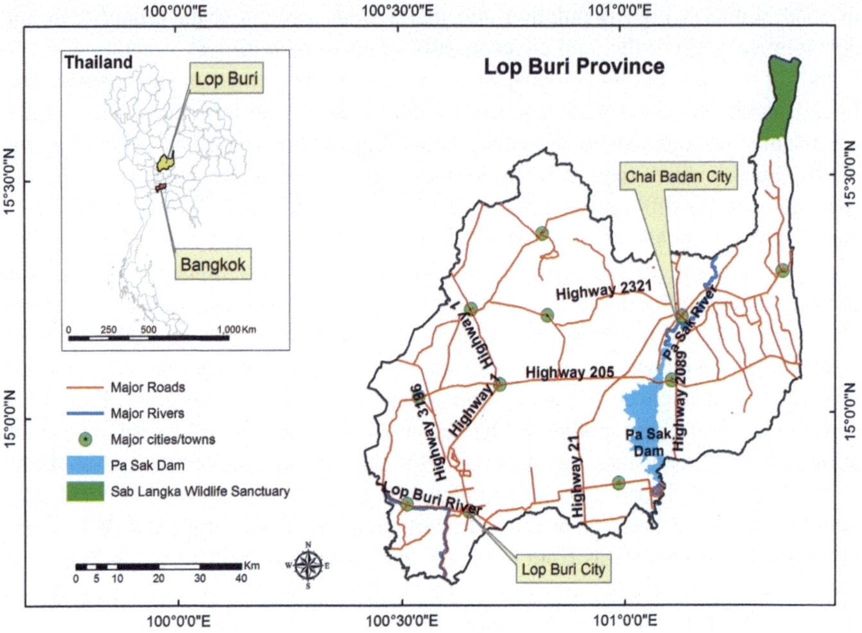

Fig. 26.1 The study area

(1) built-up areas have larger distribution along highways compared to local roads, (2) local roads have a higher distribution of agricultural landscape compared to built-up areas, and (3) larger amounts of forest cover are converted to agricultural land along local roads. The analysis relies on descriptive statistics and the use of Geographic Information System (GIS) and remote sensing technology.

26.2 Methods

26.2.1 Study Area

Lop Buri Province is located in the Central Region of Thailand (Fig. 26.1). It is approximately 150 km north of Thailand's capital city, Bangkok. The province has a land area of 6,600 km^2. Lop Buri province lies between 14°39′N and 15°35′N latitude and between 100°24′E and 101°26′E longitude. This province has a geographic advantage because it is a gateway among the Northern Thailand, Central Thailand, and Northeastern Thailand regions.

Lop Buri Province is considered as one of the wealthier provinces in Thailand (Felkner and Townsend 2004). The province's population in 2008 was 753,470. Lop Buri City is the most populated district with a population of 249,620, followed by

Chai Badan City, which has a population of 90,182. These two districts are the most important centers of economic activity in the province (Lop Buri Provincial Office 2010).

The elevation ranges from 5 to 750 m above sea level. The province has a subtropical climate influenced by monsoons. May to October is the rainy season, influenced by the southeast monsoon from the Indian Ocean. Rainfall in the province is bimodal. The greatest amount of rainfall is in September and the second greatest is in May. For the 30-year span from 1961 to 1990, the average amount of rainfall was 170 mm in May and 280 mm in September. November and April are dry months, influenced by the northeast monsoon from China. The average minimum temperature from 1961 to 1990 was 23.3 °C and the maximum was 33.3 °C. The hottest month is April and the coldest is December/January (Lop Buri Provincial Office 2006; Thai Meteorological Department 2008; Worakawin 2003).

The landscape is mainly composed of agricultural areas, with the major land cover being rice paddies and upland crops (e.g., cassava, sugarcane, maize, sunflowers). Forests, water, plantation, and built-up areas also constitute the provincial landscape. The land cover has undergone considerable alteration during the past several decades, particularly the change from forest to upland crops. This change was experienced by about 15 % of the entire province between 1989 and 2006. Other notable land cover changes in the area include rice paddies to upland crops (8 % of the study area), forest to rice (3 % of the study area), and forest to built-up areas (1 % of the study area) (Patarasuk and Binford 2012).

The road network consists of national highways, provincial highways, and local roads. The highways area paved; however, most of the local roads are not. Furthermore, the road network density is not evenly distributed throughout the province. The greatest densities are found in the southwestern and northern portions of the province. Toward the end of the 1990s, some roads were removed following the construction of the Pa Sak Dam, which was completed in 1999.

26.2.2 Land Cover Data

Land cover data used in the study were based on a previous study by Patarasuk and Binford (2012). These authors used a hybrid method classification process. Landsat images dated 20 January 1989 (TM 4), 10 March 1998 (TM 5), and 8 March 2006 (ETM+) were used to derive the land cover information. Six land cover classes were used: forest, water, plantation, rice, upland crops (e.g., cassava, sugarcane, maize, sunflowers), and built-up areas. The 2006 image has cloud cover and was excluded from the study; it comprises approximately 3.5 % of the study area. The corresponding areas for the 1989 and 1998 images were also excluded for areas of cloud cover. The overall accuracy of the 2006 land cover classification was 83 %. Land cover classification was based on the field verification in May/June/July of 2006 and 2007 and December/January of 2008.

26.2.3 Road Network Data

Road network data were obtained from three different agencies: Department of Highways, Department of Rural Roads, and Department of Environmental Quality Promotion: these were compiled into one single network. The road network dataset was geo-referenced to Landsat images with a root mean square (RMS) error <15 m. Ancillary data such as topographic maps (1:50,000 scale) and aerial photographs were used to help geo-reference the road networks. The road type used in this study was based on the classification according to the Department of Environmental Quality Promotion. The types of roads used in the study follow:

Road type 1: hard surface, two or more lanes wide (major highways, provincial highways); this type is the highest hierarchy
Road type 2: loose or light surface, two or more lanes wide (provincial highways)
Road type 3: hard surface, one lane wide (local roads)
Road type 4: loose or light surface, one lane wide (local roads); this type is the lowest hierarchy

26.2.4 Analysis

The study omitted the cart tracks in the analysis because it was difficult to geo-reference these based on Landsat images. Each road type was buffered using a Euclidean distance of 6 km. This distance was chosen because, on the basis of the previous study by Patarasuk and Binford (2012), land cover dynamics are found within 6 km of roads. Finally, land cover information was derived based on the buffered area to determine the distribution of land cover for each road type.

26.3 Results

26.3.1 General Description of Road Network

In general, roads are distributed throughout the province. Roads were most extensively developed in the northern two thirds of the province. The total road lengths from 1989 to 1998 and to 2006 were 3,347 km, 4,541 km, and 4,723 km, respectively. Hence, roads were more developed for the 1989–1998 than the 1998–2006 time periods.

Table 26.1 shows the number of roads by type described in the Sect. 26.2.3 in terms of total lengths. Road type 1, of which the majority are major highways with hard surface and two or more lanes wide, is distributed throughout the province; it accounted for 29 % of the total road lengths in 1989, 22 % in 1998, and 20 % in

Table 26.1 Total road lengths by road type

	Year 1989					Year 1998					Year 2006				
	Road type 1	Road type 2	Road type 3	Road type 4	Total 1989	Road type 1	Road type 2	Road type 3	Road type 4	Total 1998	Road type 1	Road type 2	Road type 3	Road type 4	Total 2006
Total road lengths (km)	970.75	83.47	1,371.06	922.14	3,347.42	1,006.98	115.14	1,445.79	1,973.26	4,541.16	944.54	117.82	1,471.65	2,188.97	4,722.97
Percent (%) of Total	29.00	2.49	40.96	27.55	100.00	22.17	2.54	31.84	43.45	100.00	20.00	2.49	31.16	46.35	100.00

2006. Road type 1 usually links main cities within the province as well as outside the province. For example, Highway #1 links the city of Lop Buri with Bangkok (the capital city of Thailand). Road type 2 has the least total road length in the province. This type consists of loose or light surface and is two or more lanes wide. These roads only accounted for 2–3 % of the total road lengths for the entire province. The amount of this road type was consistent throughout 1989–2006. Roads of type 2 are scattered in the northern and the eastern sections of the province. Road type 3, which classified as one lane wide with a hard surface, had the greatest total of road lengths in 1989, which proportionally accounted for 40 % of all other road types. However, the number dropped to 32 % in 1998 and remained at approximately 32 % in 2006. These roads are found mostly in the northern part of the province (i.e., north of Highway #205) (Fig. 26.1). Last, road type 4 has undergone the highest amount of development. The total road lengths accounted for approximately 28 % in 1989. The amount of roads developed was added to the total road lengths, which were 43 % in 1998 and 46 % in 2006. This type of road was mostly developed in the northern section of the province.

26.3.2 Road Type and Land Cover

The majority of the landscape in Lop Buri Province is covered by forests, water, plantation, rice paddies, upland crops, and built-up areas (Patarasuk and Binford 2012) (Fig. 26.3). Table 26.2 shows the percentage of land cover distribution within a 6-km buffer for each road type. The analysis showed that, in general, upland crops are the major land cover found along all road types. That is, upland crop cover has the highest percentage distribution within each particular road type. One exception is noted within road type 2 in 1989, where there are more forested areas than upland crops. As the year progressed, the percentage of upland crops within each road type increased. The highest increase is within road type 2 between 1989 and 1998; the percentage cover increased from 41.4 % to 56.87 %.

Forest cover is usually the second most common occurrence within the 6-km buffer of roads. Exceptions for this are within road type 2 in 1989 where it is the highest. Another exception is within road types 1 and 3, which are third highest. The results indicated that Lop Buri Province experienced forest cover decline from the 1989–1998 period to 2006. The highest amount of the forest cover decline is within the buffer of road type 2 and between 1998 and 2006 (11 % decline).

Next, rice paddy landscape fluctuates across the time period. For example, for the years 1989 and 1998, within road type 1, rice paddies have the third largest distribution, but the second largest in 2006. Across all road types, rice has the least distribution along roads of type 2 with 10.41 %, 3.73 %, and 1.60 %, respectively for 1989, 1998, and 2006. Similarly to the trend with forest cover, the analysis also revealed that the amount of rice cover within each road type had declined from 1989 to 1998 and to 2006. The biggest decline was within road type 3 in 1998 with a reduction of 7.62 %.

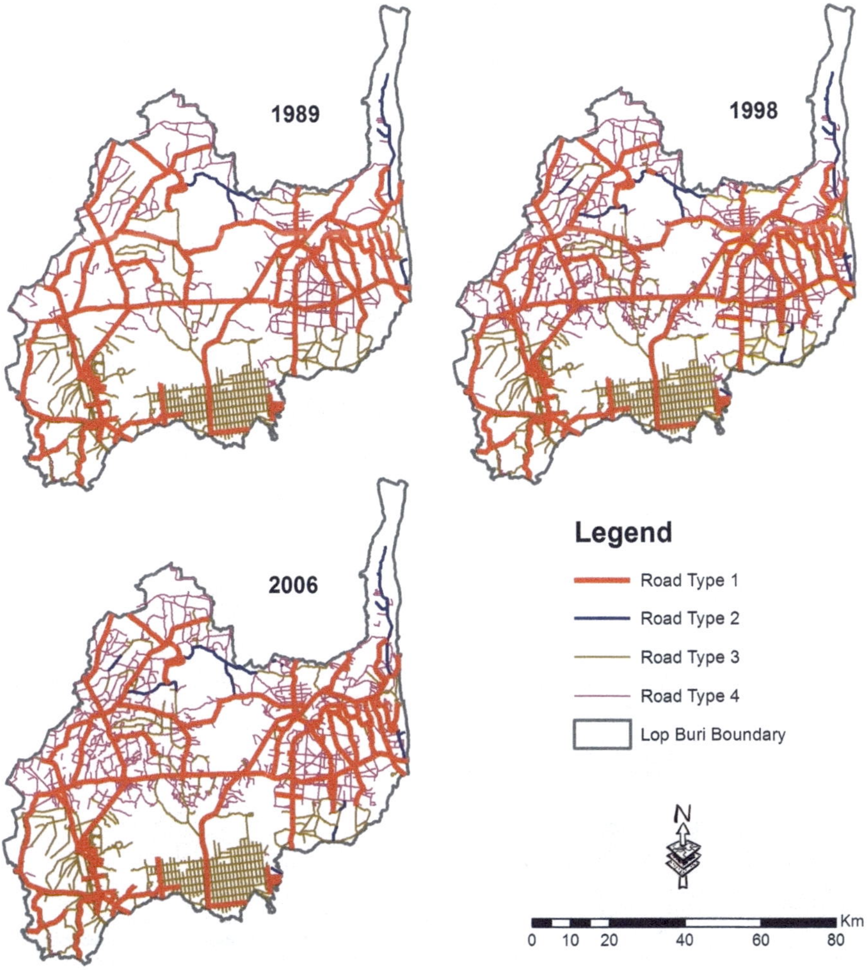

Fig. 26.2 Road network by type: type 1, hard surface, two or more lanes wide (major highways, provincial highways); type 2, loose or light surface, two or more lanes wide (provincial highways); type 3, hard surface, one lane wide (local roads); type 4, loose or light surface, one lane wide (local roads)

Plantations (teak and eucalyptus) in Lop Buri Province emerged on the landscape in 1998. However, the distribution of plantations was very small compared to other agriculture activities such as upland crops and rice. Plantations within each type of road buffer occupy less than 1 %. They are found along all road types. Note that no plantations were observed in 1989.

Water, which includes rivers, lakes, reservoirs, and ponds, contributed to less than 3 % within each buffered area. The amount of landscape covered by water has increased dramatically in 2006 because of the construction of the Pa Sak Dam/Reservoir, located in the eastern section of the province (Fig. 26.1).

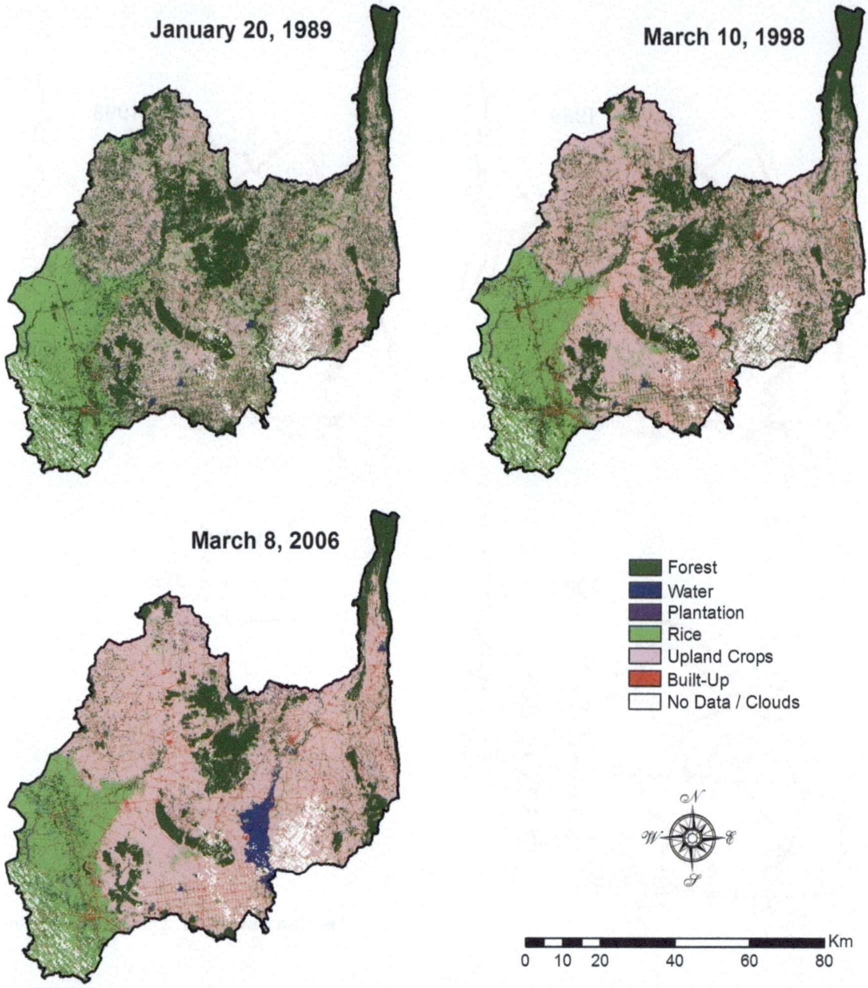

Fig. 26.3 Land cover distribution map (From Patarasuk and Binford 2012)

Last, Lop Buri Province is largely an agriculturally based economy with the majority of the landscape consisting of rice paddies and upland crop fields. Thus, only a small percentage is covered by built-up areas for human settlements. The built-up areas found within each road type are less than 7 %. The majority of built-up areas are found within the buffers of road types 1 and 3. Throughout the study time period, the amount of built-up areas has increased. The biggest increase was within road type 2 in 2006, from 4.15 % to 5.86 %.

Table 26.2 Land cover distribution within 6-km buffer of each road type

Land cover (% of the total area within 6-km buffer for each road type)	Year 1989				Year 1998				Year 2006			
	Road type 1	Road type 2	Road type 3	Road type 4	Road type 1	Road type 2	Road type 3	Road type 4	Road type 1	Road type 2	Road Type 3	Road Type 4
Forest	31.41	45.43	30.23	31.97	22.40	34.97	21.71	24.71	14.58	24.00	14.12	16.97
Water	0.55	0.17	0.67	0.50	0.44	0.17	0.47	0.33	2.24	1.46	2.95	2.70
Plantation	0.00	0.00	0.00	0.00	0.08	0.11	0.07	0.09	0.17	0.27	0.16	0.18
Rice	25.78	10.41	27.32	24.93	19.61	3.73	19.70	18.05	18.57	1.60	18.12	16.75
Upland crops	38.98	41.40	38.35	39.40	52.50	56.87	52.95	52.05	58.24	66.81	58.51	57.38
Built-up area	3.28	2.58	3.43	3.20	4.97	4.15	5.10	4.76	6.20	5.86	6.14	6.02
Total	100.00	100.00	100.00	100.00	100.00	100.00	100.00	100.00	100.00	100.00	100.00	100.00

26.4 Discussion

The results showed that agricultural activities, particularly upland crop fields, dominate the landscape. Thus, the situation in Lop Buri Province does not support the first hypothesis stated in the introduction, that built-up areas exceed agricultural areas along the highways more than along local roads. Within road types 1 and 2, rice and upland crops have contributed to more than 50 % of the landscape; built-up areas have only 3–4 % distribution. However, if compared with the built-up area land cover class, more built-up areas (by percentage) are found in road type 3 and 4 than road types 1 and 2. In contrast, the results support the second hypothesis (local roads have higher distribution of agricultural areas than built-up areas). The presence of rice paddies and upland crop fields have contributed to 50 % of the landscape along roads type 3 and 4. Last, the third hypothesis is supported as forest cover was reduced from 1989 to 2006. Within each road buffer, the amount of forest cover ranged from 30 % to 45 % in 1989; this amount decreased to 14–25 % in 2006.

The decline of forests in Lop Buri Province suggests that the land was converted to other uses within the 6-km road buffer, especially to rice and upland crops (Patarasuk and Binford 2012). More forest loss occurs adjacent to hard-surface roads (types 1 and 3) rather than loose-surface roads (types 2 and 4) when compared between highways and local roads (Table 26.2). Furthermore, rice paddies are found more often along local roads than highways, and a greater percentage of upland crops are distributed along the highway than along local roads. Highways and hard-surface roads are easily accessible all year round. They provide better travel speed, reduce travel time, and reduce travel cost in transporting agricultural products to markets (Soares-Filho et al. 2004; Verburg et al. 2004). Thus, areas along highways or hard-surface roads create more incentives for people to convert the land. An example is the National Highway No. 1, which is the main land transportation arterial between Lop Buri Province and Bangkok. Most crops grown in the province are transported to Bangkok and are exported to overseas markets such as Japan and Europe (Entwisle et al. 2008; Phantumvanit and Sathirathai 1988; Schar 2004).

Lop Buri Province is largely an agricultural area and is considered as part of Thailand's 'Rice Bowl' region (Cheyroux 2003). Rice and upland crops are largely grown in the province, and they occupy more than 50 % of the landscape. Upland crops have played an important role in the province's economy during the past several decades. They have dominated the landscape pattern by replacing forests and rice paddies (Patarasuk and Binford 2012) (Fig. 26.4). Rice paddies remain largely in the western section of the province where they are served by the irrigation systems (Figs. 26.3). Upland crops fields replaced rice paddies mostly in the upland areas, because there are no irrigation systems in the upland areas and the farmers had to rely on the availably of the rainfall (personal interview). Furthermore, the Thai government launched an agricultural diversification policy in early 1990 to encourage farmers to adopt other crops as alternatives to rice to increase their household incomes. As a result, many farmers converted rice to upland crop fields

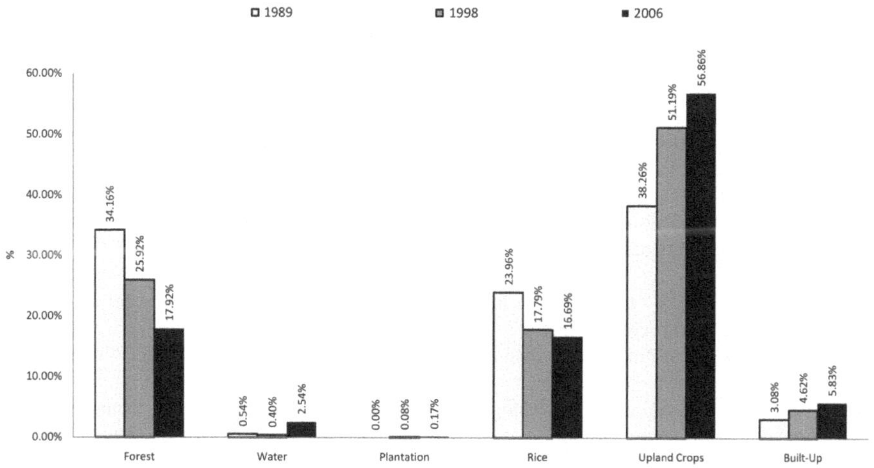

Fig. 26.4 Percentage of land cover distribution graph (From Patarasuk and Binford 2012)

(Cheyroux 2003; Schar 2004; Sirisup and Kammeier 2003). In addition, the cropping pattern has changed from rice to upland crops because of the response to demand from foreign markets, especially in Europe for the animal feeds industry (Entwisle et al. 2008; Phantumvanit and Sathirathai 1988; Schar 2004).

As roads developed, the connectivity changed. Local roads are more connected to highways, hence, improving transportation routes and access within Lop Buri Province and outside the province. The landscape pattern has also changed along all road types. Roads may have had major roles in land cover change; however, other factors can contribute to land cover change at the same time (Lambin et al. 2001), including climate pattern, household income, and population pressures (Cropper et al. 1999; Dale 1997).

This research has shown a link between land cover pattern and its distribution along different road types—hard versus loose surface and highway versus local roads—in Lop Buri Province from 1989 to 2006 by using GIS and remote sensing technology. With the development of the road network that aimed to improve social and economic development during the past two decades, the distribution of land cover along these types of roads has changed over time. Land cover change is a major contributor to local and global environmental change. The loss of forest and increase in upland crops can lead to changes in local climate and local hydrology as well as to loss of biodiversity (Dale 1997; Kummer and Turner 1994).

Acknowledgments This research was financially supported by the National Science Foundation, Geo-Informatics and Space Technology Development Agency, University of Florida Map and Imagery Library, and Altrusa International.

References

Almeida B (2005) A GIS assessment of urban sprawl in Richmond, Virginia, geography. Virginia Polytechnic Institute and State University, Blacksburg, p 141

Alves DS, Pereira JLG, Sousa CLD, Soares JV, Yamaguchi F (1999) Characterizing landscape changes in central Rondonia using Landsat TM imagery. Int J Remote Sens 20:2877–2882

Cheyroux B (2003) Fruits and vegetables in Thailand's rice bowl: the agricultural development of poldered raised bed systems in Damnoen Saduak area. In: Molle F, Srijantr T (eds) Perspectives on social and agricultural change in the Chao Phraya Delta. White Lotus, Bangkok, pp 157–176

Chomitz KM, Gray DA (1996) Roads, land use, and deforestation: a spatial model applied to Belize. World Bank Econ Rev 10:487–512

Coffin AW (2007) From roadkill to road ecology: a review of the ecological effects of roads. J Transp Geogr 15:396–406

Coffin AW (2009) Road network development and landscape dynamics in the Santa Fe River watershed, north-central Florida, 1975 to 2005. University of Florida, Department of Geography, Gainesville

Cropper M, Griffiths C, Mani M (1999) Roads, population pressures, and deforestation in Thailand, 1976–1989. Land Econ 75:58–73

Dale VH (1997) The relationship between land-use change and climate change. Ecol Appl 7:753–769

Dale VH, O'Neill RV, Southworth F, Pedlowski M (1994) Modeling effects of land management in the Brazilian Amazonian settlement of Rondonia. Conserv Biol 8:196–206

Department of Highways District 1. Lop Buri (2008) History of Lop Buri 1 District of Highway. Available from: http://www.doh.go.th/dohweb/hwyorg43100/district_history.htm. 23 October

Entwisle B, Rindfuss RR, Walsh SJ, Page PH (2008) Population growth and its spatial distribution as factors in the deforestation of Nang Rong, Thailand. Geoforum 39:879–897

Felkner JS, Townsend RM (2004) The wealth of villages: an application of GIS and spatial statistics to two structural economic models. Department of Economics, University of Chicago, Chicago

Forman RTT, Sperling D, Bissonette JA, Clevenger AP, Cutshall CD, Dale VH, Fahrig L, France R, Goldman CR, Heanue K, Jones JA, Swanson FJ, Turrentine T, Winter TC (2003) Road ecology: science and solution. Island Press, Washington, DC

Geist HJ, Lambin EF (2002) Proximate causes and underlying driving forces of tropical deforestation. Bioscience 52:143–150

Glover DR, Simon JL (1975) The effect of population density on infrastructure: the case of road building. Econ Dev Cult Change 23:453

Hafner J (1970) The impact of road development in the central plain of Thailand. University of Michigan, Ann Arbor

Hawbaker T, Radeloff V, Hammer R, Clayton M (2005) Road density and landscape pattern in relation to housing density, and ownership, land cover, and soils. Landsc Ecol 20:609–625

Hughes AC (1971) Some economic effects of road construction in rural northeast Thailand, 1958–1970. University of Hawaii

Imbernon J (2007) Population growth and deforestation: the case of Rondonia in the Brazilian Amazon. ICRAF, Nairobi

Kakizaki I (2002) Road improvement in the era of postwar rehabilitation in Thailand, 1945–1957. JAAS Aziya Kenkyu Asian Studies. Japan Association of Asian Studies, Tokyo, pp 4–31

Kermel-Torres DK, Hubert JP, Franck M (2004) Chapter 6: Tertiary sector. In: Kermel-Torres D (ed) Atlas of Thailand: spatial structures and development. Silkworm Books, Chiang Mai, pp 131–150

Kummer DM, Turner BL (1994) The human causes of deforestation in Southeast Asia. Bioscience 44:323–328

Lambin EF, Turner BL, Geist HJ, Agbola SB, Angelsen A, Bruce JW, Coomes OT, Dirzo R, Fischer G, Folke C, George PS, Homewood K, Imbernon J, Leemans R (2001) The causes of land-use and land-cover change: moving beyond the myths. Global Environ Change A 11:261–269

Laurance WF, Albernaz AKM, Schroth G, Fearnside PM, Bergen S, Venticinque EM, Costa CD (2002) Predictors of deforestation in the Brazilian Amazon. J Biogeogr 29:737–748
Layton RD (1996) Functional integrity of the highway system. Department of Civil, Construction, and Environmental Engineering, Oregon State University, Corvallis
Li X, Yeh AG-O (2004) Analyzing spatial restructuring of land use patterns in a fast growing region using remote sensing and GIS. Landsc Urban Plan 69:335–354
Lop Buri Provincial Office (2006) Lop Buri Province report, Lop Buri, Thailand
Lop Buri Provincial Office (2010) Lop Buri Province report, Lop Buri
Lugo AE, Gucinski H (2000) Function, effects, and management of forest roads. For Ecol Manag 133:249–262
Office of Information and Technology: Thepsatri Rajabhat University (2008) Lop Buri: a military city. Available from: http://library.tru.ac.th/il/lop/tour/soldier1.html
Patarasuk R, Binford MW (2012) Longitudinal analysis of the road network development and land-cover change in Lop Buri province, Thailand, 1989–2006. Appl Geogr 32:228–239
Phantumvanit D, Sathirathai KS (1988) Thailand: degradation and development in a resource-rich land. Environment 30:10–15
Puri J (2006) Factors affecting agricultural expansion in forest reserves of Thailand: the role of population and roads. University of Maryland, College Park, p 197
Queiroz C, Gautam S (1992) Road infrastructure and economic development : some diagnostic indicators. The World Bank, Washington, DC
Roberts DA, Numata I, Holmes K, Batista G, Krug T, Monteiro A, Powell B, Chadwick OA (2002) Large area mapping of land-cover change in Rondônia using multitemporal spectral mixture analysis and decision tree classifiers. J Geophys Res 107:LBA 40-41–LBA 40-17
Rojnkureesatien K (2006) Geographical accessibility and land-use and land-cover dynamics: the case of Nang Rong district, northeast Thailand. Department of Geography, University of North Carolina, Chapel Hill, p 227
Saunders SC, Mislivets MR, Chen J, Cleland DT (2002) Effects of roads on landscape structure within nested ecological units of the northern Great Lakes region, USA. Biol Conserv 103:209–225
Schar P (2004) Agriculture (Chapter 4). In: Kermel-Torres D (ed) Atlas of Thailand: spatial structures and development. Silkworm Books, Chiang Mai, pp 85–105
Sirisup S, Kammeier HD (2003) Government policy and farmers' decision making: the agricultural diversification programme for the Chao Phraya River basin (1993–1995) revisited. In: Molle F, Srijantr T (eds) Perspectives on social and agricultural change in the Chao Phraya Delta. White Lotus, Bangkok
Soares-Filho B, Alencar A, Nepstad D, Cerqueira G, del Carmen Vera Diaz M, Rivero S, Solorzano L, Voll E (2004) Simulating the response of land-cover changes to road paving and governance along a major Amazon highway: the Santarem–Cuiaba corridor. Glob Change Biol 10:745–764
Thai Meteorological Department (2008) Lop Buri Province's average temperature and rainfall from 1961–1990.Availablefrom:http://www.tmd.go.th/en/province_stat.php?StationNumber=48426. 26 November
Trombulak SC, Frissell CA (2000) Review of ecological effects of roads on terrestrial and aquatic communities. Conserv Biol 14:18–30
Verburg PH, Overmars KP, Witte N (2004) Accessibility and land-use patterns at the forest fringe in the northeastern part of the Philippines. Geogr J 170:238–255
Walsh SJ, Crawford TW, Welsh WF, Crews-Meyer KA (2001) A multiscale analysis of LULC and NDVI variation in Nang Rong district, northeast Thailand. Agric Ecosyst Environ 85:47–64
Windle J, Cramb RA (1997) Remoteness and rural development: economic impacts of rural roads on upland farmers in Sarawak, Malaysia. Asia Pac Viewpoint 38:37–53
Worakawin K (2003) Atlas of Thailand: physical geography by province. Phattana Khunnapab Vichakarn (translated from Thai language), Bangkok, Thailand

Chapter 27
Assessment of Capacities of TAOs and CBOs for Local Development in Thailand

Darin Khongsatjaviwat and Jayant K. Routray

Abstract Under the decentralized development planning process, the role and contributions of local governments as well as local institutions are becoming more important in each and every country irrespective of their overall economic development. Thailand, as a middle income and newly industrialized country, is no exception. Over the past several decades, efforts are being made to decentralize the planning and development process with a participatory approach and strong focus on people-centered development. At the local level, the Tambon Administrative Organization (TAO), as local self-government, and a number of community-based organizations (CBOs) are currently engaged in the local development process and development consequent on the introduction of decentralized development planning for more than 15 years. The development goals of these two institutions at the local levels in rural areas are primarily the same, but the roles, contributions, and processes are different. An attempt is made in this chapter to assess their capacities and contributions in catalyzing local development and also fulfilling the needs of the people within their areas of operations as well as limitations. In terms of their capacities and contributions, CBOs are better placed than the TAOs. CBOs work more closely with the people and try to meet the social, economic, and environmental as well as natural resource development needs.

Keywords Local development • Community-based organization • Tambon administrative organization • Thailand

D. Khongsatjaviwat
Department of Political Science and Public Administration, Faculty of Social Sciences, Naresuan University, Phitsanulok, Thailand

J.K. Routray (✉)
Asian Institute of Technology, PO Box 4, Klong Luang, Pathamuthani 12120, Thailand
e-mail: routray@ait.asia

A.K. Dutt et al. (eds.), *Spatial Diversity and Dynamics in Resources and Urban Development*, DOI 10.1007/978-94-017-9771-9_27

27.1 Introduction

Development planning in Thailand is under the supervision of the central government. The National Economic and Social Development Board (NESDB) is the main factor in setting up guidelines and directives for development of the nation as a whole. The first National Economic and Social Development Plan (NESDP) was implemented between 1961 and 1966, with an emphasis on infrastructure improvement for economic activities. Between NESDP 1 and NESDP 6, Thailand has been modernizing to respond to new industrialization process. The National Economic and Social Development Board (NESDB) has reinstated poverty alleviation as a major focus of its efforts under NESDP 7, 8, and 9, and has placed emphasis on developing poverty alleviation strategies based on community empowerment experiences. Plans to tackle poverty must take into account the considerable amounts of experience from projects of community empowerment during recent years. Poverty policies should endeavor to build on social capital investments and community strengths, and the importance of improving access to land and other natural resources, as key strategies of poverty alleviation.

Thailand has now implemented its 11th plan (2012–2016). For six decades, Thailand has encountered unbalanced development because of the influx of capitalism (Bello et al. 1998; Pongpaichit and Chris 2002). It is obvious that the government and market mechanisms have failed to equally distribute income and public goods to the poor, resulting in so-called government and market failures. Apparently, the government not only does empower the community and make investment in social capital but also focuses on local institutions more rigorously. The capital cities are the center of economic resources and these resources are not equally distributed. People living in rural areas are most vulnerable because they are living far away from the capital (Wattanasiri 2001). Thailand's poverty profile in 2008 showed that 5.2 million people were in the rural areas, constituting about 90 % of the total poor. The Northeast is the poorest region in the country. According to the Thailand Millennium Development Goals Report 2009, Thailand has met the MDGs target very closely in an overall perspective with only less than 10 % poor remaining to be covered in 2015. However, the poverty incidence remains highest in the northeast part of Thailand. The poorest of the poor are elderly and farmers. As a result, rural communities have lagged behind other social sectors in terms of development, employment, and income, and there has been a continuous migration of rural people to Bangkok and the large cities. At the same time, the widening income gap, poverty, and the deterioration of natural resources and the environment have contributed to increased social conflicts and tension.

The country's development may become distorted if the government concentrates only on the industrial sector. Any solution to Thailand's development problems must address the problems at the grassroots level as a large proportion of people still live in abject poverty. Balanced development and sustainable economic growth obviously require alternative models. Local institutions are therefore important to help the central government achieve its development goals.

Local institutions have been in existence for many years in Thailand. Doungkaew (1999) has clearly explained the constraints of local institutions in Thailand. She argues that (1) the majority of rural institutions are small with limited activities in scope; (2) they induce impacts on a very small section of the society; (3) funding remains a constant concern of these organizations; (4) the majority of rural institutions are not registered; and (5) only some of the initiatives undertaken by these have succeeded. Vichit-Vadakan (2001) further explains that most of the community-based organizations (CBOs) in Thailand are struggling for sustainability. Many were helped initially with "seed money" from external sources, including government agencies. Moreover, CBOs also have various problems in management and financial skills, so their management has not been effective. It has been found that their manpower skills are not properly developed and therefore they have a low level of professionalism (Boonyabancha 2004).

Tambon (sub-district) Administrative Organizations (TAOs) in Thailand have been providing certain development functions and administrative services. They are responsible for encouraging public participation and ensuring decentralized patterns of rural development in Thailand. As a form of local government at the lowest level in rural areas, TAOs promote democracy at the grassroots level among the rural Thai communities to provide public services for improving the quality of life, and implement development projects, thus fulfilling the needs of the local people. Generally, the government has full authority to make political decisions with regard to the decentralization process from the central to local levels. In the case of Thailand, the decentralization policy as stipulated in the new Constitution aims to devolve the authority as much as appropriate to local governments. In accordance with the new Constitution of 2007 and the decentralization laws, local government units such as the TAOs have become stronger and expanded their responsibilities in line with their resources, and they are functioning as local institutions. To date, most of the functions and responsibilities will be or are in the process of being devolved to the local governments. Moreover, the government has also allocated a large portion of its budget and manpower to the TAOs. The government continues this move as necessary, considering that the revenue and budget for the TAOs are essential elements for rural development. Thus, effective budget management within the TAOs would enable the people in their respective localities to initiate and implement meaningful development projects for improving their quality of life.

CBOs also complement and supplement the activities of the TAOs by promoting the people's active participation and engaging in development projects. There have been various studies on the role of institutions in rural community development in many countries by the Asian Productivity Organization over many years. However, little research is carried out on TAOs and CBOs in Thailand with reference to their effectiveness in rural or community development.

There are linkages between the central and local governments for the overall development of the country. The self-reliance of each local institution depends on its ability to sustain development projects despite limited/inadequate funding support. Effective links between development activities/services and other groups/institutions within the village are also important when capital pooling is needed for self-help groups, housewives' groups, etc. In this case, the relationships among local institutions—TAOs and CBOs vis-à-vis the central government—are absolutely essential to run the development projects efficiently. TAOs are local legal bodies financially supported by the central government, whereas CBOs are locally autonomous and are financially supported mostly by members and rarely by nongovernmental sources. The central government can also allocate some financial support to CBOs but to a limited extent. In general, financial support from the central government is spent to promote better livelihoods in the community.

To develop at the grassroots level, local institutions are required to provide strategies, programs, and activities to empower local people. Local institutions apply the Sufficiency Economy concept. Therefore, TAOs and CBOs shed light on the deep-seated problems of local people. Providing services and meeting people's expectations are two important challenges commonly faced by the TAOs and CBOs. People's expectations and the capacity of local institutions need to be analyzed by indicators covering economic, social, natural resources, and environment development aspects. The ultimate goal of such institutions in the context of rural development is the protection of people's welfare.

With this backdrop, the objective of this chapter is to assess the capacities of the TAOs and the CBOs in fulfilling the needs and expectations of people at the local level in the context of overall rural development.

27.2 Methods and Materials

The study is primarily based on secondary and primary data sets embedded with both qualitative and quantitative research methods. The descriptive type of research has been applied to describe the role and function of local institutions in the context of rural development in Thailand. To understand local institutions clearly, this research has attempted to describe the principles and activities of local institutions undertaken and linked with their achievements and contributions for rural development. The exploratory type of research was applied for the assessment of local institutions' activities in Changwat Phirsanulok. First, it has attempted to assess the socioeconomic impacts of local institutions measured through their performance. Second, it has analyzed the people's expectations and the capability of local institutions in fulfilling their needs in relationship to rural development. Eventually, this research has been conducted to answer how local institutions are able to contribute to strengthen rural development in Thailand.

The reconnaissance survey helped in gathering basic data and information about the sample Tambons chosen for the study, and provided insight into the research issues more deeply than initially conceptualized, further strengthened through key informant interviews. These two steps helped to refine the questionnaire. After pretesting, the questionnaire was finalized and employed for primary data collection for interviewing the target samples. Simple random sampling was used for the two sets of samples for data collection.

According to the sampling formula, the total sample household size of TAOs was 142 and the sample size of the members of the CBOs was 129. The selected CBOs for sampling the respondents were chosen primarily with one important consideration: that they were very active in savings and occupational, social, and microfinance group activities.

The data were analyzed by applying both qualitative and quantitative tools. The analytical techniques used were descriptive statistics and analytical statistics. The weighted average index (WAI) was the main statistical tool used following the scaling techniques. Normally, a 5-point Likert scale with a value from 0 to 1 is used for computing the WAI. The five points refer to "very high, high, moderate, low, and very low" according to the specific context, and issues that addressed this were modified according to the subject of investigation. The issues were analyzed with the WAI.

$$WAI = (W_1F_1 + W_2F_2 + W_3F_3 + \ldots. + W_nF_n) / (F_1 + F_2 + \ldots. + F_n)$$

where WAI = weighted mean index

W = assigned weight for a particular class under the 5-point scale (very low, $W_1 = 0.20$; low, $W_2 = 0.40$; moderate, $W_3 = 0.60$; high, $W_4 = 0.08$; very high, $W_5 = 1.0$)

F_1 = corresponding frequency of that class

The criteria used for interpreting the results are given below:

Criteria for assessing weighted average index:				
Very high (VH)	High (H)	Moderate (M)	Low (L)	Very low (VL)
0.81–1.0	0.61–0.8	0.41–0.6	0.21–0.4	0.00–0.20

27.3 Profile of the Study Areas

Changwat Phitsanulok was purposively chosen as the study area because it is a pilot province for decentralized planning and development. It is located between the central and northern regions of Thailand. Phitsanulok is one of the northern provinces that serves as a center of commerce, transportation, and communication between two regions. There are 82 Tambons and 374 CBOs in the province. Three Tambons in three districts were selected purposely for intensive field research. These Tambons are categorized as small, medium, and large following revenue criteria, excluding grants from the central government (more than 20 million baht is large, 6–20 million baht is medium, and less than 6 million baht is small). The purpose of selecting study areas into various classes is to understand the relative dynamics of Tambons in terms of their activities for fulfilling the needs of the people and contributing to overall development in the rural areas.

Table 27.1 General information on the Tambon Administrative Organizations (TAOs)

General information	Tambon Administrative Organization		
	Chumsang Songkram (Large)	Wang Nok-An (Medium)	Sanamkhli (Small)
Geography	Floodplain area	Hilly area	Plains area
Area (km^2)	120.39	523.55	31.12
Population (persons)	7,955	16,788	2,669
Density (persons/km^2)	66	32	86
Number of villages	11	20	6
Number of households	2,349	5,746	708
Main agricultural production	Rice, soybean, corn	Cassava, corn	Rice, corn, green bean
Commerce and service	3 gas stations	2 PTT gas stations	18 small gas stations
	14 small gas stations	4 resorts	

Source: Basic Information on TAOs and Basic Minimum Needs (2010)

Among the three Tambons taken for this study (Table 27.1), the largest area and the most populated Tambon is Wang Nok-An, but it has the least population density. On the other hand, Sanam Khli Tambon, which is the smallest in geographic area, has the lowest population but the highest population density. Being in the plains area, Chumsang Songkram and Sanam Khli have fertile soil, suitable for agriculture due to the presence of the Nan River with an irrigation canal and many swamps. Although Tambon Wang Nok-An is hilly and mountainous, it supports the cultivation of cash crops such as corn and cassava. One of the serious problems in all three Tambons is flooding in the rainy reason and lack of water during the dry season. Surface water is scarce during the dry season, but the villagers can still tap available and high-potential groundwater in the area for agricultural use.

27.3.1 Social Profile

Social information on the TAOs in the three Tambons is categorized into e*ducation, health*, and *security* (Table 27.2). Generally, the educational services provided by the three TAOs are almost equal, but the adult literacy rate of the household members in Tambon Sanam Khli is comparatively lower than the other two Tambons. Public health services rendered by the three Tambons are equal; the number of police stations provided for security purposes varies according to the number of people residing in each Tambon.

According to the Basic Minimum Needs (BMN) Report and Annual Report of the TAOs in 2010, the percentage of educational services provided by the three

Table 27.2 Social profile of the Tambon Administrative Organizations (TAOs)

Social information	Tambon Administrative Organization					
	Chumsang Songkram (Large)		Wang Nok-An (Medium)		Sanam Khli (Small)	
	Number	(%)	Number	(%)	Number	(%)
Educational service						
Children 6–15 years of age receive 9 years of compulsory education	851	100.0	1,040	100.0	244	99.6
Household members 15–60 years old are literate	3,413	99.9	4,356	99.8	1,119	99.6
Public health						
Pregnant women receive prenatal care	7	100.0	14	100.0	7	100.0
Pregnant woman receive birth-giving services and postnatal care	13	100.0	11	100.0	14	100.0
Children under 1 year of age receive the necessary vaccinations	29	100.0	41	100.0	25	100.0
Children 6–12 years old receive the necessary vaccinations	568	100.0	689	100.0	170	98.8
Households with sanitation facilities	2,181	100.0	5,312	100.0	662	100.0
Security						
Number of police substations	1		3		1	

Source: Basic Minimum Needs (2010) and Annual Report of TAOs (2010)

TAOs is quite high, considering that 100 % of the children 6–15 years of age in Chumsang Songkram and Wang Nok-An, and 99.6 % in Sanam Khli, receive 9 (nine) years of compulsory education. Generally, pupils who have completed their primary education go to secondary schools in the district or in any part of Phitsanulok Province. There are, however, a few illiterate persons in these three Tambons, but most of the 15- to 60-year-olds are literate (more than 99 %). Moreover, 100 % of pregnant women in the three Tambons have received prenatal care as well as birth-giving services and postnatal care. In addition, 100 % of the children less than 1 year of age also have received the necessary vaccinations. In addition, 100 % of children 6–12 years of age in two Tambons have received the necessary vaccinations, except in Sanam Khli, which has registered a slightly lower percentage (98.8 %). Thus, generally, the educational and public health services in the three Tambons are being carried out effectively.

It should also be noted that the TAOs in these Tambons also provide the necessary budget to set up projects to support school activities as well as those in the public health centers. Moreover, promotion of sanitation can be judged as successful, considering the fact that 100 % of the households in these three Tambons have toilets. This is one factor that could help prevent the spread of diseases such as diarrhea and food poisoning and some forms of contamination in the drinking water that causes other diseases.

27.3.2 Economic Profile

The economic information shown in Table 27.3 reveals that there is a remarkable variation in the average income per year per person between the rice farmers and the other crop farmers in the study area. As mentioned earlier, Chumsang Songkram and Sanam Khli are plain areas that are suitable for raising a variety of crops and obtaining high yields. In fact, some areas are reported to be available for rice cultivation, even two to three times per year, especially when the water resource is abundant. On the other hand, cassava and corn are the most important upland crops cultivated in Wang Nok-An. Cultivation of fruit trees and vegetables are generally undertaken either on homesteads or in fields near the farmers' homes.

Household income is one of the most important indicators for measuring the standard of living of the population, and is also an important component for economic planning. Based on the target of BMNs, the average annual income to be able to meet the basic needs of a person should not be less than 23,000 baht/person/year. In 2010, the three Tambons therefore surpassed the target and achievement of the BMN. As shown in Table 27.3, Chumsang Songkram has the highest average income person/year at 64,306 baht and Wang Nok-An had the lowest average income at 49,374 baht/person/year.

27.3.3 Infrastructure Development

The major infrastructures in the three Tambons, such as transportation, water resources, and households with e-communication, were also studied. The results (Table 27.4) show that Tambon Chumsang Songkram has improved its roads whereas the other two Tambons still have soil roads. Although all Tambons have laterite roads, the condition of these roads is subject to deterioration during the rainy season. Therefore, the TAOs need to maintain and improve the surface of the

Table 27.3 Economic profile on the Tambon Administrative Organizations (TAOs)

	Tambon Administrative Organization		
	Chumsang Songkram	Wang Nok-An	Sanam Khli
Economic information	(Large)	(Medium)	(Small)
Main occupation	Rice farmer	Upland crops	Rice farmer
Average income/person/year (baht)	64,306	49,374	55,774
Main production	Rice, soybean, corn	Cassava, corn	Rice, corn, green bean
Commerce and service	3 gas stations	2 PTT gas stations	18 small gas stations
	14 small gas stations	4 resorts	

Source: Basic Information on TAOs and Annual Report of TAOs (2010)

Table 27.4 Infrastructure profile of the Tambon Administrative Organizations (TAOs)

Infrastructure information	Tambon Administrative Organization					
	Chumsang Songkram (Large)		Wang Nok-An (Medium)		Sanam Khli (Small)	
	Number	%	Number	%	Number	%
Transportation						
Soil road	0	0	50.0 km	46.3	4.0 km	7.1
Laterite road	250 km	94.8	27.8 km	25.8	22.0 km	39.3
Asphalt road	11.68 km	4.4	15.6 km	14.4	10.0 km	17.9
Concrete road	2.11 km	0.8	14.5 km	13.5	20.0 km	35.7
Total	263.8 km	100.0	107.9 km	100.0	56.0 km	100.0
Water resources						
Reservoir/irrigation	0	0	2	1.7	1	2.9
Pool/pond	4	4.0	35	30.2	0	0
Artesian wells	30	29.7	33	28.5	19	55.9
Weir/dike	9	8.9	15	12.9	1	2.9
Village-water supply	19	18.8	26	22.4	6	17.7
Mountain-water supply	0	0	5	4.3	0	0
Natural water resources/brook/canal/marsh	39	38.6	0	0	7	20.6
Total	101	100.0	116	100.0	34	100.0
Households with electricity	1,985	98.0	4,853	95.0	643	97.0
Communication						
Post office	1	–	1	–	0	–
Public telephone	5	–	20	–	15	–
Broadcasting tower	44	–	17	–	7	–
Households with telephones	1,520	75.0	850	17.0	50	8.0

Source: Summary Report of TAOs (2010)

roads with asphalt or construct concrete roads. Based on the TAO information, infrastructure improvement has been its main task since the TAOs were established in 1997. In fact, the majority of the expenditures of the TAOs is for road maintenance and the improvement of road quality. However, because of budget limitations, it has become quite difficult for the TAOs to improve all roads, specifically by asphalting or changing to concrete roads. The asphalt and concrete roads are shorter in Chumsang Songkram and Wang Nok-An as compared to Sanam Khli, which has slightly longer roads.

All Tambons have abundant natural water resources, giving an advantage to the Tambons to grow various crops and fish for domestic use. All the villages in the Tambons get their domestic water supply from the village water systems. However, the villagers in Wang Nok-An are also able to avail themselves of the mountain water supply system. Underground water is abundant in Tambon Chumsang Songkram but it has no irrigation reservoir for agriculture.

More than 95 % of the households in the three Tambons have electricity supply, although it is much less in some mountainous remote areas in Wang Nok-An. Furthermore, as for household communication systems, the situation in Tambon Chumsang Songkram is much better than in the other two Tambons. Nevertheless, all Tambons have public phone systems, household phones, and broadcasting towers. The broadcasting tower has been an important channel for local communications, where the villagers can receive reports, announcements, and other relevant information through government news broadcasts by the village headman.

27.4 People's Expectations and the Capacity of Tambon Administrative Organizations in Rural Development

The study analyzes the degree of people's expectations and the capacity of TAOs in rural development covering three dimensions: social, economic, and environmental and natural resource development.

The people's expectations regarding the TAO rural development activities are shown in Table 27.5. The need for vocational training has received top priority. The field survey revealed that 35.9 % of the respondents need vocational training in food preservation, food processing, flower growing, dressmaking, and cattle farming. The major reason behind this is that the local people like to have supplementary occupations to increase their job opportunities and income, especially during the dry season (the farm off-season). People's expectations regarding vocational training from the TAOs were the highest (40.9 %) in Sanam Khli, as compared to the two other TAOs, Chumsang Songkram with 34.8 % and Wang Nok-An with 32.7 %. Sanam Khli is the smallest Tambon but has the highest population density. The local people there need a budget allocation from the TAO for vocational training.

The other expectation is the demand for social welfare (26.8 %). For example, TAOs must provide pensions for aging persons, the disabled, and AIDS patients. It was found that people's expectations in promoting social welfare such as pensions for aging persons, pensions for the disabled, and pensions for AIDS patients in Chumsang Songkram were dramatically higher than in the other two Tambons, as people in Chumsang Songkram have a much more advanced vision of the social safety net, particularly for vulnerable groups.

In addition, 28.8 % of the respondents also expected to receive effective support in terms of educational assistance, public health, and healthcare services such as first aid for the locality. Even though public health and healthcare activities are the responsibility of the Ministry of Health, the TAOs still have budget allocations to support relevant activities such as training the villagers to be health volunteers. In this respect, there is no difference among the three Tambons as people in each

Table 27.5 Level of people's expectations for social, economic, and environmental and natural resources development

People's expectation	Size of TAOs Large (CSSK) F	%	Medium (WNA) F	%	Small (SMK) F	%	All TAOs F	%
1. Social development								
1.1 Vocational training by the TAOs	16	34.8	17	32.7	18	40.9	51	35.9
1.2 Promoting social welfare such as pensions for aging, pensions for the disabled, and pension for AIDS patients	15	32.6	12	23.1	11	25.0	38	26.8
1.3 Supporting the basic educational system	5	10.9	7	13.5	5	11.4	17	12.0
1.4 Promotion of public health and healthcare such as first aid	7	15.2	10	19.2	7	15.9	24	16.8
1.5 Promotion of people's participation in local administration	3	6.5	6	11.5	3	6.8	12	8.5
Total	46	100.0	52	100.0	44	100	142	100.0
2. Economic development								
2.1 Infrastructure development such as construction and maintenance of roads, repair of water supply systems, irrigation for the agricultural sector	23	50.0	28	53.8	22	50.0	73	51.4
2.2 Financial support by TAOs such as providing budget/capital for the groups, creating job opportunities for the people in the locality	16	34.8	17	32.7	17	38.6	47	35.2
2.3 Income generation such as ability to generate income in the community and be able to increase employment opportunity for villagers	7	15.2	7	13.5	5	11.4	22	13.4
Total	46	100.0	52	100.0	44	100.0	142	100.0
3. Natural resources and environmental development								
3.1 In conservation of natural resources and environment including forest conservation and solving the problems of forest encroachment	18	39.1	25	48.1	20	45.5	63	44.4
3.2 In natural disaster management in the form of flood prevention, flood relief, flood recovery, and forest fire control	28	60.9	27	51.9	24	54.5	79	55.6
Total	46	100.0	52	100.0	44	100.0	142	100.0

Source: Field Survey 2006
F frequency

Tambon have been benefiting equitably from the health- and education-related services provided by each TAO, although the largest size Tambon showed the higher interest because of its size in covering more beneficiaries within the Tambon as compared to other two Tambons.

The last in the ranking of the people's expectations (8.5 %) concerned the promotion and motivation of villagers to participate in every local administrative process, such as participation in public meetings specifically regarding the formulation of guidelines for general Tambon planning as well as the monitoring and evaluation process by the local people. In this case, a similar scenario was mentioned for the public health sector and basic education system. There was no difference among the three Tambons as people in each Tambon have been benefiting equitably from the health-related services provided by each TAO. However, the larger Tambons still need to mobilize the people's participation in its own local administration system.

As far as social development is concerned, the results show that except for the component of vocational training by Tambons, all other components such as social welfare, basic education, and public health, and people's participation in local administration, are included in the higher expectations of people in general, with the exception of the large Tambon, Chumsang Songkram, in terms of population size.

The most important aspect of local people's expectations regarding economic development (51.4 %) is for the construction and maintenance of infrastructure, including the construction and maintenance of roads, water supply systems, and irrigation for the agricultural sector. The second need is for financial support (35.2 %), such as subsidizing a certain amount of budget for the groups and creating job opportunities for local people. The third need is generating income and increasing employment opportunities for the villagers (13.4 %). According to the observations of people's expectations for economic development, three Tambons have shown no significant difference, as the people were equally interested in their basic needs, linked with economic development.

Referring to natural disaster management by the TAOs, the people's perceived first priority (55.6 %) focused on flooding problems, which occur every year. The TAOs cooperate with government agencies in providing relief assistance to households affected by severe flooding. In this case, after the TAOs survey the damage, they report to the government agencies to obtain a certain amount of emergency budget to help the people affected by the disaster. At the same time, the TAOs also take responsibility for alleviating the impact of the disaster on the lives of the affected people in the Tambon by responding to basic needs such as food, medicine, and drinking water. Therefore, Chumsang Songkram, where a flood occurs almost every year, embodies the highest people's expectations, whereas the people of Wang Nok-An have expressed the highest expectations for conservation of natural forests and the environment because that area is mountainous with substantial forest coverage.

27.4.1 Capacity of Tambon Administrative Organizations in Social Development

Concerning the capacity of the TAOs in the area of social development, the study has focused on three aspects: (1) the capacity of TAOs to provide social welfare to local people, (2) the capacity of TAOs to promote people's participation, and (3) the capacity of TAOs in terms of human resource management. The degree of TAO capacity is at medium level (WAI is 0.60) but more than the average situation. Table 27.6 shows that the degree of TAO capacity in providing social welfare and in promoting human resource development/management to villagers is at a medium level, whereas the degree of capacity to promote people's participation in planning, monitoring, and evaluation is rather high (0.7).

Regarding the capacity of TAOs to provide social welfare to local people, four out of eight items exhibited medium levels with respect to the level of promotion for vocational groups, such as allocating budgets for training, field study, workshops, the ability to provide pensions for the disabled, and the ability to provide vocational training and to manage nursery centers in the Tambon. The other aspects of social development, such as the ability to promote public health and healthcare, for example, first aid, the ability of TAOs to provide pensions for aging persons, and the ability to prevent and control contagious diseases had reflected higher index value where as it was for providing pensions for AIDS patients; because the expenditure for curing these patients is very high, the TAOs are unable to supply sufficient money for them. Although overall assessment of the capacity of TAOs to provide social welfare to local people was shown at a medium level in average, Chumsang Songkram was weighted at a high level, whereas the other two were weighted at the upper medium level because the larger TAOs have greater capacity to provide social welfare to the local people than the smaller TAOs.

In terms of comparing the specific items under the capacity of each TAO in providing social welfare to local people, the capacity to promote public health and healthcare for first aid, to provide a pension for aging persons, and preventing and controlling contagious diseases were at the same high level in all three TAOs because these items were fulfilled by all TAOs as the most fundamental and important issues for the social welfare improvement of the local community.

However, regarding the capacity to promote vocational groups and to provide a pension for the disabled, it was also found that the larger-size Chumsang Songkram was weighted at a high level, while the other two were still at a medium level, as the larger Tambon had a larger budget for implementing those specific items.

The ability to promote democracy in terms of elections and public hearings had a high level of TAOs in human resource management, whereas the integration of various resources for rural development had a low level. The TAOs act as the center of several resources (such as the local media, local healthcare centers, nurseries) so that they are able to establish mutual aid groups, healthcare services, and educational and recreational activities and utilities for the people who have no chance of creating job opportunities in the society.

Table 27.6 Capacity of Tambon Administrative Organizations (TAOs) in social development

Items	Size of TAOs			Degree of capacity	
	Large (CSSK)	Medium (WNA)	Small (SNK)	Total WAI	OA
Capacity of TAOs in providing social welfare to local people					
1.1 To promote public health and healthcare for first aid	0.7	0.7	0.6	0.7	H
1.2 To provide pension for aging persons	0.7	0.7	0.7	0.7	H
1.3 Preventing and controlling contagious diseases	0.7	0.7	0.7	0.7	H
1.4 Promotion for vocational groups such as budget for training, field study, workshops	0.7		0.6	0.6	M
1.5 To provide pensions for the disabled	0.7	0.6	0.5	0.6	M
1.6 To provide vocational training	0.6	0.6	0.6	0.6	M
1.7 To manage nursery centers in Tambons	0.6	0.6	0.5	0.6	M
1.8 To provide pensions for AIDS patients	0.5	0.4	0.4	0.4	L
Aggregate (WAI)	0.7	0.6	0.6	0.6	M
Capacity of TAOs in promoting people's participation					
1.9 Promoting people's participation in planning	0.6	0.7	0.7	0.7	H
1.10 Promoting people's participation in monitoring and evaluations	0.6	0.6	0.6	0.6	H
Aggregate (WAI)	0.6	0.7	0.7	0.7	H
Capacity of TAOs in human resource management					
1.11 To promote democracy in terms of election and public hearings	0.7	0.7	0.7	0.7	H
1.12 To integrate various resources for rural development	0.5	0.6	0.6	0.6	M
1.13 To protect local arts, culture, and tradition	0.5	0.5	0.6	0.5	M
1.14 To provide community learning centers	0.5	0.5	0.5	0.5	M
Aggregate (WAI)	0.6	0.6	0.6	0.6	M
All components (WAI)	0.6	0.6	0.6	0.6	M
Criteria for assessing weighted average index					
Very high (VH)	High (H)	Medium (M)	Low (L)	Very low (VL)	
0.81–1.0	0.61–0.8	0.41–0.6	0.21–0.4	0.00–0.20	

Source: Field Survey 2006
WAI weighted average index, *OA* overall assessment

27.4.2 Capacity of TAOs for Economic Development

Based on Table 27.7, it can be observed that the ability to provide and maintain the infrastructure of the TAOs, such as constructing and maintaining basic infrastructure and repairing water supply systems, had a WAI of 0.7, which corresponds to a high-level capacity. The respondents explained that they were highly satisfied with the capacity of TAOs in providing and maintaining infrastructure services because the villagers could take advantage of the good transportation facilities and roads from their area to the city. It is evident that many roads in the Tambons have been improved, from laterite to concrete roads, using the TAO budgets. In the capacity of TAOs for economic development, all three TAOs have been assessed overall at a high level on average, but the capacity of Chumsang Songkram (large) showed significantly that the people's perception of the TAO level of capacity regarding construction and maintenance of roads was very high, although the other two were only at a high level.

Comparing the financial support capacity of each TAO, all three TAOs showed a large capacity in providing capital/budget for their members, and Chumsang Songkram (large) was significantly high in providing financial training to its members, whereas the other two were not yet satisfactory in this regard. Apart from that, the Wang Nok-An TAO (medium) showed a capacity to promote tourism dramatically better than the other two, and all three TAOs were a bit weak in promoting industrial development at the local level as well.

The ability to generate income was perceived as low with a WAI of 0.4, which means that the TAOs were unsuccessful in generating income and increasing employment opportunities for the villagers, and the people do not expect to get much assistance from the TAOs in income generation. As the overall average assessment regarding the capacity in support of income generation by each TAO is at a low level, the TAO's capacity to encourage locals in the community to propose development programs was weighted at a better level compared to the other items, and it was also found that the capacity of all TAOs to decrease the villagers' current debt was low.

27.4.3 Capacity of Tambon Administrative Organizations for Natural Resource and Environmental Development

The levels of the TAO capacity for natural resource and environmental development are shown in Table 27.8. The capacity of TAOs regarding natural disaster relief was at a high level, with a WAI of 0.7, which reveals that the TAOs attempt to respond to the real needs of the people and resolve natural disaster problems directly. The level of TAO capacity in terms of garbage collection management was low, although it should be the responsibility of the villagers to collect and properly dispose of their garbage.

Table 27.7 Capacity of Tambon Administrative Organizations for economic development

	Size of TAOs			Degree of capacity	
	Large	Medium	Small	Total	
Items	(CSSK)	(WNA)	(SNK)	WAI	OA
Infrastructure					
1.1 Construction and maintenance of roads	0.9	0.7	0.7	0.8	H
1.2 Ability to maintain, improve, and repair water supply systems	0.7	0.7	0.7	0.7	H
1.3 Watershed management for agricultural sector	0.7	0.7	0.7	0.7	H
1.4 Providing and controlling markets	0.6	0.6	0.6	0.6	M
Average (WAI)	0.7	0.7	0.7	0.7	H
Financial support					
1.5 To provide budget/capital for members	0.6	0.7	0.7	0.7	H
1.6 To create job opportunities for members	0.6	0.5	0.6	0.6	M
1.7 To provide financial training	0.7	0.5	0.5	0.6	M
1.8 To promote industrial development at local levels	0.4	0.3	0.4	0.4	L
1.9 To promote tourism	0.3	0.6	0.3	0.4	L
Average (WAI)	0.5	0.5	0.5	0.5	M
Income generation					
1.10 Encouraging locals in community to propose development programs	0.5	0.5	0.5	0.5	M
1.11 LIs are able to establish saving groups in the community	0.4	0.4	0.4	0.4	L
1.12 Villagers are able to create income opportunities after receiving capital from TAOs	0.4	0.4	0.4	0.4	L
1.13 LIs are able to increase employment opportunities for the villagers	0.4	0.5	0.4	0.4	L
1.14 Villagers are able to decrease their current debt	0.3	0.3	0.4	0.3	L
Average (WAI)	0.4	0.4	0.4	0.4	L
All components (WAI)	0.5	0.5	0.5	0.5	M
Criteria for assessing weighted average index					
Very high (VH)	High (H)	Medium (M)	Low (L)	Very low (VL)	
0.81–1.0	0.61–0.8	0.41–0.6	0.21–0.4	0.00–0.20	

Source: Field Survey 2006
WAI weighted average index, *OA* overall assessment

Table 27.8 Capacity of Tambon Administrative Organizations (TAOs) for natural resource and environmental development

Items	Size of TAOs Large (CSSK)	Medium (WNA)	Small (SNK)	Degree of capacity Total WAI	OA
Conservation of natural resources and the environment					
1.1 Natural disaster relief such as flooding	0.7	0.7	0.7	0.7	H
1.2 Conservation of natural resources and environment	0.5	0.7	0.5	0.6	H
1.3 Solving forest encroachment problems	0.5	0.5	0.5	0.5	M
1.4 Inspection of environment and regarding pollution	0.5	0.5	0.5	0.5	M
1.5 Forest fire control	0.4	0.5	0.4	0.4	L
1.6 Garbage collection management	0.5	0.4	0.4	0.4	L
Average (WAI)	0.5	0.4	0.4	0.5	M
Criteria for assessing weighted average index					
Very high (VH)	High (H)	Medium (M)	Low (L)	Very low (VL)	
0.81–1.0	0.61–0.8	0.41–0.6	0.21–0.4	0.00–0.20	

Source: Field Survey 2006
WAI weighted average index, *OA* overall assessment

The capacity of the TAOs regarding conservation of natural resources and the environment can be considered to be at a moderately important level. However, the overall capacity of the TAOs in natural resource and environment development was satisfactory at a medium level, with a WAI of 0.5. By comparing the capacity of each TAO for conservation of natural resources and of the environment, the large TAO from Chumsang Songkram was a bit better than the rest on average; however, the TAO from Wang Nok-An showed a better capacity for conservation of natural resources and the environment as that Tambon has the potential to do so.

27.5 People's Expectations from Community-Based Organizations (CBOs) in Rural Development

The results of the study also have shown that people have a very high level of expectation of the CBOs (93 %) in either facilitating or arranging vocational training programs for them. Similarly, 63.5 % of the respondents expected that the CBOs should be involved in promoting public health and healthcare-related issues (Table 27.9). The purpose of vocational training is to enhance skills in livelihood-related activities that generate alternative occupations and supplementary income.

Table 27.9 Expectations of members of community-based organizations (CBOs)

Items	Percentage of respondents
Social development	
1. Vocational training	93.0
2. Promotion of public health and healthcare services	60.5
3. Improvement of administration with transparency and financial systems	29.5
4. Promotion of people's participation in group activities	14.7
Economic development	
1. Financial support through savings and credit without interest	51.9
2. Income generation through group activities	36.4
Natural resources and environment	
1. Conservation of natural resources and protection of the environment (conserving community forests, increasing number of trees by new plantings, etc.)	36.4

Source: Field Survey 2006

Provision of vocational training is very common in Thailand through sectoral/line agencies and at the Tambon level as well. An important approach is to acquire new and upgraded skills by members through the "learning and doing" process. The members who have had vocational training are able to increase the amount of their production and to generate a marketable surplus for sale. However, the needs of the members are not only linked with vocational training but also with financial support, for which the members have a high level of expectation. For example, more than 50 % of respondents expected financial support through savings and loans without interest. Financial support was the second greatest need after vocational training. Respondents understand that these two aspects together can bring changes in economic development at the local level. More than one third of the respondents reflected on the possibility of generating income through different group activities. Similarly, an equal percentage of respondents also expressed an interest in conserving natural resources and the environment through new planting.

27.5.1 Capacity of Community-Based Organizations Regarding Social Development

The capacity of CBOs in the area of social development in the respective localities was highly correlated with different activities associated with people's participation and human resource development and management (Table 27.10). The capacity level was measured by a 5-point Likert scale and expressed through the weighted average index (WAI). This technique was also applied to other sections of this chapter. The capacity of CBOs to promote people's participation was high. A similar level of capacity was also observed for other social development activities with

Table 27.10 Capacity of community-based organizations (CBOs) in social development

Items			Capacity level	
			WAI	OA
People's participation in group activities				
1. Promoting people's participation in election of leader and committee members			0.8	High
2. Promoting people's participation in planning			0.7	High
3. Attending training and other activities			0.7	High
4. Expressing views in the meetings and public affairs			0.6	High
5. Participation in contributing labor			0.6	High
6. Promoting people's participation in monitoring and evaluation			0.8	High
Aggregate			0.7	High
Human resource management				
7. Building mutual trust and confidence among members and with the committees as well as leaders			0.8	High
8. Feeling of ownership in group activities			0.8	High
9. Promoting democratic process for group activity management system			0.7	High
10. Providing occupational training to members			0.6	High
11. Facilitating members to take a leadership role			0.5	Medium
12. Solving problems and groups conflicts			0.4	Low
Aggregate			0.6	High
Total average			0.7	High
Criteria for assessing weighted average index				
Very high (VH)	High (H)	Medium (M)	Low (L)	Very low (VL)
0.81–1.0	0.61–0.8	0.41–0.6	0.21–0.4	0.00–0.20

Source: Field Survey 2006
WAI weighted average index, *OA* overall assessment

minor variation, as the WAI varied from .6 to .8. The CBO roles and contributions seemed to be high in terms of encouraging peoples' participation in planning processes and activities, for attending training programs, expressing their views in meetings and public affairs, and contributing labor for different social development activities. The people's participation in monitoring and evaluation of socioeconomic project activities was well reflected among all areas of participation. This is quite obvious, as people always expect direct and indirect benefits from the project activities at the local level. As CBOs are people's organizations, they work very closely with the people in their localities. The functionaries are among the local people. They know each other and interact very closely. Hence, the CBOs' role and responsibility for encouraging different activities are relatively easy and also intensive, and this is why the capacity of the CBOs in this aspect as felt to be high level.

Human resource management-related activities were measured through mutual trust and confidence building, ownership in group activities, promotion of the democratic process-driven management system, providing occupational training, facilitating the leadership role of members, and solving the problems arising in group activities and also resolving group conflicts. The first four items have appeared at the high level that reflects the remarkable achievement on the part of the CBOs. In terms of resolving issues and group conflicts, the capacity of the CBOs was low, which implies that CBOs have not yet been very successful in this aspect of human resource management. Local conflicting issues were more sensitive. CBO functionaries, being local, have faced difficulties in resolving local conflicting issues. A third party or a non-local organization or higher body such as the TAO may contribute more than the CBOs. Very often the success of any development project is directly linked with the absence of group conflicts at the local level, which is an indicator of equal access and ownership of the project activities.

In comparing the component of the people's participation with the component of human resource development management, there is hardly any difference between them. Whatever reflected on the CBOs, this implies more intensive and closer interaction between the communities and the CBOs to achieve the best from the social development process and outcomes.

27.5.2 Capacity of Community-Based Organizations in Economic Development

The high level of capacity of the CBOs in economic development was reflected in the areas of ensuring benefits to the members, encouraging them to save part of their income, and increasing employment opportunities through group activities in nonagricultural activities, which was directly linked with benefiting the financial capacity of the members. Table 27.11 shows that the level of CBO capacity for

Table 27.11 Capacity of community-based organizations (CBOs) in economic development

Items			Level of capacity	
			WAI	OA
1. Ensuring benefits to the members			0.8	High
2. Encouraging the members to save through group activities			0.7	High
3. Increasing employment opportunities for members			0.7	High
4. Creating income-earning opportunities through group activities with capital support from the CBOs			0.5	Medium
5. Increasing production from group activities			0.4	Low
6. Guidance to decrease their current debt status			0.4	Low
Aggregate			0.5	Medium
Criteria for assessing weighted average index				
Very high (VH)	High (H)	Medium (M)	Low (L)	Very low (VL)
0.81–1.0	0.61–0.8	0.41–0.6	0.21–0.4	0.00–0.20

Source: Field Survey 2006
WAI weighted average index, *OA* overall assessment

creating income-earning opportunities through group activities with capital support from the CBOs was seen to be at a medium level, as they have meager resources basically raised and contributed through memberships and shares for different economic activities, sometimes supplemented by limited external support through sectoral line agencies.

The TAOs are positioned comparatively well in this respect. As a local government, TAOs have access to multiple sources of funding in addition to its own generated revenues through taxes. CBOs are composed of small groups with few members. Their resources are very small and are inadequate to meet their needs for investment in economic activities. Members of CBOs are not very successful in increasing their production level, which is linked with generating income from group activities. The CBOs' guidance for decreasing the debt level is reflected at a low level. Without a larger production and substantial income from the group activities, it is impossible to cut down the indebtedness. In these two aspects, CBOs have demonstrated a low level of capacity.

27.5.3 Capacity of CBOs in Natural Resources and Environmental Protection

The levels of the CBO capacity to educate members to conserve natural resources and convincing them to realize the importance of forestry are shown in Table 27.12. The highest level was observed for persuading members to participate in conserving national resources and the environment. Generally, CBOs are aware of the real situation of the environment in their localities and of the problems with which they are confronted. Therefore, CBOs are able to encourage the members in the locality to cooperate in solving the problems properly. For this reason, the level of CBO capacity regarding the conservation of natural resources was high, as they could fulfill the

Table 27.12 Capacity of community-based organizations (CBOs) in natural resources and environmental protection

Items			Level of capacity	
			WAI	OA
1. Participating in conservation of natural resources and environmental activities			0.6	High
2. Managing natural disasters			0.6	High
3. Establishment of community forestry management group activity			0.5	Medium
Aggregate			0.6	High
Criteria for assessing weighted average index				
Very high (VH)	High (H)	Medium (M)	Low (L)	Very low (VL)
0.81–1.0	0.61–0.8	0.41–0.6	0.21–0.4	0.00–0.20

Source: Field Survey 2006
WAI weighted average index, *OA* overall assessment

people's real felt needs and expectations. Managing natural disasters associated with floods, road accidents, and fire hazards at times at the local level was also reflected as high because the CBOs and local people have awareness as well preparedness at the local level through the decentralized planning process. Disaster prevention and management activities are the new agenda for the local government (TAO) at present, with support and linkages from the provincial to district level, and at the Tambon and village level. Local youth are engaged in volunteer services with some training and preparedness exposure. In establishing community forestry management group activities, CBO capacity was registered at a medium level, because those members have more diverse interests in harnessing benefits individually from the forests rather than attempting this collectively. In this case, CBOs have to make more effort in bringing members together for group activities with a focus on common interest.

27.6 Comparative Assessment of TAOs and CBOs

Most of the people's expectations regarding social development include the need for vocational training, which is the top priority. The major reason behind it is that local people like to have supplementary occupations to increase their job opportunities and income, especially during the dry season (the farm off-season). The highest expectation for economic development was in constructing and maintaining the infrastructure. The respondents were also highly satisfied with the capacity of TAOs in providing and maintaining infrastructure services. The results showed that there was no significant difference, as the people were equally interested in their basic needs, linked with economic development.

People's expectations and the capacity of the CBOs go hand in hand and are seen as strongly positive in the context of rural development. The CBOs are able to perform their roles and functions better than the TAOs, as reflected in the conservation of natural resources and meeting the felt needs and expectations of the people. The study has found that the capacities of the CBOs are higher than those of the TAOs. Basically, the CBOs are focused on the capacity building of the community in terms of addressing basic and socioeconomic needs in achieving sustainable development. The TAOs have remained focused as local governmental and political institutions to promote democracy in the locality.

It was also shown that the benefits of CBOs in economic aspects are greater than those of the TAOs. In other words, the performance of CBOs is more effective in rural development than the TAOs. CBOs are also able to fulfill the people's expectations much more than the TAOs because the majority of CBOs have been formed based on their interest in economic development. The common factor is that both TAOs and CBOs are weak in strengthening people's participation in local activities and development agenda, and lack administrative skills, participatory monitoring, and evaluation methods for sustainable rural development.

In Thailand, the local governments have the autonomy to manage development projects and to provide public services according to the needs of their constituents and the local communities. Local administrations are therefore authorized to formulate development plans and personnel policies as well as the budget and financial policies for their respective communities. The duties and responsibilities of the local institutions include the promotion of economic, social, cultural, and environmental development within their own authority as provided by the law. The TAO is composed of a council and executive board, which are directly elected by the local people. People have become more conscious of their people's power in terms of voting for candidates who have created and implemented good projects and policies and who truly work for their respective communities.

The TAO structure is effective in Thailand because of the anticipated outcomes and processes of decentralization, as they are a function of the distribution of power in society (Duglas 2005). The main revenues of the TAOs come from the taxes and fees collected by the central government (shared taxes). Another source of revenue is regular grants from the government. Currently, the Tambons have also generated resources within their territory by using the power and authority given to them under the decentralized policy framework. The TAOs are very much dependent on the government for meeting their financial needs, even though opportunities exist for generating internal revenue, albeit to a limited extent. The internal and external revenue ratio at present is 30:70. Most of the TAO revenue cannot cover expenditures because of several community development projects under the wide range of responsibilities assigned to them. Therefore, the central government should increase the allocated budget to support the projects of local governments. The expenditure of investments, such as infrastructure creation (roads, culverts, water resources, etc.) and maintenance, along with livelihood-related activities, are at the highest level in all the TAOs. Very limited resources are available to undertake other responsibilities at the local level and therefore many remain unattended unless linked with other external sources of funding from successive upper levels of the government line agencies. Comparing the revenue structure and achievements of the TAOs, it was observed that large and medium Tambons have more external revenue through shared taxes and grants from the central government as compared to the small Tambons.

The social development activities consist of two types: (1) support by TAOs for social support and networks, primary education, and public health; and (2) people's participation in social mobilization and community development activities. People's expectations are not in accordance with the capacity of the TAOs regarding social development. However, the people's expectations in relationship to economic development are significantly related to the capacity of the TAOs. These expectations, in conjunction with the capacity of the TAOs to conserve natural resources and protect the environment, are achieved at the moderate level.

People's expectations and the capacity of the CBOs run parallel and have been seen to be strongly positive in the context of rural development. The CBOs are able to perform their roles and functions better than the TAOs, as reflected in the conservation of natural resources and meeting the felt needs and expectations of the people. The capacities of the CBOs are higher than those of the TAOs. Basically, CBOs are focused on the capacity building of the community to address basic and socioeconomic needs in achieving sustainable development. TAOs have remained focused as local governmental and political institutions to promote democracy in the locality. It was also revealed that the benefits of the CBOs in economic aspects are higher than those of the TAOs. In other words, the performance of CBOs is more effective in rural development than that of the TAOs. CBOs are also able to fulfill the people's expectations much more than the TAOs because a majority of the CBOs have been formed based on their interest in economic issues. Obviously, many respondents were willing to participate in community development activities even if they would not gain much direct individual benefit. The reason given for their willingness to take part in community activities is because such activities responded to their social needs and that they could fully utilize their leisure time.

Generally, the CBOs work more closely with the community compared to the TAOs, but they are limited regarding financial resources. However, the TAOs focus on multi-project activities supported by the central government. Regarding rural development under a decentralized system as in Thailand, both CBOs and TAOs are expected to work together to cater to the needs of the community, which will ultimately lead to rural development. The common factor is that both TAOs and CBOs are weak in strengthening people's participation in local activities and development agenda, and lack administrative skills and participatory monitoring and evaluation methods for sustainable rural development.

27.7 Conclusion

Under the decentralized government system of Thailand, the TAOs are responsible for a wide range of activities and projects, which could allow them to generate some revenue and create job opportunities for the people at the local level. Political decentralization offers the prospect of more efficient aid disbursement, and these accord with the neoliberal values of economic advancement, rights, and democratic values (Batterbury and Fernando 2006). In addition to that, the people's expectations from the TAOs concerning social and economic development, natural recourses, and environment management have revealed that people have more expectations of programs and activities related to basic health, infrastructure, maintenance, flooding, etc. The performance of TAOs is also high in infrastructure, financial support, and income generation activities. In contrast, the CBOs perform well in their roles and functions when compared to the TAOs and at the same time their capacities are also higher than that of the TAOs. This dynamic sector working together with the people is always a major contributor to local development (Mukherjee and Zhang 2007).

One reason for this may be the close relationship of CBOs with the local people, where the local people are actively involved, and second, CBOs do have inbuilt interests in economic and community development more explicitly than the TAOs. Above all, both the CBOs and TAOs work with their own limitations and jurisdictions, and both at the same time have goals to accelerate local development. The combined efforts and strengths of both TAOs and CBOs are essentially required to work together and to support local development strategies, assisted by the central government. As Helmsing (2001) points out, local economic development (LED) is a process in which partnerships between local governments, community-based groups, and the private sector are established to manage existing resources to create jobs and to stimulate the economy of a well-defined territory; this will promote economic development at the local level and enhance the capacities of the communities to generate income and play their roles in rural development at the local levels more effectively. Therefore, the people at the community level are willing to engage in community-based activities when they believe that they can improve their livelihoods (Marschke and Berkes 2005).

Successful rural development requires that local institutions be further geared up to formulate development strategies, programs, and activities that would meet people's development needs, while simultaneously empowering the local people to be more proactive and to participate more intensively in Tambon-initiated programs and activities. In Thailand, the Sufficiency Economy approach, which has always been reiterated by King Bhumiphol, provides the strategies necessary for balanced and sustainable development. Therefore, following the King's guiding principle, the Thai government now emphasizes raising people's awareness and understanding the roots of every problem, with the main goal of eventually alleviating poverty, especially in the rural communities. The poverty level has decreased to less than 10 % at present in the country. The main factors that could contribute to achieving sustainable development should include a multi-faceted approach to rural development.

To be able to continue with the decentralization effort in Thailand, the process should be made multidimensional. First, to increase the quantity of financial allocation to the local level, the central government has to effectively allocate the central budget to support the responsibilities of the TAOs in the locality. In accordance with the decentralization Act of 1999, the share of local government revenue would have reached 35 % of total government revenue by 2006. However, at present the local revenue appears to have reached about 25 % (Chardchawarn 2010). Therefore, the TAOs are required to improve their ability to expand their own local tax bases wherever applicable, collect their own revenues, and generate more income at the local level. Krueathep (2008) further suggests that mobilizing local own-source revenues and improving local tax collection efficacy are key components to overcoming the burden of local revenues.

Second, as already initiated in Thailand, good local governance should be practiced more rigorously as a powerful mechanism for implementing rural development activities at the local level by the TAOs. Good governance means to be fully accountable to stakeholders such as workers, shareholders, persons of various social and economic strata, and so on, in the exercise of mandated activities and functions by all concerned, whether they are the central or local government, managers of state and business enterprises, or civic and civil society organizations (Lam 2003), thus providing equal opportunities and access to participation in the decision-making process. A more transparent and effective local government will provide scope to the people to effectively control and monitor its performance.

Third, people's participation has been widely recognized as an important aspect that facilitates the achievement of sustainable development. Fiscal and democratic decentralization has been coupled with participation that is supposed to translate into a local government that not only delivers better services but one that also engages citizens in the design and implementation of policies, programs, and projects (Mutebi 2003). This process takes place when the stakeholders of a community equally and democratically share ideas and visions, as well as participate and take responsibilities together to steer and implement the development activities, eventually creating a sense of ownership and partnerships in these development activities.

Local institutions are very important for strengthening local capacities; as such institutions work very closely with the local people and have autonomy in discharging their respective areas of responsibility. As discussed in the previous section, the duties and responsibilities of local institutions are quite broad and mainly cover three components: social, economic, and environmental aspects. The role of local institutions in rural community development as "service provider" is crucial, for which the central government allocates the funds to local governments for providing services in the localities such as infrastructure improvement, transportation network maintenance, healthcare, and education. However, in the case of large development projects such as irrigation, the local institutions are still dependent on the central government. Moreover, local institutions provide not only a basic infrastructure but also financial support under different micro-credit schemes to individuals or group activities.

At present the workload and responsibilities of the TAOs are increasing year by year because of the transfer of many new activities, such as construction and maintenance of roads and bridges, transport terminals, and traffic control, village water supplies, social activities for children and women, the aged, and the disadvantaged, and pre-school education, etc., to the local level. It has been recognized that the TAOs are burdened beyond their capacity and the administrative skills of the staff. If this continues, the TAOs will not be able to function effectively in the future. A need has therefore arisen to review the list of responsibilities and to readjust the activities with clear terms of reference for enabling the TAOs to be functional. As Blair (2000) points out, given sufficient political will from the central government to keep a decentralization initiative in place over time, effective local governance can be achieved, as is true in Thailand. The supply of adequate central government grants to local governments on a regular basis to meet the local needs is crucial for achieving substantial results. Despite firm efforts made by the local governments to increase their internal revenue, it remains a very meager share with reference to the total amount needed for local development. Hence, central government's commitment and responsibility should remain unchanged to achieve local development through the TAO.

The government is committed to enhancing the development potentials of both rural and urban areas to bring about a fairer distribution of benefits for development and to promote a more unified administrative system for rural development and decentralized authority in administration and decision-making processes at the local levels. Balanced development and sustained economic growth in urban and rural areas, therefore, would require meaningful and constructive models. To cope with the changing social and economic conditions, Thailand now needs new strategies and frameworks for resolving fundamental imbalances, inequalities, and poverty issues in rural areas. Focus should be placed more strongly on internal human resources and local potentials, while maintaining cultural identity. The Thai government also responds to the demand for change, articulated within the civil society. To achieve the National Economic and Social Development Plan and Royal Thai Government Policies, the role of local institutions has become equally important, as government authorities are in favor of rural development.

References

Batterbury S, Fernando J (2006) Rescaling governance and the impacts of political and environmental decentralization: an introduction. World Dev 34(11):1851–1863

Bello B, Cunningham S, Poh K (1998) A Siamese tragedy: development and disintegration in modern Thailand. Food First Books/Zed Books, London, 256 pp

Blair H (2000) Participation and accountability at the periphery: democratic local governance in six countries. World Dev 28(1):21–39

Boonyabancha S (2004) Empowering squatter citizen: local government, civil society, urban poverty reduction. In: Diana M, David S (eds) A decade of change: from the urban community development office to the Community Organization Development Institute in Thailand. Earthscan, United Kingdom, pp 25–49

Chardchawarn S (2010) Local government in Thailand: politics of decentralization and the role of bureaucrats, politicians and the people. Institute of Developing Economies, Chiba

Douglas AD (2005) The restructuring of local government in rural regions: a rural development perspective. J Rural Stud 21:231–246

Doungkaew P (1999) Role of institutions in rural community development. Asian Productivity Organization, Tokyo

Helmsing A (2001) Local economic development. Local governance and decentralization in Africa. A summary report prepared for the UNCDF symposium. Cape Town Symposium, pp 59–78

Krueathep W (2008) Measuring municipal fiscal health in Thailand: research findings and policy implications. In the Second UCLG-ASPAC Congress. Pattaya, Thailand

Lam N (2003) Development research and policy analysis division. Bulletin on Asia-Pacific Perspectives. ESCAP

Marschke M, Berkes F (2005) Local level sustainability planning for livelihoods: a Cambodian experience. World Ecol 12:21–33

Mukherjee A, Zhang X (2007) Rural industrialization in China and India: role of polices and institutions. World Dev 35(10):1621–1634

Mutebi A (2003) Challenges of participatory local management under decentralization: lessons from Chiangrai Municipality. INDES Workshop. Asia Development Bank, Bangkok

Phongpaichit P, Chris B (2002) Thailand: economy and politics. Oxford University Press, Kuala Lumpur, Revised edition

Vichit-Vadakan J (2001) Central role in development for Thai NGOs? Retrieve on July 10, 2005. From http://www.globalpolicy.org/ngos/intro/general/2002/12thai.htm

Wattanasiri C (2001) Sustainable rural development through community empowerment approach. Capacity building for local resources decentralization. In: Proceedings of Asia-Pacific International Conference. Faculty of Humanities and Social Sciences, Naresuan University, Phitsanulok, Thailand

Chapter 28
Nature of Land Use and Agricultural Change in Peripheral Regions: A Case Study of Arunachal Pradesh, India

Ravi S. Singh

Abstract Peripheral regions have typical developmental challenges. The issue becomes problematic in view of general apathy toward them revealed through their 'negligence' in development discourse. The studies on Northeast India, especially Arunachal Pradesh, which lies in the extreme northeastern part of India, are predominantly anthropological (and ethnographical). Such a treatment could be justified on various grounds; at the same time, it provides the academic rationale to attempt an analysis of land use shaped by a number of inhabiting ethnic communities while interacting with their natural surroundings in respective habitats and the intraregional pattern of agricultural change that has been mainly introduced by the governmental agencies. This state is by and large mountainous with an aggregate of 61.57 % of the geographic area being forested. The region experiences (hot-humid) subtropical conditions in the southern foothills to an alpine-type climate in the northern high-altitude mountains. Analyses presented in this study are based on secondary sources, mainly the Statistical Abstract of Arunachal Pradesh for different years. The unit of analysis considered in this exercise is district. The discussions are further supported by the author's personal field experience and observations during his fairly long stay in the region.

Keywords Agriculture • Change • Development • Land Use • Modernization • Arunachal Pradesh (India)

R.S. Singh (✉)
Department of Geography, Faculty of Science, Banaras Hindu University, Varanasi, Uttar Pradesh 221 005, India
e-mail: ravisingh.geog@gmail.com

A.K. Dutt et al. (eds.), *Spatial Diversity and Dynamics in Resources and Urban Development*, DOI 10.1007/978-94-017-9771-9_28

28.1 Introduction

Arunachal Pradesh (ArP), literally 'the land of (rising) sun,' lies in the extreme northeastern (26°28′N–29°31′N and 91°31′E–97°30′E) part of India. Before attaining full statehood in 1987, it had been one of the Union Territories of the Indian Union since 1972. This state was known in the past as the North East Frontier Tract before 1954 and the North East Frontier Agency (NEFA) during 1954–1972. Although significant in terms of geostrategic location, the state has shown a slow pace of 'development.' Before moving further, it would be appropriate to provide a geographic introduction to the study area.

28.1.1 Arunachal Pradesh (ArP): A Brief Geographic Orientation

Despite early professional interest in this part of India dating back to the nineteenth century, the absence of a complete geologic survey obscures a clear geologic picture. However, evidences of the Upper Tertiary and Lower Gondwana formations are found with traces of the (Adi) volcanoes, which are difficult to explain. ArP, being part of the Himalayan mountain system, naturally contains the major features of this mountain chain. There is apparent predominance of the west–east length over the north–south one. More than 70 % of the geographic area is above 1,350 m above mean sea level (MSL). Gradual decline in altitude from west to east is prominent. Moving from the west one finds longitudinal (erosional) valleys, ridge-like formations. Step-like variation in the relief and formation of deep gorges in the upper parts are found to the east. Moving further eastward, the antecedent drainage of Siang is found, and it forms a depositional plain (northwest–southeast). The southeastern part, being affiliated topographically to the Naga Hills, is distinctly different from the rest of ArP. Having an average annual rainfall exceeding 200 cm, Arunachal is entirely covered by a network of different order streams, and its topography is naturally sculptured by the rivers' active role similar to any other humid region. In fact, rivers have prominently defined the sub-regions and probably this is the reason almost all major administrative divisions are named after them (Singh 2005). The life cycle of an average Arunachali is closely and neatly woven around the rhythm of seasons in which rain (fall) has a distinct role. One can identify five precipitation zones forming a concentric circle pattern (Singh 2005, pp. 7–8): very high (above 400 cm), high (300–400 cm), moderately high (200–300 cm), low (100–200 cm), and very low (below 100 cm).

Three natural resources are of greatest importance across the state: forests, land, and water, which form common property resources in Arunachali village society. Life of an average Arunachali cannot be imagined in the absence of forests whose accessibility varies with altitude. Soil is the next important natural resource in this region. At the aggregate level, one finds differences in terms of soils in mountains and hills and that of the narrow river valleys. Wide intraregional variations are quite prominent. In general, soils lack well-developed soil profiles, leaching is ubiquitous, and despite a huge supply and addition of vegetal remains, the effective humus layer is too thin subsequent to washing away by widespread surface runoff. The National Soil Survey and Land Use Planning (NBSSLUP), Nagpur, has identified nine soil types in ArP that occur independently. Soil erosion and consequent loss of soil fertility are two impediments in the path of sustainable (agricultural) development. As mentioned earlier, average annual rainfall is quite high in Arunachal, but availability of water in the upper hills and mountains is very low. The situation worsens during the drier spells. Apart from domestic use, water is used for three specific development purposes: hydel power generation, irrigation, and pisciculture.

The state is currently organized into 16 districts (Fig. 28.1). As per Census 2011, the total population of ArP is provisionally estimated at 1,382,611 persons distributed over a geographic area of 83,743 km^2 (about 17 persons/km^2). There was growth of about 26 % during 2001–2011. This population could be disaggregated as scheduled tribal (64.22 %) and non-tribal (35.78 %), with 79.59 % as rural and 20.41 % as urban. ArP, similar to the whole northeast region, presents a cultural mosaic with 31 communities identified as permanent inhabitants whose sociocultural practices vary considerably. Diversity in ethnic composition could be better captured with reference to linguistic variations: 42 languages and 61 dialects are spoken and six scripts, viz., Devanagari, Assamese, Hingna, Mon, Roman, and Tibetan, are followed (Singh RS 1999). However, greater similarity is noticed in the occupational pattern, dominated by agriculture, which is complemented by collection from the forest, hunting, and fishing.

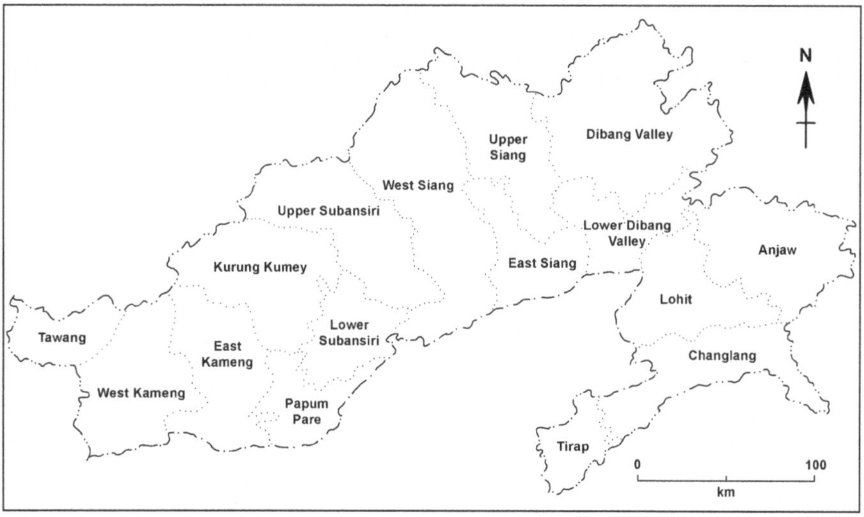

Fig. 28.1 Arunachal Pradesh: location of districts, 2010–2011 (Note: Map is notional and does not depict correct boundaries. From Statistical Abstract of Arunachal Pradesh, 2011)

The level of human development is too low (0.515) in the state. The geographic pattern presents a mixed scenario. Of the three components, income index appears better than education and health. The gap between 'performing' and 'not-performing' districts is too small. Two districts, one each in west (East Kameng) and east (Tirap), are found at the bottom where education, health, and income are in a poor state (Arunachal Pradesh Human Development Report 2005).

28.1.2 Implications of Peripherality and the Rationale of Present Study

If ArP has not been able to catch up with the pace of development of mainstream India, what could be the possible reasons? Is it locational disadvantage? At this point, the conceptual position could be briefly considered. Although periphery, and hence, 'peripheral' and 'peripherality,' are understood quite clearly in regional studies as concrete geographic terms, there is an emerging thought about the 'fuzziness' of the concepts (see Markusen 1999). Perhaps it is the result of the post-modern practice of deconstructing an ascribed and established meaning. Despite the questions raised regarding non-clarity of the meaning, a plethora of studies in both regional and development studies (Wallerstein 1974; Alonso 1964; Myrdal 1957; also see, Lomnitz 1977) indicate that 'peripherality' can be understood as 'an inherently relational condition' (Crone 2012). In a critical modern development geography perspective, which aims at understanding developmental patterns in the context of power relationships, the relational condition expressed by peripherality necessarily is an expression of 'power inequality' and 'subordinate status of the peripherality in relation to the core' (Anderson 2000). Crone (2012) points out two more important facets: multi-scalarity and typology of peripherality. Referring to Fig. 28.2, peripherality along with similar other conditions vis-à-vis the core or center could be appreciated better. No matter whether we look at the peripheral regions from a centripetal perspective or otherwise, they are generally at the receiving end but for

Fig. 28.2 Geographic positions in a system

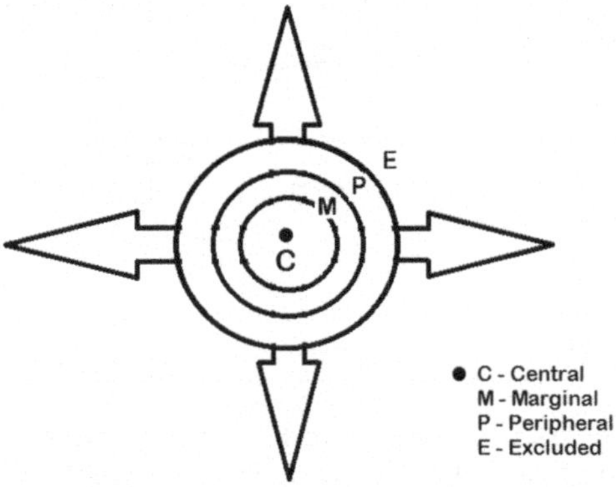

their military and defense significance. Marginality in the general sense is an issue of 'perspective,' and being a geographic fact necessarily involves 'scale' and therefore spatial resolution (Leimgruber 2004, 2010).

Leimgruber (1994, p. 8) believes that the border regions can be regarded as a manifestation of 'geometrical marginality.' There have been allegations by the peripheral states, particularly of the northeast region, in India that they are not perceived but rather seen as 'Indians.' Even in a recently published scholarly work on globalization and Indian agriculture (Bhalla and Singh 2012), I fail to locate the northeastern states (except Assam) in discussions therein. Why do they miss out? Is it the result of their peripherality, the lack of secondary data, or some other reason? When we look at the studies on ArP specifically, they are predominantly anthropological with a few from the historical perspective. Barring a few recent attempts, developmental studies have been on a backburner. One can justify such a treatment on various grounds. To me, this 'neglect' has provided the academic rationale to attempt analyses of development issues (see Singh 2000, 2001, 2005, 2006, 2008, 2010), including the present one. It seeks to understand patterns of man–nature interaction as revealed by the land use shaped (internally) by a number of inhabiting ethnic communities. These communities have been interacting with their natural surroundings in local habitats, and the intraregional pattern of agricultural change has been by and large 'forced' and 'imposed' (externally) by governmental agencies.

28.1.3 *Objective, Research Questions, Materials, and Methods*

In the backdrop of the preceding section, my objective of conducting this study is to determine the nature of land use and agriculture that has evolved in ArP in recent periods. To pursue this objective, answers to the following research questions are sought in the subsequent sections: (1) What is the nature of change across land use categories at aggregate level? (2) What is the spatial pattern of land use? (3) What are the major agricultural changes and developments taking place there? and (4) How is agricultural development distributed at the disaggregated level?

The present study being empirical, the backbone of analysis is naturally formed by secondary data sources, and the conventional geographic method of mapping also has been employed. The discussions are further substantiated by my personal field observation during 4 years of residence in the region. The unit of analysis considered in this exercise is district, and the major data source is the Statistical Abstract of Arunachal Pradesh (SAAP) of different years. Land use analysis has made use of data pertaining to the last four agricultural censuses: 1985–1986, 1990–1991, 1995–1996, and 2000–2001. The latest SAAP (2011) contains 2000–2001 agricultural census data only.

Ideally, agricultural development is measured utilizing such variables as productivity and yield, infrastructure development, and also the amount of inputs used. As a matter of fact, for different reasons discussed later in the agriculture section, one does not find sufficient progress with regard to the latter two categories of variables. Their inclusion in measuring agricultural development in ArP is likely to distort the patterns in which we are interested. For this reason, the present study employs measurement of agricultural development based on crop productivity (Bhatia 1967). Two time points are considered here, 1993–1994 and 2010–2011, to contrast and compare the development canvas. In general, the methodology of the study is descriptive, supported by some simple statistical applications.

28.1.4 Choices and Data Limitations

Except microlevel studies where compulsion of generating primary data becomes boon in disguise, dependence on secondary data is by compulsion rather than choice. In such cases, the statistical analysis part of a research has to compromise with data available instead of using what is 'appropriate.' Often would-be research is 'designed' on the basis of available data, which severely affects the 'appropriateness' of research (questions). My personal engagement and experience with research on ArP has not been satisfactory enough as the choice of data is too limited. Data generation is too sketchy. In fact, this state still waits for several crucial surveys such as geologic and cadastral in a complete fashion. In some recent surveys such as the District Level Household and Facility Survey (DLHS). coverage is satisfactory. The state government nevertheless has to do much more in this direction. Poor data generation forbids any deeper analysis. I have even consulted some older statistical sources dating back to the1950s and find that erratic data compilation is another inherent problem that worsens further with incoherency in structure and organization of data. Similarly, changing district boundaries forbid district-wise analysis of temporal changes. This problem of internal reorganization of administrative units is a limitation found in almost every state of India, and ArP is no exception. Because of such inconsistencies, assessment of the trend of performance of districts is totally forbidden. There are techniques to resolve these issues, when data are reduced to a base year level, but that again intervenes with realistic interpretation and future projection. Time points in a study focusing on spatiotemporal variations should be the same. However, it has not been possible here to adhere to a technically rigorous method for the lack of published data.

Analysis and discussion to follow in this chapter are organized into two broad heads, land use and agriculture. The following section is an analysis of land use pattern and recent changes. Here, the changes across the categories are explored first, followed by an analysis of district-level intraregional changes between different land use censuses (1985–1986 to 2000–2001). The third section, on agriculture, discusses the general nature of agricultural practice. Subsequently, the changes, from *jhooming* (also *jhuming*) to settled cultivation, availability and distribution of agricultural land, cropping pattern, and modernization (with special reference to irrigation, high yield variety seeds, and chemical fertilizers) are discussed. The fourth section pertains to the theme of progress shaping development contours since the early 1990s, a watershed in Indian development policy. The fifth section concludes.

28.2 Land Use: Change and Emerging Scenario

Land resource is crucial because it is the platform on which ultimately every development strategy rests, whether it is 'vegetative-agricultural,' residential, commercial, or industrial. Similarly, in the light of given scientific advancement, it will not be out of place to remind ourselves that food still remains basic among human needs and its supplies predominantly come from fields rather than laboratories.

Land use reflects the nature of interaction between the physical environment and human factor, characterized by the prevalent sociocultural and economic environment, and, naturally varies across the geographic environments and cultural groups. In a way it is an index of human use of land resources. Hence, the study of land use is not merely the descriptive analysis of different categories of land utilization; rather it is also an (qualitative) assessment of patterns of man–nature interaction and the extent to which human agencies are able to use the geographic area meaningfully (Singh 2005, p. 19). An understanding of land use therefore requires knowledge of the peculiarity of environmental conditions and the nature of human constraints in the region concerned. In mountainous regions such as the Himalayas, to which the larger part of ArP belongs, terrain (limitation) has tremendous role in determining the extent and efficiency of resource use, including land.

Available literature has well established that the existing land use system is the result of many 'causative factors' both natural and cultural in their wider connotation. The nature and level of land utilization consequently is determined by mainly three sets of determinants: (existing) socio-techno-economic, infrastructural facilities, and the people's living standard in the region. Of course, these three interact with each other in multiple ways, creating facilitating or impeding circumstances.

28.2.1 A Broader View

To begin with, one finds apparent positive changes with regard to categories such as operational area and cropped area. If interpreted together, they convincingly indicate agricultural development. Simultaneous considerations of the categories that have gained, and those areas not available for cultivation and cultivable waste that have declined, reflect a broad level positive correlation. Growth of total cropped area, for example, could be attributed to bringing areas not available for cultivation, cultivable waste, and (current) fallows under crops (Table 28.1).

The larger part of ArP is found to be covered by forests. Until the early 1990s a very large geographic area (more than 60 %) was covered by forests, which may be attributed to their tremendous regenerative capacity caused by the hot and humid climate complemented by difficult terrain conditions that make the forests less accessible for exploitation. During the land use census of 1985–1986 and 1990–1991, a positive change was noticeable, that is, an average annual increase of 400 ha. But, that did not continue. Excessive and unchecked commercial exploitation, on which the *kunda* business excelled to the benefit of all who were somewhere in the

Table 28.1 Land use, 1985–1986 to 2000–2001

	Area (ha)						
Land use category	1985–1986	1990–1991	Change (1985–1986 to 1990–1991)	1995–1996	Change (1990–1991 to 1995–1996)	2000–2001	Change (1995–1996 to 2000–2001)
Forests[a]	5,154,000	5,156,000	2,000	5,098,000	(−)58,000	4,759,949	(−)338,051
Operational area	344,232	349,878	5,646	405,878	56,000	393,638	(−)12,240
Total cropped area	167,469	193,193	25,724	248,064	54,871	215,064	(−)33,000
i. Net area sown	149,314	164,819	15,505	203,630	38,811	200,210	(−)3,420
ii. Area sown more than once	18,155	28,374	10,219	44,434	16,060	14,854	(−)29,580
Area not available for cultivation	48,129	42,780	(−)5,349	33,142	(−)9,638	32,386	(−)756
Miscellaneous use	28,655	44,555	15,900	14,362	(−)30,193	NA	–
Cultivable waste	44,426	33,094	(−)11,332	43,008	9,914	27,821	(−)15,187
Total fallow land	73,708	64,630	(−)9,078	88,030	23,400	87,000	(−)1,030
i. Current fallow	24,700	28,376	3,676	28,109	(−)267	22,961	(−)5,148
ii. Other than the current fallow	49,008	36,254	(−)12,754	59,921	23,667	64,039	4,118

Source: Based on Statistical Abstract of Arunachal Pradesh (SAAP) for different years

[a]Forests figures are not part of land use census

'power network,' in connivance with outsider merchants, led to a sudden decline of the 'real' forested area. The indiscriminate forest felling, legal and illegal both, came to a halt only after the December 1996 order of the Honorable Supreme Court of India, which in case of Arunachal Pradesh 'ordered saw mills to close down not only where a complete ban was directed but even within a 100-km radius of Arunachal Pradesh's state boundary' (Rosencranz and Lélé 2008). Around the same time, I had begun my lectureship in Arunachal when there were protests against this ban. I wanted to know about the whole issue and asked a local person. I remember his comments: "When this ban was not there the trucks plied all through the night making it difficult to take a proper sleep. All those who were connected to powerful people benefitted like anything. Money flowed down in the market, not only in Arunachal but also the markets down in neighboring Assam. Of course, for the common man, life was always the same—full of hardships—may there be ban (on commercial felling of forests) or not."

28.2.2 Intraregional Patterns

As mentioned earlier, changing district boundaries mean it is not possible to make a district-wise temporal comparison as that will not provide any meaningful understanding into the nature of change. However, it cannot forbid a spatial comparison and analysis of the emerging patterns (Tables 28.2 and 28.3).

28.2.2.1 Forests

Despite the Supreme Court of India ban on commercial felling of trees, forests are still central to the life of average Arunachalis who get their sustenance and livelihood from them. It is hence logical to begin with the changing intraregional distribution of forests in ArP. To present a precise picture, one can try to see with reference to the state average. In 1985–1986, all western districts had a lesser percentage of forested area than the east, except Tirap (and Changlang together). In the 1990s, the existing gap between districts reporting larger forest area (West Siang, 83.75 %) and the one with the smallest area (Changlang, 45.88 %) is considerable. In 2000–2001, as per the government statistics, not much change has taken place, at least at the state level. However, western districts appear to have increased their forested areas. The uneven distribution of forests could be explained as the outcome of degree of inaccessibility for commercial exploitation, particularly for the timber business, which devastated Arunachali forests mainly in the foothills. Location of timber mills and plywood industries in the vicinity, especially along the Arunachal–Assam border, acted as catalyst to this process; that is why districts with difficult terrain and inaccessible areas could maintain larger forest cover compared to those that were victims of rash and ruthless deforestation such as Changlang, Tirap, and Lower Subansiri.

Table 28.2 Pattern of land use by districts, 1985–1986 (ha)

State/district	Geographic area (ha)	Operational area[a]	Forest[a]	Cropped area: Total[b]	Net area sown[b]	Sown more than once[b]	Fallow land: Total[b]	Current fallow[c]	Other than current fallow[c]	Cultivable waste[b]	Area not available for cultivation[b]
(1)	(2)	(3)	(4)	(5)	(6)	(7)	(8)	(9)	(10)	(11)	(12)
Arunachal Pradesh	8,374,300	4.11	61.55	48.65	43.38	5.27	21.41	33.52	66.48	12.90	13.98
Tawang	217,200	1.79		90.23	66.67	26.10	10.72	46.64	53.36	3.87	7.96
West Kameng	742,200	2.48	51.55[d]	63.22	50.55	20.03	23.22	8.86	91.14	11.89	1.70
East Kameng	413,400	5.69	51.50	34.33	33.07	3.67	41.78	30.42	69.58	17.57	3.85
Lower Subansiri	1,013,500	4.38	51.48	46.88	46.88	Nil	16.22	56.43	43.57	4.33	22.95
Upper Subansiri	703,200	3.56	26.48	26.57	25.41	4.34	32.92	38.83	61.17	10.93	19.48
West Siang	832,500	5.50	DE	48.78	46.18	5.32	20.97	26.5	73.50	8.89	11.13
East Siang	1,019,300	5.75	50.66	54.60	45.86	16.02	11.14	54.43	45.57	6.93	31.19
Dibang Valley	1,302,900	1.00	62.63	80.10	60.69	24.24	14.99	38.62	61.38	6.33	3.47
Lohit	1,140,200	2.27	62.66	75.02	57.33	23.58	22.61	42.10	58.90	4.95	6.96
Changlang	466,200	4.63		67.99	63.93	5.97	13.73	25.98	74.02	5.52	5.60
Tirap	236,200	21.80	46.34[e]	23.29	23.26	0.16	28.78	51.82	48.18	41.42	3.50

Source: Adapted from SAAP
DE data error unverifiable
[a]Percent (%) of geographic area of the district
[b]Percent (%) of operational area
[c]Percent (%) of total fallow
[d]Includes Tawang
[e]Includes Changlang

Table 28.3 Pattern of land use by districts, 2000–2001 (ha)

				Cropped area			Fallow land				
State/district	Geographic area (ha)	Operational area[a]	Forest[a]	Total[b]	Net area sown[b]	Sown more than once[b]	Total[b]	Current fallow[c]	Other than current fallow[c]	Cultivable waste[b]	Area not available for cultivation[b]
(1)	(2)	(3)	(4)	(5)	(6)	(7)	(8)	(9)	(10)	(11)	(12)
Arunachal Pradesh	8,374,300	4.70	61.55	54.63	50.86	3.77	22.10	26.4	73.60	7.07	8.23
Tawang	217,200	2.97	49.99	85.50	69.80	18.36	7.76	78.60	21.40	1.06	18.12
West Kameng	742,200	1.07	54.60	60.44	58.39	3.39	15.84	42.65	57.35	3.14	12.69
East Kameng	413,400	4.62	75.80	58.78	55.83	5.03	32.13	8.65	91.35	4.67	1.51
Lower Subansiri	131,700	15.99	DE	85.17	78.54	7.78	12.37	48.20	51.80	1.49	5.85
Kurung Kumey	8,818	2.79	67.64	87.26	77.32	11.39	13.22	65.39	34.61	17.51	4.35
Papum Pare	287,500	6.42	87.52	74.51	70.68	4.99	12.02	64.62	35.38	2.52	8.69
Upper Subansiri	703,200	4.16	50.94	28.88	28.71	0.57	26.24	20.88	79.12	9.94	20.28
West Siang	764,300	5.41	44.53	47.43	46.79	1.35	18.28	37.91	62.09	4.04	13.20
Upper Siang	618,800	1.29	18.89	42.13	41.10	2.44	20.77	15.16	84.84	8.38	15.35
East Siang	468,700	9.75	38.80	64.90	58.09	10.50	17.97	86.82	13.18	5.55	8.70
Dibang Valley	912,900	0.07	72.90	92.64	92.16	0.52	0.18	50.00	50.00	4.80	0.16
Lower Dibang Valley	390,000	4.56	42.55	87.00	78.89	9.33	9.89	63.80	36.20	1.63	3.37
Lohit	521,200	3.06	DE	56.03	54.99	1.85	17.64	42.23	57.77	6.45	8.92
Anjaw[*]	619,000		27.50								
Changlang	466,200	9.85	88.65	59.85	59.07	1.29	21.93	12.09	87.91	5.54	4.40
Tirap	236,200	30.72	63.97	21.00	18.39	12.45	38.52	16.41	83.59	17.25	5.01

Source: Adapted from SAAP
DE data error unverifiable
[a]Percent (%) of geographic area of the district
[b]Percent (%) of operational area
[c]Percent (%) of total fallow
[*]New district carved out of (old) Lohit on 16th February 2004

28.2.2.2 Operational Area

It is true that there is an appreciable increase in the operational area. Still, a very small proportion of the geographic area is reported to be operational (Tables 28.2 and 28.3). It is a candid indication of the natural environment's dominance over the local ethnic communities in different parts of the state. Further, it also points out that there is no population pressure on the land.[1] There is a huge spatial variation but no noticeable pattern emerges as such. In 1985–1986, the operational area at state level was merely 4.11 %. Tirap had the highest at 21.80 % and Dibang Valley the lowest at 1.0 %. In the 1990s also there existed the same gap, almost 20 times, between the East Kameng highest operational area (21.83 %) district and Dibang Valley with merely 1.11 %. Similarly, during 2000–2001, despite an improvement shown by all districts, of course a marginal one, gaps remained considerably high. Here too such huge inter-district difference is basically the result of terrain complexity coupled with the lower technology possessed by local inhabitants, which together make the task of bringing a larger land area under (human) operation difficult.

28.2.2.3 Cropped Area

From the (agricultural) development aspect, the area sown is very important for all land use categories. In general, it reflects the extent to which areas have been brought under the plough and cultivation of crops. There are two subcategories under which cropped area is reported in the Indian agricultural census system, net area sown (NAS) and area sown more than once (ASMO). The former expresses extension of cropped area and the latter reveals cropping intensity.

One of the most common strategies of increasing agricultural production has been bringing more and more area under the plough. However, usability of land surface and level of demand (for such expansions) are crucial factors in this process. And, on the other hand, such extensions are at the cost of other land use/land cover; the most prominent among them has been forests that were cleared to create the first agricultural fields, and the process continues to date. In ArP one finds that neither of the aforementioned factors favors increase of cropped areas. On top of that, prevalence of community ownership of land (Roy and Kuri 2002), along with other common property resources (CPRs), probably hinders enough enthusiasm, encouragement, and incentives to extend and develop the area for cultivation (Singh 2005, p. 23). At the same time, forests are valued relatively high as the source of supplementing low productive subsistence farming and hence a common Arunachali person would not

[1]The general population density of Arunachal in 2011 was 17 persons/km^2, less than one third of the next lowest, 52, in Mizoram whereas the highest was of Bihar (1,102) among the states.

favor their destruction. This interpretation nevertheless needs further field testing as to what is the people's perception in the changed times.[2]

Tawang, Changlang, and Dibang Valley were the first three districts in terms of highest NAS in 1985–1986, which was above 60 % of the operational area (Fig. 28.3). Beside them, Lohit and West Siang too joined this group in 1990–1991. After a decade, when there was considerable reorganization of the districts, the scenario changed completely. Dibang Valley (92.16 %), Lower Dibang Valley (78.89 %), and Lower Subansiri (78.54 %) surpassed the previous figures and emerged as the top three districts in this respect in 2000–2001 while the state average remained at merely 50.8 %. The intraregional distribution hence reveals the eastern and southeastern districts doing better compared to their western counterparts, barring a few exceptions like Tawang and Lower Subansiri.

The ASMO does not provide a bright picture. In 1985–1986, a little more than 10 % of the gross cropped area was sown more than once. It improved in the next 5 years and reached around 15 %. But, again in 2000–2001, it dropped to half of the 1990–1991 mark. Tawang is the only district that has been consistently top ranking in this respect; performance of other districts in quite inconsistent (Fig. 28.4).

28.2.2.4 Fallow Land

Maintaining fallows in Arunachal, as in other parts of this region, is apparently linked with *jhooming*—the practice of shifting or slash-and-burn method of cultivation, which is why no steady trend is witnessed in this land use category. In 1985–1986, a little more than one fifth of the operational area was fallow. However, any regional pattern was difficult to discern as the top three districts were East Kameng (41.78 %), Upper Subansiri (32.92 %), and Tirap (28.78 %). In the following census of 1990–1991, clear regional trends were not deciphered. Western Arunachal, with the exception of Tawang, had larger fallows in comparison to the eastern part. The same was found again in 2000–2001. A pertinent point to be noted here is that there is a very high proportion of other fallow subcategories in the majority of the districts across the last four agricultural censuses, which confirms the wide prevalence of shifting cultivation.

28.2.2.5 Cultivable Waste

Generally, unscientific farming and poor management practices lead to unsuitability of land for continued cultivation. In the entire northeast region including ArP, deforestation, *jhooming*, and unscientific slope cultivation are together responsible for soil erosion that renders land unsuitable for cultivation. However, concerted efforts from different state government departments have yielded positive results. In

[2] There is unsubstantiated information about unauthorized private use of land that is otherwise a CPR in ArP.

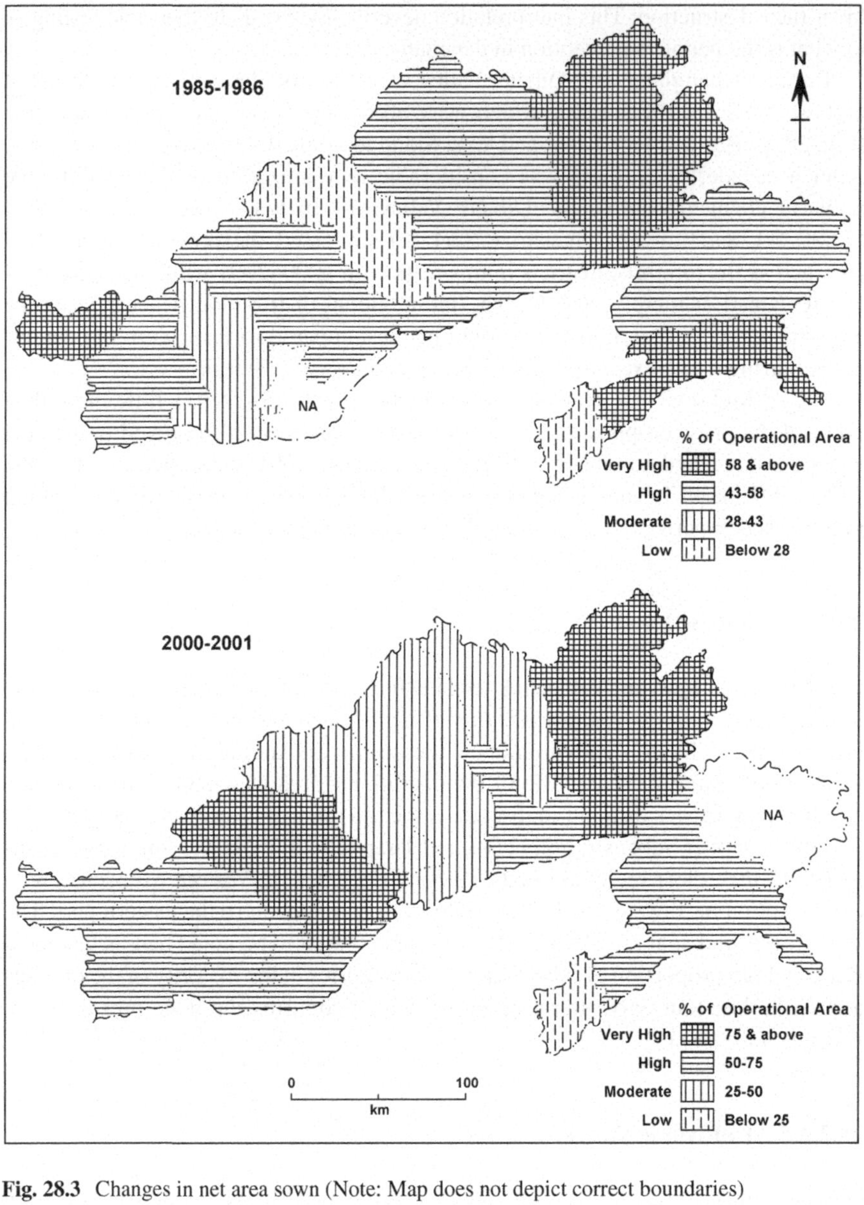

Fig. 28.3 Changes in net area sown (Note: Map does not depict correct boundaries)

1985–1986 the cultivable waste accounted for around 13 % of the operational area, which declined to 9 % in the next census of 1990–1991, and after a decade it dropped to just 7 % (2000–2001). Some older districts such as Tirap and newly carved ones such as Kurung Kumey, however, have reportedly a higher amount of wastage of cultivable land. On the other hand, Tawang shows better performance by maintaining its lowest proportion of cultivable waste across the censuses.

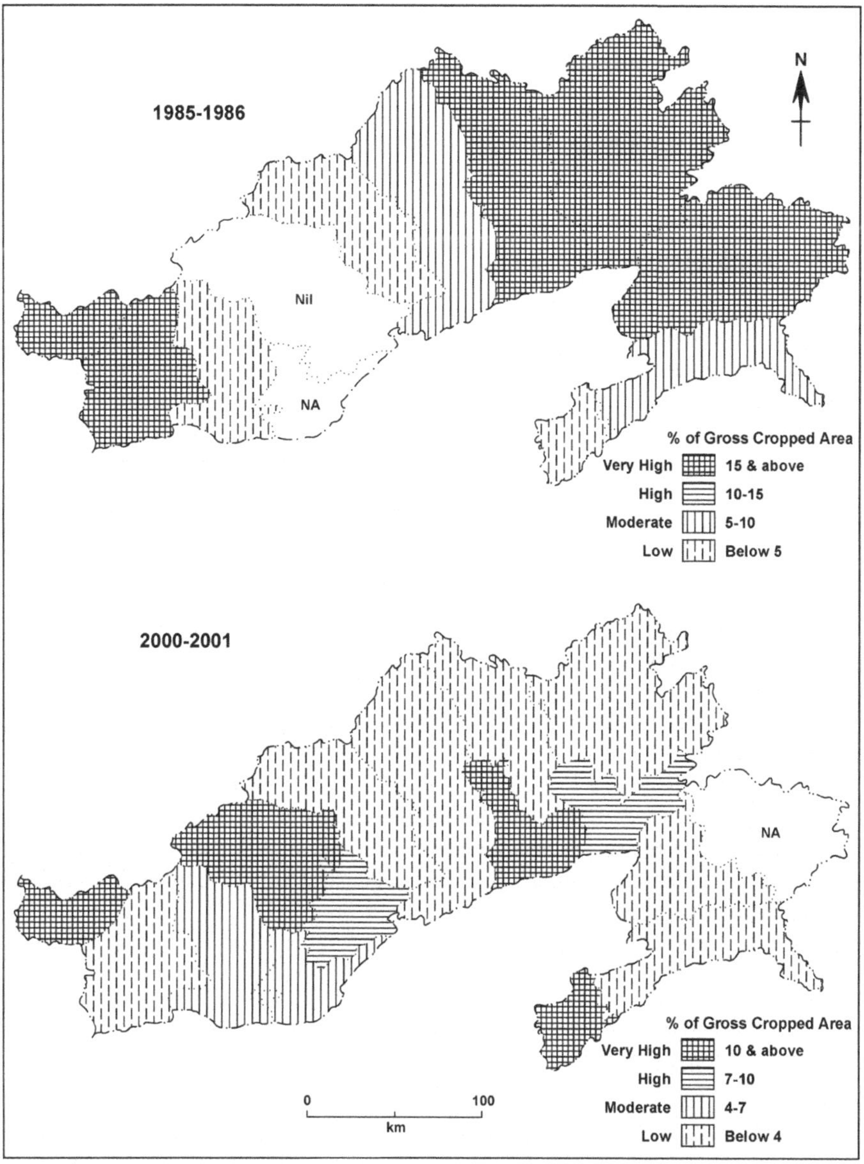

Fig. 28.4 Changes in area sown more than once (Note: Map does not depict correct boundaries)

28.2.2.6 Land Not Available for Cultivation

This category of land includes land put to nonagricultural uses such as settlements and roads and barren and uncultivable land. In the case of hilly and mountainous regions, to a considerable extent it does include highly rugged and eroded areas unfit for any kind of human utilization. As cultivable waste, in this category also signs of improvement are visible. Around 14 % of the operational area was reported under this category in 1985–1986, which declined to around 8 % in 2000–2001. It was quite high in East Siang (31.19 %), Lower Subansiri (22.95 %), and Upper Subansiri (19.48 %) in 1985–1986, whereas in 2000–2001 the top three districts were Upper Subansiri (20.28 %), Tawang (18.12 %), and Upper Siang (15.35 %).

28.3 Agriculture Practices

A majority of the population in ArP depends upon the agriculture sector for livelihood, mainly as cultivators and quite a small proportion as agricultural labor too. The people here, as do their other counterparts in northeast India, mainly practice *jhooming. Pani-kheti*, wet (settled or permanent) cultivation, is practiced by a small number of people in river valleys and plateaus where soil and water supply conditions are conducive to this. Initially, agriculture was largely restricted to the production of rice, the staple food of the people, supplemented by many minor crop products. This low-subsistence agriculture, of very meager productivity, was complemented by a number of activities, important among them being hunting, fishing, and a variety of collections from the forest. Consequently, a marketable surplus was almost unknown, preventing the emergence of monetized transactions as well as capital formation.

As noted earlier, agricultural land was held communally except for a few pockets. However, the individual cultivator enjoyed usufructuary rights during his operation of the land (Roy and Kuri 2002; Talukdar 1997). Agricultural land was not cadastral surveyed; even today a cadastral survey has not been conducted on agricultural land. After the recent enactment of a Bill on Land Settlement and Revenue in 2001, preparations are being made for a cadastral survey (Arunachal Pradesh Human Development Report 2005).

28.3.1 General Nature of Practiced Agriculture: Interplay of Nature and Culture

Agriculture is an excellent example to analyze the interplay between nature and culture, in which man, despite all his capabilities, finally depends on nature for success. Of course, the technological man has endeavored to surpass the limits posed by nature; however, the known information is yet to confirm his success conclusively. This observation becomes crucial in the context of ArP where people with limited technological access could not surmount nature's limit. They have tried to live in harmony with nature and probably because of it only they have tried to self-impose limits on their material requirements. Hence, the primary objective of agriculture practiced by the inhabitants in this mountainous and hilly state is to subsist on growing traditional crops using local practices.

Looking scientifically into the basic nature of agriculture, it is important to look into the land capability pattern. The state government's Soil and Land Use Survey Organization has conducted a survey of land resources and identified four land capability classes that are applicable in the case of Arunachal: II, III, IV, and VIII. Of these, class IV is found in the largest area across the state, whereas II is concentrated in parts of Lohit and Dibang basins, III in a narrow strip running along the Arunachal–Assam border and Siang, Subansiri, and Kameng basins, and class VIII is found along the Indo-Chinese borderline. Among them, class II is most suitable for agricultural cultivation conditionally. The rest of the capability classes suffer from severe limitations having negative impacts on cultivation. For example, class IV, which is prominent being widely distributed, suffers from profound limiting factors such as moderate to severe erosion by water, steep slopes, and soils with poor rooting zones and low water-holding capacity (Singh S 1999, pp. 77–78). Thus, the nature of agriculture in ArP needs little explanation. A cultivator here has actually limited choice rather than going in for traditional *jhooming* with the least use of modern inputs such as fertilizers, either because of their lesser utility in slope terraced farms, or too small agricultural landholdings (SAAP 2011), or their highly scattered nature, which forbids proper application of modern inputs (Tiwari and Joshi 2000, p. 83). Moreover, in such farming practice commercialization is yet to materialize, and that is why one does not find the free availability of such inputs in the local market.

28.3.2 Changes from Jhooming to (Wet) Settled Cultivation

Although *jhooming* has been found as a practice that is not environment friendly in many ways, it continues to be the dominant agricultural practice across northeast India. It needs to be recognized that in the given agro-ecological conditions, the only way is to rely on inherent soil fertility, which is not capable of permanent cultivation. And that is why keeping fallows—an ages-old practice—becomes the only option to maintain the fertility of the soil. This practice was possible, and perhaps 'sustainable,' when the land-to-man ratio was low. But now, the situations are changing. Although in comparison to other states, the northeast as a whole and ArP in particular have low population density, one cannot deny the fact that over the years this will increase. Such an anticipated increase will create pressure on *jhoom* fields in the form of gradual shortening of the *jhoom* cycle. Naturally, this will have a deteriorating impact on land quality with reference to cultivation even though the best traditional ecological knowledge of the local people is utilized to manage soil fertility (Ramakrishna 1992, 2002).

For this reason, there has been a consistent effort from the government's side to encourage people to shift to settled cultivation. There has been some success (Fig. 28.5), but such attempts have not yielded as much as desired. The academic argument put to explain it is that 'this is because the people perceive sedentary agriculture is demanding very high inputs in terms of fertilizers and pesticides'

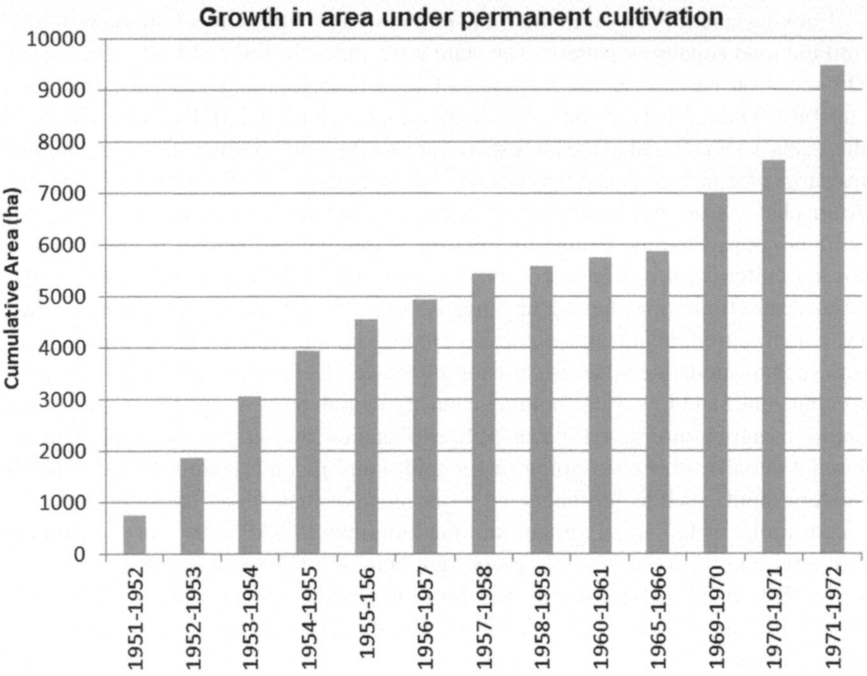

Fig. 28.5 Area under permanent cultivation during 1951–1952 to 1971–1972 reflecting slower pace of change (Based on different statistical reports)

(Arunachal Pradesh Human Development Report 2005, p. 173). Here, we tend to err. We need to remind ourselves that mere (modern) inputs cannot facilitate such a switch-over. Settled agriculture is possible only in a situation where land capability is good enough to support it, at least with the adoption of best land management practices. So, the people well understand through their indigenous traditional knowledge system that it would be disastrous to switch over to settled cultivation where nature does not permit it. Such a situation calls for considering other ways to tackle low productivity, discussed later in this chapter.

28.3.3 Agricultural Land: Availability and Distribution

Agricultural land, which is defined variably, here means 'land normally used for agricultural production.' Thus, it includes cultivated and fallow lands as well. In the context of ArP, it is needless to mention that both the area under *jhoom* and settled cultivation are considered. Singh (2008) attempted to analyze the distribution of agricultural land. Here, the main findings of that study are noted in a summarized form.

Table 28.4 Concentration of agricultural land

	Location quotient		
District	Net sown area	Gross cropped area	Total
Tawang	0.76	1.05	0.63
West Kameng	0.37	0.36	0.30
East Kameng	1.17	1,06	1.40
Papum Pare	3.17	3.25	2.52
Lower Subanasiri	0.32	0.32	0.51
Upper Subanasiri	0.72	0.71	0.57[a]
West Siang	1.42	1.26	1.49
East Siang	2.28	1.88	2.10
Upper Siang	0.55	0.48	0.49
Dibang Valley	0.92	1.00	0.67
Lohit	0.83	0.79	0.64
Changlang	1.26	1.21	1.88
Tirap	3.19	3.25	3.76

Source: Singh (2008)
[a]Land under miscellaneous tree crops or groves not included as data are not available

Table 28.4 reveals that agricultural land is highly concentrated in Tirap, Papum Pare, and East Siang. The gross cropped area (GCA) is expected to provide a better picture as ecological, techno-economic, and socio-institutional factors have direct bearing on it. The same pattern is repeated in case of the NAS and the GCA.

In view of the significance of agricultural land as the single major source of food and non-food supplies, its 'scarcity' or 'sufficiency' has economic, social, and political implications. Hence, its analyses need to take into account the people dependent on it through (agricultural) land–population ratios. With this purpose in mind, per capita availability of agricultural land with respect to total, rural, and agricultural workers population has been determined (Table 28.5).

In case of NAS, Dibang Valley has the highest per capita area, that may be with reference to the total or rural or agricultural worker population. The next highest figures are almost half of this. Lowest availability is found in Lower Subansiri, West Kameng, and Changlang (Table 28.5). It can be observed that Lower Subansiri and West Kameng had the lowest concentration of NAS (see Table 28.5). The GCA too by and large confirms the pattern noted by NAS. A degree of regionalization can be observed in the comprehensive land uses in the state. The western districts show lesser per capita availability compared to the eastern part, and hence it may be inferred that the population pressure on land is relatively high there. On comparing and contrasting district-wise values, that the inter-district gaps are not very large, which means the entire state reels under the same depressed condition, that is, lesser availability of agricultural land.

Table 28.5 Per capita distribution of agricultural land

	Per capita[a] agricultural land[b] (ha)		
District/state	Total population	Rural population	Agricultural workers
Tawang	0.144	0.166	0.503
West Kameng	0.109	0.120	0.602
East Kameng	0.368	0.498	1.099
Papum Pare	0.217	0.440	2.152
Lower Subanasiri	0.194	0.222	0.557
Upper Subanasiri	0.268[c]	0.373[c]	0.935[c]
West Siang	0.400	0.500	1.482
East Siang	0.410	0.547	1.839
Upper Siang	0.337	0.337	0.967
Dibang Valley	0.550	0.667	2.029
Lohit	0.184	0.226	0.682
Changlang	0.256	0.284	0.708
Tirap	0.322	0.380	0.867
Arunachal Pradesh	0.280	0.351	1.015

Source: Singh (2008)
[a]Population data used here are from Census of India 2001
[b]Land use data based on the latest available agricultural census
[c]Land under miscellaneous tree crops and groves not included because date are not available

28.3.4 Cropping Pattern

There has been tremendous increase in the area under major crops. During the periods 1980–1981 to 1993–1994, an increase of 59.9 thousand hectares (ha) is noticed (Singh S 1999). Similarly, during 1993–1994 to 2010–2011, another 47.6 thousand ha were added in the gross cropped area. Of all the crops, both in *jhooming* and settled cultivation, paddy emerges as the dominant crop, claiming the highest area of the GCA. Other major food grain crops have been mainly maize and millet, followed by wheat, which has been quite successful in the western districts of Tawang and West Kameng as well as East Siang in the central part. However, over the years the percentage of paddy has declined considerably, from 70 % (1993–1994) to 56 % (2010–2011) of the GCA at the state level.

There is simultaneous gain by other crops during this period. For example, oilseeds, pulses, potato, ginger, and crops such as chilies have gained significantly. Interestingly, sugarcane also is being cultivated in most of the districts, as revealed by the SAAP of different years.

It is worth mentioning here that apart from the aforementioned crops, experiments with horticultural crops, particularly fruits such as apples, orange, pineapple, banana, pears, and the recently introduced kiwi, together with spices such as big cardamom and black pepper, have picked up significantly. This trend suggests growing crop diversification and commercialization of agriculture, although many of these items are mainly meant for export outside the state. The only major hindrance in successful growth is the marketing bottleneck, arising from poor transport connectivity, as there are still many areas not connected with roads (Singh 2006; Mishra et al. 2004). In terms of total value as well as yield, these crops are at the lower end; still, they can increase poor agricultural return and provide better livelihood security to the people.

28.3.5 'Modernization' of Agriculture

The miracle called the 'Green Revolution' in agriculture that took place in India after attaining Independence during the 1960s–1970s was possible only because of the introduction and adoption of the Borlaug (high yield variety, HYV) seed, chemical, and fertilizer technology, essentially a novel paradigm in Indian agriculture. It created an upbeat mood among policy makers and planners and carved a new benchmark of agricultural growth and development. However, the gains of this miracle did not spread out evenly across the states and regions (Bhalla and Singh 2012, pp. 6–11) for a variety of reasons; important among these was its 'crop bias,' which had its natural regional fallouts. The peripheral regions such as ArP had to bear the brunt rather severely.

Looking at the nature of change from the grass roots perspective, it is not difficult to recognize that changes are not internal; rather, they are the outcome of *sarkari* ('official') interventions, which have always been external. So far, the state agencies seem to have failed in involving people.[3] For that reason, policies and programs remained more 'official' and less 'real.' Also, the union territory (UT) got statehood quite late in 1987 and for long it was administered remotely from Shillong. And it was in 1974 when the capital shifted to Itanagar, the present state capital. This administrative arrangement naturally had its implications for proper and meaningful integration of the state with the mainstream and of the implementation of development plans and programs including those related to agriculture.[4]

---

[3]There are many reasons responsible for this; the main issue has been, for a long period of time, the government officials at all the levels came from outside and did not have proper communication and interaction with local people. For this reason, perhaps, they developed apathy and lacked interest and did not take the initiative in developing the state.

[4]Fortunately, I have the Statistical Outline of North-East Frontier Agency, 1959, published from Shillong, the headquarters. The information generated is too scanty and gives an idea as to how this part of the country was treated at that time. For example, the directorate of agriculture has reported data on area brought under permanent cultivation; animals treated, castrated, and vaccinated; and farms and demonstration and upgrading centers.

Table 28.6 Growth of irrigation

Year	(Net) irrigated area (ha)
1969–1970	NA
1970–1971	NA
1971–1972	NA
1993–1994	3,138
2004–2005	664
2007–2008	44,478
2010–2011	44,478

Source: Adapted from SAAP for different years

28.3.6 Irrigation

As the traditional practice of *jhooming* was rain fed, irrigation was not required for it. In the areas where settled cultivation was practiced in a limited way, water was supplied through an indigenous method of channel irrigation in which water from rivulets and rapids is channelized to the agricultural plots. In the wake of gradual switching over to settled cultivation and governmental efforts to modernize and intensify agricultural production, the need to develop irrigation is also increasingly realized as it is one of the most important technological factors found to have direct impact on agricultural productivity. Humble beginnings in this direction were made in 1966–1967 when minor irrigation projects were started with 1,032 ha of command area (Singh S 1999). Up to the mid-1990s, limited success was noticeable (Table 28.6), but in subsequent years, appreciable progress is seen (Table 28.7). During 1993–1994, only 2 % irrigated net sown area increased more than ten times, making it a little over 20 % in 2010–2011. Earlier, Tawang had the only impressive irrigated area. The scenario has however changed after the first decade of the twenty-first century when many more districts such as Upper Siang (47.67 %), East Siang (44.83 %), Lower Subansiri (39.06 %), Papum Pare (36.64 %), Lower Dibang Valley (25.18 %), and West Siang (23.02 %) registered higher percentages of irrigated net sown area contrasted against the state average (Fig. 28.6).

28.3.7 High-Yielding Varieties (HYVs)

The application of high-yielding varieties (HYVs) has been found to increase the crop yield level and reduce the yield gaps (Singh S 1999, p. 83). Adoption of HYV seeds of wheat and paddy from the mid-1960s to 1980–1983 in northwestern India had resulted in significant increase in yield and output. During the period of 1980–1983 to 1990–1993, the 'Green Revolution' matured, which meant acceleration of growth rates at all levels including the eastern states (Bhalla and Singh 2012, pp. 214–218). However, the same did not happen in the case of most of the states of the northeastern region. For example, in ArP, a very small proportion of

Table 28.7 Concentration of irrigated area

Districts	1993–1994	2010–2011
Tawang	5.69	0.37
West Kameng	0.49	0.10
East Kameng	1.72	0.66
Upper Subansiri	0.93	0.58
Lower Subansiri	2.01	1.76
West Siang	1.18	1.04
East Siang	0.83	2.02
Upper Siang	–	2.15
Lohit	0.97	0.29
Dibang Valley	2.22	NIL
Lower Dibang Valley	–	1.13
Tirap	0.28	0.08
Changlang	0.97	0.58
Kurung Kamey	–	0.88
Papum Pare	–	1.65
Anjaw	–	–
Arunachal Pradesh	2.10	22.21

Source: Adapted from SAAP for different years

cultivated area was reported to be under the HYVs until the early 1970s (Table 28.8). It is just one illustration that the 'revolution' diffused with quite varying pace into peripheral regions, and the states had to wait longer to experiment with and experience the outcomes.

Rice being the staple food grain of the state, paddy naturally was the first crop to benefit from the HYV technology. Still, by the early 1990s, not even a quarter of the area under paddy had HYVs. In contrast, half of the area under maize and more than 90 % of wheat was under HYVs. Recently, the adoption rate of HYVs has improved only marginally, as indicated by the case of paddy, of which 36 % area was reported under HYV in 2010–2011 (SAAP 2011, p. 32). As per the latest data, of the total area under HYVs of all crops, 42 % was shared by the first three districts West Siang, Lohit, and Lower Dibang Valley only. And, among them, 15 % of the share was of West Siang alone.

28.3.8 Chemical Fertilizers

Chemical fertilizers have been considered another important component of the Green Revolution. They too have contributed in the efforts to increase agricultural productivity as well as yield by playing a complementary role in compensating for the nitrogenous-phosphoric-potashic (NPK) deficiency in soil. Research has confirmed that fertilizers have a significant role in advancing agricultural production in

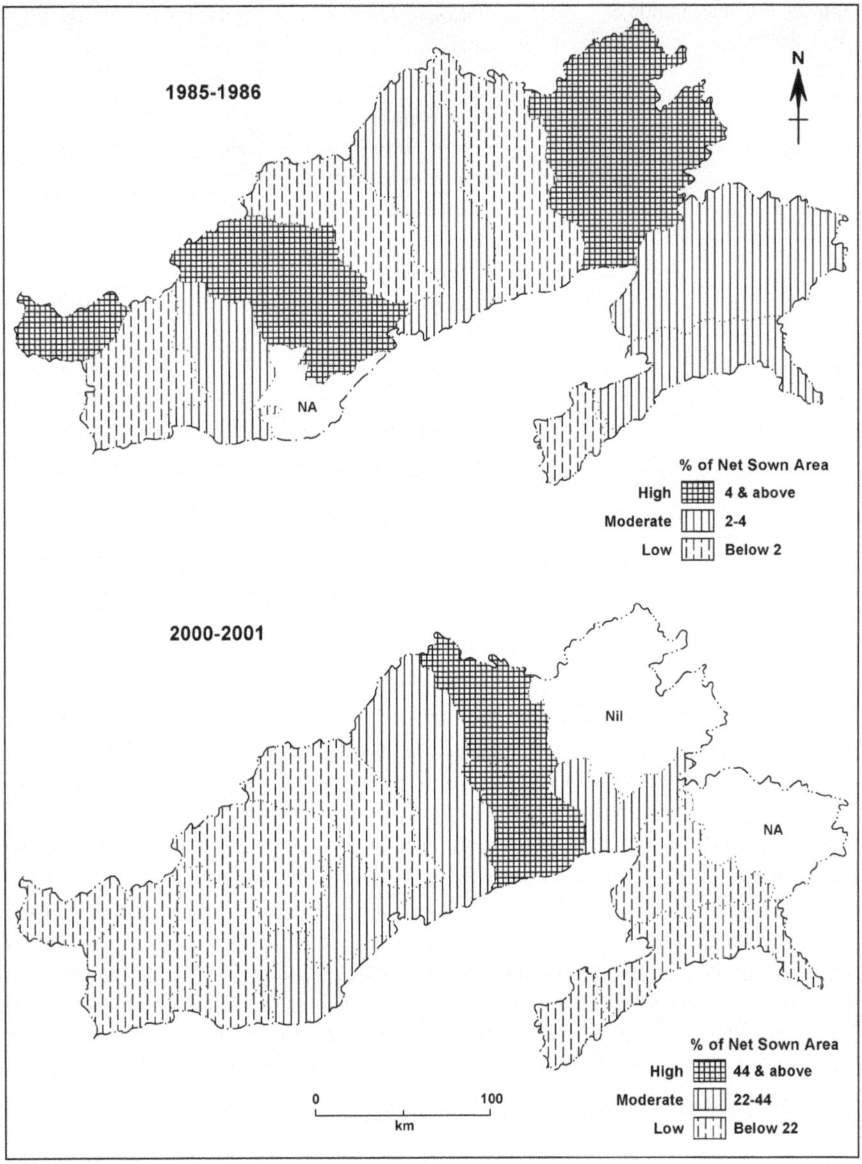

Fig. 28.6 Distribution of net irrigated area (Note: Map does not depict correct boundaries)

the lagging agricultural regions of India (Bhalla and Singh 2012, p. 86). Unfortunately, in ArP, the NPK consumption is too small and the rate of progress in this regard is very slow. In the entire state, only 750 metric tons were consumed as per the 2010–2011 statistics, which is highest so far (Table 28.9). The two

Table 28.8 Use of high-yield varieties (HYVs)

Year	Area under HYVs (ha)
1969–1970	446.72[a]
1970–1971	487.24[a]
1971–1972	450.87[a]
1993–1994	49,700
2004–2005	65,300
2007–2008	32,200
2010–2011	74,000

Source: Adapted from SAAP for different years
[a]Figures for these years are available as 'area under HYVs of seeds'

Table 28.9 Trend of fertilizer consumption

Year	Unit	Consumption
1969–1970	ha	184.91
1970–1971	ha	139.61
1971–1972	ha	54.55
1993–1994	tons	575
2004–2005	tons	695
2007–2008	tons	735
2010–2011	tons	750

Source: Adapted from SAAP for different years

westernmost districts, Tawang and West Kameng, were the leading consumers, sharing 22.13 % each of the state total. In the rest of the districts, consumption barely passes the 10 % mark.

28.4 Levels of Agricultural Development

Development both as a process and a goal has been ever desirable. With the passage of time, the need to develop agriculture has been felt in ArP too despite a comfortable land–human ratio compared to other states in India. In addition, agriculture has been the major source of people's livelihood and the state's economy depends on it substantially. There are three objectives: to meet the local demand of agricultural produce, to provide and strengthen livelihood security, and to raise the local people's income level.

It may be noted that agricultural development determinants fall broadly into two sets: physical and cultural. Often the physical challenges that are tackled by

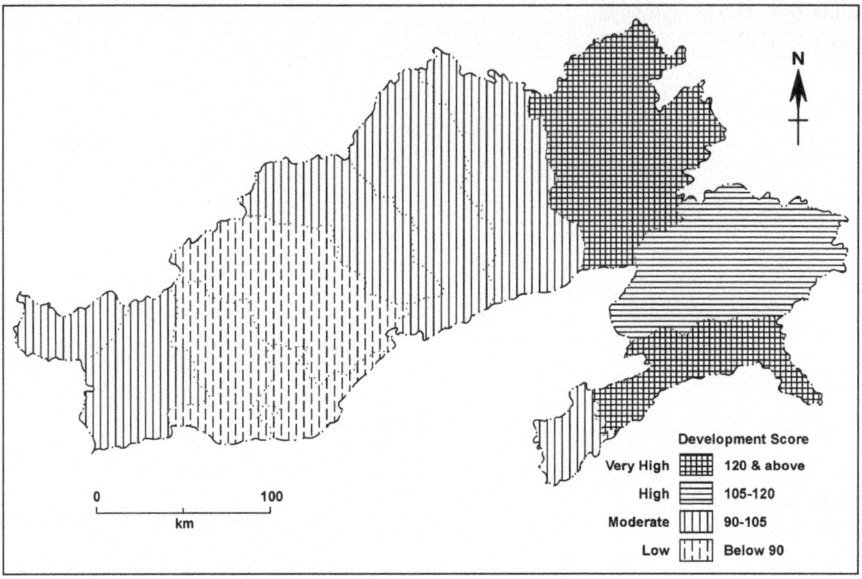

Fig. 28.7 Levels of agricultural development, 1994 (Note: Map does not depict correct boundaries)

cultural interventions are related to climate or soil. Biotechnology added a new dimension to this wherein crop plants (and therefore their seeds) were modified to make them more 'suitable' to the changing circumstances. In the context of the present discussion, farming techniques and farm technology are also considered important. They are either improvised internally in response to the societal needs of a time or acquired through the process of cultural exchange and adopted in due time. In recent times, the political economy and political ecology have emerged as crucial, especially in the context of peripheral regions. Several studies confirm the apparent development of agriculture in India after Independence (Singh 2005, p. 25; Bhalla and Singh 2012), and they have also provided alternative ways to measure the development and its (regional) distribution. From the foregoing discussions, it is easy to infer that agricultural development has taken place even in ArP; of course, not as much as in other states, especially outside the northeast region.

Now, it is time to see what kind of geographic pattern emerges with reference to development attainments. Is there uniformity in the spatial distribution of underdevelopment or not? In 1994 the average development score was 104.5. As is apparent from Fig. 28.7, most of the districts performed more poorly if the average is considered the benchmark. If an imaginary north–south line is drawn between Subansiri and Siang basins dividing ArP into east and west, the eastern part will appear to be performing better. The basic reason appears to be the difference in terms of land capability, which is apparently better in the east, and also the presence of a sizeable area under settled cultivation. As far as individual district-wise performance is

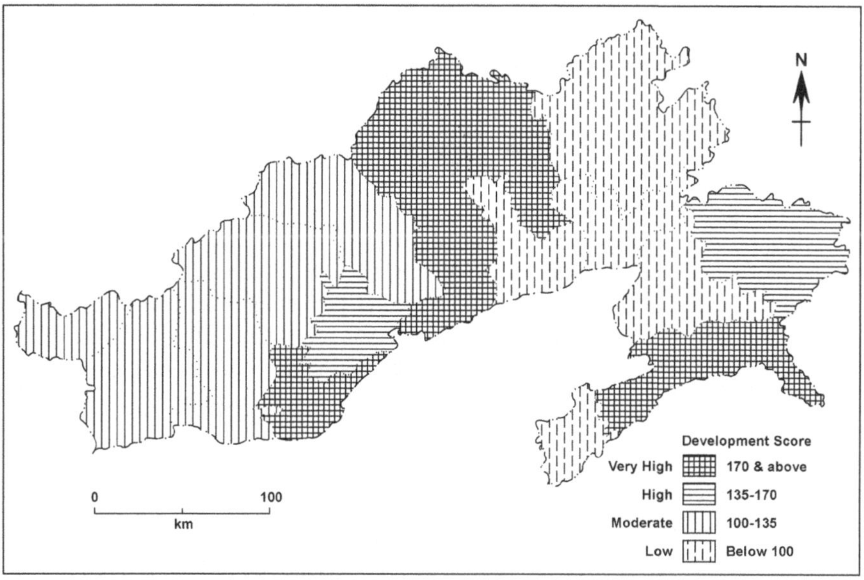

Fig. 28.8 Levels of agricultural development, 2011 (Note: Map does not depict correct boundaries)

concerned, East Kameng with a score value of 55.67 emerged as the poorest performing district and Changlang with 221.42 as the highest performing district.

The agricultural development scenario definitely improved in 2010–2011 relative to the 1990s level. The first point to note here is the average development score, which increased to 135.84. A clear 30-point improvement was attained, which attests to a positive change, but again this is too slow if evaluated with reference to time. The districts were reorganized many times during this period, and the development scores are available for 16 districts. The number of better performing districts doubled to become six and the eastern districts dominate among them. District-wise performance reflects an interesting picture that contrasts visibly compared to the previous pattern observed. Most of the eastern districts were found in the lowest category whereas the western districts perform relatively better (Fig. 28.8). Two neighboring districts, Changlang and Tirap, showed relative contrasts as compared to each other with scores of 230.92 and 75.25 as the best and worst performing districts, respectively.

28.5 Conclusions

ArP, as the other hilly states of the northeastern region, has not been able to benefit from various development initiatives because of its peripheral location and limitations imposed by the limits of the geographic environment. In spite of the odds, development as a process has marched, albeit slowly, which is why one finds encouraging changes with respect to land use and agricultural development particularly. It is said that under the neo-liberal development paradigm, state intervention has diminished over the years and the free market has assumed its role in that space. Such a change nevertheless is against regions such as ArP where the economy yet not fully monetized nor has the market penetrated the whole territory. Moreover, such underdeveloped spaces never attract the market. So, the state role remains crucial for pushing the development cart ahead in favor of this agriculture-dependent society and economy. Successful agriculture no doubt requires good external conditions such as climate, soil, topography, accessibility, and knowledge, which are not found ubiquitously. Five constraints have been identified affecting agriculture development in ArP: climatic constraints, infrastructure constraints, biophysical constraints, constraints of management, and socioeconomic constraints (Mishra et al. 2004).

Therefore, despite the slow rate of success, interventions are needed from state agencies, otherwise the development of Arunachali people will prove to be difficult. The lesson to be learnt from the past experience is to ensure people's involvement and genuine participation. No development initiative could be successful and give desirable optimum results in the absence of the stakeholders' involved contribution to it. It is rightly pointed out by a group of scholars that "Arunachal needs to plan its development in a considered and phased manner, ensuring that the development is people-centric yet decentralized; community-based but, with the Government as a facilitator; using its resources in a measured and sustainable manner, and in keeping with the aspirations of the people. The challenge is to evolve a development model that is truly sustainable and worthy of emulation." (Arunachal Pradesh Human Development Report 2005, p. 15).

References

Alonso W (1964) Location and land use. Harvard University Press, Cambridge, MA

Anderson A (2000) Paradox in the periphery: an entrepreneurial reconstruction? Entrep Reg Dev 12:91–109

Arunachal Pradesh Human Development Report (2005) Government of Arunachal Pradesh, Itanagar

Bhalla GS, Singh G (2012) Globalization and Indian agriculture: a district level analysis. Sage, New Delhi

Bhatia SS (1967) A new measure of agricultural efficiency in Uttar Pradesh, India. Econ Geogr 43(3):244–260

Crone M (2012) Re-thinking 'peripherality' in a knowledge-intensive service dominated economy. In: Danson M, De Sousa P (eds) Regional development in Northern Europe: peripherality, marginality and border issues. Regions and cities. Routledge, Abingdon

Leimgruber W (1994) Marginality and marginal regions: problems of definition. In: Chang C-YD et al (eds) Marginality and development issues in marginal regions. National Taiwan University, Taipei, pp 1–18

Leimgruber W (2004) Between global and local. Marginality and marginal regions in the context of globalization and deregulation. Ashgate, Aldershot

Leimgruber W (2010) Towards a better world? Joining forces to overcome deadlocks. In: Leimgruber W et al (eds) Geographical marginality as a global issue, vol 1 General, theoretical and methodological aspects. IGU commission on marginalization, globalization and regional and local response

Lomnitz LA (1977) Networks and marginality. Life in a Mexican Shantytown. Academic, New York

Markusen A (1999) Fuzzy concepts, scanty evidence and policy distance: the case for rigor and policy relevance in critical regional studies. Reg Stud 33:869–884

Mishra AK, Bundela DS, Satapathy KK (2004) Analysis and characterization of rice environment of Arunachal Pradesh. ENVIS Bull Himal Ecol 2(1):12–24

Myrdal G (1957) Economic theory and underdeveloped regions. Methuen, London

Ramakrishna PS (1992) Shifting agriculture and sustainable development: an interdisciplinary study from North Eastern India. Wiley Eastern, New Delhi

Ramakrishna PS (2002) Land use dynamics and sustainable development in Arunachal Pradesh: environmental planning and sustainable development, Himvikas, Occasional Publication

Rosencranz A, Lélé S (2008) Supreme Court and India's forests. Econ Pol Wkly 43(5):11–14

Roy NC, Kuri PK (2002) Land reform in Arunachal Pradesh. Classical Publishing House, New Delhi

Singh RS (1999) Ethnic minority groups of Arunachal Pradesh, India: a study in cultural geography. Natl Geogr J India 45(1–4):112–121

Singh RS (2000) Land use and level(s) of agricultural development in Arunachal Pradesh. Natl Geogr J India 46(1-4):69–80

Singh RS (2001) Gender situation in mountain environment: a case of Arunachal Pradesh. Natl Geogr J India 47(1-4):109–122

Singh RS (2005) Paths of development in Arunachal Pradesh. Northern Book Centre, New Delhi

Singh RS (2006) Infrastructure in Arunachal Pradesh, India: development and distribution. In: Singh RB (ed) Sustainable urban development. Concept Publishing, New Delhi, pp 134–148

Singh RS (2008) Geographical dimensions of land-use in Arunachal Pradesh, India. Paper presented at International Geographical Union Seminar on land use land cover change and agro-biodiversity, Department of Geography, National Post-Graduate College, University of Lucknow, Lucknow, 7–8 Mar 2008

Singh RS (2010) Recent scenario of gender patterns in Arunachal Pradesh. In: Nangia S, Jha MM, Mishra S, Ramachandran H, Velayutham M (eds) Development concerns in 21st century: essays in the honour of Dr. K.V. Sundaram. Concept Publishing, New Delhi, pp 89–110

Singh S (1999) A resource atlas of Arunachal Pradesh. Government of Arunachal Pradesh, Itanagar

Statistical Abstract of Arunachal Pradesh (2011) Government of Arunachal Pradesh, Itanagar

Talukdar AC (1997) Land tenures and land use pattern in Arunachal Pradesh: a review. In: Behera MC, Roy NC (eds) Trends in agrarian structure in the hills of North-East India. Commonwealth Publishers, New Delhi

Tiwari PC, Joshi B (2000) Optimum land use in mountains for sustainable development: a case study of Himalaya-Ganga plain. Natl Geogr J India 46(1-4):81–92

Wallerstein IM (1974) The modern world-system. Academic, New York

Web Sources

http://www.dorabjitatatrust.org/northeast/pdf/ReportonNE.pdf

http://censusindia.gov.in/2011-prov-results/data_files/india/table_1.pdf

Index

A.K. Dutt et al. (eds.), *Spatial Diversity and Dynamics in Resources and Urban Development*, DOI 10.1007/978-94-017-9771-9

The manufacturer's authorised representative in the EU is Springer Nature Customer Service Centre GmbH, Europaplatz 3, 69115 Heidelberg, Germany. If you have any concerns regarding our products, please contact ProductSafety@springernature.com

Printed and bound by CPI Group (UK) Ltd, Croydon, CR0 4YY
30/09/2026
02225448-0017